The Law of Sines

In any $\triangle ABC$,

$$\frac{a}{\sin A} = \frac{b}{\sin B} = \frac{c}{\sin C}.$$

The Law of Cosines

In any $\triangle ABC$,

$$a^2 = b^2 + c^2 - 2bc \cos A,$$
$$b^2 = a^2 + c^2 - 2ac \cos B,$$
$$c^2 = a^2 + b^2 - 2ab \cos C.$$

Trigonometric Function Values of Special Angles

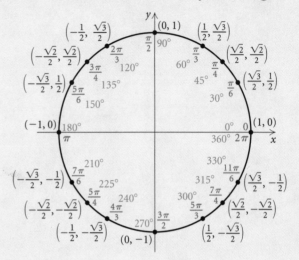

Graphs of Trigonometric Functions

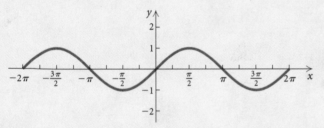

The sine function: $f(x) = \sin x$

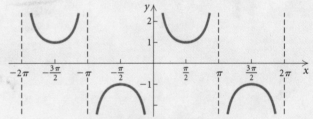

The cosecant function: $f(x) = \csc x$

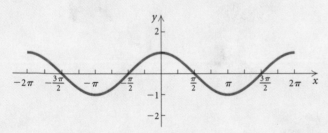

The cosine function: $f(x) = \cos x$

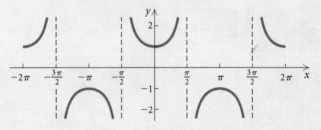

The secant function: $f(x) = \sec x$

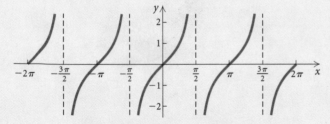

The tangent function: $f(x) = \tan x$

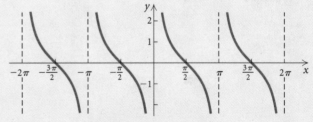

The cotangent function: $f(x) = \cot x$

Trigonometry

GRAPHS AND MODELS

Marvin L. Bittinger

Indiana University–Purdue University at Indianapolis

Judith A. Beecher

Indiana University–Purdue University at Indianapolis

David Ellenbogen

St. Michael's College

Judith A. Penna

Indiana University–Purdue University at Indianapolis

 **ADDISON-WESLEY**

An imprint of Addison Wesley Longman, Inc.

Reading, Massachusetts • Menlo Park, California • New York • Harlow, England
Don Mills, Ontario • Sydney • Mexico City • Madrid • Amsterdam

Publisher	Jason A. Jordan
Associate Editor	Christine Poolos
Managing Editor	Ron Hampton
Production Supervisor	Kathleen A. Manley
Copy Editor	Martha Morong/Quadrata, Inc.
Text Designer	Geri Davis/The Davis Group, Inc.
Marketing Manager	Brenda Bravener
Illustrators	Scientific Illustrators, J. A. K. Graphics, Parrot Graphics, and Gail Hayes
Compositor	The Beacon Group
Cover Designer	Susan Carsten
Cover Photograph	Takeshi Odawara/PHOTONICA
Manufacturing Supervisor	Ralph Mattivello

Photo Credits

63 (left), SuperStock **160,** Darryl Jones **165,** courtesy of Goss Graphics Systems **446,** SuperStock **451,** The Topps Company, Inc.

2, Graph on number of births: From THE NEW YORK TIMES, 2/12/95, p. A1. Copyright © 1995 by The New York Times Company. Reprinted with permission.

Library of Congress Cataloging-in-Publication Data

Trigonometry: graphs & models/Marvin L. Bittinger . . . [et al.].
     p. cm.
    Includes indexes.
    ISBN 0-201-33201-9
    1. Trigonometry. I. Bittinger, Marvin L.
QA531.T75 1997
516.24′2—dc21   97-41163
                                                 CIP

1 2 3 4 5 6 7 8 9 10—RNT—00999897

Contents

6

Exponential and Logarithmic Functions

Preface

Trigonometry: Graphs and Models covers college-level trigonometry and is appropriate for a one-term course in precalculus mathematics. The approach of this text is more interactive than most trigonometry texts. Our goal is to enhance the learning process through the use of technology and to provide as much support and help for students as possible in their study of trigonometry. A course in intermediate algebra is a prerequisite for the text.

Content Features

• **Integrated Technology** The technology of the graphing calculator is completely integrated throughout the text to provide a visual means of increasing understanding. In this text, we use the term "grapher" to refer to all graphing calculator technology. The use of the grapher is woven throughout the exposition, the exercise sets, and the testing program without sacrificing algebraic and trigonometric skills. We use the grapher technology to enhance, not to replace, the students' mathematical skills and to alleviate the tedium associated with certain procedures. It is assumed that each student is required to have a grapher (or at least access to one) while enrolled in this course.

• **Learning to Use the Technology** To minimize the need for valuable class time to teach students how to use a grapher, we have provided several features that shorten the learning curve while increasing the students' knowledge of the fundamentals of a grapher. The first of these is the section entitled "Introduction to Graphs and Graphers" found at the beginning of the text. It introduces students to the basic functions of the grapher. The others are the *Graphing Calculator Manual* (see p. xi) and the video series entitled *Graphing Calculator Instructional Videos* (see p. xii). In addition, a set of programs has been included in the *Graphing Calculator Manual*. All of these features have been specifically written and produced for this text.

• **Interactive Discoveries** The grapher provides an exciting teaching opportunity in which a student can discover and further investigate mathematical concepts. This unique Interactive Discovery feature is used to introduce new topics and provides a vehicle for students to "see" a concept quickly. This feature reinforces the idea that grapher technology is an integral part of the course as well as an important learning tool. It invites the student to develop analytic and reasoning skills while taking an active role in the learning process. (See pp. 50, 186, and 296.)

• **Function Emphasis** The use of technology with its immediate visualization of a concept encourages the early presentation of functions. Graphing and functions are both introduced in the first section of Chapter 1. The study of the family of functions (linear, quadratic, higher-degree polynomial, rational, exponential, logarithmic, and trigonometric) has been enhanced and streamlined with the inclusion of the grapher. Applications with graphs are incorporated throughout to amplify and add relevance to the study of functions. (See pp. 20, 104, and 174.)

• **Variety in Approaches to Solutions** Skill in solving mathematical problems is expanded when a student is exposed to a variety of approaches to finding a solution. We have carefully incorporated three solution approaches throughout the text: algebraic, graphical, and numerical. Chapter openers illustrate an application with a concurrent grapher presentation of both a table and a graph (see pp. 203 and 323). The TABLE feature on a grapher provides a numerical display or check of the solution (see pp. 123 and 207).

To highlight both the algebraic- and graphical-solution approaches in solving equations, we have used a two-column solution format in numerous examples (see pp. 248 and 434). In the algebraic/graphical side-by-side features, both methods are presented together; each method provides a complete solution. This feature emphasizes that there is more than one way to obtain a result and illustrates the comparative efficiency and accuracy of the two methods.

• **Real-Data Applications** Throughout the writing process, we conducted an energetic search for real-data applications. The result of that effort is a variety of examples and exercises that connect the mathematical content with the real world. Source lines appear with most real-data applications and charts and graphs are frequently included. Many applications are drawn from the fields of health, business and economics, life and physical sciences, social science, and areas of general interest such as sports and daily life. We encourage students to "see" and interpret the mathematics that appears around them every day. (See pp. 132–133, 165, and 441.)

• **Verifying Identities** Identities can be partially verified with a grapher using both the GRAPH and TABLE features. This use of the grapher is first seen in the Introduction to Graphs and Graphers and is continued in later chapters in discovery and verification of possible identities (see pp. 11, 207, and 424). This content feature allows a visual answer to such frequent questions as "Why isn't $(x + 2)^2$ equal to $x^2 + 4$?" This approach also provides a unique lead-in to the development of the properties of exponents and logarithms.

Pedagogical Features

- **Use of Color** The text uses full color in an extremely functional way, as seen in the design elements and artwork on nearly every page. The choice of color has been carried out in a methodical and precise manner so that its use carries a consistent meaning, which enhances the readability of the text for both student and instructor. (See pp. 29 and 191.)

- **Art Package** The text contains over 1300 art pieces including a new form of art called photorealism. Photorealism superimposes mathematics on a photograph and encourages students to "see mathematics" in familiar settings (see pp. 51 and 165). The exceptional situational art and statistical graphs throughout the text highlight the abundance of real-world applications while helping students visualize the mathematics (see pp. 129, 132, and 241). The design and use of color with the grapher windows exemplifies the impact that technology has in today's mathematical curriculum (see pp. 13, 92, and 242).

- **Annotated Examples** Over 450 examples fully prepare the student for the exercise sets. Learning is carefully guided with numerous color-coded art pieces and step-by-step annotations, with substitutions and annotations highlighted in red (see pp. 208 and 261).

- **Variety of Exercises** There are over 3500 exercises in this text. The exercise sets are enhanced not only by the inclusion of real-data applications with source lines, detailed art pieces, and technology windows that include both tables and graphs, but also by the following features.

  Technology Exercises Since the grapher is totally integrated in this text, exercise sets include both grapher and nongrapher exercises. In some cases, detailed instruction lines indicate the approach the student is expected to use. In others, the student is left to choose the approach that seems best, thereby encouraging critical thinking. (See pp. 43, 68, 194–195, and 439.)

  Skill Maintenance The exercises in this section have been specifically selected to review concepts previously taught in the text that are foundations for the material presented in the following section. They are chosen to prepare the student for the new concept(s) that will be covered next. (See pp. 33 and 127.)

Synthesis Exercises These exercises, which appear at the end of each exercise set, encourage critical thinking by requiring students to synthesize concepts from several sections or to take a concept a step further than in the regular exercises. (See pp. 90–91, 136.)

Thinking and Writing Exercises for thinking and writing, at the beginning of the synthesis exercises, are denoted with a maze icon ◈. They encourage students to both consider and write about key mathematical ideas in the chapter. Many of these exercises are open-ended, making them particularly suitable for use in class discussions or as collaborative activities. (See pp. 33, 127, 151, and 241.)

- **Chapter Openers** Each chapter opens with an application illustrated with both technology windows and situational art. The openers also include a table of contents listing section titles. (See pp. 19 and 115.)

- **Section Objectives** Content objectives are listed at the beginning of each section. These, together with subheadings throughout the section, provide a useful outline for both instructors and students. (See pp. 137, 217, and 258.)

- **Highlighted Information** Important definitions, properties, and rules are displayed in screened boxes. Summaries and procedures are listed in color-outlined boxes. Both of these design features present and organize the material for efficient learning and review. (See pp. 52, 175, and 206.)

- **Summary and Review** The Summary and Review at the end of each chapter provides an extensive set of review exercises along with a list of important properties and formulas covered in that chapter. This feature provides an excellent preparation for chapter tests and the final examination. Answers to all review exercises appear in the text along with section references that direct students to material to reexamine if they have difficulty with a particular exercise. (See pp. 198–202.)

Supplements for the Instructor

Instructor's Solutions Manual

The Instructor's Solutions Manual by Judith A. Penna contains worked-out solutions to all exercises in the exercise sets, including the thinking and writing exercises. It also includes a sample test with answers for each chapter and answers to the exercises in the appendixes. The sample tests are also included in the Student's Solutions Manual.

Printed Test Bank/Instructor's Manual

Prepared by Donna DeSpain, the Printed Test Bank/Instructor's Manual contains the following:

- 4 free-response test forms for each chapter, following the format and with the same level of difficulty as the tests in the Student's Solutions Manual.
- 2 multiple-choice test forms for each chapter.
- 6 alternate forms of the final examination, 4 with free-response questions and 2 with multiple-choice questions.
- Index to the Graphing Calculator Instructional Videos.

Testgen EQ

Testgen EQ is a computerized test generator that allows instructors to select test questions manually or randomly from selected topics or to use a ready-made test for each chapter. The test questions are algorithm-driven so that regenerated number values maintain problem types and provide a large number of test items in both multiple-choice and open-ended formats for one or more test forms. Test items can be viewed on screen, and the built-in question editor lets instructors modify existing questions or add new ones that include pictures, graphs, accurate math symbols, and variable text and numbers.

Additional features in the new Windows and Macintosh Test Generators allow the instructor to customize both the look and content of test-banks and tests. Test questions are easily transferred from the testbank to a test and can be sorted, searched, and displayed in various ways. Testgen EQ is free to adopters.

Course Management and Testing System

InterAct Math Plus for Windows and Macintosh (available from Addison Wesley Longman) combines course management and on-line testing with the features of the basic tutorial software (see "Supplements for the Student") to create an invaluable teaching resource. Consult your local Addison Wesley Longman sales consultant for details.

Supplements for the Student

Graphing Calculator Manual

The Graphing Calculator Manual by Judith A. Penna, with the assistance of John Garlow and Mike Rosenborg, contains keystroke level instruction for the Texas Instruments TI-82, TI-83, TI-85, and Hewlett Packard HP 48G graphing calculators. Modules for the Casio 9850, the Sharp 9200 and 9300, and the HP 38G are available on request. Contact your local Addison Wesley Longman sales consultant for details.

Bundled free with every copy of the text, the Graphing Calculator Manual uses actual examples and exercises from *Algebra and Trigonometry: Graphs and Models* to help teach students to use their graphing calculator. The order of topics in the Graphing Calculator Manual mirrors that of the texts, providing a just-in-time mode of instruction.

Student's Solutions Manual

The Student's Solutions Manual by Judith A. Penna contains completely worked-out solutions with step-by-step annotations for all the odd-numbered exercises in the exercise sets in the text, with the exception of the thinking and writing exercises. It also includes a self-test with answers for each chapter and a final examination.

The Student's Solutions Manual can be purchased by your students from Addison Wesley Longman.

Graphing Calculator Instructional Videos

Designed and produced specifically for *Algebra and Trigonometry: Graphs and Models*, the Graphing Calculator Instructional Videos take students through procedures on the graphing calculator using content from the text. These videos include most topics covered in the Graphing Calculator Manual, as well as several additional topics.

Every video section uses actual text examples or odd-numbered exercises from the text to help motivate students while they learn to use the graphing calculator. In the videos, an instructor shows students key-stroke level procedures that they will need to succeed with the grapher as they proceed through the course. The videos are correlated to the sections of the text.

A complete set of Graphing Calculator Instructional Videos is free to qualifying adopters.

InterAct Math Tutorial Software

InterAct Math Tutorial Software has been developed and designed by professional software engineers working closely with a team of experienced math educators.

InterAct Math Tutorial Software includes exercises that are linked with every objective in the textbook and require the same computational and problem-solving skills as their companion exercises in the text. Each exercise has an example and an interactive guided solution that are designed to involve students in the solution process and to help them identify precisely where they are having trouble. In addition, the software recognizes common student errors and provides students with appropriate customized feedback.

With its sophisticated answer recognition capabilities, InterAct Math Tutorial Software recognizes appropriate forms of the same answer for any kind of input. It also tracks student activity and scores for each section, which can then be printed out.

Available for both Windows and Macintosh computers, the software is free to qualifying adopters.

Acknowledgments

We wish to express our genuine appreciation to a number of people who contributed in special ways to the development of this textbook. Jason Jordan and Greg Tobin, our editors at Addison Wesley Longman, shared our vision and provided encouragement and motivation. In addition, the production and marketing departments of Addison Wesley Longman brought to the project their unsurpassed commitment to excellence. The unwavering support from the Higher Education Group has been a continuing source of strength for this author team. For this we are most grateful. Mike Rosenborg, Barbara Johnson, Patty Slipher, Irene Doo, and Larry Bittinger provided many constructive comments as well as accuracy checks to the manuscript.

Finally, Professor Bittinger would like to thank his MA 153 students at IUPUI for their productive response to parts of the manuscript that were class-tested in the spring of 1996 using the graphing calculator. This teaching approach resulted in the most satisfying class he has taught at IUPUI in 28 years. Further information regarding this class can be obtained from Professor Bittinger at his e-mail address, exponent@aol.com, or through his home page (see Web Connection that follows).

We would also like to thank the following reviewers for their invaluable contribution to the development of this text:

Sandra Beken, *Horry-Georgetown Technical College*

Diane Benjamin, *University of Wisconsin—Platteville*

Robert Bohac, *Northwest College*

Diane W. Burleson, *Central Piedmont Community College*

John W. Coburn, *St. Louis Community College at Florissant Valley*

Elaine N. Daniels, *Salve Regina University*

Donna DeSpain, *Benedictine University*

Eunice F. Everett, *Seminole Community College*

Rob Farinelli, *Community College of Allegheny County South Campus*

Betty P. Givan, *Eastern Kentucky University*

Allen R. Hesse, *Rochester Community College*

Heidi A. Howard, *Florida Community College at Jacksonville*

Steve Kahn, *Anne Arundel Community College*

Timothy A. Loughlin, *New York Institute of Technology*

Larry Luck, *Anoka-Ramsey Community College*

Joseph D. Mahoney, *Paducah Community College*

Peggy I. Miller, *University of Nebraska at Kearney*

John A. Oppelt, *Bellarmine College*

Tom Schaffter, *Fort Lewis College*

Eric Schulz, *Walla Walla Community College*

Judith D. Smalling, *St. Petersburg Junior College*

Craig M. Steenberg, *Lewis-Clark State College*

Kathryn C. Wetzel, *Amarillo College*

Kemble Yates, *Southern Oregon State College*

Web Connection

Students and instructors can obtain more information about books written by Professor Bittinger and his co-authors by contacting *Marv's Math Corner* on the Internet at the following address:

http://www.math.iupui.edu/~mbitting/

Included on this Web site is information about books and their supplements, as well as study tips and sample practice final examinations. Students and instructors are welcome to e-mail questions, comments, and constructive criticism.

M.L.B.
J.A.B.
D.J.E.
J.A.P.

- **Applied Chapter Openers:** Each chapter begins with a relevant application highlighting how concepts presented in the chapter can be put to use in the real world. These applications are accompanied by numerical tables, equations, and grapher windows to show students the many different ways in which problem situations can be examined.

The Trigonometric Functions 2

Musical instruments can generate complex sine waves. Consider two tones whose graphs are $y = 2 \sin x$ and $y = \sin 2x$. The combination of the two tones produces a new sound whose graph is $y = 2 \sin x + \sin 2x$.

X	Y1	Y2
0	0	0
.7854	1.4142	1
1.5708	2	0
2.3562	1.4142	−1
3.1416	0	0
3.927	−1.414	1
4.7124	−2	0

X = 3.1416

$y_1 = 2 \sin x$, $y_2 = \sin 2x$,
$y_3 = 2 \sin x + \sin 2x$

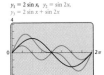

We now consider an important class of functions called *trigonometric*, or *circular*, *functions*. Historically, these functions arose from a study of triangles; hence the name trigonometric. We will begin our study with right triangles and degree measure and solve applications involving right triangles. Then we will consider trigonometric functions of angles or rotations of any size with both degree and radian measure. A circle of radius 1 (a unit circle) is then used to define the six basic trigonometric functions; hence the name circular functions. The domains and ranges of these functions consist of real numbers.

2.1 Trigonometric Functions of Acute Angles
2.2 Applications of Right Triangles
2.3 Trigonometric Functions of Any Angle
2.4 Radians, Arc Length, and Angular Speed
2.5 Circular Functions: Graphs and Properties
2.6 Graphs of Transformed Sine and Cosine Functions
SUMMARY AND REVIEW

115

b) To graph $g(x$
$g(x) = f(x +$
$f(x) = \sqrt{x}$, sh

c) To graph $g(x$
$g(x) = f(x) + 2$,
$f(x) = 1/x$, shifted

$y_1 = \frac{1}{x}$, $y_2 = \frac{1}{x} + 2$

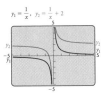

$y_1 = |x|$, $y_2 = |x - 4|$

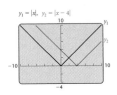

d) To graph $g(x) = |x - 4|$, think of the graph of $f(x) = |x|$. Since $g(x) = f(x - 4)$, the graph of $g(x) = |x - 4|$ is the graph of $f(x) = |x|$ shifted *right* 4 units. (See the figure on the right above.)

e) To graph $h(x) = \sqrt{x + 2} - 3$, think of the graph of $f(x) = \sqrt{x}$. In part (b), we found that the graph of $g(x) = \sqrt{x + 2}$ is the graph of $f(x) = \sqrt{x}$ shifted *left* 2 units. Since $h(x) = g(x) - 3$, we shift the graph of $g(x) = \sqrt{x + 2}$ *down* 3 units. Together, the graph of $f(x) = \sqrt{x}$ is shifted *left* 2 units and *down* 3 units.

$y_1 = \sqrt{x}$, $y_2 = \sqrt{x + 2}$, $y_3 = \sqrt{x + 2} - 3$

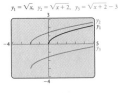

- **Grapher Integration:** The author team assumes that the student will use a grapher throughout the course and during homework or group sessions. Numerous grapher windows appear throughout the text.

• Interactive Discoveries:

Throughout the exposition, students are directed to investigate new concepts before they are formally developed. This design invites students to be actively involved with the material in order to identify a mathematical pattern or form an intuitive understanding of a new topic.

Interactive Discovery

We can graph the unit circle on a grapher. We use parametric mode with the following window and let $X_{1T} = \cos T$ and $Y_{1T} = \sin T$.

WINDOW
Tmin = 0
Tmax = 2π, or 6.2831853
Tstep = $\pi/12$, or 0.2617993
Xmin = -1.5
Xmax = 1.5
Xscl = 1
Ymin = -1
Ymax = 1
Yscl = 1

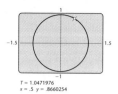

$T = 1.0471976$
$x = .5 \quad y = .8660254$

Using the trace key and an arrow key to move the cursor around the unit circle, we see the T, X, and Y values appear on the screen. What do they represent? (For more on parametric equations, see Section 5.8.)

From the definitions on page 167 and from the preceding exploration, we can relabel any point (x, y) on the unit circle as $(\cos s, \sin s)$, where s is any real number.

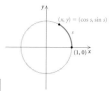

---

and Cosine Functions

can be easily observed from their graphs. We begin and cosine functions. We make a table of values, en connect those points with a smooth curve. It is unit circle and label a few points with coordinates.

---

unit circle gives us the identity
$$(\cos s)^2 + (\sin s)^2 = 1,$$
which can be expressed as
$$\mathbf{\sin^2 s + \cos^2 s = 1.}$$
It is conventional in trigonometry to use the notation $\sin^2 s$ rather than $(\sin s)^2$, which uses parentheses. Note that $\sin^2 s \neq \sin s^2$.

$y = (\sin x)^2$ $y = \sin x^2$

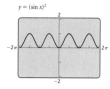   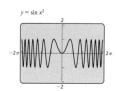

The identity $\sin^2 s + \cos^2 s = 1$ relates the sine and the cosine of any real number s. It is an important **Pythagorean identity**.

Interactive Discovery

Addition of y-values provides a unique way of developing the identity $\sin^2 x + \cos^2 x = 1$. First graph $y_1 = \sin^2 x$ and $y_2 = \cos^2 x$. By adding the y-values, visually graph the sum, $y_3 = \sin^2 x + \cos^2 x$. Then graph y_3 with a grapher and check your prediction. The resulting graph appears to be the line $y_4 = 1$, which is the graph of the expression on the right of the equals sign. These graphs do not prove the identity, but they do provide a check in the interval $[-2\pi, 2\pi]$.

$y_1 = \sin^2 x$ $y_2 = \cos^2 x$

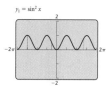   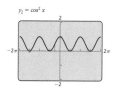

$y_3 = \sin^2 x + \cos^2 x, \quad y_4 = 1$

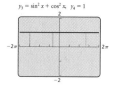

• Side-by-Side Algebraic and Graphical Solutions:

Many examples in the text are presented in a two-column format that shows simultaneous algebraic and graphical solution methods. This balanced approach allows students to compare the efficiency and appropriateness of each method.

Example 7   Solve: $\log_3 (2x - 1) - \log_3 (x - 4) = 2$.

ALGEBRAIC SOLUTION

We have

$$\log_3 (2x - 1) - \log_3 (x - 4) = 2$$

$$\log_3 \frac{2x - 1}{x - 4} = 2 \qquad \text{Using the quotient rule}$$

$$\frac{2x - 1}{x - 4} = 3^2 \qquad \text{Writing an equivalent exponential equation}$$

$$\frac{2x - 1}{x - 4} = 9$$

$$2x - 1 = 9(x - 4) \qquad \text{Multiplying by the LCD, } x - 4$$

$$2x - 1 = 9x - 36$$

$$35 = 7x$$

$$5 = x.$$

CHECK:

$$\frac{\log_3 (2x - 1) - \log_3 (x - 4) = 2}{\log_3 (2 \cdot 5 - 1) - \log_3 (5 - 4) \, ? \, 2}$$

$$\log_3 9 - \log_3 1 $$

$$2 - 0 \, \bigg| $$

$$2 \, \bigg| \, 2 \quad \text{TRUE}$$

The solution is 5.

GRAPHICAL SOLUTION

We use the change-of-base formula and graph the equations

$$y_1 = \frac{\ln (2x - 1)}{\ln 3} - \frac{\ln (x - 4)}{\ln 3}$$

and

$$y_2 = 2.$$

Then we find the point(s) of intersection.

$$y_1 = \frac{\ln (2x-1)}{\ln 3} - \frac{\ln (x-4)}{\ln 3}, \quad y_2 = 2$$

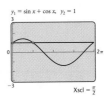

$$x = 5, \, y = 2$$

The solution is 5.

Example 9   Solve $\sin x + \cos x = 1$ in $[0, 2\pi)$.

ALGEBRAIC SOLUTION

We have

$$\sin x + \cos x = 1$$

$$\sin^2 x + 2 \sin x \cos x + \cos^2 x = 1 \qquad \text{Squaring both sides}$$

$$2 \sin x \cos x + 1 = 1 \qquad \text{Using } \sin^2 x + \cos^2 x = 1$$

$$2 \sin x \cos x = 0$$

$$\sin 2x = 0. \qquad \text{Using } 2 \sin x \cos x = \sin 2x$$

We are looking for solutions x to the equation for which $0 \le x < 2\pi$. Multiplying by 2, we get $0 \le 2x < 4\pi$, which is the interval we consider to solve $\sin 2x = 0$. These values of $2x$ are 0, π, 2π, and 3π. Thus the desired values of x in $[0, 2\pi)$ satisfying this equation are 0, $\pi/2$, π, and $3\pi/2$. Now we check these in the original equation:

$$\sin 0 + \cos 0 = 0 + 1 = 1,$$

$$\sin \frac{\pi}{2} + \cos \frac{\pi}{2} = 1 + 0 = 1,$$

$$\sin \pi + \cos \pi = 0 + (-1) = -1,$$

$$\sin \frac{3\pi}{2} + \cos \frac{3\pi}{2} = (-1) + 0 = -1.$$

We find that π and $3\pi/2$ do not check, but the other values do. Thus the solutions are

$$0 \quad \text{and} \quad \frac{\pi}{2}.$$

When the solution process involves squaring both sides, values are often obtained that are not solutions of the original equation. As we saw in this example, it is important to check the possible solutions.

GRAPHICAL SOLUTION

We can graph the left side and then the right side of the equation as seen in the first window below. Then we look for points of intersection. We could also rewrite the equation as $\sin x + \cos x - 1 = 0$, graph the left side, and look for the zeros of the function, as illustrated in the second window below. In each window, we see the solutions in $[0, 2\pi)$ as 0 and $\pi/2$.

$$y_1 = \sin x + \cos x, \quad y_2 = 1$$

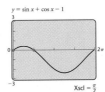

$$Xscl = \frac{\pi}{2}$$

$$y = \sin x + \cos x - 1$$

$$Xscl = \frac{\pi}{2}$$

This example illustrates a valuable advantage of the grapher—that is, with a grapher, extraneous solutions do not appear.

---

encounter equations for which an algebraic solution impossible.

$$\ldots^{.5x} - 7.3 = 2.08x + 6.2.$$

an equation for which an algebraic solution seems ble.

ons

$$\ldots 7.3 \quad \text{and} \quad y_2 = 2.08x + 6.2$$

of intersection. (See the figure at left.) nsider the equation

$$\ldots .08x - 6.2, \quad \text{or} \quad y = e^{0.5x} - 2.08x - 13.5,$$

sing TRACE and ZOOM or a SOLVE feature. The ap- are -6.471 and 6.610.

- **Art:** Generous amounts of color-coded technical and situational art appear throughout the text to enhance understanding of an example or exercise, to interest students, and to aid in the visualization of concepts.

17. *Safety Line to Raft.* Each spring Madison uses his vacation time to ready his lake property for the summer. He wants to run a new safety line from point *B* on the shore to the corner of the anchored diving raft. The current safety line, which runs perpendicular to the shore line to point *A*, is 40 ft long. He estimates the angle from *B* to the corner of the raft to be 50°. Approximately how much rope does he need for the new safety line if he allows 5 ft of rope at each end to fasten the rope?

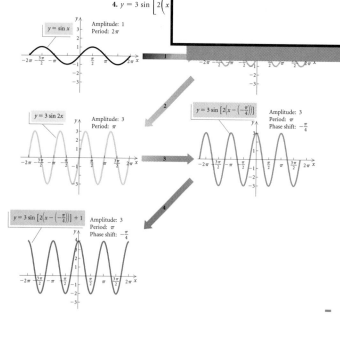

18. *Enclosing an Area.* Glynis is enclosing a triangular area in a corner of her fenced rectangular backyard for her Labrador retriever. In order for a certain tree to be included in this pen, one side needs to be 14.5 ft and make a 53° angle with the new side. How long is the new side?

19. *Easel Display.* A marketing group is designing an easel to display posters advertising their newest products. They want the easel to be 6 ft tall and fit flush against a wall. For optimal eye contact, the best angle between the front and back legs of the easel is 23°. How far from the wall should the front legs be placed in order to obtain this angle?

20. *Height of a Tree.* A supervisor must train a new team of loggers to estimate the heights of trees. As an example, she walks off 40 ft from the base of a tree and estimates the angle of elevation to the

tree's peak to be 70°. Approximately how tall is the tree?

21. *Sand Dunes National Park.* While visiting the Sand Dunes National Park in Colorado, Ian approximated the angle of elevation to the top of a sand dune to be 20°. After walking 800 ft closer, he guessed that the angle of elevation had increased by 15°. Approximately how tall is the dune he was observing?

22. *Tee-Shirt Design.* A new tee-shirt design is to have a regular pentagon inscribed in a circle, as shown in the figure. Each side of the pentagon is to be 3.5 in. long. Find the radius of the circumscribed circle.

23. *Inscribed Octagon.* A regular octagon is inscribed in a circle of radius 15.8 cm. Find the perimeter of the octagon.

SECTION 2.6

Amplitude = | |

Period = | $\dfrac{2\pi}{C}$ |

Phase shift =

To create the final gra...
Then we sketch gra...

1. $y = \sin 2x$
2. $y = 3 \sin 2x$
3. $y = 3 \sin \left[2\left(x \right.\right.$
4. $y = 3 \sin \left[2\left(x \right.\right.$

$y = \sin x$ Amplitude: 1
Period: 2π

$y = 3 \sin 2x$ Amplitude: 3
Period: π

$y = 3 \sin \left[2\left(x - \left(-\frac{\pi}{4}\right)\right)\right]$ Amplitude: 3
Period: π
Phase shift: $-\frac{\pi}{4}$

$y = 3 \sin \left[2\left(x - \left(-\frac{\pi}{4}\right)\right)\right] + 1$ Amplitude: 3
Period: π
Phase shift: $-\frac{\pi}{4}$

- To assist students with understanding, **graphs with multiple curves** use different colors for each curve. Movement on the graph may be indicated by a gradual shift in color on an accompanying shift arrow.

- **Direction lines** in the exercise sets instruct students to use both algebraic and graphical methods to solve problems. Sometimes the student is asked to solve with a grapher and check algebraically. In other problems, the student is directed to solve algebraically and check the work with a grapher. Many exercises do not specify a solution method and allow the student to determine the method, thus encouraging critical thinking and analytical skills.

6.4 Exercise Set

Solve each exponential equation algebraically. Then check on a grapher.

1. $3^x = 81$ **2.** $2^x = 32$

3. $2^{2x} = 8$ **4.** $3^{7x} = 27$

5. $2^x = 33$ **6.** $2^x = 40$

7. $5^{4x-7} = 125$ **8.** $4^{3x-5} = 16$

9. $27 = 3^{5x} \cdot 9^{x^2}$ **10.** $3^{x^2+4x} = \frac{1}{27}$

11. $84^x = 70$ **12.** $28^x = 10^{-3x}$

13. $e^t = 1000$ **14.** $e^{-t} = 0.04$

15. $e^{-0.03t} = 0.08$ **16.** $1000e^{0.09t} = 5000$

17. $3^x = 2^{x-1}$ **18.** $5^{x+2} = 4^{1-x}$

19. $(3.9)^x = 48$ **20.** $250 - (1.87)^x = 0$

21. $e^x + e^{-x} = 5$ **22.** $e^x - 6e^{-x} = 1$

23. $\dfrac{e^x + e^{-x}}{e^x - e^{-x}} = 3$ **24.** $\dfrac{5^x - 5^{-x}}{5^x + 5^{-x}} = 8$

Solve each logarithmic equation algebraically. Then check on a grapher.

25. $\log_5 x = 4$ **26.** $\log_2 x = -3$

27. $\log x = -4$ **28.** $\log x = 1$

29. $\ln x = 1$ **30.** $\ln x = -2$

31. $\log_2 (10 + 3x) = 5$ **32.** $\log_5 (8 - 7x) = 3$

33. $\log x + \log (x - 9) = 1$

34. $\log_2 (x + 1) + \log_2 (x - 1) = 3$

35. $\log_8 (x + 1) - \log_8 x = 2$

36. $\log x - \log (x + 3) = -1$

37. $\log_4 (x + 3) + \log_4 (x - 3) = 2$

38. $\ln (x + 1) - \ln x = \ln 4$

39. $\log (2x + 1) - \log (x - 2) = 1$

40. $\log_5 (x + 4) + \log_5 (x - 4) = 2$

Use only a grapher. Find approximate solutions of each equation or approximate the point(s) of intersection of a pair of equations.

41. $e^{7.2x} = 14.009$ **42.** $0.082e^{0.05x} = 0.034$

43. $xe^{3x} - 1 = 3$ **44.** $5e^{5x} + 10 = 3x + 40$

45. $4 \ln (x + 3.4) = 2.5$ **46.** $\ln x^2 = -x^2$

47. $\log_8 x + \log_8 (x + 2) = 2$

48. $\log_3 x + 7 = 4 - \log_5 x$

49. $\log_5 (x + 7) - \log_5 (2x - 3) = 1$

50. $y = \ln 3x, \quad y = 3x - 8$

51. $2.3x + 3.8y = 12.4, \quad y = 1.1 \ln (x - 2.05)$

52. $y = 2.3 \ln (x + 10.7), \quad y = 10e^{-0.07x^2}$

53. $y = 2.3 \ln (x + 10.7), \quad y = 10e^{-0.007x^2}$

Skill Maintenance

54. Solve $K = \frac{1}{2}mv^2$ for v.

Solve.

55. $x^4 + 5x^2 = 36$ **56.** $t^{2/3} - 10 = 3t^{1/3}$

57. *Total Sales of Goodyear.* The following table shows factual data regarding total sales of The Goodyear Tire and Rubber Company.

YEAR, x	TOTAL SALES, y (IN MILLIONS)
1. 1991	\$10,906.8
2. 1992	11,784.9
3. 1993	11,643.4
4. 1994	12,288.2

Source: The Goodyear Tire and Rubber Company Annual Report.

a) Use linear regression on a grapher to fit an equation $y = mx + b$, where $x = 1$ corresponds to 1991, to the data points. Predict total sales in 1999. (See Appendix A.3.)

b) Use quadratic regression on a grapher to fit an equation $y = ax^2 + bx + c$ to the data points. Predict total sales in 1999. (See Appendix A.3.)

Synthesis

58. ◆ In Example 3, we took the natural logarithm on both sides. What would have happened had we taken the common logarithm? Explain which seems best to you and why.

59. ◆ Explain how Exercises 29 and 30 could be solved using the graph of $f(x) = \ln x$.

Any pair of sides and the included a also have

$$K = \frac{1}{2}ab \sin C \quad \text{and} \quad K =$$

The Area of a Triangle

The area K of any $\triangle ABC$ is one h of two sides and the sine of the in

$$K = \frac{1}{2}bc \sin A = \frac{1}{2} ab \sin$$

Example 6 *Area of the Peace Monume* the City Program sponsored by Marian C children have turned a vacant downtown lot into a monument for peace.* This community project brought together neighborhood volunteers, businesses, and government in hopes of showing children how to develop positive nonviolent ways of dealing with conflict. A landscape architect[†] used the children's drawings and ideas to design a triangular-shaped peace garden. Two sides of the property, formed by Indiana Avenue and Senate Avenue, measure 182 ft and 230 ft, respectively, and together form a 44.7° angle. The third side of the garden, formed by an apartment building, measures 163 ft. What is the area of this property?

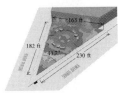

SOLUTION   Since we do not know a height of the triangle, we use the area formula:

$$K = \frac{1}{2}bc \sin A$$
$$K = \frac{1}{2} \cdot 182 \text{ ft} \cdot 230 \text{ ft} \cdot \sin 44.7°$$
$$K \approx 14{,}722 \text{ ft}^2.$$

The area of the property is approximately 14,722 ft².

*The Indianapolis Star, August 6, 1995, p. J8.
[†]Alan Day, a landscape architect with Browning Day Mullins Dierdorf, Inc., donated his time to this project.

- **Examples/Exercises:** The examples and exercises reflect the text's focus. Many use situational or grapher art and many also incorporate recent, source-based data to illustrate applications and concepts. An Index of Applications is included at the end of the text.

• Synthesis/Skill Maintenance exercises: Skill Maintenance exercises prepare students for the next section by reviewing and reinforcing skills that are used in the following section. Synthesis exercises require additional thought and encourage students to take the material one step further by combining concepts. Exercises marked with the maze icon are designed to foster critical thinking and writing skills; they may also be completed by small groups.

Skill Maintenance

51. Add: $\dfrac{2x}{x^2 - 1} + \dfrac{4}{x - 1}$.

52. Multiply: $(x^2 - 4)(3x + 7)$.

53. Simplify: $\dfrac{2x^2 - 3x - 20}{2x^2 + 3x - 5}$.

54. Subtract: $\dfrac{1}{x - 6} - 4$.

Synthesis

55. ◆ In the equations $y = A \sin (Cx - D) + B$ and $y = A \cos (Cx - D) + B$, which constants translate the graphs and which constants stretch and shrink the graphs? Describe in your own words the effect of each constant.

56. ◆ In the transformation steps listed in this section, why must step (1) precede step (3)? Give an example that illustrates this.

Some graphers can use regression to fit a trigonometric function to a set of data.

57. *Sales.* Sales of certain products fluctuate in cycles. The data in the following table show the total sales of skis per month for a business in a northern climate.

x, DAY	NUMBER OF DAYLIGHT HOURS, y
10. January 10	5.0
50. February 19	9.1
62. March 3	10.4
118. April 28	16.4
134. May 14	18.2
162. June 11	20.7
198. July 17	19.5
234. August 22	15.7
262. September 19	12.7
274. October 1	11.4
318. November 14	6.7
362. December 28	4.3

Source: The Astronomical Almanac, 1995, Washington: U.S. Government Printing Office.

a) Using the SINE REGRESSION feature on a grapher, model these data with an equation of the form $y = A \sin (Cx - D) + B$.

b) Approximate the number of daylight hours in Kajaani for April 22, July 4, and December 15.

The transformational techniques that we learned in this section for graphing the sine and cosine functions can also be applied to the other trigonometric functions. Sketch a graph of each of the following. Then check your work with a grapher.

59. $y = -\tan x$ **60.** $y = \tan (-x)$

61. $y = -2 + \cot x$ **62.** $y = -\dfrac{3}{2} \csc x$

63. $y = 2 \tan \dfrac{1}{2} x$ **64.** $y = \cot 2x$

65. $y = 2 \sec (x - \pi)$ **66.** $y = 4 \tan \left(\dfrac{1}{4}x + \dfrac{\pi}{8}\right)$

67. $y = 2 \csc \left(\dfrac{1}{2}x - \dfrac{3\pi}{4}\right)$ **68.** $y = 4 \sec (2x - \pi)$

Use a grapher to graph each function.

69. $f(x) = e^{-x/2} \cos x$ **70.** $f(x) = x \sin x$

71. $f(x) = 0.6x^2 \cos x$ **72.** $f(x) = 2^{-x} \cos x$

73. $f(x) = |\tan x|$ **74.** $f(x) = (\tan x)(\csc x)$

CHAPTER

6 Summary and Review

Important Properties and Formulas

Exponential Function:	$f(x) = a^x$
The Number $e = 2.7182818284\ldots$	
Logarithmic Function:	$f(x) = \log_a x$
A Logarithm is an Exponent:	$\log_a x = y \leftrightarrow x = a^y$
The Change-of-Base Formula:	$\log_b M = \dfrac{\log_a M}{\log_a b}$
The Product Rule:	$\log_a MN = \log_a M + \log_a N$
The Power Rule:	$\log_a M^p = p \log_a M$
The Quotient Rule:	$\log_a \dfrac{M}{N} = \log_a M - \log_a N$
Other Properties:	$\log_a a = 1, \qquad \log_a 1 = 0,$
	$\log_a a^x = x, \qquad a^{\log_a x} = x$
Base–Exponent Property:	$a^x = a^y \leftrightarrow x = y$
Exponential Growth Model:	$P(t) = P_0 e^{kt}$
Exponential Decay Model:	$P(t) = P_0 e^{-kt}$
Interest Compounded Continuously:	$P(t) = P_0 e^{kt}$
Limited Growth:	$P(t) = \dfrac{a}{1 + be^{-kt}}$

REVIEW EXERCISES

In Exercises 1–6, match the equation with one of figures (a)–(f), which follow. If needed, use a grapher.

a)   b)

e)   f)

c)   d)

1. $f(x) = e^{x-3}$ **2.** $f(x) = \log_3 x$
3. $y = -\log_3 (x + 1)$ **4.** $y = \left(\dfrac{1}{2}\right)^x$
5. $f(x) = 3(1 - e^{-x}), \ x \geq 0$
6. $f(x) = |\ln (x - 4)|$

• End-of-Chapter material: End-of-Chapter material includes a summary and review of properties and formulas along with a complete set of Review Exercises. Review Exercises also include Synthesis exercises, as well as critical thinking and writing exercises. The answers to the Review Exercises, which appear at the back of the text, have text section references to further aid students.

Trigonometry

GRAPHS AND MODELS

Introduction to Graphs and Graphers

- *Review hand-drawn graphs of equations.*
- *Use a grapher to create graphs with various viewing windows, to create x, y tables for equations, and to determine whether an equation is an identity.*
- *Use a grapher to find coordinates of points on a graph, to solve equations, and to find the points of intersection of two graphs.*

Graphing calculators and computers equipped with graphing software can alleviate much of the toil of graph making. We will henceforth refer to all such graphing utilities simply as **graphers**. As the equations we encounter become more complicated, it becomes increasingly difficult to produce accurate hand-drawn graphs. It can take considerable time to calculate just a few ordered pairs that are solutions. In addition, many ordered pairs are often required to produce an accurate graph. The use of a grapher can not only shorten this process but also perform many other mathematical procedures efficiently. Also, many equations arise that are difficult to analyze without machine assistance. Throughout this text, we will use graphers to enhance the learning process. The following is our philosophy regarding the use of graphers.

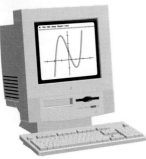

The Use of the Grapher

The grapher creates a visual presentation that *increases* understanding and *saves* time. It is used to enhance the mathematics, not to replace it!

Most of our discussion of the grapher is presented in a relatively generic form. Expanded discussion specific to certain graphers is included in the Graphing Calculator Manual that accompanies this book. To determine specific procedures, you should consult this manual, your user's manual, or your instructor.

Graphs

Hand-drawn graphs are reviewed first. With this groundwork established, we then consider the use of graphers. This introduction to graphs and graphers provides the foundation for the remainder of the text.

Graphs provide a natural way to link algebra and geometry. It is not uncommon to open a newspaper or magazine and encounter graphs. Shown below are examples of bar, circle, and line graphs.

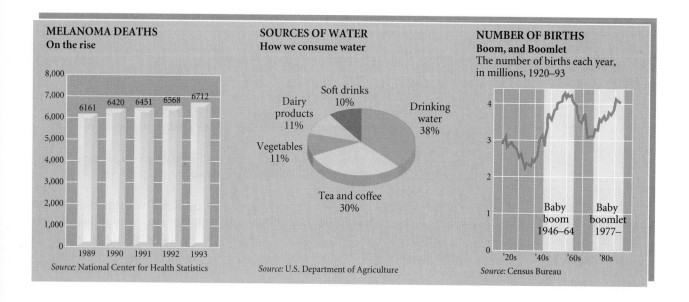

MELANOMA DEATHS
On the rise

Source: National Center for Health Statistics

SOURCES OF WATER
How we consume water

Soft drinks 10%
Dairy products 11%
Vegetables 11%
Tea and coffee 30%
Drinking water 38%

Source: U.S. Department of Agriculture

NUMBER OF BIRTHS
Boom, and Boomlet
The number of births each year, in millions, 1920–93

Baby boom 1946–64
Baby boomlet 1977–

Source: Census Bureau

Many real-world situations can be modeled using equations in which two variables appear. We use a plane to graph a pair of numbers. To locate points on a plane, we use two perpendicular number lines, called **axes**, which intersect at 0. We call this point the **origin**. The horizontal axis is called the **x-axis**, and the vertical axis is called the **y-axis**. (Other variables can be used.) The axes divide the plane into four regions, called **quadrants**, denoted by Roman numerals, numbered counterclockwise from the upper right. Arrows show the positive direction of each axis.

Each point (a, b) in the plane is called an **ordered pair**. The first number, a, indicates the point's horizontal location with respect to the y-axis, and the second number, b, indicates the point's vertical location with respect to the x-axis. We call a the **first coordinate**, **x-coordinate**, or **abscissa**. We call b the **second coordinate**, **y-coordinate**, or **ordinate**. Such a representation is called the **Cartesian coordinate system** in honor of the great French mathematician and philosopher René Descartes (1596–1650).

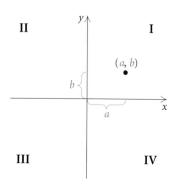

Example 1 Graph and label the points in the set

$$\{(-3, 5), (4, 3), (3, 4), (-4, -2), (3, -4), (0, 4), (-3, 0), (0, 0)\}.$$

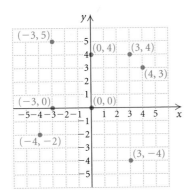

SOLUTION   To graph or **plot** $(-3, 5)$, we note that the x-coordinate tells us to move from the origin 3 units to the left of the y-axis. Then we move 5 units up from the x-axis. To graph the other points, we proceed in a similar manner. (See the graph at left.) Note that the origin, $(0, 0)$, is located at the intersection of the axes and that the point $(4, 3)$ is different from the point $(3, 4)$.

A graph of a set of points, as in Example 1, is called a **scatterplot**.

Solutions of Equations

Equations in two variables, like $2x + 3y = 18$, have solutions that are ordered pairs such that when the first coordinate is substituted for x and the second coordinate is substituted for y, the result is a true equation.

Example 2   Determine whether each ordered pair is a solution of $2x + 3y = 18$.

a) $(-5, 7)$ b) $(3, 4)$

SOLUTION   We check as follows.

a)
$$2x + 3y = 18$$

$2(-5) + 3(7) \; ? \; 18$ We substitute -5 for x and 7 for y (alphabetical order).

$-10 + 21$

$\quad\quad 11 \; | \; 18$ FALSE

The equation $11 = 18$ is false, so $(-5, 7)$ is not a solution.

b)
$$2x + 3y = 18$$

$2(3) + 3(4) \; ? \; 18$ We substitute 3 for x and 4 for y.

$6 + 12$

$\quad\quad 18 \; | \; 18$ TRUE

The equation $18 = 18$ is true, so $(3, 4)$ is a solution.

Graphs of Equations

The equation considered in Example 2 actually has an infinite number of solutions. Since we cannot list all of the solutions, we will make a drawing, called a **graph**, that represents them.

To Graph an Equation

To graph an equation is to make a drawing that represents the solutions of that equation.

At left are some suggestions for making hand-drawn graphs.

Suggestions for Hand-Drawn Graphs

1. Use graph paper.
2. Draw axes and label them with the variables.
3. Use arrows on the axes to indicate positive directions.
4. Scale the axes, that is, mark numbers on the axes.
5. Calculate solutions and list the ordered pairs in a table.
6. Plot solutions, look for patterns, and complete the graph. Label the graph with the equation being graphed.

Example 3   Graph: $2x + 3y = 18$.

SOLUTION   To find ordered pairs that are solutions of this equation, we can replace either x or y with any number and then solve for the other variable. For instance, if x is replaced with 0, then

$$2 \cdot 0 + 3y = 18$$
$$3y = 18$$
$$y = 6. \qquad \text{Dividing by 3}$$

Thus $(0, 6)$ is a solution. If x is replaced with 5, then

$$2 \cdot 5 + 3y = 18$$
$$10 + 3y = 18$$
$$3y = 8 \qquad \text{Subtracting 10}$$
$$y = \tfrac{8}{3}. \qquad \text{Dividing by 3}$$

Thus $\left(5, \tfrac{8}{3}\right)$ is a solution. If y is replaced with 0, then

$$2x + 3 \cdot 0 = 18$$
$$2x = 18$$
$$x = 9. \qquad \text{Dividing by 2}$$

Thus $(9, 0)$ is a solution.

We continue choosing values for one variable and finding the corresponding values of the other. We list the solutions in a table, and then plot the points. Note that the points appear to lie on a straight line.

x	y	(x, y)
0	6	$(0, 6)$
5	$\frac{8}{3}$	$\left(5, \frac{8}{3}\right)$
9	0	$(9, 0)$
-1	$\frac{20}{3}$	$\left(-1, \frac{20}{3}\right)$

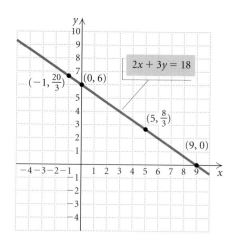

In fact, were we to graph additional solutions of $2x + 3y = 18$, they would be on the same straight line. Thus, to complete the graph, we use a straightedge to draw a line as shown in the figure. That line represents all solutions of the equation. ▬

Graphs of lines are examined in detail in Section 1.3.

Example 4   Graph: $y = x^2 - 5$.

SOLUTION   Since y is expressed in terms of x, the easiest way to find solutions of this equation is to replace x with various values and then calculate the corresponding values for y. In cases like this, we say that x is the **independent variable** (because we *choose* its value) and y is the **dependent variable** (because its value is then calculated).

x	y	(x, y)
0	-5	$(0, -5)$
-1	-4	$(-1, -4)$
1	-4	$(1, -4)$
-2	-1	$(-2, -1)$
2	-1	$(2, -1)$
-3	4	$(-3, 4)$
3	4	$(3, 4)$

① Select values for x.
② Compute values for y.

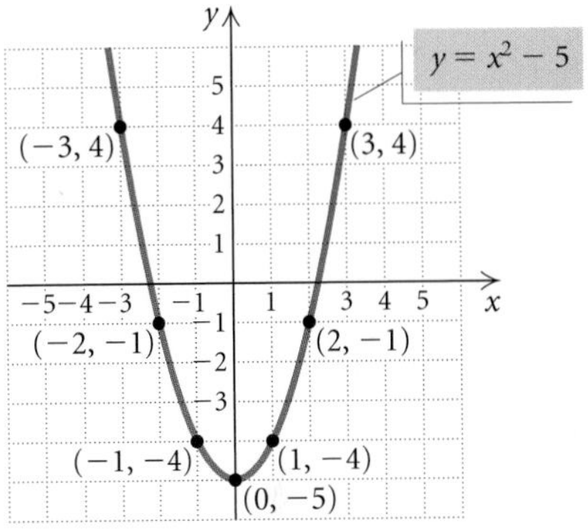

Next, we plot these points and connect them. We note that as the absolute value of x increases, $x^2 - 5$ also increases. Thus the graph is a curve that rises on either side of the y-axis.

Graphs of curves similar to the one in Example 4 are examined in more detail in Section 1.6.

Example 5   Graph: $x = y^2 + 1$.

SOLUTION   Since x is expressed in terms of y, we select values for y and then find the corresponding values for x. In this case, y is the independent variable and x the dependent variable.

x	y	(x, y)
1	0	$(1, 0)$
2	-1	$(2, -1)$
2	1	$(2, 1)$
5	-2	$(5, -2)$
5	2	$(5, 2)$

① Select values for y.
② Compute values for x.

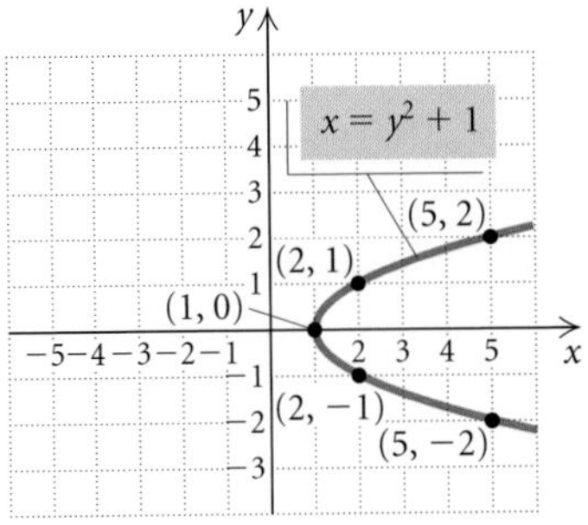

When plotting, we must remember that x is the first coordinate and y is the second coordinate. Note in the figure that the graph moves farther from the x-axis as it extends farther to the right.

Example 6 Graph: $xy = 12$.

SOLUTION   To find numbers that satisfy the equation, it helps to solve for y (that is, $y = 12/x$) or solve for x (that is, $x = 12/y$). Typically, we solve for y if it is convenient. Then we make substitutions.

x	y	(x, y)
2	6	$(2, 6)$
-2	-6	$(-2, -6)$
3	4	$(3, 4)$
-3	-4	$(-3, -4)$
4	3	$(4, 3)$
-4	-3	$(-4, -3)$
6	2	$(6, 2)$
-6	-2	$(-6, -2)$

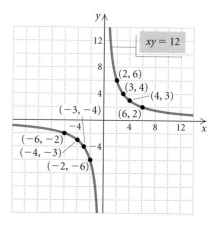

We plot these points and connect them. As the absolute value of x becomes small, the absolute value of y becomes large, and vice versa. Since neither x nor y can be 0, the graph does not cross either axis. Thus the graph consists of two branches. ▬

Example 7 Graph: $y = |x|$.

SOLUTION   We find numbers that satisfy the equation. For example, when $x = -3$, $y = |-3| = 3$. When $x = 2$, $y = |2| = 2$, and when $x = 0$, $y = |0| = 0$.

x	y	(x, y)
0	0	$(0, 0)$
-1	1	$(-1, 1)$
1	1	$(1, 1)$
-2	2	$(-2, 2)$
2	2	$(2, 2)$
-3	3	$(-3, 3)$
3	3	$(3, 3)$

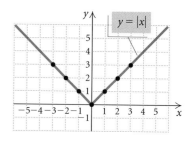

We plot these points and connect them. As the points fall farther from the origin, to the left or right, the absolute value of x increases. Thus the graph is a V-shaped curve that rises to the left and right of the y-axis. ▬

Graphers and Viewing Windows

We now consider the use of graphers. One feature common to all graphers is the **viewing window**. This refers to the rectangular portion of the xy-plane that appears on the screen of a grapher. The notation we

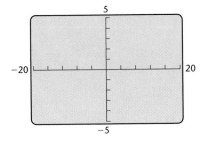

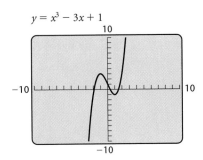

will use to describe viewing windows consists of four numbers, [L, R, B, T] in brackets, which represent the *Left* and *Right* endpoints of the x-axis and the *Bottom* and *Top* endpoints of the y-axis. A WINDOW feature might be used to set these dimensions, but the method varies for each grapher. The screen at top left is a window setting of $[-20, 20, -5, 5]$ with axis scaling denoted as Xscl = 4 and Yscl = 1, which means that there are 4 units between tick marks on the x-axis and 1 unit between tick marks on the y-axis. The screen at bottom left shows the results from this setting.

Axis scaling must be chosen with care, because tick marks become blurred and indistinguishable when too many appear. On some graphers, a setting of $[-10, 10, -10, 10]$, Xscl = 1, Yscl = 1 is considered the **standard window**. There is usually a procedure to obtain a standard setting quickly. Consult your manual.

The primary use of graphers is to graph equations. As an example, let's graph the equation $y = x^3 - 3x + 1$. The equation might be entered using the notation $y = x\text{\textasciicircum}3 - 3x + 1$. Some software use BASIC notation, in which case the equation would be entered as $y = x\text{\textasciicircum}3 - 3*x + 1$. We obtain the following graph.

You may need to change the viewing window in order to clearly reveal the curvature of a graph. For example, each of the following is a graph of $y = 3x^5 - 20x^3$. The graph on the right best represents the curve.

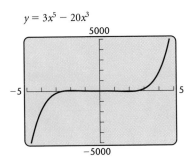

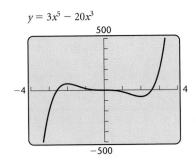

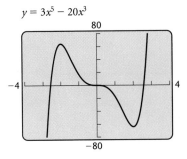

To graph an equation like $2x + 3y = 18$, most graphers require that the equation be solved for y, that is, "$y = \ldots$". When we solve $2x + 3y = 18$ for y, we obtain $y = (18 - 2x)/3$, or $y = 6 - \frac{2}{3}x$. As another example, we can write an equation like $x = y^2 + 1$ as

$y = \pm\sqrt{x-1}$ and graph the individual equations $y_1 = \sqrt{x-1}$ and $y_2 = -\sqrt{x-1}$. The combination of the two graphs yields the graph of $x = y^2 + 1$.

Just as the form in which an equation is entered varies among graphers, so too do other types of notation. Some examples of commonly used grapher notation follow.

EQUATION	POSSIBLE GRAPHER NOTATION
$y = 4x^3 - 7x^2 + 5x - 10$	$y = 4x\char94 3 - 7x^2 + 5x - 10$ or $4*x\char94 3 - 7*x\char94 2 + 5*x - 10$
$y = \sqrt{x-1}$	$y = (x-1)\char94(1/2)$ or $\sqrt{\,}(x-1)$
$y = |x-5|$	$y = abs\,(x-5)$
$y = \sqrt[3]{x}$	$y = x\char94(1/3)$ or $\sqrt[3]{\,}x$
$y = \dfrac{7.8x^2 - 1}{x^5 + 3}$	$y = (7.8x^2 - 1)/(x\char94 5 + 3)$ or $(7.8*x\char94 2 - 1)/(x\char94 5 + 3)$

Example 8   Graph each of the following equations choosing a viewing window that best reveals the curvature of the graph. Answers may vary.

a) $2x + 3y = 18$ **b)** $y = x^2 - 5$ **c)** $x = y^2 + 1$
d) $xy = 12$ **e)** $y = |x|$ **f)** $y = x^4 - 2x^2 - 3$

SOLUTION

a) $2x + 3y = 18$ **b)** $y = x^2 - 5$

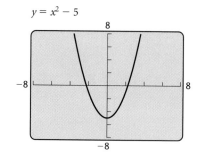

c) $x = y^2 + 1$ **d)** $xy = 12$

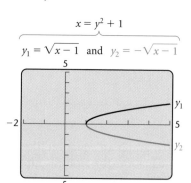

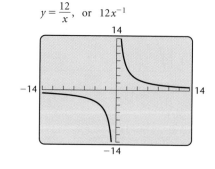

e) $y = |x|$

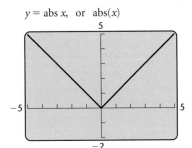

f) $y = x^4 - 2x^2 - 3$

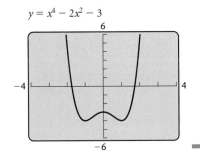

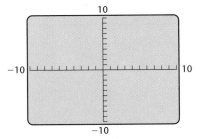

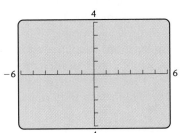

Squaring a Viewing Window

Consider the $[-10, 10, -10, 10]$ viewing window shown first at left. Note that the distance between units is not visually the same on both axes. In this case, the length of the interval shown on the y-axis is about $\frac{2}{3}$ of the length of the interval shown on the x-axis. If we change to the window shown second at left, with dimensions $[-6, 6, -4, 4]$, we get a graph for which the units are visually about the same on both axes. We choose these dimensions so that the length of the y-axis is $\frac{2}{3}$ the length of the x-axis. The window has been *squared*. Any dimensions in this ratio will produce the desired effect with this grapher.

On another grapher, the ratio might be $\frac{8}{15}$, and the window $[-7.5, 7.5, -4, 4]$ would show the units visually the same on both axes.* Creating such a window is called **squaring**. On some graphers, there is a ZSQUARE feature for automatic window squaring. Consult your manual about how to square a window.

Each of the following is a graph of the line $y = 2x - 3$, but the viewing windows are different.

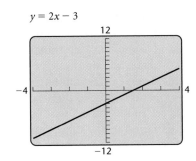

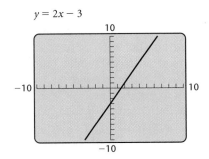

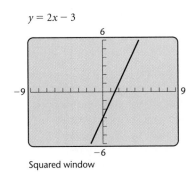

Squared window

When the x-units and the y-units are visually the same, we get an accurate representation of the *slope* of the line. We will study slope in Section 1.3.

---

*The exact ratio is 63/95 on a TI-82 or a TI-83, 63/127 on a TI-85, and 103/239 on a TI-92.

A squared window also eliminates distortion of the graph. Compare the graph of the circle $x^2 + y^2 = 4$ shown here in both a nonsquared and a squared window. (See Appendix A.2, for review of equations of circles.)

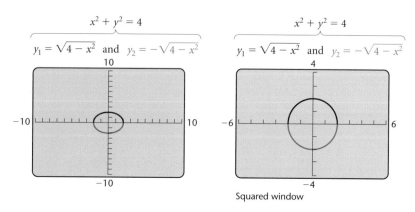

$$x^2 + y^2 = 4$$
$$y_1 = \sqrt{4 - x^2} \quad \text{and} \quad y_2 = -\sqrt{4 - x^2}$$

$$x^2 + y^2 = 4$$
$$y_1 = \sqrt{4 - x^2} \quad \text{and} \quad y_2 = -\sqrt{4 - x^2}$$

Squared window

The Table Feature

TblMin = 2, ΔTbl = .25

X	Y₁
2	−1
2.25	−1.333
2.5	−2
2.75	−4
3	ERROR
3.25	4
3.5	2

X = 2.25

Many graphers can display a table of values similar to the tables we used for hand-drawn graphs. For an equation entered on the "$y =$" screen, we can use the table set-up screen to choose a minimum x-value along with a step-value or increment. The grapher then produces a table of x- and y-values for the given equation. For example, if we enter the equation $y = 1/(x - 3)$ along with a minimum x-value of 2 and an increment of 0.25, the table at left is produced. We see that $y = -1$ when $x = 2$, $y = -1.3333$ when $x = 2.25$, and so on. The ERROR entry indicates that 3 is not an acceptable value for x.

We can use the arrow keys to scroll through more x, y-values for this equation. Some graphers can display both a graph and a table on a split screen.

The TABLE feature can also be used to evaluate an expression in one variable for particular values of the variable. Again the expression is entered on the "$y =$" screen and the table is set so that the independent variable is in ASK mode. This allows us to enter x-values one at a time and see the corresponding y-values for the given expression. Suppose we want to evaluate $2x^5 - 6x^3 + 7$ for $x = -8$ and for $x = 5$. First we enter $y = 2x^5 - 6x^3 + 7$; then, with the table set in ASK mode, we enter -8. The corresponding y-value $-62,457$ is displayed in the table. Similarly, we enter 5 and read the y-value 5507. We can continue to enter x-values as desired.

Program

The Graphing Calculator Manual that accompanies this text contains programs that can be entered in the TI-82, TI-83, TI-85, or HP38G graphing calculators. For a complete list of these program references, see the Program heading in the index. The first program is designed for the TI-85.

TABLE: This program provides the TI-85 with a TABLE feature.

X	Y₁
−8	−62457
5	5507

X =

Identities

Consider the equation

$$(x + 1)^2 = x^2 + 2x + 1.$$

We know from the laws of algebra that this equation is true for every possible substitution of x by a real number. Let's look at this equation graphically. We consider two equations of the form "$y = \ldots$", one using the expression on the left side of the equation and one using the expression on the right:

$$y_1 = (x + 1)^2 \quad \text{and} \quad y_2 = x^2 + 2x + 1.$$

If we graph these two equations using the same viewing window, $[-5, 5, -1, 6]$, we see that the graphs seem to coincide. You might try other viewing windows to see that they do not differ. The TABLE feature will also confirm that y_1 and y_2 appear to have the same value for a given value of x.

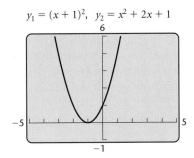

The equation $(x + 1)^2 = x^2 + 2x + 1$ is called an **identity**. As illustrated above, it is true for every possible substitution. The graphs are the same for the values of x for which both expressions are defined. One way to get a partial check of a potential identity, or to make a conjecture that an equation is an identity, is to graph each side and see if the graphs coincide. Although this procedure can be fruitful, one must be aware of its shortcomings. First, we rarely can draw all or even most of a graph. Thus there is always the possibility that somewhere outside the window, the graphs differ. The second limitation is that the size and scale of the window may deceive us about whether graphs really do coincide. An alternate partial check can also be obtained using the TABLE feature or other function evaluator. Many values of a pair of expressions can be compared very quickly.

Example 9 Determine graphically whether each of the following seems to be an identity.

a) $(x^2)^3 = x^6$ **b)** $\sqrt{x + 4} = \sqrt{x} + 2$

SOLUTION

a) We leave it to the student to graph $y_1 = (x^2)^3$ and $y_2 = x^6$. The graphs appear to coincide no matter what the viewing window. Thus, $(x^2)^3 = x^6$ seems to be an identity.

b) We graph $y_1 = \sqrt{x + 4}$ and $y_2 = \sqrt{x} + 2$ using the viewing window $[-6, 8, -1, 6]$. We see that the graphs differ, so the equation is *not* an identity. Comparing y-values using the TABLE feature also illustrates that this equation is *not* an identity.

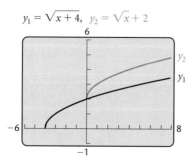

$y_1 = \sqrt{x + 4}, \quad y_2 = \sqrt{x} + 2$

TblMin $= -2$, ΔTbl $= 1$

X	Y₁	Y₂
−2	1.4142	ERROR
−1	1.7321	ERROR
0	2	2
1	2.2361	3
2	2.4495	3.4142
3	2.6458	3.7321
4	2.8284	4

X = 1

The Trace Feature

Once a graph has been created, we can investigate some of its points by using the TRACE feature that most graphers offer. Consider the graph of $y = x^3 - 3x + 1$ in the viewing window $[-10, 10, -10, 10]$ with Xscl $= 2$ and Yscl $= 2$. When the TRACE key is pressed, a cursor (often blinking) appears somewhere on the graph, while the x- and y-coordinates are shown elsewhere on the screen. These coordinates will change as the cursor is moved along the graph using the arrow keys ($\triangleleft$, $\triangleright$). Some points on the graph are shown in the figures below.

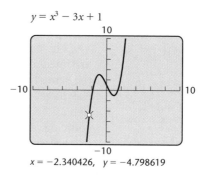

$y = x^3 - 3x + 1$

$x = -2.340426, \quad y = -4.798619$

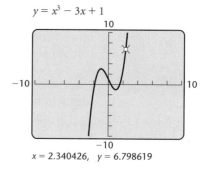

$y = x^3 - 3x + 1$

$x = 2.340426, \quad y = 6.798619$

$y = x^3 - 3x + 1$

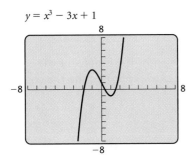

Equation Solving

Example 10   Solve $x^3 - 3x + 1 = 0$ graphically. Approximate solutions to three decimal places.

SOLUTION   The solutions of the equation $x^3 - 3x + 1 = 0$ are the first coordinates of the points where the graph of $y = x^3 - 3x + 1$ crosses the x-axis. These points are called the **x-intercepts** of the graph. An x-intercept has the general form $(a, 0)$ and occurs at an x-value a for which $y = 0$. It appears from the graph shown at left that there are three such x-values, one near -2, one near 0, and one near 1.5.

To acquire more precision, we can change the viewing window. A fast way to do this on many graphers is to use a feature called ZOOM. The zoom-in feature allows a portion of the graph to be magnified; that is, a small section of the graph is enlarged in the viewing window. Let's examine the graph near $x = -2$. We trace to the y-value closest to 0 and zoom in on the graph at that point. We continue to zoom in until the desired accuracy is achieved.

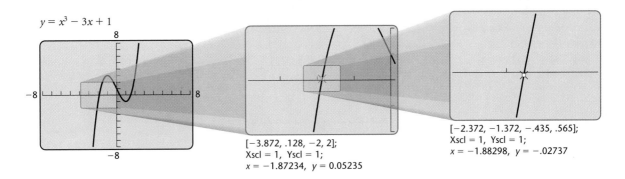

$y = x^3 - 3x + 1$

$[-3.872, .128, -2, 2]$;
Xscl = 1, Yscl = 1;
$x = -1.87234$, $y = 0.05235$

$[-2.372, -1.372, -.435, .565]$;
Xscl = 1, Yscl = 1;
$x = -1.88298$, $y = -.02737$

To find the solution to the nearest thousandth, we trace and zoom until the cursor's x-value just to the left of the intercept and the cursor's x-value just to the right of the intercept are the same, when rounded to the nearest thousandth. By using the TRACE and ZOOM features three more times, we find the solution to be about -1.879. In a similar manner, we find that the other solutions of the equation $x^3 - 3x + 1 = 0$ are about 0.347 and 1.532. If available, we might also use the SOLVE or ROOT features to approximate solutions.

There are many ways in which the ZOOM feature can be used. For example, we can zoom in for more precision, as in Example 10, or zoom out on a graph—say, to reveal more of its curvature—adjusting the factors of the zoom in any way we choose. In Example 10, we used zoom factors of 4. We may also be able to use a ZOOM-BOX feature to zoom in on a boxed region of our choosing. All such details can be found by consulting the manual for your particular grapher or the Graphing Calculator Manual that accompanies this book.

Finding Points of Intersection

There are many situations in which we want to determine the point(s) of intersection of two graphs.

Example 11   Find the points of intersection of the graphs of the equations $y_1 = 3x^5 - 20x^3$ and $y_2 = 34.7 - 1.28x^2$. Approximate the coordinates to three decimal places.

SOLUTION

Method 1. We use a grapher, trying to create graphs that clearly show the curvature. Then we look for the coordinates of the points at which the graphs cross each other. It appears that there are three points of intersection. We can use TRACE and ZOOM to approximate their coordinates, or we can use the INTERSECT feature, if available. Experiment to see which you prefer.

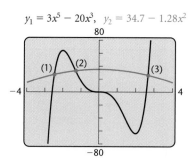

The values found can be checked by substitution. The points of intersection are about $(-2.463, 26.936)$, $(-1.296, 32.551)$, and $(2.668, 25.591)$.

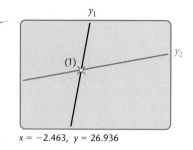

$x = -2.463,\ y = 26.936$

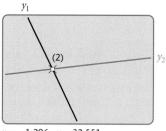

$x = -1.296,\ y = 32.551$

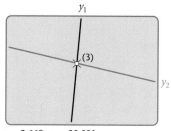

$x = 2.668,\ y = 25.591$

Method 2. Another method is to use the SOLVE feature. We find the x-values that are solutions of the equation $y_1 = y_2$. To do so, we solve

$$y_1 - y_2 = 0, \quad \text{or} \quad 3x^5 - 20x^3 + 1.28x^2 - 34.7 = 0.$$

On some graphers there is actually a POLY or ROOT feature, which allows you to enter this polynomial and gives all the solutions of this equation directly. We round each x-value to three decimal places. Then, to compute the second coordinate, we substitute the x-value into either y_1 or y_2.

Example 12 Solve: $\frac{2}{3}x - 7 = 5$.

SOLUTION

Method 1. One way to solve this equation is to ask ourselves, "For what *x*-values will the expression $\frac{2}{3}x - 7$ equal 5?" This can be visualized by asking, "Where will the graph of $y = \frac{2}{3}x - 7$ cross the horizontal line $y = 5$?" Thus we graph the equations $y_1 = \frac{2}{3}x - 7$ and $y_2 = 5$ and look for the *first coordinate* of their point of intersection. The INTERSECT feature provides us with an efficient means of finding this point. TRACE and ZOOM could also be used.

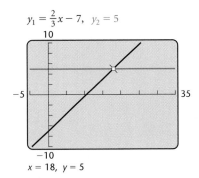

$y_1 = \frac{2}{3}x - 7, \quad y_2 = 5$

$x = 18, \ y = 5$

X	Y1	Y2
17.7	4.8	5
17.8	4.8667	5
17.9	4.9333	5
18	5	5
18.1	5.0667	5
18.2	5.1333	5
18.3	5.2	5

X = 18

We see that the point of intersection is (18, 5). Thus the solution is 18.

We can also confirm the solution numerically, using the table shown at left. The step-value in the table is 0.1. We choose the minimum *x*-value to be 17.7 and observe the *y*-values as the *x*-values increase to 18. The solution can also be checked using the table set in ASK mode.

Method 2. Another way to solve this equation is to solve

$$y_1 - y_2 = 0, \quad \text{or} \quad \frac{2}{3}x - 7 - 5 = 0, \quad \text{or} \quad \frac{2}{3}x - 12 = 0.$$

We can do this in a manner similar to Example 10. We graph the equation $y = \frac{2}{3}x - 12$ and see that it intersects the *x*-axis at (18, 0). Thus the solution is 18. We could also use the SOLVE or ROOT features.

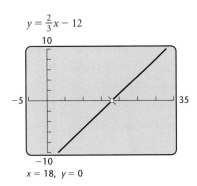

$y = \frac{2}{3}x - 12$

$x = 18, \ y = 0$

A Special Case

Graphers, for all their advantages, also have their disadvantages, one of which can occur with certain kinds of rational equations. Consider the equation $y = 8/(x^2 - 4)$. The numbers -2 and 2 cannot be substituted into the equation because they result in division by 0. Thus no point with either of these numbers as a first coordinate can be part of the graph. Now look at the two graphs of $y = 8/(x^2 - 4)$ shown below.

CONNECTED MODE | DOT MODE

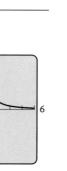
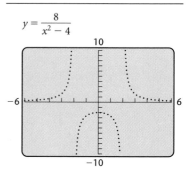

In CONNECTED mode, a grapher connects plotted points with line segments. In DOT mode, it simply plots unconnected points. In the graph on the left, the grapher has connected the points plotted on either side of the x-values -2 and 2 with vertical lines. These lines, however, cannot be part of the graph because substitutions of -2 and 2 are not allowed in this equation. Thus this is not a correct graph of this equation. The graph on the right is more accurate. When graphing equations in which a variable appears in a denominator, use DOT mode if you have a choice.

This kind of difficulty should not deter you from using a grapher. As we will see throughout the text, we will often need to apply our math knowledge to make the best use of a grapher.

Exercise Set

Graph and label each set of points. Make a hand-drawn graph.

1. $\{(4, 0), (-3, -5), (-1, 4), (0, 2), (2, -2)\}$

2. $\{(1, 4), (-4, -2), (-5, 0), (2, -4), (4, 0)\}$

3. $\{(-5, 1), (5, 1), (2, 3), (2, -1), (0, 1)\}$

4. $\{(4, 0), (4, -3), (-5, 2), (-5, 0), (-1, -5)\}$

Determine whether the given ordered pairs are solutions of the given equation.

5. $(1, -1), (0, 3);\ y = 2x - 3$

6. $(2, 5), (-2, -5);\ y = 3x - 1$

7. $\left(-\frac{1}{2}, -\frac{4}{5}\right), \left(0, \frac{3}{5}\right);\ 2a + 5b = 3$

8. $\left(0, \frac{3}{2}\right), \left(\frac{2}{3}, 1\right);\ 3m + 4n = 6$

9. $(-0.75, 2.75), (2, -1);\ x^2 - y^2 = 3$

10. $(2, -4), (4, -5);\ 5x + 2y^2 = 70$

Make a hand-drawn graph of each equation. Use the indicated x-values along with any others you are curious about.

11. $y = 3 - 2x;\ x = -3, -2, -1, 0, 1, 2, 3$

12. $x + y = 4;\ x = -3, -2, -1, 0, 1, 2, 3$

13. $3x - 4y = 12;\ x = -3, -2, -1, 0, 1, 2, 3$

14. $y = x^2$; $x = -3, -2, -1, 0, 1, 2, 3$

15. $y = -x^2$; $x = -3, -2, -1, 0, 1, 2, 3$

16. $y = \sqrt{x}$; $x = 0, 1, 4, 9$

17. $y = |x| - 2$; $x = -3, -2, -1, 0, 1, 2, 3$

18. $y = x^3$; $x = -2, -1, 0, 1, 2$

19. $y = \sqrt{x} - 4$; $x = 0, 1, 4, 9$

20. $y = |x| + 2$; $x = -3, -2, -1, 0, 1, 2, 3$

21. $y = -\dfrac{2}{x}$; $x = -4, -2, -1, -\dfrac{1}{2}, \dfrac{1}{2}, 1, 2, 4$

22. $y = 4 - x^2$; $x = -3, -2, -1, 0, 1, 2, 3$

In Exercises 23–30, use a grapher to match the equations with graphs (a)–(h).

a)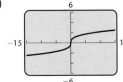

b)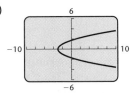

c)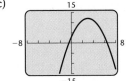

d)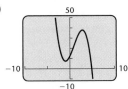

e)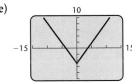

f)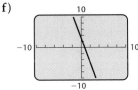

g)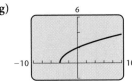

h)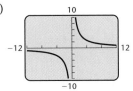

23. $y = 3 - 4x$ **24.** $y = |x| - 5$

25. $y = 3 + 6x - x^2$ **26.** $y = \sqrt{x + 4}$

27. $x = y^2 - 3$ **28.** $y = \sqrt[3]{x}$

29. $y = 12.4 + 9.1x + 3.07x^2 - 1.1x^3$

30. $xy = 10$

Graph each of the following equations, using the standard viewing window, $[-10, 10, -10, 10]$. Then try other windows until you find one that best represents

the graph. Answers may vary.

31. $y = \frac{1}{5}x + 2$ **32.** $4x - 5y = 20$

33. $x = y^2 + 2$ **34.** $y = |x - 1|$

35. $xy = -18$ **36.** $y = (x + 2)^2$

37. $y = x^2 - 4x + 3$ **38.** $x = 4 - y^2$

39. $y = \sqrt{x} - 2$

40. $y = x^3 + 2x^2 - 4x - 13$

41. $y^2 = 4 + x$

42. $y = \sqrt[3]{x + 4}$

In Exercises 43–46, an equation is given together with four choices of viewing windows and scaling conditions. Choose the best option. If you reject a situation, explain why you did so.

43. $y = 2x^3 - x^4$

   a) $[-1, 1, -1, 1]$, Xscl $= 1$, Yscl $= 1$

   b) $[-2, 3, -4, 3]$, Xscl $= 1$, Yscl $= 1$

   c) $[-2, 3, -4, 3]$, Xscl $= 0.01$, Yscl $= 0.1$

   d) $[-20, 30, -40, 30]$, Xscl $= 1$, Yscl $= 1$

44. $y = x^4 - 8x^3 + 18x^2$

   a) $[-10, 10, -10, 10]$, Xscl $= 1$, Yscl $= 1$

   b) $[-1, 1, -10, 10]$, Xscl $= 1$, Yscl $= 1$

   c) $[-3, 6, -10, 60]$, Xscl $= 1$, Yscl $= 10$

   d) $[-10, 60, -3, 6]$, Xscl $= 1$, Yscl $= 10$

45. $y = \dfrac{8x}{x^2 + 1}$

   a) $[-1, 1, -10, 10]$, Xscl $= 1$, Yscl $= 1$

   b) $[-5, 5, -7, 7]$, Xscl $= 1$, Yscl $= 1$

   c) $[-3, 6, -10, 60]$, Xscl $= 1$, Yscl $= 10$

   d) $[-100, 100, -100, 100]$, Xscl $= 10$, Yscl $= 10$

46. $y = 4x - x^3$

   a) $[-0.1, 10, -10, 10]$, Xscl $= 1$, Yscl $= 1$

   b) $[-10, 0.1, -10, 10]$, Xscl $= 1$, Yscl $= 1$

   c) $[-5, 5, -0.1, 1]$, Xscl $= 1$, Yscl $= 1$

   d) $[-3, 3, -10, 10]$, Xscl $= 1$, Yscl $= 1$

47. Using a grapher, complete this table for the equations

$$y_1 = \sqrt{10 - x^2} \quad \text{and} \quad y_2 = \frac{16}{x + 2.8}.$$

Then extend the table from -2.6 to 3.3.

X	Y1	Y2
−3.3		
−3.2		
−3.1		
−3		
−2.9		
−2.8		
−2.7		

X = −3.3

Determine with a grapher whether each of the following seems to be an identity.

48. $7x = x \cdot 7$

49. $(x - 5)^2 = x^2 - 10x + 25$

50. $(x + 4)^2 = x^2 + 16$

51. $x^3 + x^4 = x^7$

52. $(x - 9)^2 = (x + 3)(x - 3)$

53. $x^3 - 1 = (x - 1)(x^2 + x + 1)$

54. $(x + 2)^3 = x^3 + 8$

55. $\sqrt{x^2 - 16} = x - 4$

56. $(x^2)^3 = x^6$

57. $\dfrac{x^5}{x^2} = x^3$

58. $\sqrt{1 + x} = 1 + \dfrac{x}{2} - \dfrac{x^2}{8} + \dfrac{x^3}{16}$

Solve with a grapher.

59. $\frac{3}{4}x + 2 = -4$

60. $-5 = -\frac{3}{2}x - 3$

61. $3x + 4 = -\frac{2}{5}x + 1$

62. $15 - 2x = \dfrac{2x + 5}{3}$

63. $1.4x + 0.7 = 0.9x - 2.2$

64. $-2.6(x - 8.4) + 1.92 = 23x - 0.93(8x + 11.3)$

65. $x - 7.4 = 2.8\sqrt{x + 1.1}$

66. $\sqrt{x - 3.2} + \sqrt{x + 5.03} = 4.91$

67. $x^3 - 6x^2 = -9x - 1$

68. $2.1x^4 - 4.3x^2 = 5$

69. $1.09x^2 - 0.8x^4 = -7.6$

70. $144x + 140 = x^3 - 3x^2$

71. $x^4 + 4x^3 + 300 = 36x^2 + 160x$

72. $x^3 = 2x^2 + 2$

73. $|x + 1| + |x - 2| = 5$

74. $|x + 1| + |x - 2| = 0$

Find the points of intersection of the graphs of each pair of equations using a grapher.

75. $y_1 = x^3 + 3x^2 - 9x - 13, \quad y_2 = 0$

76. $y_1 = \sqrt{7 - x^2}, \quad y_2 = 2.5$

77. $y = x^3 - 3x^2, \quad 4x - 7y = 20$

78. $y = x^4 - 2x^3, \quad y = -0.7x + 3$

79. $y = \dfrac{8x}{x^2 + 1}, \quad y = 0.9x$

80. $y = x\sqrt{8 - x^2}, \quad 93x - 100y = 1$

81. *A Hole in a Graph.*   Consider the graph of
$$y = \dfrac{x^2 - 4}{x - 2}.$$

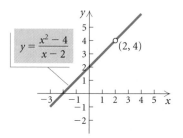

We cannot substitute $x = 2$ into the equation. This creates a "hole" in the graph. The point $(2, 4)$ is not part of the graph. On some graphers it may be very difficult to find the hole. Graph this function and try to find the hole. Experiment with some of the grapher's features.

In Exercises 82 and 83, make a hand-drawn graph. Look for a pattern to find the missing data. Try to determine an equation that fits the data.

82.

x	y
1	4
2	7
3	10
4	13
5	
6	
	31
	34

83.

x	y
0	1
1	2
2	5
3	10
4	17
5	
	82
	122

Graphs, Functions, and Models

1

During a thunderstorm, it is possible to calculate how far away, y (in miles), lightning is when the sound of thunder arrives x seconds after the lightning has been sighted. The relationship between x and y can be described by a function, $y = f(x) = \frac{1}{5}x$.

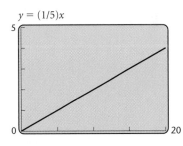

X	Y₁	
0	0	
1	.2	
2	.4	
5	**1**	
10	2	
X = 5		

$y = (1/5)x$

This weather application is just one example of the widespread use of graphs in today's society. In this chapter, we will focus attention on functions and equations with graphs that can be used to model real-life situations mathematically. Such models are invaluable when doing analyses and making predictions in fields ranging from astronomy to zoology.

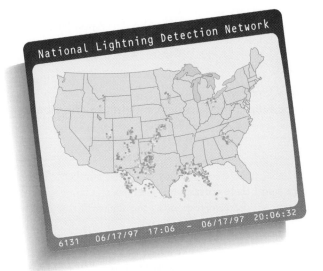

National Lightning Detection Network

6131 06/17/97 17:06 — 06/17/97 20:06:32

1.1

Functions, Graphs, and Graphers

- *Determine whether a correspondence or a relation is a function.*
- *Find function values, or outputs, using a formula.*
- *Find the domain and the range of a function.*
- *Determine whether a graph is that of a function.*
- *Solve application problems using functions.*

We now focus attention on a concept that is fundamental to many areas of mathematics—the idea of a *function*.

Functions

We first consider an application.

Lightning–Time–Thunder Distance. During a thunderstorm, it is possible to calculate how far away, *y* (in miles), lightning is when the sound of thunder arrives *x* seconds after the lightning has been sighted. We can examine the relationship between *x* and *y* in several ways:

x	y	ORDERED PAIRS: (x, y)	CORRESPONDENCE
0	0	$(0, 0)$	$0 \longrightarrow 0$
1	$\frac{1}{5}$	$\left(1, \frac{1}{5}\right)$	$1 \longrightarrow \frac{1}{5}$
2	$\frac{2}{5}$	$\left(2, \frac{2}{5}\right)$	$2 \longrightarrow \frac{2}{5}$
5	1	$(5, 1)$	$5 \longrightarrow 1$
10	2	$(10, 2)$	$10 \longrightarrow 2$

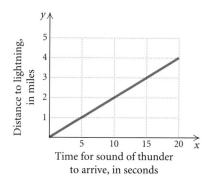

The ordered pairs imply a relationship, or correspondence, between the first and second coordinates. We can see this relationship in the graph as well. There is also an equation that describes the correspondence. It is

$$y = \frac{1}{5}x.$$

This is an example of a *function*.

Let's consider some other functions before giving a definition.

DOMAIN		RANGE
To each registered student	there corresponds	an I. D. number.
To each mountain bike sold	there corresponds	its price.
To each number between -3 and 3	there corresponds	the square of that number.

In each example, the first set is called the **domain** and the second set is called the **range**. For each member, or **element**, in the domain, there is *exactly one* member of the range to which it corresponds. Thus each registered student has exactly *one* I. D. number, each mountain bike has exactly *one* price, and each number between −3 and 3 has exactly *one* square. Each correspondence is a *function*.

> **Function**
>
> A *function* is a correspondence between a first set, called the *domain*, and a second set, called the *range*, such that each member of the domain corresponds to *exactly one* member of the range.

It is important to note that not every correspondence between two sets is a function.

Example 1 Determine whether each of the following correspondences is a function.

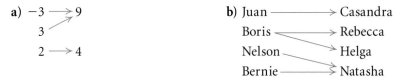

a) −3 ⟶ 9
     3
     2 ⟶ 4

b) Juan ⟶ Casandra
    Boris ⟶ Rebecca
    Nelson ⟶ Helga
    Bernie ⟶ Natasha

SOLUTION

a) This correspondence *is* a function because each member of the domain corresponds to exactly one member of the range.

b) This correspondence *is not* a function because there is a member of the domain (Boris) that is paired with two different members of the range.

Example 2 Determine whether each of the following correspondences is a function.

DOMAIN	CORRESPONDENCE	RANGE
a) Years in which a president is elected	The person elected	A set of presidents
b) The integers	Each number's cube root	A set of real numbers
c) All states in the United States	A senator from that state	The set of all U.S. senators

SOLUTION

a) This correspondence *is* a function, because in each presidential election *exactly one* president is elected.

b) This correspondence *is* a function, because each integer has *exactly one* cube root.

c) This correspondence *is not* a function, because each state can be paired with *two* different senators.   ▬

When a correspondence between two sets is not a function, it is still an example of a **relation**.

> **Relation**
>
> A *relation* is a correspondence between a first set, called the *domain,* and a second set, called the *range,* such that each member of the domain corresponds to *at least one* member of the range.

All the correspondences in Examples 1 and 2 are relations, but, as we have seen, not all are functions. Relations are sometimes written as sets of ordered pairs in which elements of the domain are the first part (coordinate) of each ordered pair and elements of the range are the second part (coordinate). For example, instead of writing $-3 \longrightarrow 9$, we write the ordered pair $(-3, 9)$.

Example 3   Determine whether each of the following relations is a function. Identify the domain and the range.

a) $\{(-2, 5), (5, 7), (0, 1), (4, -2)\}$

b) $\{(9, -5), (9, 5), (2, 4)\}$

SOLUTION

a) The relation *is* a function because no two ordered pairs have the same first coordinate and different second coordinates.

The domain is the set of all first coordinates: $\{-2, 5, 0, 4\}$.

The range is the set of all second coordinates: $\{5, 7, 1, -2\}$.

b) The relation *is not* a function because the ordered pairs $(9, -5)$ and $(9, 5)$ have the same first coordinate and different second coordinates.

The domain is the set of all first coordinates: $\{9, 2\}$.

The range is the set of all second coordinates: $\{-5, 5, 4\}$.   ▬

Notation for Functions

Functions used in mathematics are often given by equations. They generally require that certain calculations be performed in order to determine which member of the range is paired with each member of the domain. For example, in the Introduction to Graphs and Graphers, we graphed the function $y = x^2 - 5$ by doing calculations like the following:

for $x = 3$, $y = 3^2 - 5 = 4$,
for $x = 2$, $y = 2^2 - 5 = -1$,
for $x = 1$, $y = 1^2 - 5 = -4$,   and so on.

A more concise notation is often used. For $y = x^2 - 5$, the **inputs** (members of the domain) are values of x substituted into the equation. The **outputs** (members of the range) are the resulting values of y. If we call the function f, we can use x to represent an arbitrary *input* and $f(x)$—read "f of x", or "f at x", or "the value of f at x"—to represent the corresponding *output*. In this notation, the function given by $y = x^2 - 5$ is written as $f(x) = x^2 - 5$ and the above calculations would be

$$f(3) = 3^2 - 5 = 4,$$
$$f(2) = 2^2 - 5 = -1,$$
$$f(1) = 1^2 - 5 = -4. \qquad \text{Keep in mind that } f(x) \text{ } does \text{ } not \text{ mean}$$
$$f \cdot x.$$

Thus, instead of writing "when $x = 3$, the value of y is 4," we can simply write "$f(3) = 4$," which can also be read as "f of 3 is 4" or "for the input 3, the output of f is 4."

Example 4   A function f is given by $f(x) = 2x^2 - x + 3$. Find each of the following.

a) $f(0)$ **b)** $f(-7)$

c) $f(5a)$ **d)** $f(a - 4)$

SOLUTION   One way to find function values when a formula is given is to think of the formula as follows:

$$f(\blacksquare) = 2(\blacksquare)^2 - (\blacksquare) + 3.$$

Then to find an output for a given input we think: "Whatever goes in the blank on the left goes in the blank(s) on the right." This gives us a "recipe" for finding outputs. We can now solve the example.

a) $f(0) = 2 \cdot (0)^2 - 0 + 3 = 0 - 0 + 3 = 3$

b) $f(-7) = 2(-7)^2 - (-7) + 3 = 2 \cdot 49 + 7 + 3 = 108$

c) $f(5a) = 2(5a)^2 - 5a + 3 = 2 \cdot 25a^2 - 5a + 3 = 50a^2 - 5a + 3$

d) $f(a - 4) = 2(a - 4)^2 - (a - 4) + 3 = 2(a^2 - 8a + 16) - a + 4 + 3$
$$= 2a^2 - 16a + 32 - a + 4 + 3$$
$$= 2a^2 - 17a + 39 \qquad \blacksquare$$

It is important for students preparing for calculus to be able to simplify rational expressions like

$$\frac{f(a + h) - f(a)}{h}.$$

Example 5   For the function f given by $f(x) = 2x^2 - x - 3$, construct and simplify the expression

$$\frac{f(a + h) - f(a)}{h}.$$

SOLUTION   We find $f(a + h)$ and $f(a)$:

$$f(a + h) = 2(a + h)^2 - (a + h) - 3$$
$$= 2[a^2 + 2ah + h^2] - a - h - 3$$
$$= 2a^2 + 4ah + 2h^2 - a - h - 3,$$
$$f(a) = 2a^2 - a - 3.$$

Then

$$\frac{f(a + h) - f(a)}{h}$$

$$= \frac{[2a^2 + 4ah + 2h^2 - a - h - 3] - [2a^2 - a - 3]}{h}$$

$$= \frac{2a^2 + 4ah + 2h^2 - a - h - 3 - 2a^2 + a + 3}{h} = \frac{4ah + 2h^2 - h}{h}$$

$$= \frac{h(4a + 2h - 1)}{h \cdot 1} = \frac{h}{h} \cdot \frac{4a + 2h - 1}{1} = 4a + 2h - 1.$$ ▬

Finding Domains of Functions

When a function is given by a formula, we find function values by making substitutions of numbers or inputs into the formula. When a function f, whose inputs and outputs are real numbers, is given by a formula, the *domain* is understood to be the set of all inputs for which the expression is defined as a real number. When a substitution results in an expression that is not defined as a real number, we say that the function value *does not exist* and that the number being substituted *is not* in the domain of the function.

Example 6   Find the indicated function values. Simplify, if possible.

a) $f(1)$ and $f(3)$, for $f(x) = \dfrac{1}{x - 3}$

b) $g(16)$ and $g(-7)$, for $g(x) = \sqrt{x} + 5$

SOLUTION

a) $f(1) = \dfrac{1}{1 - 3} = \dfrac{1}{-2} = -\dfrac{1}{2};$

$f(3) = \dfrac{1}{3 - 3} = \dfrac{1}{0}$

Since division by 0 is not defined, the number 3 is not in the domain of f. Thus, $f(3)$ does not exist.

b) $g(16) = \sqrt{16} + 5 = 4 + 5 = 9$;

$g(-7) = \sqrt{-7} + 5$

Since $\sqrt{-7}$ is not defined as a real number, the number -7 is not in the domain of g. Thus, $g(-7)$ does not exist.   ▬

We can see from Example 7 that inputs that make a denominator 0 or a square-root radicand negative are not in the domain of a function.

Example 7   Find the domain of each of the following functions.

a) $f(x) = \dfrac{1}{x - 3}$

b) $g(x) = \sqrt{x} + 5$

c) $h(x) = \dfrac{3x^2 - x + 7}{x^2 + 2x - 3}$

d) $p(x) = \sqrt{4 - x^2}$

e) $F(x) = x^3 + |x|$

SOLUTION

a) The input 3 results in a denominator of 0. The domain is $\{x \,|\, x \neq 3\}$, or $(-\infty, 3) \cup (3, \infty)$.

b) We can substitute any number for which the radicand is nonnegative, that is, for which $x \geq 0$. Thus the domain is $\{x \,|\, x \geq 0\}$, or the interval $[0, \infty)$.

c) Although we can substitute any real number in the numerator, we must avoid inputs that make the denominator 0. To find those inputs, we solve $x^2 + 2x - 3 = 0$, or $(x + 3)(x - 1) = 0$. Thus the domain consists of the set of all real numbers except -3 and 1, or $\{x \,|\, x \neq -3$ *and* $x \neq 1\}$, or $(-\infty, -3) \cup (-3, 1) \cup (1, \infty)$.

d) We must avoid inputs for which the radicand is negative. Thus the domain is all real numbers for which $4 - x^2 \geq 0$, or $x^2 \leq 4$. These inputs are all numbers in the interval $[-2, 2]$. You can check this by making some substitutions or, on a grapher, by looking at the graphs of $y_1 = x^2$ and $y_2 = 4$.

e) All substitutions are suitable. The domain is the set of all real numbers, $\mathbb{R}$.   ▬

Function Values and Tables on a Grapher

A grapher can be used in many different ways to find function values. We can simply do a calculation. The process is the same as that described in the Introduction to Graphs and Graphers.

With a TABLE feature, function values can be displayed for a string of numbers like $x = -3, -2.75, -2.5, -2.25, -2, -1.75$, and so on. Consider the function $f(x) = \sqrt{4 - x^2}$. Shown at left is part of a table for this function. Note that we entered the function as $y = \sqrt{(4 - x^2)}$, or $(4 - x^\wedge 2)^\wedge 0.5$. An ERROR message tells us when a number is not in the domain.

X	Y₁
−3	ERROR
−2.75	ERROR
−2.5	ERROR
−2.25	ERROR
−2	0
−1.75	.96825
−1.50	1.3229

X = −3

It is not easy, maybe not even possible, to determine a domain using a table, but it may point out certain numbers that are not in the domain.

Graphs of Functions

Most of the functions we study in this course and in calculus are given by formulas rather than sets of ordered pairs. We graph functions in much the same way as we do equations. We find ordered pairs (x, y) or $(x, f(x))$, look for patterns, and complete the graph.

Example 8   Graph each of the following functions.

a) $f(x) = x^2 - 5$ **b)** $f(x) = x^3 - x$ **c)** $f(x) = \sqrt{x + 4}$

SOLUTION   Most graphers do not use function notation "$f(x) = \ldots$" to enter a function formula. Instead, we must enter the function using "$y = \ldots$". The graphs follow.

a) $f(x) = x^2 - 5$

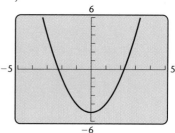

$y = x^2 - 5$

b) $f(x) = x^3 - x$

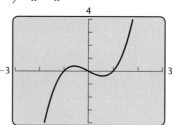

$y = x^3 - x$

c) $f(x) = \sqrt{x + 4}$

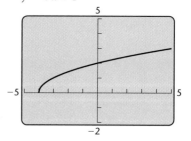

$y = \sqrt{x + 4}$

To find a function value, like $f(3)$, from a graph, we locate the input 3 on the horizontal axis, move vertically to the graph of the function, and then horizontally to find the output on the vertical axis. See the graph on the left below.

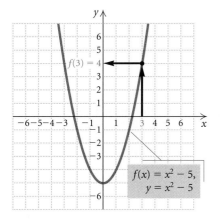

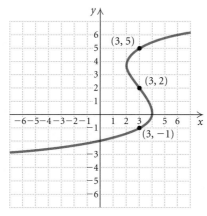

Since 3 is paired with more than one member of the range, the graph does not represent a function.

When one member of the domain is paired with two or more different members of the range, the correspondence *is not* a function. Thus, when a graph contains two or more different points with the same first coordinate, the graph cannot represent a function (see the graph on the right above). Points sharing a common first coordinate are vertically above or below each other. This leads us to the *vertical-line test*.

The Vertical-Line Test

If it is possible for a vertical line to cross a graph more than once, then the graph is not the graph of a function.

To apply the vertical-line test, we try to find a vertical line that crosses the graph more than once. If we do, then the graph is not that of a function. If we do not, then the graph is that of a function.

Example 9   Which of graphs (a) through (f) (in red) are graphs of functions?

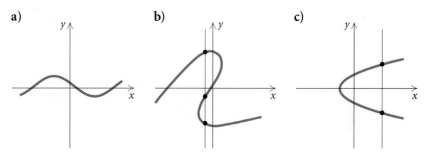

d) **e)** **f)**

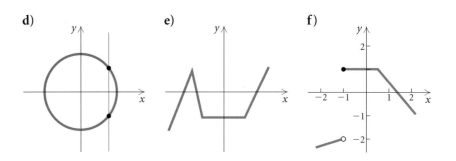

In graph (f), the solid dot shows that $(-1, 1)$ belongs to the graph. The open circle shows that $(-1, -2)$ does *not* belong to the graph.

SOLUTION   Graphs (a), (e), and (f) are graphs of functions because we cannot find a vertical line that crosses any of them more than once. In (b) the vertical line crosses the graph in three points and so it is not a function. Also, in (c) and (d), we can find a vertical line that crosses the graph more than once.   ▬

The Domain and the Range Using a Grapher

Keep the following in mind regarding the *graph* of a function:

> **Domain** = the set of a function's inputs, found on the horizontal axis;
>
> **Range** = the set of a function's outputs, found on the vertical axis.

By carefully examining the graph of a function, we may be able to determine the function's domain as well as its range. Consider the graph of $f(x) = \sqrt{4 - (x - 1)^2}$, shown below. We look for the inputs on the x-axis that correspond to a point on the graph. We see that they extend from -1 to 3, inclusive. Thus the domain is $\{x \mid -1 \leq x \leq 3\}$, or the interval $[-1, 3]$.

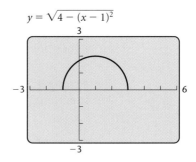

$$y = \sqrt{4 - (x - 1)^2}$$

To find the range, we look for the outputs on the y-axis. We see that they extend from 0 to 2, inclusive. Thus the range of this function is $\{y \mid 0 \leq y \leq 2\}$, or the interval $[0, 2]$. We can confirm our results using the TRACE feature, moving the cursor from left to right along the curve. We can also confirm our results using a TABLE feature. (See Exercise 63 in Exercise Set 1.1.)

Example 10   Use a grapher to graph each of the following functions. Then estimate the domain and the range of each.

a) $f(x) = \sqrt{x + 4}$

b) $f(x) = x^3 - x$

c) $f(x) = \dfrac{12}{x}$

d) $f(x) = x^4 - 2x^2 - 3$

SOLUTION

a)

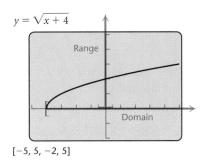

$$y = \sqrt{x + 4}$$

$[-5, 5, -2, 5]$

Domain $= [-4, \infty)$;
range $= [0, \infty)$

b)

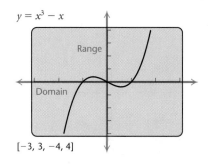

$$y = x^3 - x$$

$[-3, 3, -4, 4]$

Domain $=$ all real numbers, $\mathbb{R}$;
range $=$ all real numbers, $\mathbb{R}$

c)

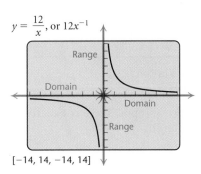

$$y = \frac{12}{x}, \text{ or } 12x^{-1}$$

$[-14, 14, -14, 14]$

Since the graph does not cross the y-axis, 0 is excluded as an input.
Domain $= (-\infty, 0) \cup (0, \infty)$;
range $= (-\infty, 0) \cup (0, \infty)$

d)

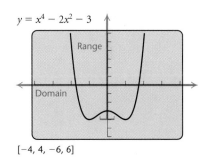

$$y = x^4 - 2x^2 - 3$$

$[-4, 4, -6, 6]$

Domain $=$ all real numbers, $\mathbb{R}$;
range $= [-4, \infty)$

Always consider adding the reasoning of Example 7 to a graphical analysis. Think, "What can I substitute?" to find the domain. Think, "What do I get out?" to find the range. Thus, in Example 10(d), it might not look like the domain is all real numbers because the graph rises steeply, but by reexamining the formula we see that we can indeed substitute any real number.

Applications of Functions

Example 11 *Speed of Sound in Air.* The speed S of sound in air is a function of the temperature t, in degrees Fahrenheit, and is given by

$$S(t) = 1087.7\sqrt{\frac{5t + 2457}{2457}},$$

where S is in feet per second.

a) Graph the function using the viewing window $[-50, 150, 1000, 1250]$ with Xscl $= 50$ and Yscl $= 50$.

b) Find the speed of sound in air when the temperature is $0°$, $32°$, and $70°$ Fahrenheit.

SOLUTION

a) The graph is shown at left. Note that $S(t)$ must be changed to y and t must be changed to x.

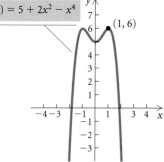

b) We use a grapher to compute the function values. We find that

$$S(0) = 1087.7 \text{ ft/sec},$$
$$S(32) \approx 1122.6 \text{ ft/sec}, \quad \text{and}$$
$$S(70) \approx 1162.6 \text{ ft/sec}.$$

We can also use the TRACE feature to approximate these function values, or we can use the TABLE feature, set in ASK mode, to find them.

The following is a review of several of the function concepts considered in this section.

FUNCTION CONCEPTS	GRAPH
Formula for f: $f(x) = 5 + 2x^2 - x^4$. For every input, there is exactly one output. $(1, 6)$ is on the graph. 1 is an input. 6 is an output. $f(1) = 6$ Domain: set of all inputs $= \mathbb{R}$. Range: set of all outputs $= (-\infty, 6]$.	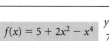

1.1 Exercise Set

Determine whether each correspondence is a function.

1. $a \longrightarrow w$
$b \longrightarrow y$
$c \longrightarrow z$

2. $m \longrightarrow q$
$n \quad r$
$o \quad s$

3. $-6 \longrightarrow 36$
$-2 \longrightarrow 4$
2

4. $-3 \longrightarrow 2$
$1 \longrightarrow 4$
$5 \longrightarrow 6$
$9 \longrightarrow 8$

5. $m \longrightarrow A$
$n \quad B$
$r \quad C$
$s \quad D$

6. $a \longrightarrow r$
$b \quad s$
$c \quad t$
d

7.

NAME OF SPACE MISSION	YEAR OF LAUNCH
Sputnik 1	1957
Mercury	1962
Mariner 2	1969
Apollo 11	

(*Source: Cambridge Factfinder*, 1993)

8.

ANIMAL	SPEED ON THE GROUND (IN KILOMETERS/HOUR)
Cheetah	110
Lion	80
Giraffe	50
Ostrich	

(*Source: Cambridge Factfinder*, 1993)

DOMAIN	CORRESPONDENCE	RANGE
9. A set of cars in a parking lot	Each car's license number	A set of numbers
10. A set of people in a town	A doctor a person uses	A set of doctors
11. A set of members of a family	Each person's eye color	A set of colors
12. A set of members of a rock band	An instrument each person plays	A set of instruments
13. A set of students in a class	A student sitting in a neighboring seat	A set of students
14. A set of bags of chips on a shelf	Each bag's weight	A set of weights

Determine whether each relation is a function. Identify the domain and the range.

15. $\{(2, 10), (3, 15), (4, 20)\}$

16. $\{(3, 1), (5, 1), (7, 1)\}$

17. $\{(-7, 3), (-2, 1), (-2, 4), (0, 7)\}$

18. $\{(1, 3), (1, 5), (1, 7), (1, 9)\}$

19. $\{(-2, 1), (0, 1), (2, 1), (4, 1), (-3, 1)\}$

20. $\{(5, 0), (3, -1), (0, 0), (5, -1), (3, -2)\}$

21. A graph of a function f is shown below. Find $f(-1)$, $f(0)$, and $f(1)$.

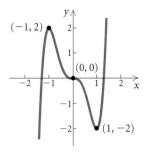

22. Given that $f(x) = 5x^2 + 4x$, find:
a) $f(0)$;
b) $f(-1)$;
c) $f(3)$;
d) $f(t)$;
e) $f(t - 1)$;
f) $\dfrac{f(a + h) - f(a)}{h}$.

23. Given that $g(x) = 3x^2 - 2x + 1$, find:
a) $g(0)$;
b) $g(-1)$;
c) $g(3)$;
d) $g(-x)$;
e) $g(1 - t)$;
f) $\dfrac{g(a + h) - g(a)}{h}$.

24. Given that $f(x) = 2|x| + 3x$, find:
a) $f(1)$;
b) $f(-2)$;
c) $f(-x)$;
d) $f(2y)$;
e) $f(2 - h)$;
f) $\dfrac{f(a + h) - f(a)}{h}$.

25. Given that $g(x) = x^3$, find:
a) $g(2)$;
b) $g(-2)$;
c) $g(-x)$;
d) $g(3y)$;
e) $g(2 + h)$;
f) $\dfrac{g(a + h) - g(a)}{h}$.

26. Given that $f(x) = \dfrac{x}{2-x}$, find:

a) $f(2)$;   **b)** $f(1)$;
c) $f(-16)$;   **d)** $f(-x)$;
e) $f\left(-\dfrac{2}{3}\right)$ **f)** $\dfrac{f(x+h) - f(x)}{h}$.

27. Given that $g(x) = \dfrac{x-4}{x+3}$, find:

a) $g(5)$;   **b)** $g(4)$;
c) $g(-3)$;   **d)** $g(-16.25)$;
e) $g(x+h)$;   **f)** $\dfrac{g(x+h) - g(x)}{h}$.

28. Given that $f(x) = x^2$, find

$$\dfrac{f(a+h) - f(a)}{h}.$$

29. Find $g(0)$, $g(-1)$, $g(5)$, and $g\left(\tfrac{1}{2}\right)$ for

$$g(x) = \dfrac{x}{\sqrt{1-x^2}}.$$

30. Find $h(0)$, $h(2)$, and $h(-x)$ for

$$h(x) = x + \sqrt{x^2 - 1}.$$

Find the domain of each function. Do not use a grapher.

31. $f(x) = 7x + 4$ **32.** $f(x) = |3x - 2|$

33. $f(x) = 4 - \dfrac{2}{x}$ **34.** $f(x) = \dfrac{1}{x^4}$

35. $f(x) = \sqrt{7x + 4}$ **36.** $f(x) = \sqrt{x - 3}$

37. $f(x) = \dfrac{1}{x^2 - 4x - 5}$ **38.** $f(x) = \dfrac{x^4 - 2x^3 + 7}{3x^2 - 10x - 8}$

39. $f(x) = \sqrt{9 - x^2}$ **40.** $f(x) = \dfrac{6x}{\sqrt{9 - x^2}}$

Use a grapher to graph each function. Then visually estimate the domain and the range.

41. $f(x) = |x|$ **42.** $f(x) = |x| - 10.3$
43. $f(x) = \sqrt{9 - x^2}$ **44.** $f(x) = -\sqrt{25 - x^2}$
45. $f(x) = (x - 1)^3 + 2$ **46.** $f(x) = (x - 2)^4 + 1$
47. $f(x) = \sqrt{7 - x}$ **48.** $f(x) = \sqrt{x + 8}$
49. $f(x) = \dfrac{18}{x}$ **50.** $f(x) = 2x^2 - x^4 + 5$

Determine whether each graph is that of a function. An open dot indicates that the point does not belong to the graph.

51.

52.

53.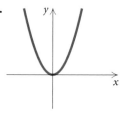

54.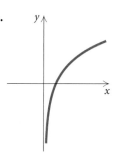

55.

56.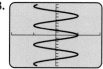

57.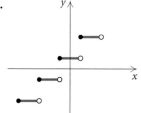

58.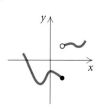

In Exercises 59–62, determine whether the graph is that of a function. Find the domain and the range.

59.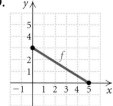

60.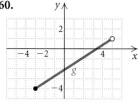

61.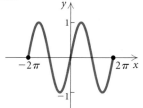

62.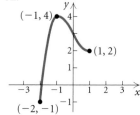

63. Use the TABLE feature to complete the input–output table for the equations

$$y_1 = \sqrt{4 - (x - 1)^2} \quad \text{and} \quad y_2 = \dfrac{23x}{x - 3.2}.$$

Then use the TABLE feature to consider inputs from -2.0 to 4.8.

X	Y1	Y2
2.2		
2.4		
2.6		
2.8		
3.0		
3.2		
3.4		
X = 2.2		

a) What does the table tell you about the domain of each function? the range?

b) What seems to be the largest output of each function? the smallest?

64. *Average Price of a Movie Ticket.*   The average price of a movie ticket, in dollars, can be estimated by the function P given by

$$P(x) = 0.1522x - 298.592$$

where $x =$ the year. Thus, $P(1998)$ is the average price of a movie ticket in 1998. The price is lower than what might be expected due to lower prices for matinees, senior citizens' discounts, and so on.

a) Use the function to predict the average price in 1998, 2000, and 2010.

b) When will the average price be $8.00?

65. *Boiling Point and Elevation.*   The elevation E, in meters, above sea level at which the boiling point of water is t degrees Celsius is given by the function

$$E(t) = 1000(100 - t) + 580(100 - t)^2.$$

At what elevation is the boiling point $99.5°$? $100°$?

66. *Territorial Area of an Animal.*   The territorial area of an animal is defined to be its defended, or exclusive region. For example, a lion has a certain region over which it is considered ruler. It has been shown that the territorial area I, in acres, of predatory animals is a function of body weight w, in pounds, and is given by the function

$$T(w) = w^{1.31}.$$

Find the territorial area of animals whose body weights are 0.5 lb, 10 lb, 20 lb, 100 lb, and 200 lb.

Skill Maintenance

To the student and the instructor: The skill maintenance exercises review skills covered previously in the text and anticipate the learning in the next section. You can expect these kinds of exercises in almost every exercise set.

Solve.

67. $\frac{2}{3}x + 7 = 12$

68. $\frac{4}{5}x + 1 = -\frac{2}{3}x + 7$

69. $\sqrt{2x - 4} = 7$

70. $3x^2 - 13x = 10$

Synthesis

To the student and the instructor: The synthesis exercises, found at the end of every exercise set, challenge students to combine concepts or skills developed in that section or in preceding parts of the text. Writing Exercises, denoted by the ◆ icon, are meant to be answered with one or more sentences.

71. ◆ Explain in your own words what a function is.

72. ◆ Explain in your own words the difference between the domain of a function and the range.

73. Construct

$$\frac{f(x + h) - f(x)}{h}$$

for $f(x) = \sqrt{x}$ and rationalize the numerator.

74. Make a hand-drawn graph of a function for which the domain is $[-4, 4]$ and the range is $[1, 2] \cup [3, 5]$. Answers may vary.

75. Give an example of two different functions that have the same domain and the same range, but have no pairs in common. Answers may vary.

76. Make a hand-drawn graph of a function for which the domain is $[-3, -1] \cup [1, 5]$ and the range is $\{1, 2, 3, 4\}$. Answers may vary.

77. Suppose that for some function f, $f(x - 1) = 5x$. Find $f(6)$.

1.2
Functions and Applications

- *Find zeros of functions.*
- *Graph functions, looking for intervals on which the function is increasing, decreasing, or constant, and determine relative maxima and minima.*
- *Graph functions defined piecewise.*
- *Given an application, find a function formula that models the application; find the domain of the function and function values, and then graph the function.*

Because functions occur in so many real-world situations, it is important to be able to analyze them carefully. In this section, we examine a variety of functions and formulate some mathematical models.

Finding Zeros on a Grapher

An input for which a function's output is 0 is called a **zero** of the function.

Zeros of Functions

An input c of a function f is called a *zero* of the function, if the output for c is 0. That is, $f(c) = 0$.

For the function given by $f(x) = x^4 - 4x^2 + 3$, the zeros are $-\sqrt{3}$, -1, 1, and $\sqrt{3}$.

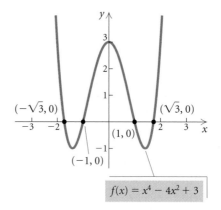

$$f(x) = x^4 - 4x^2 + 3$$

Note that these numbers are the first coordinates of the x-intercepts, $(-\sqrt{3}, 0)$, $(-1, 0)$, $(1, 0)$, and $(\sqrt{3}, 0)$. The **x-intercepts** are the points where the graph crosses the x-axis.

Finding exact values of the zeros of a function can be difficult. We can find approximations using a grapher in much the same way that we solve equations.

Example 1 Find the zeros of the function f given by
$$f(x) = 0.1x^3 - 0.6x^2 - 0.1x + 2.$$
Approximate the zeros to three decimal places.

We have not developed an algebraic procedure that will yield the zeros.

We use a grapher, trying to create a graph that clearly shows the curvature. Then we look for points where the graph crosses the x-axis. It appears that there are three zeros, one near -2, one near 2, and one near 6. We use TRACE and ZOOM to approximate these zeros, or we can use a SOLVE or ROOT feature.

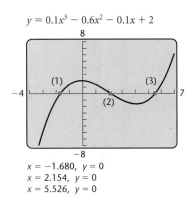

$$y = 0.1x^3 - 0.6x^2 - 0.1x + 2$$

$x = -1.680,\ y = 0$
$x = 2.154,\ y = 0$
$x = 5.526,\ y = 0$

The zeros are about -1.680, 2.154, and 5.526.

Increasing, Decreasing, and Constant Functions

On a given interval, if the graph of a function rises from left to right, it is said to be **increasing** on that interval. If the graph drops from left to right, it is said to be **decreasing**. If the function stays the same from left to right, it is said to be **constant**.

Interactive Discovery

Graph each of the following functions on a grapher using the given viewing window:

$$y = |x + 1| + |x - 2| - 5, \quad [-6, 6, -6, 6];$$
$$y = 5 - (x + 2)^2, \quad [-2, 4, -10, 6];$$
$$y = 4, \quad [-20, 20, -3, 6].$$

Then using the TRACE feature, move the cursor along the graph from left to right, observing what happens to the second coordinate. Confirm the results with a TABLE feature.

We are led to the following definitions.

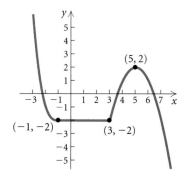

> **Increasing, Decreasing, and Constant Functions**
>
> A function f is said to be *increasing* on an interval if for all a and b in that interval, $a < b$ implies $f(a) < f(b)$.
>
> A function f is said to be *decreasing* on an interval if for all a and b in that interval, $a < b$ implies $f(a) > f(b)$.
>
> A function f is said to be *constant* on an interval if for all a and b in that interval, $f(a) = f(b)$.

Example 2   Determine the intervals on which the function is each of the following.

a) Increasing   **b)** Decreasing   **c)** Constant

SOLUTION

a) The function is increasing on the interval $[3, 5]$.

b) The function is decreasing on the intervals $(-\infty, -1]$ and $[5, \infty)$.

c) The function is constant on the interval $[-1, 3]$.

Relative Maximum and Minimum Values

Consider the graph shown below. Note the "peaks" and "valleys" at the points c_1, c_2, and c_3. The function value $f(c_2)$ is called a **relative maximum** (plural, **maxima**). Each of the function values $f(c_1)$ and $f(c_3)$ is called a **relative minimum** (plural, **minima**).

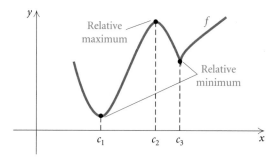

> **Relative Maxima and Minima**
>
> Suppose that f is a function for which $f(c)$ exists for some c in the domain of f. Then:
>
> $f(c)$ is a *relative maximum* if there exists an open interval I containing c in the domain such that $f(c) \geq f(x)$, for all x in I; and
>
> $f(c)$ is a *relative minimum* if there exists an open interval I containing c in the domain such that $f(c) \leq f(x)$, for all x in I.

Simply stated, $f(c)$ is a relative maximum if $f(c)$ is a "high point" in some interval, and $f(c)$ is a relative minimum if $f(c)$ is a "low point" in some interval.

We can use a grapher to approximate relative maxima or minima. On some graphers, this is done using the TRACE and ZOOM features. On other graphers, there may be a MAX or MIN feature that can be used. If you take a calculus course, you will learn a method for determining exact values of relative maxima and minima.

Example 3 Use a grapher to determine any relative maxima or minima of the function $f(x) = 0.1x^3 - 0.6x^2 - 0.1x + 2$, and to determine intervals on which the function is increasing or decreasing.

SOLUTION   We first graph the function, experimenting with the window as needed. The curvature is seen fairly well with $[-4, 6, -3, 3]$.

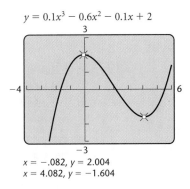

$$y = 0.1x^3 - 0.6x^2 - 0.1x + 2$$

$x = -.082, y = 2.004$
$x = 4.082, y = -1.604$

The graph starts rising, or increasing, from the left. We use the TRACE feature and move the cursor from left to right, noting whether the second coordinate increases. We note that the graph tends to stop increasing when x is somewhere around 0. Using the ZOOM feature, we get a good approximation of the relative maximum and where it occurs: 2.004 at $x = -0.082$.

From this relative maximum, the graph decreases to a point near $x = 4$ before it starts increasing again. We zoom in and get a good approximation of the relative minimum: -1.604 at $x = 4.082$.

The function is increasing on the intervals $(-\infty, -0.082]$ and $[4.082, \infty)$. It is decreasing on the interval $[-0.082, 4.082]$. ▬

Interactive Discovery

If your grapher has a TABLE feature, try to discover a way to use it to check the results of Example 3. Consider a small interval around each input and a small step value. Are there any difficulties?

Functions Defined Piecewise

Sometimes functions are defined **piecewise** using different output formulas for different parts of the domain.

Example 4   Make a hand-drawn graph of the function defined as

$$f(x) = \begin{cases} 4, & \text{for } x \leq 0, \\ 4 - x^2, & \text{for } 0 < x \leq 2, \\ 2x - 6, & \text{for } x > 2. \end{cases}$$

SOLUTION   We create the graph in three parts, as shown and described below.

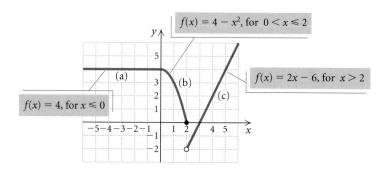

a) We graph $f(x) = 4$ *only* for inputs x less than or equal to 0 (that is, on the interval $(-\infty, 0]$).

b) We graph $f(x) = 4 - x^2$ *only* for inputs x greater than 0 and less than or equal to 2 (that is, on the interval $(0, 2]$).

c) We graph $f(x) = 2x - 6$ *only* for inputs x greater than 2 (that is, on the interval $(2, \infty)$).

To graph a function defined piecewise on a grapher, consult your manual. For piecewise-defined functions, you should select the DOT feature. Although some do not have the capability, you might try the following way to enter the function formula for Example 4. It incorporates parenthetic descriptions of the intervals:

$$f(x) = 4 \; (x \leq 0) \; + (4 - x^2) \; (x > 0)(x \leq 2) + (2x - 6) \; (x > 2).$$

If you wish, you might enter this equation as three separate functions:

$$y_1 = 4 \; (x \leq 0) \;, \qquad y_2 = (4 - x^2) \; (x > 0)(x \leq 2) \;,$$

$$y_3 = (2x - 6) \; (x > 2) \;.$$

Example 5   Make a hand-drawn graph of the function defined as

$$f(x) = \begin{cases} \dfrac{x^2 - 4}{x + 2}, & \text{for } x \neq -2, \\ 3, & \text{for } x = -2. \end{cases}$$

SOLUTION   When $x \neq -2$, the denominator of $(x^2 - 4)/(x + 2)$ is nonzero, so we can simplify:

$$\frac{x^2 - 4}{x + 2} = \frac{(x + 2)(x - 2)}{x + 2} = x - 2.$$

Thus,

$$f(x) = x - 2, \quad \text{for } x \neq -2.$$

The graph of this part of the function consists of a line with a "hole" at the point $(-2, -4)$, indicated by the open dot. At $x = -2$, we have $f(-2) = 3$, so the point $(-2, 3)$ is plotted above the open dot.

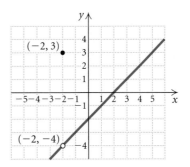

When $y = (x^2 - 4)/(x + 2)$ is graphed using a grapher, the hole may not be visible. If a TABLE feature is available, the following table can be created. The ERROR message indicates that -2 is not in the domain of the function g given by $g(x) = (x^2 - 4)/(x + 2)$. However, -2 *is* in the domain of f because $f(-2)$ is defined to be 3.

X	Y₁
−2.3	−4.3
−2.2	−4.2
−2.1	−4.1
−2	**ERROR**
−1.9	−3.9
−1.8	−3.8
−1.7	−3.7

X = −2

A function with importance in calculus and computer programming is the *greatest integer function*, f, denoted as $f(x) = \text{INT}(x)$, or $[\![x]\!]$.

Greatest Integer Function

$f(x) = \text{INT}(x) = $ the greatest integer less than or equal to x.

The greatest integer function pairs each input with the greatest integer less than or equal to that input. To check this, note that 1, 1.2, and 1.9 are all paired with the y-value 1. Similarly, 2, 2.3, and 2.8 are all paired with 2, and -3.4, -3.06, and -3.99027 are all paired with -4.

Example 6   Graph: $f(x) = \text{INT}(x)$.

SOLUTION   The greatest integer function can also be defined by a piecewise function with an infinite number of statements:

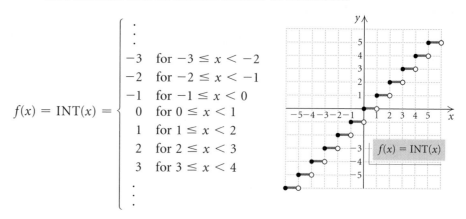

$$f(x) = \text{INT}(x) = \begin{cases} \vdots \\ -3 & \text{for } -3 \le x < -2 \\ -2 & \text{for } -2 \le x < -1 \\ -1 & \text{for } -1 \le x < 0 \\ 0 & \text{for } 0 \le x < 1 \\ 1 & \text{for } 1 \le x < 2 \\ 2 & \text{for } 2 \le x < 3 \\ 3 & \text{for } 3 \le x < 4 \\ \vdots \end{cases}$$

On a grapher, we would see the graph on the left below if we used CON-NECTED mode, which connects points (or rectangles called **pixels** on a grapher) with line segments. We would see the graph on the right if we used DOT mode. The DOT mode is preferable, though even it may not show the open dots at the endpoints of the segments.

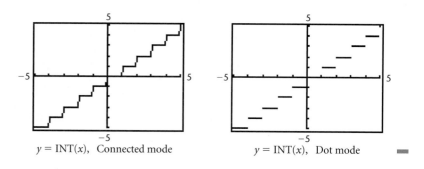

$y = \text{INT}(x)$, Connected mode $y = \text{INT}(x)$, Dot mode

Applications of Functions

Many real-world situations can be modeled by functions.

Example 7 *Car Distance.* Jamie and Ruben drive away from a restaurant at right angles to each other. Jamie's speed is 65 mph and Ruben's is 55 mph.

a) Express the distance between the cars as a function of time.

b) Find the domain of the function.

c) Graph the function.

SOLUTION

a) Suppose 1 hr goes by. At that time, Jamie has traveled 65 mi and Ruben has traveled 55 mi. We can use the Pythagorean theorem then to find the distance between them. This distance would be the length of the hypotenuse of a triangle with legs measuring 65 mi and 55 mi. After 2 hr, the triangle's legs would measure 130 mi and 110 mi.

Observing that the distances will always be changing, we make a sketch and let $t = $ the time, in hours, that Jamie and Ruben have been driving since leaving the restaurant.

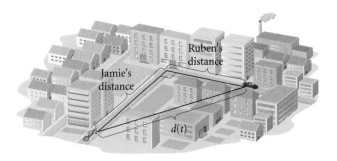

After t hours, Jamie has traveled $65t$ miles and Ruben $55t$ miles. We can use the Pythagorean theorem again:

$$[d(t)]^2 = (65t)^2 + (55t)^2.$$

Because distance must be nonnegative, we need only consider the positive square root when solving for $d(t)$:

$$
\begin{aligned}
d(t) &= \sqrt{(65t)^2 + (55t)^2} \\
&= \sqrt{4225t^2 + 3025t^2} \\
&= \sqrt{7250t^2} \\
&= \sqrt{7250}\,\sqrt{t^2} \\
&\approx 85.15|t| \qquad \text{Approximating the root to two decimal places} \\
&\approx 85.15t. \qquad \text{Since } t \geq 0,\ |t| = t.
\end{aligned}
$$

Thus, $d(t) = 85.15t,\ t \geq 0$.

b) Since the time traveled must be nonnegative, the domain is the set of nonnegative real numbers $[0, \infty)$.

c) Because of the ease of using a grapher, we can almost always visualize a problem by making a graph. Making such graphs should become a habit as you do applications and problem solving. ▬

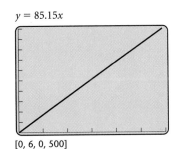

$y = 85.15x$

$[0, 6, 0, 500]$

Example 8 *Storage Area.* The Sound Shop has 20 ft of dividers with which to set off a rectangular area for the storage of overstock. If a corner of the store is used, the partition need only form two sides of a rectangle.

a) Express the floor area as a function of the length of the partition.

b) Find the domain of the function.

c) Graph the function.

d) Find the dimensions that maximize the area.

SOLUTION

a) Note that the dividers will form two sides of a rectangle. If, for example, 14 ft of dividers are used for the length of the rectangle, that would leave $20 - 14$, or 6 ft of dividers for the width. Thus if $x =$ the length, in feet, of the rectangle, then $20 - x =$ the width. We represent this information in a sketch.

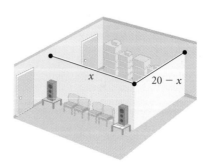

The area, $A(x)$, is given by

$$A(x) = x(20 - x) \qquad \text{Area = length · width}$$
$$= 20x - x^2.$$

The function $A(x) = 20x - x^2$ can be used to express the rectangle's area as a function of the length.

b) Because the rectangle's length must be positive and less than 20 ft (why?), we restrict the domain of A to $\{x \mid 0 < x < 20\}$, that is, the interval $(0, 20)$.

c) The graph is shown at left with the viewing window $[-5, 25, 0, 120]$.

d) We use the TRACE and ZOOM features or the MAX–MIN feature as in Example 3. The maximum value (overall largest value) of the area function is 100 when $x = 10$. Thus the dimensions that maximize the area are

Length $= x = 10$ ft   and

Width $= 20 - x = 20 - 10 = 10$ ft.

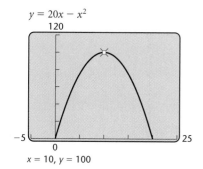

$y = 20x - x^2$
120

-5 0   25
$x = 10, y = 100$

Summary

Let's summarize our analysis of functions by considering the function f given by $f(x) = x^4 - 6x^3 - 4x^2 + 53.2x - 42.6$.

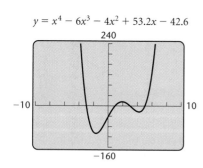

$y = x^4 - 6x^3 - 4x^2 + 53.2x - 42.6$
240

-10 10

-160

Function:	$f(x) = x^4 - 6x^3 - 4x^2 + 53.2x - 42.6$
Zeros:	$-2.975, 0.950, 3, 5.025$
Relative maximum:	15.921 at $x = 1.914$
Relative minima:	-106.907 at $x = -1.643$, -23.101 at $x = 4.229$
Increasing on:	$[-1.643, 1.914], [4.229, \infty)$
Decreasing on:	$(-\infty, -1.643], [1.914, 4.229]$
Domain:	All real numbers, $\mathbb{R}$
Range:	$[-106.907, \infty)$

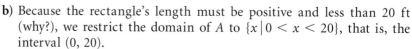

1.2 Exercise Set

For each function, find the zeros.

1. $f(x) = x^3 - 3x - 1$

2. $f(x) = x^3 + 3x^2 - 9x - 13$

3. $f(x) = x^4 - 2x^2$

4. $f(x) = x^4 - 2x^3 - 5.6$

5. $f(x) = x^3 - x$

6. $f(x) = 2x^3 - x^2 - 14x - 10$

7. $f(x) = \frac{1}{2}(|x - 4| + |x - 7|) - 4$

8. $f(x) = \sqrt{7 - x^2}$

9. $f(x) = |x + 1| + |x - 2| - 5$

10. $f(x) = |x + 1| + |x - 2|$

11. $f(x) = |x + 1| + |x - 2| - 3$

12. $f(x) = x^8 + 8x^7 - 28x^6 - 56x^5 + 70x^4 + 56x^3 - 28x^2 - 8x + 1$

Determine intervals on which each function is (a) increasing, (b) decreasing, and (c) constant.

13.

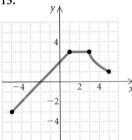

14.

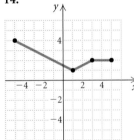

15.

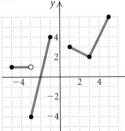

16.

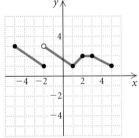

Graph each function using the given viewing window. Find where the function is increasing or decreasing and any relative maxima or minima. Change viewing windows if it seems appropriate for further analysis.

17. $f(x) = x^2$, $[-4, 4, -1, 10]$

18. $f(x) = 4 - x^2$, $[-5, 5, -5, 5]$

19. $f(x) = 5 - |x|$, $[-10, 10, -10, 10]$

20. $f(x) = |x + 3| - 5$, $[-10, 10, -10, 10]$

21. $f(x) = x^2 - 6x + 10$, $[-4, 10, -1, 20]$

22. $f(x) = -x^2 - 8x - 9$, $[-10, 2, -10, 10]$

23. $f(x) = -x^3 + 6x^2 - 9x - 4$, $[-3, 7, -20, 15]$

24. $f(x) = 0.2x^3 - 0.2x^2 - 5x - 4$, $[-10, 10, -30, 20]$

25. $f(x) = 1.1x^4 - 5.3x^2 + 4.07$, $[-4, 4, -4, 8]$

26. $f(x) = 1.2(x + 3)^4 + 10.3(x + 3)^2 + 9.78$, $[-9, 3, -40, 100]$

27. *Temperature During an Illness.* The temperature of a patient during an illness is given by the function

$$T(t) = -0.1t^2 + 1.2t + 98.6, \quad 0 \le t \le 12,$$

where T = the temperature, in degrees Fahrenheit, at time t, in days, after the onset of the illness.

a) Graph the function using a grapher.

b) Find the relative maximum.

c) At what time was the patient's temperature the highest? What was the highest temperature?

28. *Advertising Effect.* A software firm estimates that it will sell N units of a new CD-ROM video game after spending a dollars on advertising, where

$$N(a) = -a^2 + 300a + 6, \quad 0 \le a \le 300,$$

and a is measured in thousands of dollars.

a) Graph the function using a grapher.

b) Find the relative maximum.

c) For what advertising expenditure will the greatest number of games be sold? How many games will be sold for that amount?

Use a grapher. Find where each function is increasing or decreasing. Consider the entire set of real numbers if no domain is given.

29. $f(x) = \dfrac{8x}{x^2 + 1}$

30. $f(x) = \dfrac{-4}{x^2 + 1}$

31. $f(x) = x\sqrt{4 - x^2}$, for $-2 \le x \le 2$

32. $f(x) = -0.8x\sqrt{9 - x^2}$, for $-3 \le x \le 3$

Make a hand-drawn graph of each of the following. Check your results on a grapher.

33. $f(x) = \begin{cases} \frac{1}{2}x, & \text{for } x < 0, \\ x + 3, & \text{for } x \ge 0 \end{cases}$

34. $f(x) = \begin{cases} -\frac{1}{3}x + 2, & \text{for } x \le 0, \\ x - 5, & \text{for } x > 0 \end{cases}$

35. $f(x) = \begin{cases} -\frac{3}{4}x + 2, & \text{for } x < 4, \\ -1, & \text{for } x \ge 4 \end{cases}$

36. $f(x) = \begin{cases} 4, & \text{for } x \le -2, \\ x + 1, & \text{for } -2 < x < 3, \\ -x, & \text{for } x \ge -3 \end{cases}$

37. $f(x) = \begin{cases} x + 1, & \text{for } x \le -3, \\ -1, & \text{for } -3 < x < 4, \\ \frac{1}{2}x, & \text{for } x \ge 4 \end{cases}$

38. $f(x) = \begin{cases} \dfrac{x^2 - 9}{x + 3}, & \text{for } x \ne -3, \\ 5, & \text{for } x = -3 \end{cases}$

39. $f(x) = \begin{cases} 2, & \text{for } x = 5, \\ \dfrac{x^2 - 25}{x - 5}, & \text{for } x \ne 5 \end{cases}$

40. $f(x) = \begin{cases} \dfrac{x^2 + 3x + 2}{x + 1}, & \text{for } x \ne -1, \\ 7, & \text{for } x = -1 \end{cases}$

41. $f(x) = \text{INT}(x)$ **42.** $f(x) = 1 + \text{INT}(x)$

43. $f(x) = \text{INT}(x) - 1$ **44.** $f(x) = [\![x]\!] - 2$

Graph each of the following with a grapher, if possible.

45. $f(x) = \begin{cases} \sqrt[3]{x}, & \text{for } x \le -1, \\ x^2 - 3x, & \text{for } -1 < x < 4, \\ \sqrt{x - 4}, & \text{for } x \ge 4 \end{cases}$

46. $f(x) = \begin{cases} \left| \dfrac{x}{3} + 2 \right|, & \text{for } x < 5, \\ \sqrt[3]{x - 8}, & \text{for } x \ge 5 \end{cases}$

47. $f(x) = \begin{cases} \sqrt[3]{x + 27}, & \text{for } x < 1, \\ \left| 2 - \dfrac{x}{5} \right|, & \text{for } x \ge 1 \end{cases}$

48. *Airplane Distance.*   An airplane is flying at an altitude of 3700 ft. The slanted distance directly to the airport is d feet. Express the horizontal distance h as a function of d.

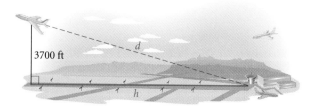

49. *Rising Balloon.*   A hot-air balloon rises straight up from the ground at a rate of 120 ft/min. The balloon is tracked from a rangefinder at point P, which is 400 ft from the release point Q of the balloon. Let d = the distance from the balloon to the rangefinder at point P and t = the time, in minutes, since the balloon was released. Express d as a function of t.

50. *Triangular Flag.*   A scout troop is designing a triangular flag so that the length of its base, in inches, is 7 less than twice the height, h. Express the area of the flag as a function of the height.

51. *Garden Area.*   Yardbird Landscaping has 48 m of fencing with which to enclose a rectangular garden. If the garden is x meters long, express the garden's area as a function of the length.

52. *Tablecloth Area.*   A tailor uses 16 ft of lace to trim the edges of a rectangular tablecloth. If the tablecloth is w feet wide, express its area as a function of the width.

53. *Inscribed Rhombus.*   A rhombus is inscribed in a rectangle that is w meters wide with a perimeter of 40 m. Each vertex of the rhombus is a midpoint of a side of the rectangle. Express the area of the rhombus as a function of the rectangle's width. (*Hint:* Consider the area of the rectangle.)

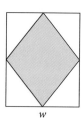

54. *Gas Tank Volume.* A gas tank has ends that are hemispheres of radius r feet. The cylindrical midsection is 6 ft long. Express the volume of the tank as a function of r.

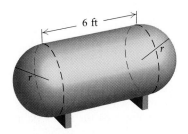

55. *Golf Distance Finder.* A device used in golf to estimate the distance d, in yards, to a hole measures the size s, in inches, that the 7-ft pin appears to be in a viewfinder. Express the distance d as a function of s.

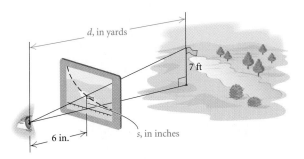

56. *Play Space.* A daycare center has 30 ft of dividers with which to enclose a rectangular play space in a corner of a large room. The sides against the wall require no partition. Suppose the play space is x feet long.

a) Express the area of the play space as a function of x.
b) Find the domain of the function.
c) Graph the function.
d) What dimensions yield the maximum area?

57. *Volume of a Box.* From a 12-cm by 12-cm piece of cardboard, square corners are cut out so that the sides can be folded up to make a box.

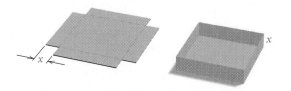

a) Express the volume of the box as a function of the length, x, in centimeters, of a cut-out square.
b) Find the domain of the function.
c) Graph the function.
d) What dimensions yield the maximum volume?

58. *Cost of Material.* A rectangular box with volume 320 ft^3 is built with a square base and top. The cost is \$1.50/ft^2 for the bottom, \$2.50/ft^2 for the sides, and \$1/ft^2 for the top. Let $x =$ the length of the base, in feet.

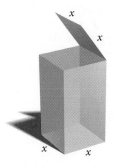

a) Express the cost of the box as a function of x.
b) Find the domain of the function.
c) Graph the function.
d) What dimensions minimize the cost of the box?

59. *Area of an Inscribed Rectangle.* A rectangle that is x feet wide is inscribed in a circle of radius 8 ft.

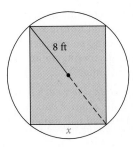

a) Express the area of the rectangle as a function of x.
b) Find the domain of the function.
c) Graph the function.
d) What dimensions maximize the area of the rectangle?

60. *Area of an Inscribed Rectangle.* A rectangle that is x meters wide is inscribed in a circle of diameter 20 m.

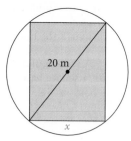

a) Express the area of the rectangle as a function of x.

b) Find the domain of the function.

c) Graph the function.

d) What dimensions maximize the area of the rectangle?

Skill Maintenance

If $f(x) = 7x + 2$, find each of the following.

61. $\dfrac{f(b) - f(a)}{b - a}$ **62.** $\dfrac{f(a + h) - f(a)}{h}$

Synthesis

63. ◆ Describe a real-world situation that could be modeled by a function that is, in turn, increasing, then constant, and finally decreasing.

64. ◆ Simply stated, a *continuous function* is a function whose graph can be drawn without lifting the pencil from the paper. Examine several functions in this exercise set to see if they are continuous. Then explore the continuous functions to find the relative maxima and minima. For continuous functions, how can you connect the ideas of increasing and decreasing on an interval to relative maxima and minima?

Use a grapher. Estimate where each function is increasing or decreasing and any relative maxima or minima. Consider the entire set of real numbers if no domain is given.

65. $f(x) = 20.17x^3 - 3.24x^5$

66. $f(x) = 3.22x^5 - 5.208x^3 - 11$

67. $f(x) = x^4 + 4x^3 - 36x^2 - 160x + 400$

68. $f(x) = -x^6 - 4x^5 + 54x^4 + 160x^3 - 641x^2 - 828x + 1200$

69. *Parking Costs.* A parking garage charges $2 for up to (but not including) 1 hr of parking, $4 for up to 2 hr of parking, $6 for up to 3 hr of parking, and so on. Let $C(t) =$ the cost of parking for t hours.

a) Graph the function.

b) Write an equation for $C(t)$ using the greatest integer notation INT.

70. If $\text{INT}(x + 2) = -3$, what are the possible inputs for x?

71. If $[\text{INT}(x)]^2 = 25$, what are the possible inputs for x?

72. *Minimizing Power Line Costs.* A power line is constructed from a power station at point A to an island at point C, which is 1 mi directly out in the water from a point B on the shore. Point B is 4 mi downshore from the power station at A. It costs $5000 per mile to lay the power line under water and $3000 per mile to lay the power line under ground. The line comes to the shore at point S downshore from A. Note that S could very well be A or B. Let $x =$ the distance from B to S.

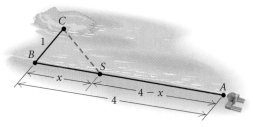

a) Express the cost C of laying the line as a function of x.

b) At what distance x should the line come to shore in order to minimize cost?

73. *Volume of an Inscribed Cylinder.* A right circular cylinder of height h and radius r is inscribed in a right circular cone with a height of 10 ft and a base with a radius of 6 ft.

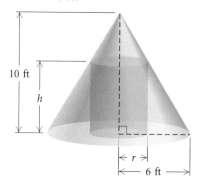

a) Express the height h of the cylinder as a function of r.

b) Express the volume V as a function of r.

c) Express the volume V as a function of h.

1.3
Linear Functions and Applications

- *Graph linear functions and equations, finding the slope and the y-intercept.*
- *Determine equations of lines.*
- *Solve applied problems involving linear functions.*

In real-life situations, we often need to make decisions on the basis of limited information. When the given information is used to formulate an equation or inequality that at least approximates the situation mathematically, we have created a **model**. The most frequently used mathematical models are *linear*—the graphs of these models are straight lines.

Linear Functions

Let's begin to examine the connections among equations, functions, and graphs that are straight lines.

Interactive Discovery

Graph each of the following equations on a grapher. Which have graphs that are lines? Look for patterns.

$$4x + 5y = 20, \qquad y = 5 - x^2,$$

$$y = \frac{x}{2} + 12, \qquad y = 1.5x - 0.05,$$

$$y = -2x - 5, \qquad y = \frac{1}{x},$$

$$x = -3, \qquad y = 6.2,$$

$$y = \sqrt{x}, \qquad x = 4.9.$$

(Some graphers are not able to graph equations of the type $x = a$. Consult your manual.) Which of these equations have graphs that are lines and are also functions?

We have the following results and related terminology.

Linear Functions

A function f is a *linear function* if it is given by $f(x) = mx + b$, where m and b are constants.

If $m = 0$, the function simplifies to the *constant function* $f(x) = b$. If $m = 1$ and $b = 0$, the function simplifies to the *identity function* $f(x) = x$.

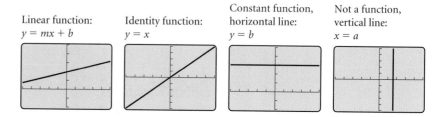

Horizontal and Vertical Lines

Horizontal lines are given by equations of the type $y = b$ or $f(x) = b$. (They are functions.)

Vertical lines are given by equations of the type $x = a$. (They are not functions.)

| Linear function: $y = mx + b$ | Identity function: $y = x$ | Constant function, horizontal line: $y = b$ | Not a function, vertical line: $x = a$ |

The Linear Function $f(x) = mx + b$ and Slope

To attach meaning to the constant m in the equation $f(x) = mx + b$, we first consider an application. FaxMax is an office machine business. Its total costs for two different time periods are given by two functions shown in the tables and graphs that follow. The variable x represents time, in months. The variable y represents total costs, in thousands of dollars, over that amount of time. Look for a pattern.

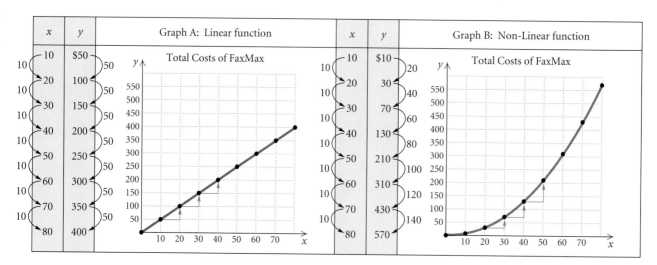

x	y
10	$50
20	100
30	150
40	200
50	250
60	300
70	350
80	400

Graph A: Linear function — Total Costs of FaxMax

x	y
10	$10
20	30
30	70
40	130
50	210
60	310
70	430
80	570

Graph B: Non-Linear function — Total Costs of FaxMax

We see in graph A that every change of 10 months results in a $50 thousand change in total costs. But in graph B, changes of 10 months do not result in constant changes in total costs. This is a way to distinguish linear from nonlinear functions. The rate at which a linear function changes, or the steepness of its graph, is constant.

Mathematically, we define a line's steepness, or **slope**, as the ratio of its vertical change (rise) to the corresponding horizontal change (run).

Slope

The *slope m* of a line containing points (x_1, y_1) and (x_2, y_2) is given by

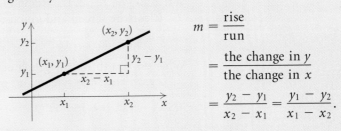

$$m = \frac{\text{rise}}{\text{run}}$$

$$= \frac{\text{the change in } y}{\text{the change in } x}$$

$$= \frac{y_2 - y_1}{x_2 - x_1} = \frac{y_1 - y_2}{x_1 - x_2}.$$

Example 1 Make a hand-drawn graph of the function $f(x) = -\frac{2}{3}x + 1$ and determine its slope.

SOLUTION Since the equation for f is in the form $f(x) = mx + b$, we know it is a linear function. We can graph it by connecting two points on the graph with a straight line. We calculate two ordered pairs, plot the points, graph the function, and determine the slope:

$$f(3) = -\frac{2}{3} \cdot 3 + 1 = -1;$$

$$f(9) = -\frac{2}{3} \cdot 9 + 1 = -5;$$

Pairs: $(3, -1)$, $(9, -5)$;

$$\text{Slope} = m = \frac{y_2 - y_1}{x_2 - x_1}$$

$$= \frac{-5 - (-1)}{9 - 3}$$

$$= \frac{-4}{6} = -\frac{2}{3}.$$

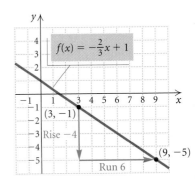

The slope is the same for any two points on a line. Thus, to check our work, note that $f(6) = -\frac{2}{3} \cdot 6 + 1 = -3$. Using the points $(6, -3)$ and $(3, -1)$, we have

$$m = \frac{-3 - (-1)}{6 - 3} = \frac{-2}{3} = -\frac{2}{3}.$$

We can also use the points in the opposite order when computing slope, so long as we are consistent. Note also that the slope of the line is indeed the number m in the equation for the function $f(x) = -\frac{2}{3}x + 1$. ▬

Let's explore the effect of the slope m in linear equations of the type $f(x) = mx$.

Interactive Discovery

With a square viewing window (see the Introduction to Graphs and Graphers), graph the following equations:

$$y_1 = x, \qquad y_2 = 2x, \qquad y_3 = 5x, \quad \text{and} \quad y_4 = 10x.$$

What do you think the graph of $y = 128x$ will look like?

Clear the screen and graph the following equations:

$$y_1 = x, \qquad y_2 = \tfrac{3}{4}x, \qquad y_3 = 0.48x, \quad \text{and} \quad y_4 = \tfrac{3}{25}x.$$

What do you think the graph of $y = 0.000029x$ will look like?

Again clear the screen and graph each set of equations:

$$y_1 = -x, \qquad y_2 = -2x, \qquad y_3 = -4x, \quad \text{and} \quad y_4 = -10x$$

and

$$y_1 = -x, \qquad y_2 = -\tfrac{2}{3}x, \qquad y_3 = -0.35x, \quad \text{and} \quad y_4 = -\tfrac{1}{10}x.$$

From your observations, what do you think the graphs of $y = -200x$ and $y = -0.000017x$ will look like?

If a line slants up from left to right, the change in x and the change in y have the same sign, so the line has a positive slope. The larger the slope is, the steeper the line. If a line slants down from left to right, the change in x and the change in y are of opposite signs, so the line has a negative slope. The larger the absolute value of the slope, the steeper the line.

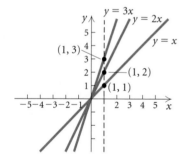

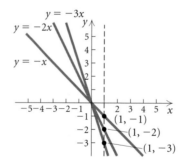

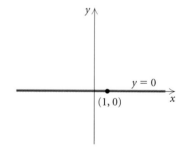

When $m = 0$, $y = 0x$, or $y = 0$, as shown in the graph on the right above. Note that this is both the x-axis and a horizontal line.

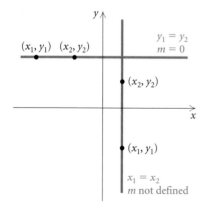

Horizontal and Vertical Lines

If a line is horizontal, the change in y for any two points is 0. Thus a horizontal line has slope 0.

If a line is vertical, the change in x for any two points is 0. Thus the slope is *not defined* because we cannot divide by 0.

Note that zero slope and an undefined slope are two very different concepts.

Applications of Slope

Slope has many real-world applications. For example, numbers like 2%, 4%, and 7% are often used to represent the **grade** of a road. Such a number is meant to tell how steep a road is on a hill or mountain. For example, a 4% grade means that the road rises 4 ft for every horizontal distance of 100 ft if a vehicle is going up; and −4% means that the road is dropping 4 ft for every 100 ft, if the vehicle is going down.

The concept of grade is also used in cardiology when a person runs on a treadmill. A physician may change the slope, or grade, of a treadmill to measure its effect on heart rate. Another example occurs in hydrology. When a river flows, the strength or force of the river depends on how far the river falls vertically compared to how far it flows horizontally.

Example 2 *Ramps for the Handicapped.* Construction laws regarding access ramps for the handicapped state that every vertical rise of 1 ft requires a horizontal run of 12 ft. What is the grade, or slope, of such a ramp?

SOLUTION   The grade, or slope, is given by

$$m = \tfrac{1}{12} \approx 0.083 \approx 8.3\%.$$

Slope–Intercept Equations of Lines

Let's explore the effect of the constant b in linear equations of the type $f(x) = mx + b$.

Begin with the graph of $y = x$ and a square viewing window. Now graph the lines $y = x + 3$ and $y = x - 4$ in the same viewing window. How do the lines differ from $y = x$? What do you think the line $y = x - 6$ will look like?

Try graphing $y = -0.5x$, $y = -0.5x - 4$, and $y = -0.5x + 3$ in the same viewing window. Describe what happens to the graph of $y = -0.5x$ when a number b is added.

Compare the graphs of the equations

$$y = 3x \quad \text{and} \quad y = 3x - 2.$$

Note that the graph of $y = 3x - 2$ is a shift down of the graph of $y = 3x$, and that $y = 3x - 2$ has y-intercept $(0, -2)$. That is, the graph crosses the y-axis at $(0, -2)$.

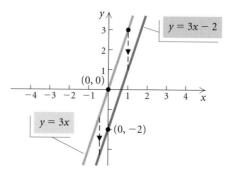

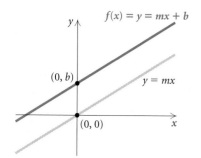

> **The Slope–Intercept Equation**
>
> The linear function f given by
>
> $$f(x) = mx + b$$
>
> has a graph that is a straight line parallel to $y = mx$. The constant m is called the slope, and the y-intercept is $(0, b)$, or b.

We know that a nonvertical line has slope m and y-intercept $(0, b)$ and that its equation is given by $y = mx + b$. The advantage of $y = mx + b$ is that we can read the slope m and the y-intercept b directly from the equation.

Example 3   Find the slope and the y-intercept of the line with equation $y = -0.25x - 3.8$.

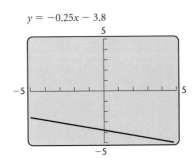

$y = -0.25x - 3.8$

SOLUTION

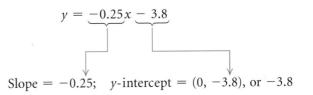

$$y = \underbrace{-0.25}x \; \underbrace{- \; 3.8}$$

Slope $= -0.25$; y-intercept $= (0, -3.8)$, or -3.8 ▬

Example 4   Find the slope and the y-intercept of the line with equation $y = 8$.

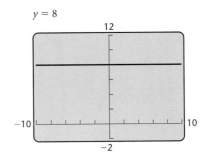

$y = 8$

SOLUTION   We can rewrite this equation as $y = 0x + 8$. We then see that the slope is 0 and the y-intercept is $(0, 8)$, or 8. The graph is a horizontal line.   ▬

Any equation whose graph is a straight line is a **linear equation.** To find the slope and the y-intercept of the graph of a linear equation, we can solve for y, and then read the information from the equation.

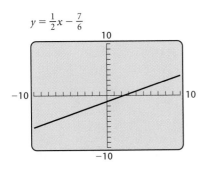

$y = \frac{1}{2}x - \frac{7}{6}$

Example 5   Find the slope and the y-intercept of the line with equation $3x - 6y - 7 = 0$.

SOLUTION   We solve for y, obtaining $y = \frac{1}{2}x - \frac{7}{6}$. Thus the slope is $\frac{1}{2}$ and the y-intercept is $\left(0, -\frac{7}{6}\right)$, or $-\frac{7}{6}$.   ▬

Example 6   A line has slope $-\frac{7}{9}$ and y-intercept $(0, 16)$. Find an equation of the line.

SOLUTION   We use the slope–intercept equation and substitute $-\frac{7}{9}$ for m and 16 for b:

$$y = mx + b$$
$$y = -\frac{7}{9}x + 16.$$

Point–Slope Equations of Lines

Suppose that we have a nonvertical line and that the coordinates of point P_1 are (x_1, y_1). We can think of P_1 as fixed and imagine a movable point P on the line with coordinates (x, y). Thus the slope is given by

$$\frac{y - y_1}{x - x_1} = m.$$

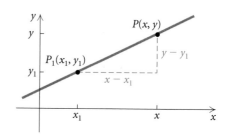

Multiplying on both sides by $x - x_1$, we get the *point–slope equation*.

> **Point–Slope Equation**
>
> The *point–slope equation* of the line with slope m passing through (x_1, y_1) is
>
> $$y - y_1 = m(x - x_1).$$

Thus if we know the slope of a line and the coordinates of one point on the line, we can find an equation of the line.

Example 7   Find an equation of the line containing the point $\left(\frac{1}{2}, -1\right)$ and with slope 5.

SOLUTION   If we substitute in $y - y_1 = m(x - x_1)$, we get

$$y - (-1) = 5\left(x - \tfrac{1}{2}\right),$$

which simplifies as follows:

$$y + 1 = 5x - \tfrac{5}{2}$$
$$y = 5x - \tfrac{5}{2} - 1$$
$$y = 5x - \tfrac{7}{2}. \quad \text{Slope–intercept equation}$$

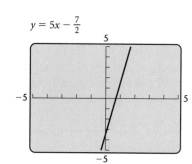

$y = 5x - \frac{7}{2}$

The advantage of having a grapher when doing a problem like the one in Example 7 is that you have a quick, visual way of seeing your result.

Two-Point Equations of Lines

Suppose that a nonvertical line contains the points $P_1(x_1, y_1)$ and $P_2(x_2, y_2)$. The slope of the line is

$$m = \frac{y_2 - y_1}{x_2 - x_1}.$$

If we substitute $\frac{y_2 - y_1}{x_2 - x_1}$ for m in the point–slope equation,

$$y - y_1 = m(x - x_1),$$

we get the *two-point equation.*

Two-Point Equation

The *two-point equation* of the line passing through the points (x_1, y_1) and (x_2, y_2) is

$$y - y_1 = \frac{y_2 - y_1}{x_2 - x_1}(x - x_1).$$

Example 8   Find an equation of the line containing the points $(2, 3)$ and $(1, -4)$.

SOLUTION   If we take $(2, 3)$ as P_1 and $(1, -4)$ as P_2 and use the two-point equation, we get

$$y - 3 = \frac{-4 - 3}{1 - 2}(x - 2),$$

which simplifies to $y = 7x - 11$.

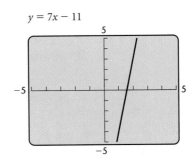

$y = 7x - 11$

Parallel Lines

How can we examine a pair of equations to determine whether their graphs are parallel—that is, they do not intersect? We can explore this with a grapher.

Graph each of the following pairs of lines. Try to determine whether they are parallel. You will probably need to change viewing windows and either zoom in or zoom out to decide for sure. Complete the table and look for a pattern.

	SLOPES		
PAIRS OF EQUATIONS	m_1	m_2	PARALLEL?
$y = -0.38x + 4.2$, $y = -0.37x - 5.1$			
$y = -0.38x + 4.2$, $50y + 19x = 21$			
$2x + 1 = y$, $y + 3x = 4$			
$y = 8.02x - 11.3$, $y = 7.98x - 11.3$			
$y = -3$, $y = 4.2$			
$x = -5$, $x = 2$			

Can you find a way to know whether two lines are parallel without graphing?

If two different lines are vertical, then they are parallel. Thus two equations such as $x = a_1$, $x = a_2$, where $a_1 \neq a_2$, have graphs that are *parallel lines*. Two nonvertical lines such as $y = mx + b_1$, $y = mx + b_2$, where $b_1 \neq b_2$, also have graphs that are *parallel lines*.

> **Parallel Lines**
>
> Vertical lines are *parallel*. Nonvertical lines are *parallel* if and only if they have the same slope and different y-intercepts.

Perpendicular Lines

How can we examine a pair of equations to determine whether they are perpendicular? Let's explore this with a grapher.

Graph each of the following pairs of lines (see the table at the top of the following page). Try to determine whether they are perpendicular. You will probably need to change viewing windows and either zoom in or zoom out to decide for sure. However, you must use square viewing windows so that you can see or measure whether angles between lines seem to be 90°. Complete the table and look for a pattern.

(continued)

PAIRS OF EQUATIONS	SLOPES		PRODUCT OF SLOPES	PERPENDICULAR?
	m_1	m_2	$m_1 \cdot m_2$	
$y = -\frac{2}{5}x + 3, \ y = \frac{5}{2}x + 3$				
$y = 0.1875x - 2, \ y = -\frac{16}{3}x - 5.1$				
$3y + 4x = -21, \ 4y - 3x = 8$				
$y = -3x + 4, \ y = 2x - 7$				
$x = -5, \ y = 2$				

Can you find a way to determine whether two lines are perpendicular without graphing?

Perpendicular Lines

Two lines with slopes m_1 and m_2 are *perpendicular* if and only if the product of their slopes is -1:

$$m_1 m_2 = -1.$$

Lines are also *perpendicular* if one is vertical ($x = a$) and the other is horizontal ($y = b$).

If a line has slope m_1, the slope m_2 of a line perpendicular to it is $-1/m_1$ (the slope of one line is the opposite of the reciprocal of the other).

Example 9   Determine whether each of the following pairs of lines is parallel, perpendicular, or neither.

a) $y + 2 = 5x, \ 5y + x = -15$
b) $2y + 4x = 8, \ 5 + 2x = -y$
c) $2x + 1 = y, \ y + 3x = 4$

SOLUTION   We use an algebraic procedure to know for sure. We can create graphs on a grapher as a check.

a) We solve each equation for y:

$$y = 5x - 2, \qquad y = -\frac{1}{5}x - 3.$$

The slopes are 5 and $-\frac{1}{5}$. Their product is -1, so the lines are perpendicular (see the figure at left).

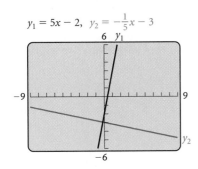

$y_1 = 5x - 2, \quad y_2 = -\frac{1}{5}x - 3$

b) Solving each equation for y, we get

$$y = -2x + 4, \qquad y = -2x - 5.$$

We see that $m_1 = -2$ and $m_2 = -2$. Since the slopes are the same and the y-intercepts, 4 and -5, are different, the lines are parallel (see the figure on the left below).

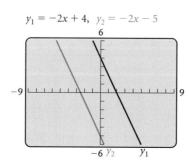

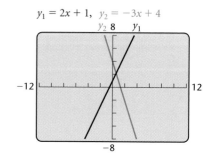

c) Solving each equation for y, we get

$$y = 2x + 1, \qquad y = -3x + 4.$$

Since $m_1 = 2$ and $m_2 = -3$, we know that $m_1 \neq m_2$ and that $m_1 m_2 \neq -1$. It follows that the lines are neither perpendicular nor parallel (see the figure on the right above).

Example 10 Write equations of the lines (a) parallel and (b) perpendicular to the graph of the line $4y - x = 20$ and both containing the point $(2, -3)$.

SOLUTION   We first solve $4y - x = 20$ for y to get $y = \frac{1}{4}x + 5$. Thus the slope is $\frac{1}{4}$.

a) The line parallel to the given line will have slope $\frac{1}{4}$. We use the point–slope equation for a line with slope $\frac{1}{4}$ and containing the point $(2, -3)$:

$$y - y_1 = m(x - x_1)$$
$$y - (-3) = \tfrac{1}{4}(x - 2)$$
$$y = \tfrac{1}{4}x - \tfrac{7}{2}.$$

b) The slope of the perpendicular line is -4. Now we use the point–slope equation to write an equation for a line with slope -4 and containing the point $(2, -3)$:

$$y - y_1 = m(x - x_1)$$
$$y - (-3) = -4(x - 2)$$
$$y = -4x + 5.$$

We can visualize our work on a grapher.

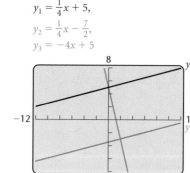

Regression

Using linear regression or curve fitting to model data is presented in Appendix A.3.

Summary of Terminology About Lines

TERMINOLOGY	MATHEMATICAL INTERPRETATION
Slope	$m = \dfrac{y_2 - y_1}{x_2 - x_1}$
Slope–intercept equation	$y = mx + b$
Point–slope equation	$y - y_1 = m(x - x_1)$
Two-point equation	$y - y_1 = \dfrac{y_2 - y_1}{x_2 - x_1}(x - x_1)$
Horizontal lines	$y = b$
Vertical lines	$x = a$
Parallel lines	$m_1 = m_2,\ b_1 \neq b_2$
Perpendicular lines	$m_1 m_2 = -1$

Applications of Linear Functions

We now consider an application of linear functions.

Example 11 *Anthropology Estimates.* An anthropologist can use certain linear functions to estimate the height of a male or a female, given the length of certain bones. The humerus is the bone from the elbow to the shoulder. Let $x =$ the length of the humerus, in centimeters. Then the height, in centimeters, of an adult male with a humerus of length x is given by the function

$$M(x) = 2.89x + 70.64.$$

The height, in centimeters, of an adult female with a humerus of length x is given by the function

$$F(x) = 2.75x + 71.48.$$

a) A 26-cm humerus was uncovered in a ruins. Assuming it was from a female, how tall was she?

b) Graph F.

c) Find the domain of F.

SOLUTION

a) We substitute into the function:

$$F(26) = 2.75(26) + 71.48 = 142.98.$$

Thus the female was 142.98 cm tall.

b) The graph is shown at left.

c) Theoretically, the domain of the function is the set of all real numbers. However, the context of the problem dictates a different domain. One could not find a bone with a length of 0 or less. Thus the domain consists of positive real numbers, that is, the interval $(0, \infty)$.

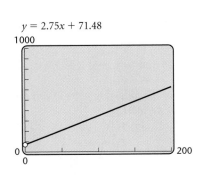

$y = 2.75x + 71.48$

Humerus

1.3 Exercise Set

Each table of data contains input–output values for a function. Answer the following questions for each table.

a) *Is the change in the inputs x the same?*
b) *Is the change in the outputs y the same?*
c) *Is the function linear?*

1.

x	y
−3	7
−2	10
−1	13
0	16
1	19
2	22
3	25

2.

x	y
20	12.4
30	24.8
40	49.6
50	99.2
60	198.4
70	396.8
80	793.6

3.

x	y
11	3.2
26	5.7
41	8.2
56	9.3
71	11.3
86	13.7
101	19.1

4.

x	y
2	−8
4	−12
6	−16
8	−20
10	−24
12	−28
14	−36

Find the slope of the line containing the given points.

5.

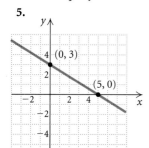

6.

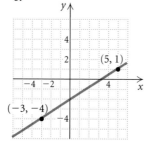

7. $(9, 4)$ and $(−1, 2)$

8. $(−3, 7)$ and $(5, −1)$

9. $(4, −9)$ and $(−5, 6)$

10. $(−6, −1)$ and $(2, −13)$

11. $(\pi, −3)$ and $(\pi, 2)$

12. $(\sqrt{2}, −4)$ and $(0.56, −4)$

13. (a, a^2) and $(a + h, (a + h)^2)$

14. $(a, 3a + 1)$ and $(a + h, 3(a + h) + 1)$

15. *Road Grade.* Using the following figure, find the road grade and an equation giving the height y as a function of the horizontal distance x.

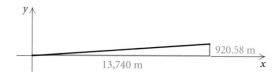

16. *Heart Arrhythmia.* A treadmill is 5 ft long and is set at an 8% grade when a heart arrhythmia occurs. How high is the end of the treadmill?

Find the slope and the y-intercept of the equation.

17. $y = \frac{3}{5}x − 7$

18. $f(x) = −2x + 3$

19. $f(x) = 5 − \frac{1}{2}x$

20. $y = 2 + \frac{3}{7}x$

21. $3x + 2y = 10$

22. $2x − 3y = 12$

23. $4x − 3f(x) − 15 = 6$

24. $9 = 3 + 5x − 2f(x)$

Write a slope–intercept equation for a line with the given characteristics.

25. $m = \frac{2}{9}$, y-intercept $(0, 4)$

26. $m = −\frac{3}{8}$, y-intercept $(0, 5)$

27. $m = −4$, y-intercept $(0, −7)$

28. $m = \frac{2}{7}$, y-intercept $(0, −6)$

29. $m = −4.2$, y-intercept $(0, \frac{3}{4})$

30. $m = −4$, y-intercept $(0, −\frac{3}{2})$

31. $m = \frac{2}{9}$, passes through $(3, 7)$

32. $m = −\frac{3}{8}$, passes through $(5, 6)$

33. $m = 3$, passes through $(1, −2)$

34. $m = −2$, passes through $(−5, 1)$

35. $m = −\frac{3}{5}$, passes through $(−4, −1)$

36. $m = \frac{2}{3}$, passes through $(−4, −5)$

37. Passes through $(−1, 5)$ and $(2, −4)$

38. Passes through $(2, −1)$ and $(7, −11)$

39. Passes through $(7, 0)$ and $(−1, 4)$

40. Passes through $(−3, 7)$ and $(−1, −5)$

Determine whether each of the following pairs of lines is parallel, perpendicular, or neither.

41. $x + 2y = 5$,
     $2x + 4y = 8$

42. $2x − 5y = −3$,
     $2x + 5y = 4$

43. $y = 4x − 5$,
     $4y = 8 − x$

44. $y = 7 − x$,
     $y = x + 3$

Write a slope–intercept equation for a line passing through the given point that is parallel to the given line. Then write a second equation for a line passing through the given point that is perpendicular to the given line.

45. $(3, 5)$, $y = \frac{2}{7}x + 1$

46. $(-1, 6)$, $f(x) = 2x + 9$

47. $(-7, 0)$, $y = -0.3x + 4.3$

48. $(-4, -5)$, $2x + y = -4$

49. $(3, -2)$, $3x + 4y = 5$

50. $(8, -2)$, $y = 4.2(x - 3) + 1$

51. $(3, -3)$, $x = -1$

52. $(4, -5)$, $y = -1$

53. *Ideal Weight.* One formula to estimate the ideal weight of a woman is to multiply her height by 3.5 and subtract 110. Let W = the ideal weight, in pounds, and h = height, in inches.

a) Express W as a linear function of h.
b) Find the ideal weight of a woman whose height is 62 in.
c) Graph W.
d) Find the domain of the function.

54. *Pressure at Sea Depth.* The function P, given by

$$P(d) = \frac{1}{33}d + 1,$$

gives the pressure, in atmospheres (atm), at a depth d, in feet, under the sea.

a) Find $P(0)$, $P(5)$, $P(10)$, $P(33)$, and $P(200)$.
b) Graph P.
c) Find the domain of the function.

55. *Stopping Distance on Glare Ice.* The stopping distance (at some fixed speed) of regular tires on glare ice is a function of the air temperature F, in degrees Fahrenheit. This function is estimated by

$$D(F) = 2F + 115,$$

where $D(F)$ = the stopping distance, in feet, when the air temperature is F, in degrees Fahrenheit.

a) Find $D(0°)$, $D(-20°)$, $D(10°)$, and $D(32°)$.
b) Graph D.
c) Explain why the domain should be restricted to $[-57.5°, 32°]$.

56. *Anthropology Estimates.* Consider Example 10 and the function

$$M(x) = 2.89x + 70.64$$

for estimating the height of a male.

a) Find the height of a male if a 26-cm humerus is found in an archeological dig.
b) Graph M.
c) Find the domain of M.

57. *Reaction Time.* While driving a car, you suddenly see a school crossing guard. Your brain registers the information and sends a signal to your foot to hit the brake. The car travels a distance D, in feet, during this time, where D is a function of the speed r, in miles per hour, that the car is traveling when you see the crossing guard. That reaction distance is a linear function given by

$$D(r) = \frac{11r + 5}{10}.$$

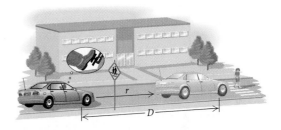

a) Find the slope of this line and interpret its meaning in this application.
b) Find $D(5)$, $D(10)$, $D(20)$, $D(50)$, and $D(65)$.
c) Graph D.
d) What is the domain of this function? Explain.

58. *Straight-line Depreciation.* A company buys a new color printer for $5200 to print banners for a sales campaign. The printer is purchased on January 1 and is expected to last 8 yr, at the end of which time its *trade-in*, or *salvage value*, will be $1100. If the company figures the decline or depreciation in value to be the same each year, then the salvage value V, after t years, is given by the linear function

$$V(t) = \$5200 - \$512.50t, \quad \text{for } 0 \leq t \leq 8.$$

a) Find $V(0)$, $V(1)$, $V(2)$, $V(3)$, and $V(8)$.
b) Graph V.
c) Find the domain and the range of this function.

For Exercises 59 and 60, express the slope as a ratio of two quantities.

59.

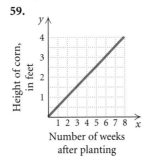

Number of weeks after planting

60.

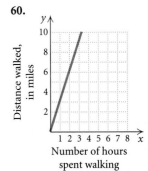

Number of hours spent walking

61. *Total Cost.*   The Cellular Connection charges $60 for a phone and $40 per month under its economy plan. Write and graph an equation that can be used to determine the total cost, $C(t)$, of operating a Cellular Connection phone for t months.

62. *Total Cost.*   Twin Cities Cable Television charges a $35 installation fee and $30 per month for "deluxe" service. Write and graph an equation that can be used to determine the total cost, $C(t)$, for t months of deluxe cable television service. Find the total cost for 8 months of service.

In Exercises 63 and 64, the term **fixed costs** *refers to the start-up costs of operating a business. This includes machinery and building costs. The term* **variable costs** *refers to what it costs a business to produce or service one item.*

63. Kara's Custom Tees experienced fixed costs of $800 and variable costs of $3 per shirt. Write an equation that can be used to determine the total expenses encountered by Kara's Custom Tees. Then graph the equation.

64. It's My Racquet experienced fixed costs of $500 and variable costs of $2 for each tennis racquet that is restrung. Write an equation that can be used to determine the total expenses encountered by It's My Racquet. Then graph the equation.

Direct Variation.   Many applications can be modeled by linear functions like

$$y = kx, \quad k > 0,$$

where the variables are nonnegative: These are functions of **direct variation.** *The constant k is called a* **variation constant.** *Note that the function is increasing over the interval* $[0, \infty)$.

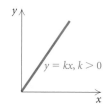

65. *House of Representatives.*   The number of representatives N that each state has varies directly as the number of people P living in the state. If New York, with 18,119,416 residents, has 31 representatives, how many representatives does Michigan, with a population of 9,436,628, have?

66. *Hooke's Law.*   Hooke's law states that the distance d that a spring will stretch varies directly as the mass m of an object hanging from the spring.

Suppose that a 3-kg mass stretches a spring 40 cm. Find a function of direct variation. Then use the equation to predict how far the spring will stretch with a 5-kg mass.

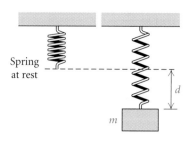

Skill Maintenance

If $f(x) = x^2 - 3x$, *find each of the following.*

67. $f(-5)$ **68.** $f(5)$

69. $f(-a)$ **70.** $f(a + h)$

Synthesis

71. ◆ Discuss why the graph of a vertical line $x = a$ cannot be that of a function.

72. ◆ Explain as you would to a fellow student how the idea of slope can be used to describe the slant and the steepness of a line using a number.

73. Find k so that the line containing the points $(-3, k)$ and $(4, 8)$ is parallel to the line containing the points $(5, 3)$ and $(1, -6)$.

74. Find an equation of the line passing through the point $(4, 5)$ perpendicular to the line passing through the points $(-1, 3)$ and $(2, 9)$.

75. *Fahrenheit and Celsius Temperatures.*   Fahrenheit temperature F is a linear function of Celsius temperature C. When C is 0, F is 32; and when C is 100, F is 212. Use these data to express C as a function of F and to express F as a function of C.

Suppose that f is a linear function. Then $f(x) = mx + b$. *Determine whether each of the following is true or false.*

76. $f(c + d) = f(c) + f(d)$

77. $f(cd) = f(c)f(d)$

78. $f(kx) = kf(x)$

79. $f(c - d) = f(c) - f(d)$

Let $f(x) = mx + b$. *Find a formula for* $f(x)$ *given each of the following.*

80. $f(3x) = 3f(x)$ **81.** $f(x + 2) = f(x) + 2$

1.4
Symmetry

- *Determine whether a graph is symmetric with respect to the x-axis, the y-axis, and the origin.*
- *Determine whether a function is even, odd, or neither even nor odd.*

Symmetry

Symmetry occurs often in nature and in art. For example, when viewed from the front, the bodies of most animals are at least approximately symmetric. This means that each eye is the same distance from the center of the bridge of the nose, each shoulder is the same distance from the center of the chest, and so on. Architects have used symmetry for thousands of years to enhance the beauty of buildings.

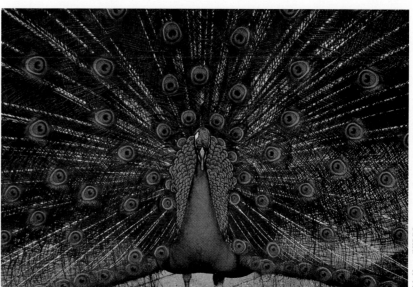

M. Hyett/VIREO

Although symmetry has important uses throughout mathematics, our present discussion is included to better enable us to graph and analyze equations and functions.

Consider the points (4, 2) and (4, −2) that appear in the graph of $x = y^2$.

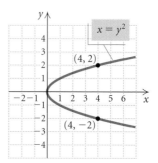

Points like these have the same x-value but opposite y-values, and are **reflections** of each other across the x-axis. Any graph that contains the reflection of every point across the x-axis is said to be **symmetric with respect to the x-axis**. If we fold the graph on the x-axis, the parts above and below the x-axis will coincide.

Consider the points $(3, 4)$ and $(-3, 4)$ that appear in the graph of $y = x^2 - 5$.

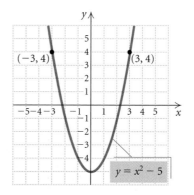

Points like these have the same y-value but opposite x-values, and are **reflections** of each other across the y-axis. Any graph that contains the reflection of every point across the y-axis is said to be **symmetric with respect to the y-axis**. If we fold the graph on the y-axis, the parts to the left and right of the y-axis will coincide.

Consider the points $(3, \sqrt{7})$ and $(-3, -\sqrt{7})$ that appear in the graph of $x^2 = y^2 + 2$. (You can create such a graph on a grapher by solving for y, obtaining $y_1 = \sqrt{x^2 - 2}$ and $y_2 = -\sqrt{x^2 - 2}$.)

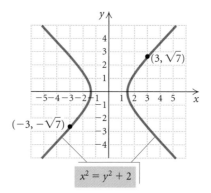

Note that if we take the opposites of the coordinates of one pair, we get the other pair. If for any point (x, y) on a graph the point $(-x, -y)$ is also on the graph, then the graph is said to be **symmetric with respect to the origin**. Visually, if we rotate the graph 180° about the origin, the resulting figure coincides with the original.

> **Algebraic Tests of Symmetry**
>
> *x-axis*: If replacing y by $-y$ produces an equivalent equation, then the graph is *symmetric with respect to the x-axis.*
>
> *y-axis*: If replacing x by $-x$ produces an equivalent equation, then the graph is *symmetric with respect to the y-axis.*
>
> *Origin*: If replacing x by $-x$ and y by $-y$ produces an equivalent equation, then the graph is *symmetric with respect to the origin.*

Example 1   Test $y = x^2 + 2$ for symmetry with respect to the x-axis, the y-axis, and the origin.

ALGEBRAIC SOLUTION

X-AXIS

Replace y by $-y$:

$$y = x^2 + 2$$
$$\downarrow$$
$$-y = x^2 + 2$$
$$y = -x^2 - 2. \qquad \text{Multiplying both sides by } -1$$

Is the resulting equation equivalent to the original? The answer is no, so the graph *is not* symmetric with respect to the x-axis.

Y-AXIS

Replace x by $-x$:

$$y = x^2 + 2$$
$$\searrow$$
$$y = (-x)^2 + 2$$
$$y = x^2 + 2. \qquad \text{Simplifying}$$

Is the resulting equation equivalent to the original? The answer is yes, so the graph *is* symmetric with respect to the y-axis.

ORIGIN

Replace x by $-x$ and y by $-y$:

$$y = x^2 + 2$$
$$\downarrow \qquad \downarrow$$
$$-y = (-x)^2 + 2$$
$$-y = x^2 + 2 \qquad \text{Simplifying}$$
$$y = -x^2 - 2.$$

Is the resulting equation equivalent to the original? The answer is no, so the graph *is not* symmetric with respect to the origin.

GRAPHICAL SOLUTION

We use a grapher to graph the equation. Note that if the graph were folded on the x-axis, the parts above and below the x-axis would not coincide. If it were folded on the y-axis, the parts to the left and right of the y-axis would coincide. If we rotated it 180°, the resulting graph would not coincide with the original graph.

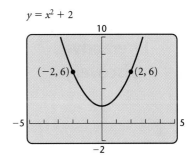

Thus we see that the graph *is not* symmetric with respect to the x-axis or the origin. The graph *is* symmetric with respect to the y-axis.

The algebraic method is often easier to apply than the graphical, especially with equations that we may not be able to graph easily.

Example 2   Test $x^2 + y^4 = 5$ for symmetry with respect to the x-axis, the y-axis, and the origin.

---

r-- *ALGEBRAIC SOLUTION*

X-AXIS

Replace y by $-y$:

$$x^2 + y^4 = 5$$

$$x^2 + (-y)^4 = 5$$

$$x^2 + y^4 = 5.$$

The resulting equation is equivalent to the original equation. Thus the graph is symmetric with respect to the x-axis.

  We leave it to the student to verify algebraically that the graph is also symmetric with respect to the y-axis and the origin.

---

r-- *GRAPHICAL SOLUTION*

With a grapher, we see symmetry with respect to both axes and with respect to the origin.

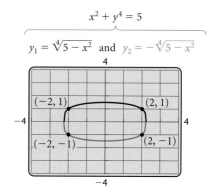

$$x^2 + y^4 = 5$$

$$y_1 = \sqrt[4]{5 - x^2} \text{ and } y_2 = -\sqrt[4]{5 - x^2}$$

Even and Odd Functions

Now we relate symmetry to graphs of functions.

Interactive Discovery

Consider the function f given by $f(x) = x^2$. Find and graph

$$y_1 = f(x) = x^2, \qquad y_2 = f(-x) = (-x)^2, \quad \text{and} \quad y_3 = -f(x) = -x^2.$$

Use the graphs to determine whether $f(x) = f(-x)$ is an identity. Use the graphs to determine whether $f(x) = -f(x)$ is an identity.

  Finding $f(-x)$ is the same as replacing x with $-x$ in its equation, and finding $-f(x)$ is the same as replacing y with $-y$ in its equation. Can you make a connection between symmetry and the properties $f(x) = f(-x)$ and $f(-x) = -f(x)$?

  Repeat this procedure for $g(x) = x^3$.

The preceding Interactive Discovery leads us to the following definitions and procedure.

Even and Odd Functions

If the graph of a function f is symmetric with respect to the y-axis, we say that it is an *even function*. That is, for each x in the domain of f, $f(x) = f(-x)$.

If the graph of a function f is symmetric with respect to the origin, we say that it is an *odd function*. That is, for each x in the domain of f, $f(-x) = -f(x)$.

Algebraic Procedure for Determining Even and Odd Functions

Given the function $f(x)$:

1. Find $f(-x)$ and simplify.

2. Find $-f(x)$ and simplify.

3. If $f(x) = f(-x)$, then f is even. If $f(-x) = -f(x)$, then f is odd.

Except for the function $f(x) = 0$, a function cannot be *both* even and odd. Thus if we see in step (1) that $f(x) = f(-x)$ (that is, f is even), we need not complete the process.

Example 3 Determine whether each of the following functions is even, odd, or neither.

a) $f(x) = 5x^7 - 6x^3 - 2x$ **b)** $g(x) = x^4 - 2x$

c) $h(x) = 5x^6 - 3x^2 - 7$

a)

r-- ALGEBRAIC SOLUTION

$f(x) = 5x^7 - 6x^3 - 2x$

1. $f(-x) = 5(-x)^7 - 6(-x)^3 - 2(-x)$

$\qquad = 5(-x^7) - 6(-x^3) + 2x$

$\qquad\qquad (-x)^7 = (-1 \cdot x)^7 = (-1)^7 x^7 = -x^7$

$\qquad = -5x^7 + 6x^3 + 2x$

2. $-f(x) = -(5x^7 - 6x^3 - 2x)$

$\qquad\quad = -5x^7 + 6x^3 + 2x$

3. We see that $f(x) \neq f(-x)$. Thus, f is *not* even. We see that $f(-x) = -f(x)$. Thus, f is odd.

r-- GRAPHICAL SOLUTION

$y = 5x^7 - 6x^3 - 2x$

We see visually that the graph is symmetric with respect to the origin. Thus the function is odd.

b)

r-- ALGEBRAIC SOLUTION

$g(x) = x^4 - 2x$

1. $g(-x) = (-x)^4 - 2(-x)$

$\qquad = x^4 + 2x$

2. $-g(x) = -(x^4 - 2x)$

$\qquad\quad = -x^4 + 2x$

3. We see that $g(x) \neq g(-x)$. The function is *not* even. We see that $g(-x) \neq -g(x)$. The function is *not* odd. Thus g is neither even nor odd.

r-- GRAPHICAL SOLUTION

$y = x^4 - 2x$

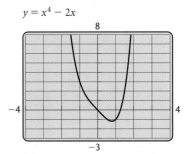

We see visually that the graph is symmetric with respect to neither the y-axis nor the origin. Thus the function is neither even nor odd.

c)

r-- *ALGEBRAIC SOLUTION*

$h(x) = 5x^6 - 3x^2 - 7$

1. $h(-x) = 5(-x)^6 - 3(-x)^2 - 7$
$= 5x^6 - 3x^2 - 7$

We see that $h(x) = h(-x)$. Thus the function is even.

r-- *GRAPHICAL SOLUTION*

$y = 5x^6 - 3x^2 - 7$

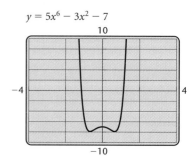

We see visually that the graph is symmetric with respect to the y-axis. Thus the function is even.

1.4 Exercise Set

Determine visually whether each graph is symmetric with respect to the x-axis, the y-axis, and the origin.

1.

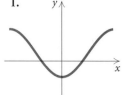

2.

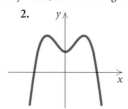

3.

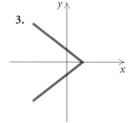

4.

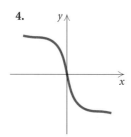

5.

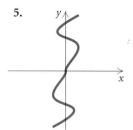

6.

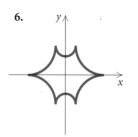

First, graph each equation and determine visually whether it is symmetric with respect to the x-axis, the y-axis, and the origin. Then verify your assertion algebraically.

7. $y = |x| - 2$ **8.** $y = |x + 5|$

9. $5y = 4x + 5$ **10.** $2x - 5 = 3y$

11. $5y = 2x^2 - 3$ **12.** $x^2 + 4 = 3y$

13. $y = \dfrac{1}{x}$ **14.** $y = -\dfrac{4}{x}$

Test algebraically whether each graph is symmetric with respect to the x-axis, the y-axis, and the origin. Then check your work graphically, if possible, using a grapher.

15. $5x - 5y = 0$ **16.** $6x + 7y = 0$

17. $3x^2 - 2y^2 = 3$ **18.** $5y = 7x^2 - 2x$

19. $y = |2x|$ **20.** $y^3 = 2x^2$

21. $2x^4 + 3 = y^2$ **22.** $2y^2 = 5x^2 + 12$

23. $3y^3 = 4x^3 + 2$ **24.** $3x = |y|$

25. $xy = 12$ **26.** $xy - x^2 = 3$

Determine visually whether each function is even, odd, or neither even nor odd.

27.

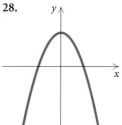

28.

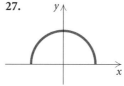

29. 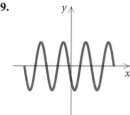 **30.**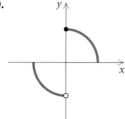

31. 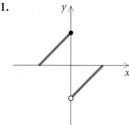 **32.**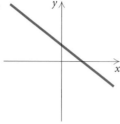

Test algebraically whether each function is even, odd, or neither even nor odd. Then check your work graphically, where possible, using a grapher.

33. $f(x) = -3x^3 + 2x$ **34.** $f(x) = 2x^2 + 4x$

35. $f(x) = 5x^2 + 2x^4 - 1$

36. $f(x) = 3x^4 - 4x^2$

37. $f(x) = 4x$ **38.** $f(x) = |3x|$

39. $f(x) = x^{24}$ **40.** $f(x) = 7x^3 + 4x - 2$

41. $f(x) = x^{17}$ **42.** $f(x) = x + \dfrac{1}{x}$

43. $f(x) = x - |x|$ **44.** $f(x) = \sqrt{x}$

45. $f(x) = \sqrt[3]{x}$ **46.** $f(x) = 8$

47. $f(x) = \sqrt[3]{x - 2}$ **48.** $f(x) = -\dfrac{2}{x}$

49. $f(x) = \dfrac{1}{x^2}$ **50.** $f(x) = \sqrt{x^2 + 1}$

Skill Maintenance

Given that $f(x) = 3x^2 - 2x$, find and simplify each of the following.

51. $f(a - 5)$ **52.** $f(a + 1)$

53. $f(a + 4)$ **54.** $f(a - 1)$

Synthesis

55. ◆ Consider the constant function $f(x) = 0$. Determine whether the graph of this function is symmetric with respect to the x-axis, the y-axis,

and the origin. Determine whether this function is even or odd. In general, can a function be symmetric with respect to the x-axis? Explain.

56. ◆ Describe conditions under which you would know whether a polynomial function

$$f(x) = a_n x^n + a_{n-1} x^{n-1} + \cdots + a_2 x^2 + a_1 x + a_0$$

is even or odd without using an algebraic or graphical procedure. Explain.

Determine whether each function is even, odd, or neither even nor odd. Use any method.

57. $f(x) = x|x^3|$ **58.** $f(x) = x^2(5 - |x|)$

59. $f(x) = \dfrac{1 - x^4}{x^3 + 1}$ **60.** $f(x) = \dfrac{x^2 + 1}{x^3 - 1}$

61. $f(x) = x\sqrt{10 - x^2}$ **62.** $f(x) = \dfrac{-8x}{x^2 + 1}$

Determine whether each graph is symmetric with respect to the x-axis, the y-axis, and the origin. Use any method.

63. **64.**

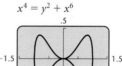

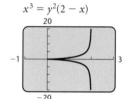

65. **66.**

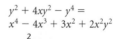

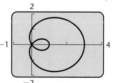

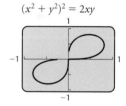

67. Given $f(x) = \frac{1}{2}x^4 - 5x^3 + 2$, let $g(x) = f(-x)$. Graph both f and g using the same set of axes. How do the graphs compare?

68. Consider symmetries with respect to the x-axis, the y-axis, and the origin. Prove that symmetry with respect to any two of these implies symmetry with respect to the other.

1.5

Transformations of Functions

• *Given the graph of a function, graph its transformation under translations, reflections, stretchings, and shrinkings.*

Throughout this chapter, we have considered many kinds of functions. Let's review some of them below.

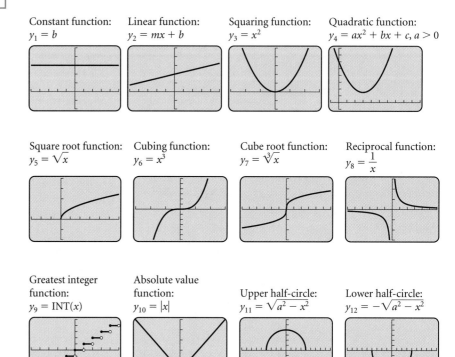

Constant function:
$y_1 = b$

Linear function:
$y_2 = mx + b$

Squaring function:
$y_3 = x^2$

Quadratic function:
$y_4 = ax^2 + bx + c, a > 0$

Square root function:
$y_5 = \sqrt{x}$

Cubing function:
$y_6 = x^3$

Cube root function:
$y_7 = \sqrt[3]{x}$

Reciprocal function:
$y_8 = \dfrac{1}{x}$

Greatest integer function:
$y_9 = \text{INT}(x)$

Absolute value function:
$y_{10} = |x|$

Upper half-circle:
$y_{11} = \sqrt{a^2 - x^2}$

Lower half-circle:
$y_{12} = -\sqrt{a^2 - x^2}$

These functions can be considered building blocks for many other functions. We can create graphs of new functions by shifting them horizontally or vertically, stretching or shrinking them, and reflecting them across an axis. In this section, we consider these *transformations*.

Vertical and Horizontal Translations

Suppose we have a function given by $y_1 = f(x)$. Let's explore the graph of a new function $y_2 = f(x) + b$ or $y_3 = f(x) - b$, for $b > 0$.

Interactive Discovery

Consider the function $y_1 = \frac{1}{5}x^4$. Graph it and the functions $y_2 = \frac{1}{5}x^4 - 5$ and $y_3 = \frac{1}{5}x^4 + 3$ using the same viewing window, $[-10, 10, -8, 10]$. What pattern do you see? Test it with some other graphs.

The effect is a shift of $f(x)$ up or down. Such a shift is called a **vertical translation**.

> ### *Vertical Translation*
>
> For $b > 0$,
>
> > the graph of $y_2 = f(x) + b$ is the graph of $y_1 = f(x)$ shifted *up* b units;
> >
> > the graph of $y_3 = f(x) - b$ is the graph of $y_1 = f(x)$ shifted *down* b units.

Suppose we have a function given by $y_1 = f(x)$. Let's explore the graph of a new function $y_2 = f(x + d)$ or $y_3 = f(x - d)$, for $d > 0$.

Interactive Discovery

Consider the function $y_1 = \frac{1}{5}x^4$. Graph it and the functions $y_2 = \frac{1}{5}(x - 3)^4$ and $y_3 = \frac{1}{5}(x + 7)^4$ using the same viewing window, $[-10, 10, -2, 10]$. What pattern do you see? Test it with some other graphs.

The effect is a shift of $f(x)$ to the right or left. Such a shift is called a **horizontal translation**.

> ### *Horizontal Translation*
>
> For $d > 0$;
>
> > the graph of $y_2 = f(x - d)$ is the graph of $y_1 = f(x)$ shifted *right* d units;
> >
> > the graph of $y_3 = f(x + d)$ is the graph of $y_1 = f(x)$ shifted *left* d units.

Example 1   Graph each of the following. Before doing so, describe how each graph can be obtained from one of the basic graphs shown on the preceding page.

a) $g(x) = x^2 - 6$ 　　　　　　　　**b)** $g(x) = \sqrt{x} + 2$

c) $g(x) = \dfrac{1}{x} + 2$ 　　　　　　**d)** $g(x) = |x - 4|$

e) $h(x) = \sqrt{x + 2} - 3$

SOLUTION

a) To graph $g(x) = x^2 - 6$, think of the graph of $f(x) = x^2$. Since $g(x) = f(x) - 6$, the graph of $g(x) = x^2 - 6$ is the graph of $f(x) = x^2$, shifted, or translated, *down* 6 units. (See the figure at left.)

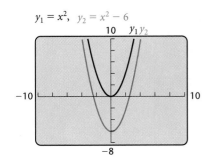

$y_1 = x^2, \quad y_2 = x^2 - 6$

b) To graph $g(x) = \sqrt{x + 2}$, think of the graph of $f(x) = \sqrt{x}$. Since $g(x) = f(x + 2)$, the graph of $g(x) = \sqrt{x + 2}$ is the graph of $f(x) = \sqrt{x}$, shifted *left* 2 units.

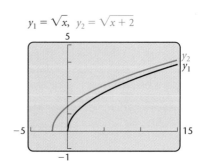

$y_1 = \sqrt{x}, \ y_2 = \sqrt{x + 2}$

c) To graph $g(x) = 1/x + 2$, think of the graph of $f(x) = 1/x$. Since $g(x) = f(x) + 2$, the graph of $g(x) = 1/x + 2$ is the graph of $f(x) = 1/x$, shifted *up* 2 units. (See the figure on the left below.)

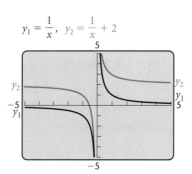

$y_1 = \dfrac{1}{x}, \ y_2 = \dfrac{1}{x} + 2$

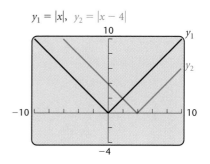

$y_1 = |x|, \ y_2 = |x - 4|$

d) To graph $g(x) = |x - 4|$, think of the graph of $f(x) = |x|$. Since $g(x) = f(x - 4)$, the graph of $g(x) = |x - 4|$ is the graph of $f(x) = |x|$ shifted *right* 4 units. (See the figure on the right above.)

e) To graph $h(x) = \sqrt{x + 2} - 3$, think of the graph of $f(x) = \sqrt{x}$. In part (b), we found that the graph of $g(x) = \sqrt{x + 2}$ is the graph of $f(x) = \sqrt{x}$ shifted left 2 units. Since $h(x) = g(x) - 3$, we shift the graph of $g(x) = \sqrt{x + 2}$ *down* 3 units. Together, the graph of $f(x) = \sqrt{x}$ is shifted *left* 2 units and *down* 3 units.

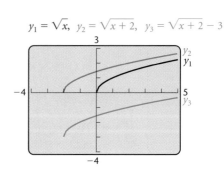

$y_1 = \sqrt{x}, \ y_2 = \sqrt{x + 2}, \ y_3 = \sqrt{x + 2} - 3$

Reflections

Suppose we have a function given by $y = f(x)$. Let's explore the graph of a new function $g(x) = -f(x)$ or $g(x) = f(-x)$.

Consider the functions $y_1 = \frac{1}{5}x^4$ and $y_2 = -\frac{1}{5}x^4$. Graph y_1 in the viewing window $[-10, 10, -10, 10]$. Then graph y_2 in the same viewing window. What pattern do you see? Test it with some other functions in which $y_2 = -y_1$.

Consider the functions $y_1 = 2x^3 - x^4 + 5$ and $y_2 = 2(-x)^3 - (-x)^4 + 5$. Graph y_1 in the viewing window $[-4, 4, -10, 10]$. Then graph y_2 in the same viewing window. What pattern do you see? Test it with some other functions in which x is replaced with $-x$.

Given the graph of $y_1 = f(x)$, we can reflect each point across the x-axis to obtain the graph of $y_2 = -f(x)$. We can reflect each point of y_1 across the y-axis to obtain the graph of $y_2 = f(-x)$. The new graphs are called **reflections** of $f(x)$.

Reflections

The graph of $y_2 = -f(x)$ is the *reflection* of the graph of $y_1 = f(x)$ across the x-axis.

The graph of $y_2 = f(-x)$ is the *reflection* of the graph of $y_1 = f(x)$ across the y-axis.

Example 2   Graph each of the following. Before doing so, describe how each graph can be obtained from the graph of $f(x) = x^3 - 4x^2$.

a) $g(x) = (-x)^3 - 4(-x)^2$ **b)** $g(x) = 4x^2 - x^3$

SOLUTION

a) We first note that

$$f(-x) = (-x)^3 - 4(-x)^2 = g(x).$$

Thus the graph of g is a reflection of the graph of f across the y-axis.

$$y_1 = x^3 - 4x^2, \ \ y_2 = (-x)^3 - 4(-x)^2$$

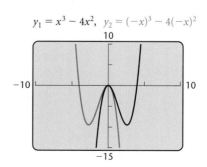

b) We first note that

$$-f(x) = -(x^3 - 4x^2)$$
$$= -x^3 + 4x^2$$
$$= 4x^2 - x^3$$
$$= g(x).$$

Thus the graph of g is a reflection of the graph of f across the x-axis.

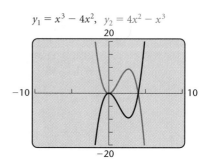

$y_1 = x^3 - 4x^2, \quad y_2 = 4x^2 - x^3$

Vertical and Horizontal Stretchings and Shrinkings

Suppose we have a function given by $y_1 = f(x)$. Let's explore the graph of a new function $y_2 = af(x)$ or $y_3 = f(cx)$.

Interactive Discovery

Graph the function $y_1 = x^3 - x$ using the viewing window $[-3, 3, -1, 1]$. Then graph $y_2 = \frac{1}{10}(x^3 - x)$ using the same viewing window. Clear y_2 and graph $y_3 = 2(x^3 - x)$. Then clear y_3 and graph $y_4 = -2(x^3 - x)$. What pattern do you see? Test it with some other graphs.

Now graph y_1 and $y_2 = (2x)^3 - (2x)$ using the same viewing window. Clear y_2 and graph $y_3 = \left(\frac{1}{2}x\right)^3 - \left(\frac{1}{2}x\right)$. Then clear y_3 and graph $y_4 = \left(-\frac{1}{2}x\right)^3 - \left(-\frac{1}{2}x\right)$. What pattern do you see? Test it with some other graphs.

Consider any function f given by $y = f(x)$. Multiplying on the right by any constant a, where $|a| > 1$, to obtain $g(x) = af(x)$ will *stretch* the graph vertically away from the x-axis. If $0 < |a| < 1$, then the graph will be flattened or *shrunk* vertically toward the x-axis. If $a < 0$, the graph is also reflected across the x-axis.

Vertical Stretching and Shrinking

The graph of $y_2 = af(x)$ can be obtained from the graph of $y_1 = f(x)$ by

stretching vertically for $|a| > 1$, or

shrinking vertically for $0 < |a| < 1$.

For $a < 0$, the graph is also reflected across the x-axis.

The constant c in the equation $g(x) = f(cx)$ will *stretch* the graph of $y = f(x)$ horizontally away from the y-axis if $0 < |c| < 1$. If $|c| > 1$, the graph will be *shrunk* horizontally toward the y-axis. If $c < 0$, the graph is also reflected across the y-axis.

Horizontal Stretching and Shrinking

The graph of $y_2 = f(cx)$ can be obtained from the graph of $y_1 = f(x)$ by

> shrinking horizontally for $|c| > 1$, or
>
> stretching horizontally for $0 < |c| < 1$.

For $c < 0$, the graph is also reflected across the y-axis.

It is instructive to use these concepts now to create hand-drawn graphs from a given graph.

Example 3 Shown below is a graph of $y = f(x)$ for some function f. No formula for f is given. Make a hand-drawn graph of each of the following.

a) $g(x) = f(2x)$

b) $h(x) = f\!\left(\tfrac{1}{2}x\right)$

c) $t(x) = f\!\left(-\tfrac{1}{2}x\right)$

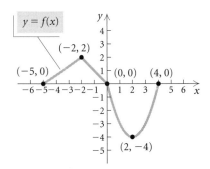

SOLUTION

a) Since $|2| > 1$, the graph of $g(x) = f(2x)$ is a horizontal shrinking of the graph of $y = f(x)$. We can consider the key points $(-5, 0)$, $(-2, 2)$, $(0, 0)$, $(2, -4)$, and $(4, 0)$. The transformation divides each x-coordinate by 2 to obtain the key points $(-2.5, 0)$, $(-1, 2)$, $(0, 0)$, $(1, -4)$, and $(2, 0)$ of the graph of $g(x) = f(2x)$. The graph is shown at the top of the following page.

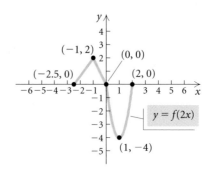

b) Since $\left|\frac{1}{2}\right| < 1$, the graph of $h(x) = f\left(\frac{1}{2}x\right)$ is a horizontal stretching of the graph of $y = f(x)$. We can consider the key points $(-5, 0)$, $(-2, 2)$, $(0, 0)$, $(2, -4)$, and $(4, 0)$. The transformation divides each x-coordinate by $\frac{1}{2}$ (which is the same as multiplying by 2) to obtain the key points $(-10, 0)$, $(-4, 2)$, $(0, 0)$, $(4, -4)$, and $(8, 0)$ of the graph of $h(x) = f\left(\frac{1}{2}x\right)$. The graph is shown below.

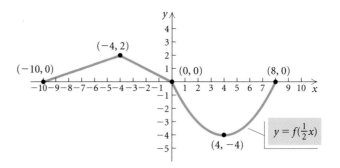

c) The graph of $t(x) = f\left(-\frac{1}{2}x\right)$ can be obtained by reflecting the graph in part (b) across the y-axis. It is shown below.

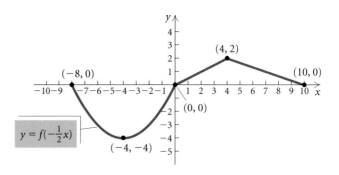

Example 4 Use the graph of $y = f(x)$ given in Example 3. Make a hand-drawn graph of

$$g(x) = -2f(x - 3) + 1.$$

SOLUTION

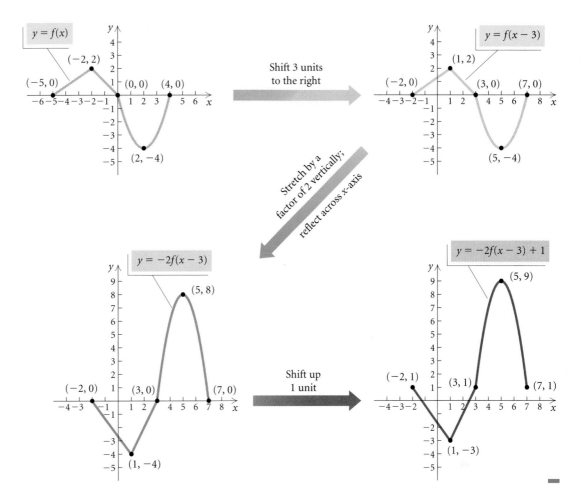

Vertical or Horizontal Translation

For $b > 0$,

> the graph of $y_2 = f(x) + b$ is the graph of $y_1 = f(x)$ shifted *up b* units;
>
> the graph of $y_2 = f(x) - b$ is the graph of $y_1 = f(x)$ shifted *down b* units.

For $d > 0$,

> the graph of $y_2 = f(x - d)$ is the graph of $y_1 = f(x)$ shifted *right d* units;
>
> the graph of $y_2 = f(x + d)$ is the graph of $y_1 = f(x)$ shifted *left d* units.

Reflections

The graph of $y_2 = -f(x)$ is the reflection of the graph of $y_1 = f(x)$ across the x-axis.

The graph of $y_2 = f(-x)$ is the reflection of the graph of $y_1 = f(x)$ across the y-axis.

Vertical Stretching or Shrinking

The graph of $y_2 = af(x)$ can be obtained from the graph of $y_1 = f(x)$ by

stretching vertically for $|a| > 1$, or

shrinking vertically for $0 < |a| < 1$.

For $a < 0$, the graph is also reflected across the x-axis.

Horizontal Stretching or Shrinking

The graph of $y_2 = f(cx)$ can be obtained from the graph of $y_1 = f(x)$ by

shrinking horizontally for $|c| > 1$, or

stretching horizontally for $0 < |c| < 1$.

For $c < 0$, the graph is also reflected across the y-axis.

1.5 Exercise Set

Graph each of the following on a grapher. Before doing so, describe how each graph can be obtained from one of the basic graphs at the beginning of this section.

1. $f(x) = x^2 + 1$

2. $f(x) = x^3 - 4$

3. $f(x) = \dfrac{1}{x} + 3$

4. $f(x) = \sqrt{x} + 2$

5. $f(x) = \sqrt{x} - 5$

6. $f(x) = -\sqrt{x}$

7. $f(x) = -x^2$

8. $f(x) = (x - 4)^2$

9. $f(x) = |x - 3|$

10. $g(x) = |3x|$

11. $g(x) = (x + 5)^3$

12. $g(x) = \dfrac{1}{x - 5}$

13. $g(x) = (x + 1)^2$

14. $f(x) = 2x^2$

15. $f(x) = \frac{1}{2}\sqrt{x}$

16. $f(x) = \frac{1}{2}x^3$

17. $g(x) = \dfrac{2}{x}$

18. $f(x) = |x - 3| - 4$

19. $f(x) = 3\sqrt{x} - 5$

20. $f(x) = 5 - \dfrac{1}{x}$

21. $f(x) = \frac{1}{2}(x - 3)^2$

22. $f(x) = \sqrt{x - 3} + 2$

23. $g(x) = \left|\frac{1}{3}x\right| - 4$

24. $f(x) = \frac{2}{3}x^3 - 4$

25. $f(x) = (x + 5)^2 - 4$

26. $f(x) = (-x)^3 - 5$

27. $f(x) = -\frac{1}{4}(x - 5)^2$

28. $g(x) = \sqrt{-x} - 2$

29. $f(x) = \dfrac{1}{x + 3} + 2$

30. $g(x) = (x - 2)^3 - 5$

31. $f(x) = (x + 4)^3 + 3$

32. $f(x) = \dfrac{1}{-x} + 2$

33. $f(x) = \sqrt{-x} + 5$

34. $f(x) = \frac{4}{5}(x - 4)^2 - 5$

35. $f(x) = 3(x + 4)^2 - 3$

36. $g(x) = 2(x + 1)^3 + 4$

37. $f(x) = 0.43(x - 3)^3 + 2.4$

38. $f(x) = 0.3|x - 5.2| + 2.8$

39. $g(x) = 2.8(x + 5.2)^2 + 1.1$

40. $f(x) = 1.8\sqrt{x - 3.4} - 4.8$

Write an equation for a function that has a graph with the given characteristics. Check your answer on a grapher.

41. The shape of $y = x^2$, but upside-down and shifted right 8 units

42. The shape of $y = \sqrt{x}$, but shifted left 6 units and down 5 units

43. The shape of $y = |x|$, but shifted left 7 units and up 2 units

44. The shape of $y = x^3$, but upside-down and shifted right 5 units

45. The shape of $y = 1/x$, but shrunk vertically by a factor of $\frac{1}{2}$ and shifted down 3 units

46. The shape of $y = x^2$, but shifted right 6 units and up 2 units

47. The shape of $y = x^2$, but upside-down and shifted right 3 units and up 4 units

48. The shape of $y = |x|$, but stretched horizontally by a factor of 2 and shifted down 5 units

49. The shape of $y = \sqrt{x}$, but reflected across the y-axis and shifted left 2 units and down 1 unit

50. The shape of $y = 1/x$, but reflected across the x-axis and shifted up 1 unit

51. The shape of $y = x^3$, but shifted left 4 units and shrunk vertically by a factor of 0.83.

A graph of $y = f(x)$ follows. No formula for f is given. Make a hand-drawn graph of each of the following.

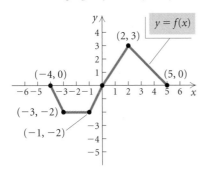

52. $g(x) = \frac{1}{2}f(x)$ **53.** $g(x) = -2f(x)$

54. $g(x) = f(2x)$ **55.** $g(x) = f\left(-\frac{1}{2}x\right)$

56. $g(x) = -3f(x + 1) - 4$ **57.** $g(x) = -\frac{1}{2}f(x - 1) + 3$

The graph of the function f is shown in figure (a). Match each function g in Exercises 58–65 with the appropriate graph from (a)–(h). Some graphs may be used more than once.

a)

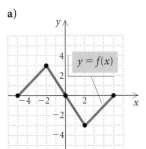

b)

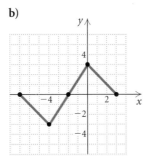

c)

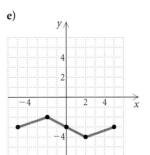

d)

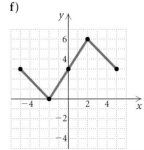

e)

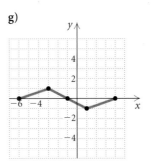

f)

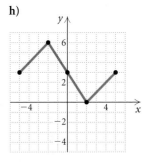

g)

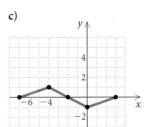

h)

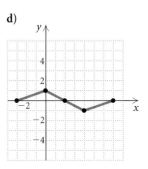

58. $g(x) = f(x) + 3$ **59.** $g(x) = f(-x) + 3$

60. $g(x) = -f(-x)$ **61.** $g(x) = -f(x) + 3$

62. $g(x) = \frac{1}{3}f(x) - 3$ **63.** $g(x) = \frac{1}{3}f(x - 2)$

64. $g(x) = -f(x + 2)$ **65.** $g(x) = \frac{1}{3}f(x + 2)$

Use a grapher to graph each set of functions using a SIMULTANEOUS feature. (With this feature, you cannot tell which function is being graphed first.) After the curves are displayed, match each curve with the appropriate function. Check your answers by setting the WINDOW format to LabelOn. Then when you use the TRACE feature, the label of the curve being traced will appear.

66. $y_1 = (x + 1)^2$,   **67.** $y_1 = (x - 5)^3 + 4$,
    $y_2 = (x - 3)^2 + 5$, $y_2 = (x + 5)^3 - 4$,
    $y_3 = (x + 3)^2 - 5$, $y_3 = (x - 2)^3$,
    $y_4 = x^2 + 1$ $y_4 = x^3 - 2$

68. $y_1 = \dfrac{1}{x + 5}$,

$y_2 = \dfrac{1}{x - 3} + 4$,

$y_3 = \dfrac{1}{x + 3} - 4$,

$y_4 = \dfrac{1}{x} + 5$

69. $y_1 = \sqrt{9 - x^2}$,

$y_2 = -\sqrt{9 - x^2}$,

$y_3 = \dfrac{2}{3}\sqrt{9 - x^2}$,

$y_4 = 4 - \sqrt{9 - x^2}$

70. Using a $[-10, 10, -200, 800]$ window with Yscl $= 100$, show that

$$y_1 = x^4 - 12x^3 + 34x^2 + 12x - 35$$

and

$$y_2 = x^4 + 12x^3 + 34x^2 - 12x - 35$$

are reflections of each other across the y-axis. Verify this algebraically.

For each pair of functions, determine if $g(x) = f(-x)$ using algebra. Then, using the TABLE *feature on a grapher, check your answers by looking at y_1 and y_2 for x-values near 0. You can then check graphically, but be careful to use a suitable window.*

71. $f(x) = 2x^4 - 35x^3 + 3x - 5$,
$g(x) = 2x^4 + 35x^3 - 3x - 5$

72. $f(x) = \frac{1}{4}x^4 + \frac{1}{5}x^3 - 81x^2 - 17$,
$g(x) = \frac{1}{4}x^4 + \frac{1}{5}x^3 + 81x^2 - 17$

A graph of the function $f(x) = x^3 - 3x^2$ is shown below. Exercises 73–76 show graphs of functions transformed from this one. Find a formula for each function.

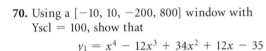

73.

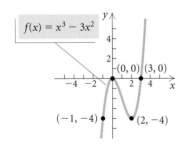

74.

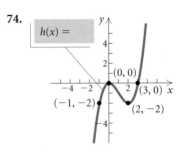

75.

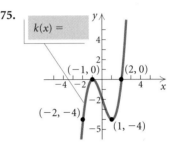

76.

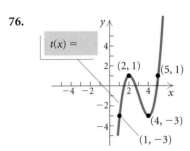

Skill Maintenance

Graph.

77. $4x - 3y = 12$

78. $y = x^2 + 1$

79. $y = x^3 + 2x - 5$

80. $y = -3x^2 - x + 3$

Synthesis

81. ◆ Explain in your own words why the graph of $y = f(-x)$ is a reflection of the graph of $y = f(x)$ across the y-axis.

82. ◆ Without drawing the graph, describe what the graph of $f(x) = |x^2 - 9|$ looks like.

83. The graph of $f(x) = \sqrt{x}$ passes through the points $(0, 0)$, $(1, 1)$, and $(4, 2)$. Transform this function to one whose graph passes through the points $(4, 7)$, $(5, 6)$, and $(8, 5)$. Check your answer on a grapher.

84. The graph of $f(x) = |x|$ passes through the points $(-3, 3)$, $(0, 0)$, and $(3, 3)$. Transform this function to one whose graph passes through the points $(5, 1)$, $(8, 4)$, and $(11, 1)$. Check your answer on a grapher.

Graph each of the following on a grapher. Before doing so, describe how each graph can be obtained from a more basic graph. Give the domain and the range of each function.

85. $f(x) = \text{INT}\left(x - \frac{1}{2}\right)$

86. $g(x) = \sqrt{7 - x}$

87. $f(x) = 3 \cdot \text{INT}(x + 2) + 1$

88. $f(x) = |x^2 - 4|$

89. $g(x) = \left| \dfrac{1}{x} \right|$

90. $f(x) = \left| \sqrt{x} - 1 \right|$

91. $f(x) = \left| |x| - 5 \right|$

92. Find the roots, or zeros, of $f(x) = 3x^5 - 20x^3$. Then without using a grapher, state the roots of $f(x - 3)$ and $f(x + 8)$.

93. If $(3, 4)$ is a point on the graph of $y = f(x)$, what point do you know is on the graph of $y = 2f(x)$? of $y = 2 + f(x)$? of $y = f(2x)$?

94. If $(-1, 5)$ is a point on the graph of $y = f(x)$, find b such that $(2, b)$ is on the graph of $y = f(x - 3)$.

1.6
Quadratic Functions and Applications

- *Find the vertex, the line of symmetry, and the maxima or minima of the graph of a quadratic function using the method of completing the square.*
- *Graph quadratic functions.*
- *Solve applications using quadratic models.*

We define quadratic functions as follows.

Quadratic Functions

A *quadratic function f* is a second-degree polynomial function

$$f(x) = ax^2 + bx + c, \quad a \neq 0,$$

where a, b, and c are real numbers.

We have already graphed some quadratic functions in this chapter. Now we develop methods for analyzing the graph of any quadratic function. These techniques will allow us to find the exact value of a maximum or a minimum.

Graphing Quadratic Functions of the Type $f(x) = a(x - h)^2 + k$

The graph of a quadratic function is called a **parabola**. We will see that the graph of every parabola evolves from the graph of the squaring function $f(x) = x^2$ using the transformations that we discussed in Section 1.5.

Interactive Discovery

Think of transformations and look for patterns. Consider the following functions:

$$y_1 = x^2, \qquad y_2 = -0.4x^2,$$
$$y_3 = -0.4(x - 2)^2, \qquad y_4 = -0.4(x - 2)^2 + 3.$$

Graph y_1 and y_2. How do you get from the graph of y_1 to y_2?
Graph y_2 and y_3. How do you get from the graph of y_2 to y_3?
Graph y_3 and y_4. How do you get from the graph of y_3 to y_4?
  Consider the following functions:

$$y_1 = x^2, \qquad y_2 = 2x^2,$$
$$y_3 = 2(x + 3)^2, \qquad y_4 = 2(x + 3)^2 - 5.$$

Graph y_1 and y_2. How do you get from the graph of y_1 to y_2?
Graph y_2 and y_3. How do you get from the graph of y_2 to y_3?
Graph y_3 and y_4. How do you get from the graph of y_3 to y_4?

We get the graph of $f(x) = a(x - h)^2 + k$ from the graph of $f(x) = x^2$ as follows:

$$f(x) = x^2$$
$$\downarrow$$
$$f(x) = ax^2 \qquad \text{Vertical stretching or shrinking with a reflection across the } x\text{-axis if } a < 0$$
$$\downarrow$$
$$f(x) = a(x - h)^2 \qquad \text{Horizontal translation}$$
$$\downarrow$$
$$f(x) = a(x - h)^2 + k. \qquad \text{Vertical translation}$$

Consider the following graphs of the form $f(x) = a(x - h)^2 + k$. The point at which the graph turns is called the **vertex**. If $f(x)$ has a maximum or a minimum value, it occurs at the vertex. Each graph has a line $x = h$ that is called the **line of symmetry**.

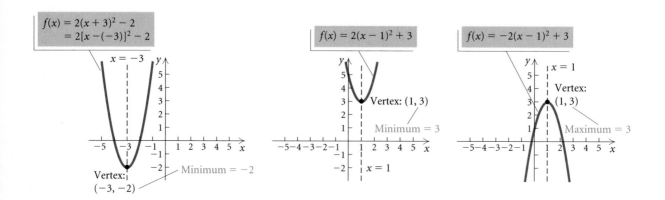

We summarize as follows.

The graph of $f(x) = a(x - h)^2 + k$:

> opens up if $a > 0$ and down if $a < 0$:
>
> has (h, k) as a vertex;
>
> has $x = h$ as the line of symmetry;
>
> has k as a minimum value (output) if $a > 0$ and has k as a maximum value if $a < 0$.

If the equation is in the form $f(x) = a(x - h)^2 + k$, we can determine a great deal of information about the graph without graphing.

FUNCTION	$\begin{aligned} f(x) &= 3\left(x - \tfrac{1}{4}\right)^2 - 2 \\ &= 3\left(x - \tfrac{1}{4}\right)^2 + (-2) \end{aligned}$	$\begin{aligned} g(x) &= -3(x + 5)^2 + 7 \\ &= -3[x - (-5)]^2 + 7 \end{aligned}$
VERTEX	$\left(\tfrac{1}{4}, -2\right)$	$(-5, 7)$
LINE OF SYMMETRY	$x = \tfrac{1}{4}$	$x = -5$
MAXIMUM	No: $3 > 0$, graph opens up.	Yes, 7: $-3 < 0$, graph opens down.
MINIMUM	Yes, -2; $3 > 0$, graph opens up.	No: $-3 < 0$, graph opens down.

Note that the vertex (h, k) is used to find the maximum or the minimum value of the function. The maximum or minimum is the number k, *not* the ordered pair (h, k).

Graphing Quadratic Functions of the Type $f(x) = ax^2 + bx + c, a \neq 0$

We now use a modification of the method of completing the square as an aid in graphing and analyzing quadratic functions of the form $f(x) = ax^2 + bx + c$, $a \neq 0$. (See Appendix A.1 for review of completing the square.)

Example 1   Complete the square to find the vertex, the line of symmetry, and the maximum or minimum value of $f(x) = x^2 + 10x + 23$. Then graph the function.

SOLUTION   To express $f(x) = x^2 + 10x + 23$ in the form $f(x) = a(x - h)^2 + k$, we proceed as follows. We construct a trinomial square. To do so, we take half the coefficient of x and square it. The number is $(10/2)^2$, or 25. We now add and subtract that number on the right-hand side. We can think of this as adding $25 - 25$, which is 0.

$$f(x) = x^2 + 10x + 23$$

Note that 25 completes the square for $x^2 + 10x$.

$$= x^2 + 10x + 25 - 25 + 23$$

Adding $25 - 25$, or 0, to the right side

$$= (x^2 + 10x + 25) - 25 + 23$$

$$= (x + 5)^2 - 2$$

Factoring and simplifying

$$= [x - (-5)]^2 + (-2)$$

Writing in the form $f(x) = a(x - h)^2 + k$

From this form of the function, we know the following:

Vertex: $(-5, -2)$;

Line of symmetry: $x = -5$;

Minimum $= -2$.

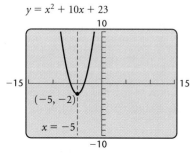

$y = x^2 + 10x + 23$

$(-5, -2)$

$x = -5$

The graph of f is a shift of the graph of $y = x^2$ left 5 units and down 2 units.   —

Keep in mind that a line of symmetry is not part of the graph; it is a characteristic of the graph. If you fold the graph on its line of symmetry, the two halves of the graph will coincide.

Example 2   Complete the square to find the vertex, the line of symmetry, and the maximum or minimum value of $g(x) = x^2/2 - 4x + 8$. Then graph the function.

SOLUTION   To complete the square, we factor $\frac{1}{2}$ out of the first two terms. This makes the coefficient of x^2 inside the parentheses 1:

$$g(x) = \frac{x^2}{2} - 4x + 8$$

$$= \frac{1}{2}(x^2 - 8x) + 8.$$

Factoring $\frac{1}{2}$ out of the first two terms

Now we complete the square inside the parentheses: Half of -8 is -4, and $(-4)^2 = 16$. We add and subtract 16 inside the parentheses:

$$g(x) = \frac{1}{2}(x^2 - 8x + 16 - 16) + 8$$

$$= \frac{1}{2}(x^2 - 8x + 16) - \frac{1}{2} \cdot 16 + 8$$

Using the distributive law to remove -16 from within the parentheses

$$= \frac{1}{2}(x - 4)^2.$$

Factoring and simplifying

We know the following:

Vertex: $(4, 0)$;

Line of symmetry: $x = 4$;

Minimum $= 0$.

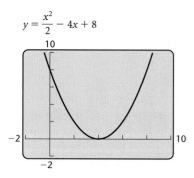

$y = \dfrac{x^2}{2} - 4x + 8$

The graph of g is a shrinking of the graph of $y = x^2$ vertically followed by a shift to the right 4 units.   —

Example 3 Complete the square to find the vertex, the line of symmetry, and the maximum or minimum value of $f(x) = -2x^2 + 10x - \frac{23}{2}$. Then graph the function.

SOLUTION

$$f(x) = -2x^2 + 10x - \frac{23}{2}$$

$$= -2(x^2 - 5x) - \frac{23}{2} \qquad \text{Factoring } -2 \text{ out of the first two terms}$$

$$= -2\left(x^2 - 5x + \frac{25}{4} - \frac{25}{4}\right) - \frac{23}{2} \qquad \text{Completing the square inside the parentheses}$$

$$= -2\left(x^2 - 5x + \frac{25}{4}\right) - 2\left(-\frac{25}{4}\right) - \frac{23}{2} \qquad \text{Removing } -\frac{25}{4} \text{ from within the parentheses}$$

$$= -2\left(x - \frac{5}{2}\right)^2 + \frac{25}{2} - \frac{23}{2}$$

$$= -2\left(x - \frac{5}{2}\right)^2 + 1.$$

This form of the function yields the following:

Vertex: $\left(\frac{5}{2}, 1\right)$:

Line of symmetry: $x = \frac{5}{2}$;

Maximum $= 1$.

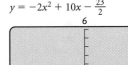

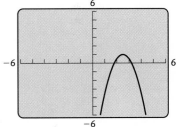

The graph is found by stretching the graph of $f(x) = x^2$ vertically, reflecting the graph across the x-axis, shifting the graph to the right $\frac{5}{2}$ units, and shifting that curve up 1 unit. ▬

In many situations, we want to find the coordinates of the vertex directly from the equation $f(x) = ax^2 + bx + c$ using a formula. One way to develop such a formula is to consider the x-coordinate of the vertex as centered between the x-intercepts, or zeros, of the function. The zeros of a quadratic function $f(x) = ax^2 + bx + c$ are the solutions of the quadratic equation $ax^2 + bx + c = 0$.

Quadratic Equations

A *quadratic equation* is an equation equivalent to

$$ax^2 + bx + c = 0, \quad a \neq 0,$$

where the coefficients a, b, and c are real numbers.

The quadratic formula can be used to solve any quadratic equation. (See Appendix A.1 for review of solving quadratic equations.)

> **The Quadratic Formula**
>
> The solutions of $ax^2 + bx + c = 0$, $a \neq 0$, are given by
> $$x = \frac{-b \pm \sqrt{b^2 - 4ac}}{2a}.$$

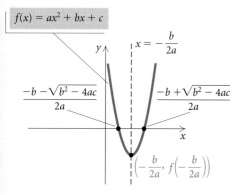

By averaging the two solutions of $ax^2 + bx + c = 0$, we find a formula for the x-coordinate of the vertex:

$$x\text{-coordinate of vertex} = \frac{\dfrac{-b - \sqrt{b^2 - 4ac}}{2a} + \dfrac{-b + \sqrt{b^2 - 4ac}}{2a}}{2}$$

$$= \frac{\dfrac{-2b}{2a}}{2} = \frac{-\dfrac{b}{a}}{2} = -\frac{b}{2a}.$$

We use this value of x to find the y-coordinate of the vertex, $f\left(-\dfrac{b}{2a}\right)$.

> **The Vertex of a Parabola**
>
> The *vertex* of the graph of $f(x) = ax^2 + bx + c$ is
> $$\left(-\frac{b}{2a}, f\left(-\frac{b}{2a}\right)\right).$$
>
> $\uparrow$ $\uparrow$
>
> We calculate the   We substitute to
> x-coordinate.   find the y-coordinate.

Example 4   For the function $f(x) = -x^2 + 14x - 47$:

a) Find the vertex.

b) Determine whether there is a maximum or minimum value and find that value.

c) Find the range.

d) On what intervals is the function increasing? decreasing?

SOLUTION   There is no need to graph the function.

a) The x-coordinate of the vertex is
$$\frac{-b}{2a}, \quad \text{or} \quad \frac{-14}{2(-1)}, \quad \text{or} \quad 7.$$
Since
$$f(7) = -7^2 + 14 \cdot 7 - 47 = 2,$$
the vertex is $(7, 2)$.

b) Since $a = -1$ and is negative, the graph opens down so the second coordinate of the vertex, 2, is the maximum value of the function.

c) The range is $(-\infty, 2]$.

d) Since the graph opens down, function values increase to the left of the vertex and decrease to the right of the vertex. Thus the function is increasing on the interval $(-\infty, 7]$ and decreasing on $[7, \infty)$.

Quadratic Applications

Quadratic functions have many uses as models in science, engineering, and business. The simplest use of quadratic functions in applications occurs when we merely evaluate a quadratic function. In such cases, a model has already been developed.

Example 5 *Vertical Leap.* A formula relating an athlete's vertical leap V, in inches, to hang time T, in seconds, is

$$V(T) = 48T^2.$$

Anfernee Hardaway of the Orlando Magic has a vertical leap of 36 in. What is his hang time? (Source: National Basketball Association)

SOLUTION   Since the model has already been developed, we need only evaluate. Note that the function V expresses vertical leap, in inches, as a function of hang time, in seconds.

$$
\begin{aligned}
V &= 48T^2 \\
36 &= 48T^2 \qquad \text{Substituting} \\
0.75 &= T^2 \\
0.866 &\approx T^2 \qquad \text{We only want the positive solution.}
\end{aligned}
$$

Anfernee Hardaway's hang time is 0.866 sec.

Example 6 *Maximizing Area.* A stone mason has enough stones to enclose a rectangular patio with 60 ft of stone wall. If the attached house forms one side of the rectangle, what is the maximum area that the mason can enclose? What should the dimensions of the patio be in order to yield this area?

SOLUTION   In this case, it is helpful to make use of a five-step problem-solving strategy.

1. Familiarize. Suppose the patio were 10 ft wide. It would then be $60 - 2 \cdot 10 = 40$ ft long. The area would be $(10 \text{ ft})(40 \text{ ft}) = 400 \text{ ft}^2$. To see if this is the maximum area possible, we would need to check other possible combinations of length and width. Instead, we will try to find an area function and determine where a maximum occurs.

We make a sketch of the situation, using w to represent the patio's width, in feet. Since only 60 ft of stone wall is available and the house serves as one side, then $(60 - 2w)$ feet of stone is available for the length.

2. **Translate.** Since the area of a rectangle is length times width, we have

$$A(w) = (60 - 2w)w$$
$$= -2w^2 + 60w,$$

where $A(w) = $ the area of the patio, in square feet, as a function of the width, w.

3. **Carry out.** To solve this problem, we need to determine the maximum value of $A(w)$ and find the dimensions for which that maximum occurs. Since A is a quadratic function and w^2 has a negative coefficient, we know that the function has a maximum value that occurs at the vertex of the graph of the function. The first coordinate of the vertex is

$$w = -\frac{b}{2a} = -\frac{60}{2(-2)} = 15 \text{ ft.}$$

Thus, if $w = 15$ ft, then the length $l = 60 - 2 \cdot 15 = 30$ ft; and the area is $15 \cdot 30$, or 450 ft^2.

4. **Check**. As a partial check, we note that 450 ft$^2 > 400$ ft^2, which is the area we found in a guess in the familiarize step. A more complete check, assuming that the function $A(w)$ is correct, would examine a table of values for $A(w) = (60 - 2w)w$ and/or examine its graph.

> **Regression**
>
> Using quadratic regression or curve fitting to model data is presented in Appendix A.3.

X	Y₁
14.7	449.82
14.8	449.92
14.9	449.98
15	450
15.1	449.98
15.2	449.92
15.3	449.82
X = 15	

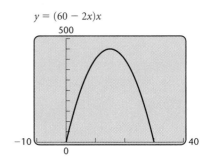

$y = (60 - 2x)x$

5. **State**. A maximum area of 450 ft^2 will occur if the patio is 15 ft wide and 30 ft long.

1.6 Exercise Set

Complete the square to:

a) *find the vertex;*
b) *find the line of symmetry;*
c) *determine whether there is a maximum or minimum value and find that value.*

Then check your answers with a grapher.

1. $f(x) = x^2 - 8x + 12$

2. $g(x) = x^2 + 7x - 8$

3. $f(x) = x^2 - 7x + 12$

4. $g(x) = x^2 - 5x + 6$

5. $g(x) = 2x^2 + 6x + 8$

6. $f(x) = 2x^2 - 10x + 14$

7. $g(x) = -2x^2 + 2x + 1$

8. $f(x) = -3x^2 - 3x + 1$

In Exercises 9–16, match the equation with figures (a)–(h). Use a grapher only as a check.

a)

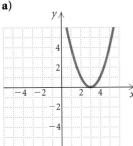

b)

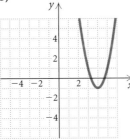

c)

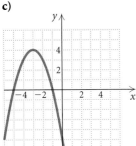

d)

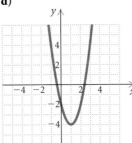

e)

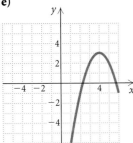

f)

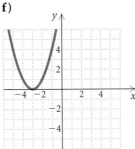

g)

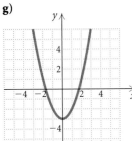

h)

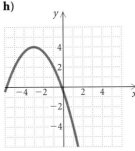

9. $y = (x + 3)^2$

10. $y = -(x - 4)^2 + 3$

11. $y = 2(x - 4)^2 - 1$

12. $y = x^2 - 3$

13. $y = -\frac{1}{2}(x + 3)^2 + 4$

14. $y = (x - 3)^2$

15. $y = -(x + 3)^2 + 4$

16. $y = 2(x - 1)^2 - 4$

In Exercises 17–24:

a) *Find the vertex.*

b) *Determine whether there is a maximum or minimum value and find that value.*

c) *Find the range.*

d) *Find the intervals on which the function is increasing and the function is decreasing.*

Then check your answers with a grapher.

17. $f(x) = x^2 - 6x + 5$

18. $f(x) = x^2 + 4x - 5$

19. $f(x) = 2x^2 + 4x - 16$

20. $f(x) = \frac{1}{2}x^2 - 3x + \frac{5}{2}$

21. $f(x) = -\frac{1}{2}x^2 + 5x - 8$

22. $f(x) = -2x^2 - 24x - 64$

23. $f(x) = 3x^2 + 6x + 5$

24. $f(x) = -3x^2 + 24x - 49$

25. *Games in a Sports League.* If there are x teams in a sports league and all the teams play each other twice, a total of $N(x)$ games are played, where

$$N(x) = x^2 - x.$$

A softball league has 9 teams, each of which plays the others twice. If the league pays \$45 per game for the field and umpires, how much will it cost to play the entire schedule?

26. *Projectile Motion.* A stone thrown downward with an initial velocity of 34.3 m/sec will travel a distance of s meters, where

$$s(t) = 4.9t^2 + 34.3t$$

and t is in seconds. If a stone is thrown downward at 34.3 m/sec from a height of 294 m, how long will it take the stone to hit the ground?

27. *Maximizing Volume.* A plastics manufacturer plans to produce a one-compartment vertical file by bending the long side of an 8-in. by 14-in. sheet of plastic along two lines to form a U-shape. How tall should the file be in order to maximize the volume that the file can hold?

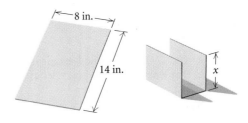

28. *Maximizing Area.*   The sum of the base and the height of a parallelogram is 69 cm. Find the dimensions for which the area is a maximum.

29. *Maximizing Area.*   The sum of the base and the height of a triangle is 20 cm. Find the dimensions for which the area is a maximum.

30. *Maximizing Area.*   A fourth-grade class decides to enclose a rectangular garden, using the side of the school as one side of the rectangle. What is the maximum area that the class can enclose with 32 ft of fence? What should the dimensions of the garden be in order to yield this area?

31. *Maximizing Area.*   A rancher needs to enclose two adjacent rectangular corrals, one for sheep and one for cattle. If a river forms one side of the corrals and 240 yd of fencing is available, what is the largest total area that can be enclosed?

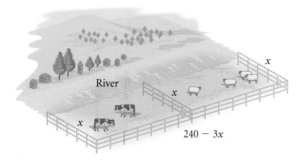

32. *Minimizing Cost.*   Aki's Bicycle Designs has determined that when *x* hundred bicycles are built, the average cost per bicycle is given by

$$C(x) = 0.1x^2 - 0.7x + 2.425,$$

where $C(x)$ is in hundreds of dollars. How many bicycles should be built in order to minimize the average cost per bicycle?

Maximizing Profit.   *In business, profit is the difference between revenue and cost, that is,*

$$Total\ profit = Total\ revenue - Total\ cost,$$
$$P(x) = R(x) - C(x),$$

where x is the number of units. Find the maximum profit and the number of units that must be sold in order to yield the maximum profit for each of the following.

33. $R(x) = 50x - 0.5x^2$, $C(x) = 10x + 3$

34. $R(x) = 5x$, $C(x) = 0.001x^2 + 1.2x + 60$

35. *Norman Window.*   A Norman window is a rectangle with a semicircle on top. Sky Blue Windows is designing a Norman window that will require 24 ft of trim. What dimensions will allow the maximum amount of light to enter a house?

A Norman window

Skill Maintenance

Given that $f(x) = x^3 - 7x$ *and* $g(x) = x^2 + 1$, *find each of the following.*

36. $g(3)$ **37.** $f(3)$

38. $f(4) - g(4)$ **39.** $f(3) + g(3)$

Synthesis

40. ◈ Write a problem for a classmate to solve. Design the problem so that it is a maximum or minimum problem using a quadratic function.

41. ◈ The graph of a quadratic function can have 0, 1, or 2 *x*-intercepts. How can you predict the number of *x*-intercepts without drawing the graph or (completely) solving an equation?

42. Find *b* such that $f(x) = -4x^2 + bx + 3$ has a maximum value of 50.

43. Find c such that $f(x) = -0.2x^2 - 3x + c$ has a maximum value of -225.

44. Find a quadratic function that has $(4, -5)$ as a vertex and contains the point $(-3, 1)$.

45. Graph: $f(x) = (|x| - 5)^2 - 3$.

46. *Minimizing Area.* A 24-in. piece of string is cut into two pieces. One piece is used to form a circle while the other is used to form a square. How should the string be cut so that the sum of the areas is a minimum?

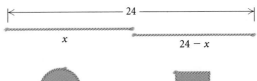

1.7

The Algebra of Functions

- *Compute function values for the sum, the difference, the product, and the quotient of two functions, and determine the domains.*
- *Find the composition of two functions and the domain of the composition; and decompose a function as a composition of two functions.*

We now consider five methods of combining two functions to obtain a new function. Among these are addition, subtraction, multiplication, and division.

Sums, Differences, Products, and Quotients

Consider the following two functions f and g:

$$f(x) = x + 2 \quad \text{and} \quad g(x) = x^2 + 1.$$

Since $f(3) = 3 + 2 = 5$ and $g(3) = 3^2 + 1 = 10$, we have

$$f(3) + g(3) = 5 + 10 = 15, \qquad f(3) - g(3) = 5 - 10 = -5,$$

$$f(3) \cdot g(3) = 5 \cdot 10 = 50, \quad \text{and} \quad \frac{f(3)}{g(3)} = \frac{5}{10} = \frac{1}{2}.$$

In fact, so long as x is in the domain of both f and g, we can easily compute $f(x) + g(x)$, $f(x) - g(x)$, $f(x) \cdot g(x)$, and, assuming $g(x) \neq 0$, $f(x)/g(x)$. Notation has been developed to facilitate this work.

> **Sums, Differences, Products, and Quotients of Functions**
>
> If f and g are functions and x is in the domain of each function, then
>
> $(f + g)(x) = f(x) + g(x), \qquad (f - g)(x) = f(x) - g(x),$
>
> $(fg)(x) = f(x) \cdot g(x), \qquad (f/g)(x) = f(x)/g(x),$
>
> $\qquad\qquad\qquad\qquad\quad$ provided $g(x) \neq 0$.

Example 1   Given that $f(x) = x + 1$ and $g(x) = \sqrt{x + 3}$, find each of the following.

a) $(f + g)(x)$ **b)** $(f + g)(6)$ **c)** $(f + g)(-4)$

SOLUTION

a) $(f + g)(x) = f(x) + g(x)$
$$= (x + 1) + \sqrt{x + 3} \quad \text{This cannot be simplified.}$$

b) We can find $(f + g)(6)$ provided 6 is in the domain of each function. The domain of f is all real numbers. The domain of g is all real numbers x for which $x + 3 \geq 0$, or $x \geq -3$. This is the interval $[-3, \infty)$. Because 6 is in both domains, we have

$$f(6) = 6 + 1 = 7, \quad g(6) = \sqrt{6 + 3} = \sqrt{9} = 3,$$
$$(f + g)(6) = f(6) + g(6) = 7 + 3 = 10.$$

Another method is to use the formula found in part (a):

$$(f + g)(6) = (6 + 1) + \sqrt{6 + 3} = 7 + \sqrt{9} = 7 + 3 = 10.$$

c) To find $(f + g)(-4)$, we must first determine whether -4 is in the domain of each function. We note that -4 is not in the domain of g, $[-3, \infty)$. That is, $\sqrt{-4 + 3}$ is not a real number. Thus, $(f + g)(-4)$ does not exist.   ▬

It is useful to view the concept of the sum of two functions graphically. In the graph on the left below, we see the graphs of two functions f and g and their sum, $f + g$. Consider finding $(f + g)(4) = f(4) + g(4)$. We can locate $g(4)$ on the graph of g and use a compass to measure it. Then we move that setting on top of $f(4)$ and add. The sum gives us $(f + g)(4)$.

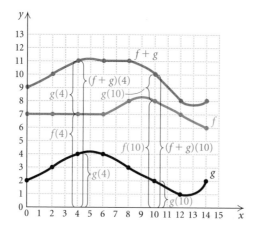

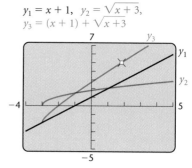

$y_1 = x + 1, \quad y_2 = \sqrt{x + 3},$
$y_3 = (x + 1) + \sqrt{x + 3}$

With this in mind, let's view Example 1 from a graphical perspective. See the graph on the right above. Use a grapher to graph

$$y_1 = x + 1, \quad y_2 = \sqrt{x + 3}, \quad \text{and} \quad y_3 = (x + 1) + \sqrt{x + 3}.$$

(On some graphers, you can enter y_1 and y_2 and then key the grapher to enter $y_3 = y_1 + y_2$ directly.) Note that because the domain of y_2 is $[-3, \infty)$, the graph of the sum exists only over $[-3, \infty)$. By using the TRACE feature and jumping the cursor vertically among the three graphs, we can confirm that the y-coordinates of the graph of $(f + g)(x)$ are the sums of the corresponding y-coordinates of the graphs of $f(x)$ and $g(x)$.

Example 2 Given that $f(x) = x^2 - 4$ and $g(x) = x + 2$, find each of the following.

a) $(f + g)(x)$ **b)** $(f - g)(x)$ **c)** $(fg)(x)$

d) $(f/g)(x)$ **e)** $(gg)(x)$

f) The domain of $f + g, f - g, fg$, and f/g

SOLUTION

a) $(f + g)(x) = f(x) + g(x) = (x^2 - 4) + (x + 2) = x^2 + x - 2$

b) $(f - g)(x) = f(x) - g(x) = (x^2 - 4) - (x + 2) = x^2 - x - 6$

c) $(fg)(x) = f(x) \cdot g(x) = (x^2 - 4)(x + 2) = x^3 + 2x^2 - 4x - 8$

d) $(f/g)(x) = \dfrac{f(x)}{g(x)}$

$\qquad = \dfrac{x^2 - 4}{x + 2}$ Note that $x \neq -2$.

$\qquad = \dfrac{(x + 2)(x - 2)}{x + 2}$ Factoring

$\qquad = x - 2$ Removing a factor of 1: $\dfrac{x + 2}{x + 2} = 1$

Thus, $(f/g)(x) = x - 2$ with the added stipulation that $x \neq -2$.

e) $(gg)(x) = [g(x)]^2 = (x + 2)^2 = x^2 + 4x + 4$

f) The domain of f is the set of all real numbers. The domain of g is the set of all real numbers. The domain of $f + g, f - g$, and fg is the set of numbers in the intersection of the domains—that is, the set of numbers in both domains, which is again the set of real numbers. For f/g, we must exclude -2, since $g(-2) = 0$. Thus the domain of f/g is the set of real numbers excluding -2, or $(-\infty, -2) \cup (-2, \infty)$. ▬

The Composition of Functions

In the real world, it is not uncommon for a function's output to depend on some input that is itself an output of some function. For instance, the amount a person pays as state income tax usually depends on the amount of adjusted gross income on the person's federal tax return, which, in turn, depends on his or her annual earnings. Such functions are called *composite functions*.

To illustrate how composite functions work, suppose a chemistry student needs a formula to convert Fahrenheit temperatures to Kelvin units. The formula $c(t) = \frac{5}{9}(t - 32)$ gives the Celsius temperature $c(t)$ that corresponds to the Fahrenheit temperature t. The formula $k(c) = c + 273$ gives the Kelvin temperature $k(c)$ that corresponds to the Celsius temperature c. Thus, 50° Fahrenheit corresponds to

$$c(50) = \tfrac{5}{9}(50 - 32) = \tfrac{5}{9}(18) = 10° \text{ Celsius}$$

and since 10° Celsius corresponds to

$$k(10) = 10 + 273 = 283° \text{ Kelvin,}$$

we see that 50° Fahrenheit is the same as 283° Kelvin. This two-step procedure can be used to convert any Fahrenheit temperature to Kelvin units.

In the table shown at left, we use a grapher to convert Fahrenheit temperatures, x, to Celsius temperatures, y_1, using $y_1 = \tfrac{5}{9}(x - 32)$. We also convert Celsius temperatures to Kelvin units, y_2, using $y_2 = y_1 + 273$.

A student making numerous conversions might look for a formula that converts directly from Fahrenheit to Kelvin. Such a formula can be found by substitution:

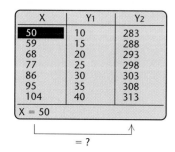

X	Y1	Y2
50	10	283
59	15	288
68	20	293
77	25	298
86	30	303
95	35	308
104	40	313

X = 50

= ?

$$
\begin{aligned}
y_2 &= y_1 + 273 \\
&= \left[\frac{5}{9}(x - 32) \right] + 273 \qquad \text{Substituting} \\
&= \frac{5}{9}x - \frac{160}{9} + 273 \\
&= \frac{5}{9}x - \frac{160}{9} + \frac{2457}{9} \\
&= \frac{5x + 2297}{9}. \qquad \text{Simplifying}
\end{aligned}
$$

We can show on a grapher that the same values that appear in the table for y_2 will appear when y_2 is entered as

$$y_2 = \frac{5x + 2297}{9}.$$

In the more commonly used function notation, we have

$$
\begin{aligned}
k(c(t)) &= c(t) + 273 \\
&= \frac{5}{9}(t - 32) + 273 \qquad \text{Substituting} \\
&= \frac{5t + 2297}{9}. \qquad \text{Simplifying as above}
\end{aligned}
$$

Since the last equation expresses the Kelvin temperature as a new function, K, of the Fahrenheit temperature, t, we can write

$$K(t) = \frac{5t + 2297}{9},$$

where $K(t) =$ the Kelvin temperature corresponding to the Fahrenheit temperature, t. Here we have $K(t) = k(c(t))$. The new function K is called the **composition** of k and c and can be denoted $k \circ c$ (read "k composed with c," "the composition of k and c," or "k circle c").

Composition of Functions

The *composite function f ∘ g*, the *composition* of f and g, is defined as

$$(f \circ g)(x) = f(g(x)),$$

where x is in the domain of g and $g(x)$ is in the domain of f.

Example 3 Given that $f(x) = 2x - 5$ and $g(x) = x^2 - 3x + 8$, find each of the following.

a) $(f \circ g)(7)$ and $(g \circ f)(7)$ **b)** $(f \circ g)(x)$ and $(g \circ f)(x)$

SOLUTION Consider each function separately:

$f(x) = 2x - 5$ This function multiplies each input by 2 and subtracts 5.

and

$g(x) = x^2 - 3x + 8$. This function squares an input, subtracts 3 times the input from the result, and then adds 8.

a) To find $(f \circ g)(7)$, we first find $g(7)$. Then we use $g(7)$ as an input for f:

$$\begin{aligned}(f \circ g)(7) = f(g(7)) &= f(7^2 - 3 \cdot 7 + 8) \\ &= f(36) = 2 \cdot 36 - 5 \\ &= 67.\end{aligned}$$

To find $(g \circ f)(7)$, we first find $f(7)$. Then we use $f(7)$ as an input for g:

$$\begin{aligned}(g \circ f)(7) = g(f(7)) &= g(2 \cdot 7 - 5) \\ &= g(9) = 9^2 - 3 \cdot 9 + 8 \\ &= 62.\end{aligned}$$

b) To find $(f \circ g)(x)$, we substitute $g(x)$ for x in the equation for $f(x)$:

$$\begin{aligned}(f \circ g)(x) = f(g(x)) &= f(x^2 - 3x + 8) \qquad \text{Substituting } x^2 - 3x + 8 \text{ for } g(x)\\ &= 2(x^2 - 3x + 8) - 5 \\ &= 2x^2 - 6x + 16 - 5 \\ &= 2x^2 - 6x + 11.\end{aligned}$$

To find $(g \circ f)(x)$, we substitute $f(x)$ for x in the equation for $g(x)$:

$$\begin{aligned}(g \circ f)(x) = g(f(x)) &= g(2x - 5) \qquad \text{Substituting } 2x - 5 \text{ for } f(x)\\ &= (2x - 5)^2 - 3(2x - 5) + 8 \\ &= 4x^2 - 20x + 25 - 6x + 15 + 8 \\ &= 4x^2 - 26x + 48.\end{aligned}$$

Note in Example 3 that, as a rule, $(f \circ g)(x) \neq (g \circ f)(x)$. We can check this graphically.

Example 4   Given that $f(x) = \sqrt{x}$ and $g(x) = x - 3$:

a) Find $h(x)$ and $k(x)$ if $h = f \circ g$ and $k = g \circ f$.

b) Graph h and k.

c) Find the domains of h and k.

SOLUTION

a) $h(x) = (f \circ g)(x) = f(g(x)) = f(x - 3) = \sqrt{x - 3}$

$k(x) = (g \circ f)(x) = g(f(x)) = g(\sqrt{x}) = \sqrt{x} - 3$

b) We actually used function composition when we worked with transformations in Section 1.5. For example, we know that the graph of $h(x) = \sqrt{x - 3}$ has the same shape as $y = \sqrt{x}$ shifted right 3 units. This occurs because g subtracts 3 units from each input before f takes the square root. When this sequence is reversed, as in the graph of $k(x) = \sqrt{x} - 3$, the subtraction of 3 occurs *after* the square root is taken. Thus the graph of k has the same shape as the graph of $y = \sqrt{x}$ shifted *down* 3 units.

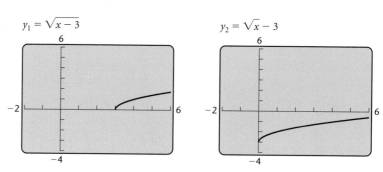

c) The domain of f is $\{x | x \geq 0\}$, or the interval $[0, \infty)$. The domain of g is $\mathbb{R}$.

To find the domain of $h = f \circ g$, we must consider that the outputs for g will serve as inputs for f. Since the inputs for f cannot be negative, we must have

$$g(x) = x - 3, \quad \text{where } x - 3 \geq 0,$$

$$\text{or}\quad g(x) = x - 3, \quad \text{where } x \geq 3.$$

Thus the domain of $h = \{x | x \geq 3\}$, or the interval $[3, \infty)$, as the graph confirms.

To find the domain of $k = g \circ f$, we must consider that the outputs for f will serve as inputs for g. Since g can accept *any* real number as an input, any output from f is acceptable. Thus the domain of $k = \{x | x \geq 0\}$, as the graph confirms.   ▬

Decomposing a Function as a Composition

In calculus, one needs to recognize how a function can be expressed as a composition. In this way, we are "decomposing" the function.

Example 5 Find $f(x)$ and $g(x)$ such that $h(x) = (f \circ g)(x)$: $h(x) = (2x - 3)^5$.

SOLUTION   We have expressed $h(x)$ as $(2x - 3)$ to the 5th power. Two functions that can be used for the composition are

$$f(x) = x^5 \quad \text{and} \quad g(x) = 2x - 3.$$

We can check by forming the composition:

$$h(x) = (f \circ g)(x) = f(g(x)) = f(2x - 3) = (2x - 3)^5.$$

This is the most "obvious" answer to the question. There can be other less obvious answers. For example, if

$$f(x) = (x + 7)^5 \quad \text{and} \quad g(x) = 2x - 10,$$

then

$$h(x) = (f \circ g)(x) = f(g(x))$$
$$= f(2x - 10) = [(2x - 10) + 7]^5 = (2x - 3)^5.$$

1.7 Exercise Set

Given that $f(x) = x^2 - 3$ and $g(x) = 2x + 1$, find each of the following, if it exists.

1. $(f + g)(5)$

2. $(f - g)(3)$

3. $(f \circ g)(4)$

4. $(g \circ f)(4)$

5. $(fg)(2)$

6. $(fg)(-2)$

7. $(f/g)(-1)$

8. $(f/g)\left(-\frac{1}{2}\right)$

9. $(fg)\left(-\frac{1}{2}\right)$

10. $(f/g)(-\sqrt{3})$

11. $(f/g)(\sqrt{3})$

12. $(f - g)(0)$

For each pair of functions in Exercises 13–16:

a) Find the domain of f, g, $f + g$, $f - g$, fg, ff, f/g, g/f, $f \circ g$, and $g \circ f$.

b) Find $(f + g)(x)$, $(f - g)(x)$, $(fg)(x)$, $(ff)(x)$, $(f/g)(x)$, $(g/f)(x)$, $(f \circ g)(x)$, and $(g \circ f)(x)$.

13. $f(x) = x - 3$, $g(x) = \sqrt{x + 4}$

14. $f(x) = x^2 - 1$, $g(x) = 2x + 5$

15. $f(x) = x^3$, $g(x) = 2x^2 + 5x - 3$

16. $f(x) = x^2$, $g(x) = \sqrt{x}$

For each pair of functions in Exercises 17–20, find $(f + g)(x)$, $(f - g)(x)$, $(fg)(x)$, $(f/g)(x)$, and $(f \circ g)(x)$.

17. $f(x) = x^2 - 4$, $g(x) = x^2 + 2$

18. $f(x) = x^2 + 2$, $g(x) = 4x - 7$

19. $f(x) = \sqrt{x - 7}$, $g(x) = x^2 - 25$

20. $f(x) = \sqrt{x - 1}$, $g(x) = \sqrt{3x - 8}$

For Exercises 21–26, consider the functions F and G as shown in the following graph.

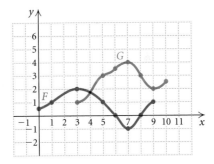

21. Find the domain of F, the domain of G, and the domain of $F + G$.

22. Find the domain of $F - G$, FG, and F/G.

23. Find the domain of G/F.

24. Graph $F + G$.

25. Graph $G - F$.

26. Graph $F - G$.

Find $(f \circ g)(x)$ and $(g \circ f)(x)$.

27. $f(x) = x + 3$, $g(x) = x - 3$

28. $f(x) = \frac{4}{5}x$, $g(x) = \frac{5}{4}x$

29. $f(x) = 3x - 7$, $g(x) = \dfrac{x + 7}{3}$

30. $f(x) = \frac{2}{3}x - \frac{4}{5}$, $g(x) = 1.5x + 1.2$

31. $f(x) = 20$, $g(x) = 0.05$

32. $f(x) = x^4$, $g(x) = \sqrt[4]{x}$

33. $f(x) = \sqrt{x + 5}$, $g(x) = x^2 - 5$

34. $f(x) = x^5 - 2$, $g(x) = \sqrt[5]{x + 2}$

35. $f(x) = \dfrac{1 - x}{x}$, $g(x) = \dfrac{1}{1 + x}$

36. $f(x) = \dfrac{x^2 - 1}{x^2 + 1}$, $g(x) = \dfrac{3x - 4}{5x - 2}$

37. $f(x) = x^3 - 5x^2 + 3x + 7$, $g(x) = x + 1$

38. $f(x) = x - 1$, $g(x) = x^3 + 2x^2 - 3x - 9$

Find $f(x)$ and $g(x)$ such that $h(x) = (f \circ g)(x)$. Answers may vary, but try to select the most obvious answer.

39. $h(x) = (4 + 3x)^5$ **40.** $h(x) = \sqrt[3]{x^2 - 8}$

41. $h(x) = \dfrac{1}{(x - 2)^4}$ **42.** $h(x) = \dfrac{1}{\sqrt{3x + 7}}$

43. $h(x) = \dfrac{x^3 - 1}{x^3 + 1}$ **44.** $h(x) = \left| 9x^2 - 4 \right|$

45. $h(x) = \left(\dfrac{2 + x^3}{2 - x^3} \right)^6$ **46.** $h(x) = (\sqrt{x} - 3)^4$

47. $h(x) = \sqrt{\dfrac{x - 5}{x + 2}}$

48. $h(x) = \sqrt{1 + \sqrt{1 + x}}$

49. $h(x) = (x + 2)^3 - 5(x + 2)^2 + 3(x + 2) - 1$

50. $h(x) = 2(x - 1)^{5/3} + 5(x - 1)^{2/3}$

51. *Total Cost, Revenue, and Profit.* In economics, functions that involve revenue, cost, and profit are used. For example, suppose that $R(x)$ and $C(x)$ denote the total revenue and the total cost, respectively, of producing a new kind of tool for King Hardware Stores. Then the difference

$$P(x) = R(x) - C(x)$$

represents the total profit for producing x tools. Given

$$R(x) = 60x - 0.4x^2 \quad \text{and} \quad C(x) = 3x + 13,$$

find each of the following.

a) $P(x)$
b) $R(100)$, $C(100)$, and $P(100)$
c) Use a grapher and graph the three functions using the viewing window $[0, 160, 0, 3000]$.

52. *Dress Sizes.* A dress that is size x in France is size $s(x)$ in the United States, where $s(x) = x - 32$. A dress that is size x in the United States is size $y(x)$

in Italy, where $y(x) = 2(x + 12)$. Find a function that will convert French dress sizes to Italian dress sizes.

53. *Ripple Spread.* A stone is thrown into a pond. A circular ripple is spreading over the pond in such a way that the radius is increasing at the rate of 3 ft/sec.

a) Find a function $r(t)$ for the radius in terms of t.
b) Find a function $A(r)$ for the area of the ripple in terms of the radius r.
c) Find $(A \circ r)(t)$. Explain the meaning of this function.

54. *Airplane Distance.* An airplane is 300 ft from the control tower at the end of the runway. It takes off at a speed of 250 mph.

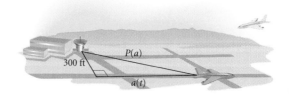

a) Let $a(t) = $ the distance that the airplane travels down the runway in time t. Find a formula for the function.

b) Let $P = $ the distance of the airplane from the control tower. Find a formula for P in terms of the distance a—that is, find an expression for $P(a)$.

c) Find $(P \circ a)(t)$. Explain the meaning of this function.

Skill Maintenance

Refer to the graph accompanying Exercises 21–26.

55. Find the range of F.

56. Find the range of G.

57. Evaluate $5x^3 - 7x^2 + 2x - 7$ for $x = -1$.

58. Factor: $9x^5 - 64x^4 + 72x^3$.

Synthesis

59. ◆ Does $(f + g)(x) = (f \circ g)(x)$ for any functions f and g? Why or why not?

60. ◆ Explain how the function f given by $f(x) = (x - 5)^2 + 3$ can be regarded as a composition of two functions, each of which is a translation.

In Exercises 61–64, use a grapher to graph f, g, and $f + g$ using the same viewing window.

61. $f(x) = x^2$, $g(x) = 3 - 2x$

62. $f(x) = \dfrac{2}{x}$, $g(x) = x$

63. $f(x) = 5 - x^2$, $g(x) = x^2 - 5$

64. $f(x) = |x|$, $g(x) = \text{INT}(x)$

Refer to the graph accompanying Exercises 21–26.

65. Find the domain of $F \circ G$, $G \circ G$, and $F \circ F$.

66. Find the range of $F \circ G$.

67. Consider $f(x) = 3x + b$ and $g(x) = 2x - 1$. Find b such that $(f \circ g)(x) = (g \circ f)(x)$ for all real numbers x.

68. Consider $f(x) = 3x - 4$ and $g(x) = mx + b$. Find m and b such that

$$(f \circ g)(x) = (g \circ f)(x) = x$$

for all real numbers x.

69. For $f(x) = 1/(1 - x)$, find $(f \circ f)(x)$ and $(f \circ (f \circ f))(x)$.

State whether each of the following is true or false.

70. The composition of two even functions is even.

71. The sum of two even functions is even.

72. The product of two odd functions is odd.

73. The composition of two odd functions is odd.

74. The product of an even function and an odd function is odd.

75. Show that if f is *any* function, then the function E defined by

$$E(x) = \frac{f(x) + f(-x)}{2}$$

is even.

76. Show that if f is *any* function, then the function O defined by

$$O(x) = \frac{f(x) - f(-x)}{2}$$

is odd.

77. Consider the functions E and O of Exercises 75 and 76.

a) Show that $f(x) = E(x) + O(x)$. This means that every function can be expressed as the sum of an even and an odd function.

b) Let $f(x) = 4x^3 - 11x^2 + \sqrt{x} - 10$. Express f as a sum of an even function and an odd function.

1.8
Inverse Functions

- *Determine whether a function is one-to-one, and if it is, find a formula for its inverse.*
- *Simplify expressions of the type $(f \circ f^{-1})(x)$ and $(f^{-1} \circ f)(x)$.*

When we go from an output of a function back to its input or inputs, we get an inverse relation. When that relation is a function, we have an inverse function. We now study inverse functions and how to find their formulas when the original function has a formula. We do this to understand the relationship between exponential and logarithmic functions.

Inverses

Consider the relation h given as follows:

$$h = \{(-8, 5), (4, -2), (-7, 1), (3.8, 6.2)\}.$$

Suppose we *interchange* the first and second coordinates. The relation we obtain is called the **inverse** of the relation h and is given as follows:

$$\text{Inverse of } h = \{(5, -8), (-2, 4), (1, -7), (6.2, 3.8)\}.$$

> ### Inverse Relation
>
> Interchanging the first and second coordinates of each ordered pair in a relation produces the *inverse relation*.

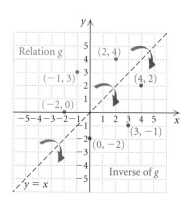

Example 1   Consider the relation g given by

$$g = \{(2, 4), (-1, 3), (-2, 0)\}.$$

Graph the relation in blue. Find the inverse and graph it in red.

SOLUTION   The relation g is shown in blue in the figure at left. The inverse of the relation is

$$\{(4, 2), (3, -1), (0, -2)\}$$

and is shown in red. The pairs in the inverse are reflections across the line $y = x$.   ▬

> ### Inverse Relation
>
> If a relation is defined by an equation, interchanging the variables produces an equation of the *inverse relation*.

Example 2   Find an equation for the inverse of the relation:

$$y = x^2 - 5x.$$

SOLUTION   We interchange x and y and obtain an equation of the inverse:

$$x = y^2 - 5y.$$

Interactive Discovery

Graph each of the following relations:

$$y = 3x + 2, \qquad y = x, \qquad xy = 2,$$
$$x^2 + 3y^2 = 4, \qquad y^2 = 4x - 5.$$

Then find the inverse of each and graph it using the same set of axes. What pattern do you see? Be sure to use a square viewing window.

If a relation is given by an equation, the solutions of the inverse can be found from those of the original equation by interchanging the first and second coordinates of each ordered pair. Thus the graphs of a relation and its inverse are always reflections of each other across the line $y = x$.

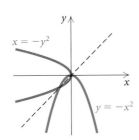

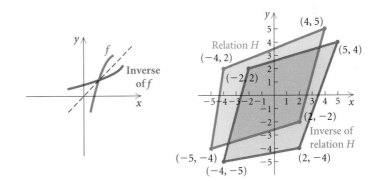

Inverses and One-to-One Functions

Let's consider the following two functions.

NUMBER (DOMAIN)	CUBING (RANGE)
-3	$\longrightarrow -27$
-2	$\longrightarrow -8$
-1	$\longrightarrow -1$
0	$\longrightarrow 0$
1	$\longrightarrow 1$
2	$\longrightarrow 8$
3	$\longrightarrow 27$

YEAR (DOMAIN)	FIRST-CLASS POSTAGE COST, IN CENTS (RANGE)
1978	$\longrightarrow 15$
1983	$\longrightarrow 20$
1984	
1989	$\longrightarrow 25$
1991	$\longrightarrow 29$
1995	$\longrightarrow 32$

Source: U.S. Postal Service.

Suppose we reverse the arrows. Are these inverse relations functions?

NUMBER (RANGE)	CUBING (DOMAIN)
−3 ←	−27
−2 ←	−8
−1 ←	−1
0 ←	0
1 ←	1
2 ←	8
3 ←	27

YEAR (RANGE)	FIRST-CLASS POSTAGE COST, IN CENTS (DOMAIN)
1978 ←	15
1983 ←	20
1984 ←	
1989 ←	25
1991 ←	29
1995 ←	32

We see that the inverse of the cubing function is a function, but the inverse of the postage function is not a function. Like all functions, each input in the postage function has exactly one output. However, the output for both 1983 and 1984 is the same number, 20. Thus in the inverse of the postage function, the input 20 has *two* outputs, 1983 and 1984. When the same output comes from two or more different inputs, the inverse cannot be a function. In the cubing function, each output corresponds to exactly one input, so its inverse is also a function. The cubing function is an example of a **one-to-one function**. If the inverse of a function f is also a function, it is named f^{-1} (read "f-inverse").

The −1 in f^{-1} is *not* an exponent!

One-to-One Function and Inverses

A function f is *one-to-one* if different inputs have different outputs—that is,

  if $a \neq b$, then $f(a) \neq f(b)$. Or,

A function f is *one-to-one* if when the outputs are the same, the inputs are the same—that is,

  if $f(a) = f(b)$, then $a = b$.

If a function is one-to-one, then its inverse is a function.

The domain of a one-to-one function f is the range of the inverse f^{-1}.

The range of a one-to-one function f is the domain of the inverse f^{-1}.

Example 3 Given the function f described by $f(x) = 2x - 3$, prove that f is one-to-one (that is, it has an inverse that is a function).

SOLUTION To show that f is one-to-one, we show that if $f(a) = f(b)$, then $a = b$. Assume that $f(a) = f(b)$ for any numbers a and b in the domain of f. Then

$$2a - 3 = 2b - 3$$
$$2a = 2b \qquad \text{Adding 3}$$
$$a = b. \qquad \text{Dividing by 2}$$

Thus, if $f(a) = f(b)$, then $a = b$. This shows that f is one-to-one.   ▬

Example 4   Given the function g described by $g(x) = x^2$, prove that g is not one-to-one.

SOLUTION   To prove that g is not one-to-one, we need to find two numbers a and b for which $a \neq b$ and $g(a) = g(b)$. Two such numbers are -3 and 3, because $-3 \neq 3$ and $g(-3) = g(3) = 9$. Thus g is not one-to-one.

The graph below shows a function, in blue, and its inverse, in red. To determine whether the inverse is a function, we can apply the vertical-line test to its graph. By reflecting each such vertical line back across the line $y = x$, we obtain an equivalent **horizontal-line test** for the original function.

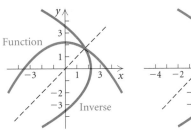

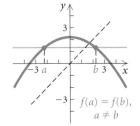

The vertical-line test shows that the inverse is not a function.

The horizontal-line test shows that the function is not one-to-one.   ▬

Horizontal-Line Test

If it is possible for a horizontal line to intersect the graph of a function more than once, then the function is *not* one-to-one and its inverse is *not* a function.

Example 5   Graph each of the following functions. Then determine whether each is one-to-one and thus has an inverse that is a function.

a) $f(x) = 4 - x$
b) $f(x) = x^2$
c) $f(x) = \sqrt[3]{x + 2} + 3$
d) $f(x) = 3x^5 - 20x^3$

SOLUTION   We graph each function using a grapher. Then we apply the horizontal-line test.

a) $y = 4 - x$

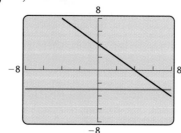

b) $y = x^2$

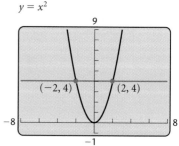

c) $y = \sqrt[3]{x + 2} + 3$

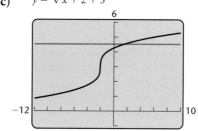

d) $y = 3x^5 - 20x^3$

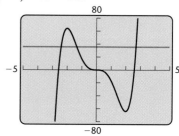

RESULT	REASON
a) One-to-one; inverse is a function	No horizontal line crosses the graph more than once.
b) Not one-to-one; inverse is not a function	There are many horizontal lines that cross the graph more than once. In particular, the line $y = 4$ does so. Note that where the line crosses, the first coordinates are -2 and 2. Although these are different inputs, they have the same output—that is, $-2 \neq 2$, but $$f(-2) = (-2)^2$$ $$= 4$$ $$= 2^2 = f(2).$$
c) One-to-one; inverse is a function	No horizontal line crosses the graph more than once.
d) Not one-to-one; inverse is not a function	There are many horizontal lines that cross the graph more than once. ▬

Finding Formulas for Inverses

Suppose that a function is described by a formula. If it has an inverse that is a function, we proceed as follows to find a formula for f^{-1}.

> **Obtaining a Formula for an Inverse**
>
> If a function f is one-to-one, a formula for its inverse can generally be found as follows:
>
> **1.** Replace $f(x)$ with y.
> **2.** Interchange x and y.
> **3.** Solve for y.
> **4.** Replace y with $f^{-1}(x)$.

Example 6   Determine whether the function $f(x) = 2x - 3$ is one-to-one, and if it is, find a formula for $f^{-1}(x)$.

SOLUTION   The graph of f is shown at left. It passes the horizontal-line test. Thus it is one-to-one and its inverse is a function.
    We also proved that f is one-to-one in Example 3.

$y = 2x - 3$

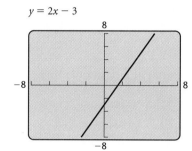

1. Replace $f(x)$ with y: $y = 2x - 3$

2. Interchange x and y: $x = 2y - 3$

3. Solve for y: $x + 3 = 2y$

$$\frac{x + 3}{2} = y$$

4. Replace y with $f^{-1}(x)$: $f^{-1}(x) = \dfrac{x + 3}{2}.$

Consider

$$f(x) = 2x - 3 \quad \text{and} \quad f^{-1}(x) = \frac{x + 3}{2}$$

from Example 6. For the input 5, we have

$$f(5) = 2 \cdot 5 - 3 = 10 - 3 = 7.$$

The output is 7. Now we use 7 for the input in the inverse:

$$f^{-1}(7) = \frac{7 + 3}{2} = \frac{10}{2} = 5.$$

The function f takes the number 5 to 7. The inverse function f^{-1} takes the number 7 back to 5.

Example 7   Graph

$$f(x) = 2x - 3 \quad \text{and} \quad f^{-1}(x) = \frac{x + 3}{2}$$

using the same set of axes. Then compare the two graphs.

$y_1 = f(x) = 2x - 3,$

$y_2 = f^{-1}(x) = \dfrac{x + 3}{2}$

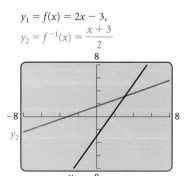

SOLUTION

Method 1. The graphs of f and f^{-1} are shown at left. Note that the graph of f^{-1} can be drawn by reflecting the graph of f across the line $y = x$. That is, if we were to graph $f(x) = 2x - 3$ in wet ink and fold along the line $y = x$, the graph of $f^{-1}(x) = (x + 3)/2$ would be formed by the ink transferred from f.

When we interchange x and y in finding a formula for the inverse of $f(x) = 2x - 3$, we are in effect reflecting the graph of that function across the line $y = x$. For example, when the coordinates of the y-intercept of the graph of f, $(0, -3)$, are reversed, we get the x-intercept of the graph of f^{-1}, $(-3, 0)$.

Method 2. Some graphers have a feature that graphs inverses automatically. If we were to start with $y_1 = 2x - 3$, the graphs of both y_1 and its inverse $y_2 = (x + 3)/2$ would be drawn. Be sure to square the viewing window. Consult your manual. ▬

> The graph of f^{-1} is a reflection of the graph of f across the line $y = x$.

Example 8   Consider $g(x) = x^3 + 2$.

a) Determine whether the function is one-to-one.

b) If it is one-to-one, find a formula for its inverse.

c) Graph the function and its inverse. Use a square viewing window.

SOLUTION

a) The graph of $g(x) = x^3 + 2$ is shown below. It passes the horizontal-line test and thus has an inverse that is a function.

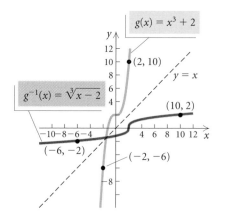

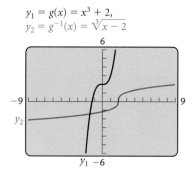

b) We follow the procedure for finding an inverse.

   1. Replace $g(x)$ with y: $y = x^3 + 2$

   2. Interchange x and y: $x = y^3 + 2$

3. Solve for y:

$$x - 2 = y^3$$
$$\sqrt[3]{x - 2} = y$$

4. Replace y with $g^{-1}(x)$: $g^{-1}(x) = \sqrt[3]{x - 2}$.

c) To find the graph, we reflect the graph of $g(x) = x^3 + 2$ across the line $y = x$. This can be done by using a grapher or by plotting points. See the graph on the left above. Be sure to square the viewing window.

Inverse Functions and Composition

Suppose that we were to use some input a for a one-to-one function f and find its output, $f(a)$. The function f^{-1} would then take that output back to a. Similarly, if we began with an input b for the function f^{-1} and found its output, $f^{-1}(b)$, the original function f would then take that output back to b. This is summarized as follows.

If a function f is one-to-one, then f^{-1} is the unique function such that each of the following holds:

$$(f^{-1} \circ f)(x) = f^{-1}(f(x)) = x, \quad \text{for each } x \text{ in the domain of } f$$
$$\text{(range of } f^{-1}\text{),} \quad \text{and}$$

$$(f \circ f^{-1})(x) = f(f^{-1}(x)) = x, \quad \text{for each } x \text{ in the domain of}$$
$$f^{-1} \text{ (range of } f\text{).}$$

Example 9   Given that $f(x) = 5x + 8$, use composition of functions to show that $f^{-1}(x) = (x - 8)/5$.

SOLUTION   We find $(f^{-1} \circ f)(x)$ and $(f \circ f^{-1})(x)$ and check to see that each is x:

$$(f^{-1} \circ f)(x) = f^{-1}(f(x)) = f^{-1}(5x + 8) = \frac{(5x + 8) - 8}{5} = \frac{5x}{5} = x;$$

$$(f \circ f^{-1})(x) = f(f^{-1}(x))$$

$$= f\left(\frac{x - 8}{5}\right)$$

$$= 5\left(\frac{x - 8}{5}\right) + 8$$

$$= x - 8 + 8 = x.$$

Restricting a Domain

In the case in which the inverse of a function is not a function, the domain can be restricted to allow the inverse to be a function. We saw in Example 5 that $f(x) = x^2$ is not one-to-one. The graph is shown at the top of the following page.

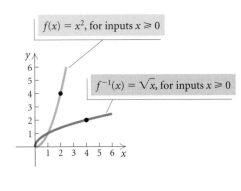

$f(x) = x^2$, for inputs $x \geq 0$

$f^{-1}(x) = \sqrt{x}$, for inputs $x \geq 0$

$f(x) = x^2$

Suppose we had not known this and had tried to find a formula for the inverse as follows:

$$y = x^2 \qquad \text{Replacing } f(x) \text{ with } y$$
$$x = y^2. \qquad \text{Interchanging } x \text{ and } y$$

We cannot solve for y and get only one value, since most real numbers have two square roots:

$$\pm\sqrt{x} = y.$$

This is not the equation of a function. An input of, say, 4 would yield two outputs, -2 and 2. In such cases, it is convenient to consider "part" of the function by restricting the domain of $f(x)$. For example, if we restrict the domain of $f(x) = x^2$ to nonnegative numbers, then its inverse is a function. See the graphs of $f(x) = x^2$, $x \geq 0$, and $f^{-1}(x) = \sqrt{x}$, $x \geq 0$ on the right above.

1.8 Exercise Set

Find the inverse of each relation.

1. $\{(7, 8), (-2, 8), (3, -4), (8, -8)\}$

2. $\{(0, 1), (5, 6), (-2, -4)\}$

3. $\{(-1, -1), (-3, 4)\}$

4. $\{(-1, 3), (2, 5), (-3, 5), (2, 0)\}$

Find an equation of each inverse relation.

5. $y = 4x - 5$

6. $2x^2 + 5y^2 = 4$

7. $x^3y = -5$

8. $y = 3x^2 - 5x + 9$

Graph each equation by substituting and plotting points. Then reflect the graph across the line $y = x$ to obtain the graph of its inverse.

9. $x = y^2 - 3$

10. $y = x^2 + 1$

11. $y = |x|$

12. $x = |y|$

Using the horizontal-line test, determine whether each function is one-to-one.

13. $f(x) = 2.7^x$

14. $f(x) = 2^{-x}$

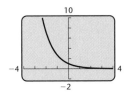

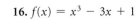

15. $f(x) = 4 - x^2$

16. $f(x) = x^3 - 3x + 1$

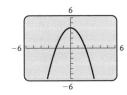

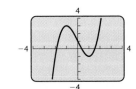

17. $f(x) = \dfrac{8}{x^2 - 4}$

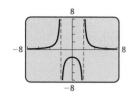

18. $f(x) = \sqrt{\dfrac{10}{4 + x}}$

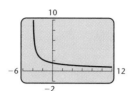

19. $f(x) = \sqrt[3]{x + 2} - 2$

20. $f(x) = \dfrac{8}{x}$

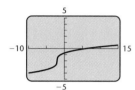

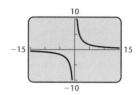

Determine whether each of the following is one-to-one. Use a grapher and the horizontal-line test or use an algebraic procedure.

21. $f(x) = 5x - 8$ **22.** $f(x) = 3 + 4x$

23. $f(x) = 1 - x^2$ **24.** $f(x) = |x| - 2$

25. $f(x) = |x + 2|$ **26.** $f(x) = -0.8$

27. $f(x) = -\dfrac{4}{x}$ **28.** $f(x) = \dfrac{2}{x + 3}$

Graph each function and its inverse using a grapher. Use an inverse drawing feature, if available. Find the domain and the range of f. Find the domain and the range of the inverse f^{-1}.

29. $f(x) = 0.8x + 1.7$ **30.** $f(x) = 2.7 - 1.08x$

31. $f(x) = \frac{1}{2}x - 4$ **32.** $f(x) = x^3 - 1$

33. $f(x) = \sqrt{x - 3}$ **34.** $f(x) = -\dfrac{2}{x}$

35. $f(x) = x^2 - 4, \; x \geq 0$

36. $f(x) = 3 - x^2, \; x \geq 0$

37. $f(x) = (3x - 9)^3$

38. $f(x) = \sqrt[3]{\dfrac{x - 3.2}{1.4}}$

Given each function:

a) *Determine whether it is one-to-one, using a grapher if desired.*

b) *If it is one-to-one, find a formula for the inverse.*

39. $f(x) = x + 4$ **40.** $f(x) = 7 - x$

41. $f(x) = 2x$ **42.** $f(x) = 5x + 8$

43. $f(x) = \dfrac{4}{x + 7}$ **44.** $f(x) = -\dfrac{3}{x}$

45. $f(x) = \dfrac{x + 4}{x - 3}$ **46.** $f(x) = \dfrac{5x - 3}{2x + 1}$

47. $f(x) = x^3 - 1$ **48.** $f(x) = (x + 5)^3$

49. $f(x) = x\sqrt{4 - x^2}$

50. $f(x) = 4x^5 - 20x^3 + 2x^2 - 5x + 1$

51. $f(x) = 5x^2 - 2, \; x \geq 2$

52. $f(x) = 4x^2 + 3, \; x \geq 0$

53. $f(x) = \sqrt{x + 1}$

54. $f(x) = \sqrt[3]{x - 8}$

Find each inverse by thinking about the operations of the function and then reversing, or undoing, them. Check your work algebraically.

FUNCTION	INVERSE
55. $f(x) = 3x$	$f^{-1}(x) = $ ▮
56. $f(x) = \frac{1}{4}x + 7$	$f^{-1}(x) = $ ▮
57. $f(x) = -x$	$f^{-1}(x) = $ ▮
58. $f(x) = \sqrt[3]{x} - 5$	$f^{-1}(x) = $ ▮
59. $f(x) = \sqrt[3]{x - 5}$	$f^{-1}(x) = $ ▮
60. $f(x) = x^{-1}$	$f^{-1}(x) = $ ▮

For each function f, use composition of functions to show that f^{-1} is as given.

61. $f(x) = \dfrac{7}{8}x, \; f^{-1}(x) = \dfrac{8}{7}x$

62. $f(x) = \dfrac{(x + 5)}{4}, \; f^{-1}(x) = 4x - 5$

63. $f(x) = \dfrac{(1 - x)}{x}, \; f^{-1}(x) = \dfrac{1}{x + 1}$

64. $f(x) = \sqrt[3]{x + 4}, \; f^{-1}(x) = x^3 - 4$

65. Find $f(f^{-1}(5))$ and $f^{-1}(f(a))$:
$$f(x) = x^3 - 4.$$

66. Find $f^{-1}(f(p))$ and $f(f^{-1}(1253))$:
$$f(x) = \sqrt[5]{\dfrac{2x - 7}{3x + 4}}.$$

67. *Dress Sizes in the United States and Italy.* A function that will convert dress sizes in the United States to those in Italy is
$$g(x) = 2(x + 12).$$

a) Find the dress sizes in Italy that correspond to sizes 6, 8, 10, 14, and 18 in the United States.

b) Find a formula for the inverse of the function.

c) Use the inverse function to find the dress sizes in the United States that correspond to 36, 40, 44, 52, and 60 in Italy.

68. *Bus Chartering.* An organization determines that the cost per person of chartering a bus is given by

the formula

$$C(x) = \frac{100 + 5x}{x},$$

where x = the number of people in the group and $C(x)$ is in dollars. Determine $C^{-1}(x)$ and explain what it represents.

69. *Reaction Distance.* You are driving a car when a deer suddenly darts across the road in front of you. Your brain registers the emergency and sends a signal to your foot to hit the brake. The car travels a distance D, in feet, during this time, where D is a function of the speed r, in miles per hour, that the car is traveling when you see the deer. That reaction distance D is a linear function given by

$$D(r) = \frac{11r + 5}{10}.$$

a) Find $D(0)$, $D(10)$, $D(20)$, $D(50)$, and $D(65)$.
b) Graph $D(r)$.
c) Find $D^{-1}(r)$ and explain what it represents.
d) Graph the inverse.

70. *Bread Consumption.* The number N of 1-lb loaves of bread consumed per person per year t years after 1995 is given by the function

$$N(t) = 0.6514t + 53.1599.$$

(*Source:* Department of Agriculture)

a) Find the consumption of bread per person in 1998 and 2000.
b) Use a grapher to graph the function and its inverse.
c) Explain what the inverse represents.

Skill Maintenance

Graph each of the following.

71. $y = x^3 - x$

72. $x = y^3 - y$

73. $f(x) = \sqrt[3]{x}$

74. $f(x) = \dfrac{8}{x^2 - 4}$

Synthesis

75. ◆ Suppose that you have graphed a function using a grapher and you see that it is one-to-one. How could you then use the TRACE feature to make a hand-drawn graph of the inverse?

76. ◆ The following formulas for the conversion between Fahrenheit and Celsius temperatures have been considered several times in this text:

$$C = \tfrac{5}{9}(F - 32)$$

and

$$F = \tfrac{9}{5}C + 32.$$

Discuss these formulas from the standpoint of inverses.

Using only a grapher, determine whether the functions are inverses of each other.

77. $f(x) = \sqrt[3]{\dfrac{x - 3.2}{1.4}}$, $g(x) = 1.4x^3 + 3.2$

78. $f(x) = \dfrac{2x - 5}{4x + 7}$, $g(x) = \dfrac{7x - 4}{5x + 2}$

79. $f(x) = \dfrac{2}{3}$, $g(x) = \dfrac{3}{2}$

80. $f(x) = x^4$, $x \geq 0$; $g(x) = \sqrt[4]{x}$

81. Find three examples of functions that are their own inverses, that is, $f = f^{-1}$.

82. Consider the function f given by

$$f(x) = \begin{cases} x^3 + 2, & x \leq -1, \\ x^2, & -1 < x < 1, \\ x + 1, & x \geq 1. \end{cases}$$

Does f have an inverse that is a function? Why or why not?

Summary and Review

Important Properties and Formulas

Terminology about Lines

Slope: $m = \dfrac{y_2 - y_1}{x_2 - x_1}$

The Slope–intercept Equation:
$$y = mx + b$$

The Point–slope Equation:
$$y - y_1 = m(x - x_1)$$

The Two-point Equation:
$$y - y_1 = \frac{y_2 - y_1}{x_2 - x_1}(x - x_1)$$

Horizontal Lines: $y = b$

Vertical Lines: $x = a$

Parallel Lines: $m_1 = m_2, \ b_1 \neq b_2$

Perpendicular Lines:
$$m_1 m_2 = -1, \text{ or}$$
$$x = a, y = b$$

Tests for Symmetry

x-axis: If replacing y by $-y$ produces an equivalent equation, then the graph is symmetric with respect to the x-axis.

y-axis: If replacing x by $-x$ produces an equivalent equation, then the graph is symmetric with respect to the y-axis.

Origin: If replacing x by $-x$ and y by $-y$ produces an equivalent equation, then the graph is symmetric with respect to the origin.

Even Function: $f(-x) = f(x)$

Odd Function: $f(-x) = -f(x)$

Transformations

Vertical Translation: $y = f(x) \pm b$

Horizontal Translation: $y = f(x \pm d)$

Reflection across the x-axis: $y = -f(x)$

Reflection across the y-axis: $y = f(-x)$

Vertical Stretching or Shrinking:
$$y = af(x)$$

Horizontal Stretching or Shrinking:
$$y = f(ax)$$

The Algebra of Functions

The Sum of Two Functions:
$$(f + g)(x) = f(x) + g(x)$$

The Difference of Two Functions:
$$(f - g)(x) = f(x) - g(x)$$

The Product of Two Functions:
$$(fg)(x) = f(x) \cdot g(x)$$

The Quotient of Two Functions:
$$(f/g)(x) = f(x)/g(x), \ g(x) \neq 0$$

The Composition of Two Functions:
$$(f \circ g)(x) = f(g(x))$$

REVIEW EXERCISES

Determine whether each relation is a function. Find the domain and the range.

1. $\{(3, 1), (5, 3), (7, 7), (3, 5)\}$

2. $\{(2, 7), (-2, -7), (7, -2), (0, 2), (1, -4)\}$

Determine whether each graph is that of a function

3. 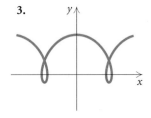   **4.**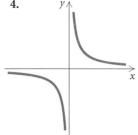

Find the domain of each function. Do not use a grapher.

5. $f(x) = 4 - 5x + x^2$ **6.** $f(x) = \sqrt{7 - 3x}$

7. $f(x) = \dfrac{1}{x^2 - 6x + 5}$ **8.** $f(x) = \dfrac{-5x}{|16 - x^2|}$

Use a grapher to graph each function. Then visually estimate the domain and the range.

9. $f(x) = \sqrt{16 - (x + 1)^2}$

10. $f(x) = |x - 3| + |x + 4| - 12$

11. $f(x) = x^3 - 5x^2 + 2x - 7$

12. $f(x) = x^4 - 8x^2 - 3$

13. Given that $f(x) = x^2 - x - 3$, find:

    a) $f(0)$;   **b)** $f(-3)$;

    c) $f(a - 1)$;   **d)** $\dfrac{f(x + h) - f(x)}{h}$.

Use a grapher to determine graphically whether each of the following seems to be an identity.

14. $\sqrt{x + 9} = \sqrt{x} + 3$ **15.** $x^2 \cdot x^3 = x^5$

16. $x^2 + x^3 = x^5$ **17.** $x + 3 = 3 + x$

Make a hand-drawn graph of each of the following. Check your results on a grapher.

18. $f(x) = \begin{cases} x^3, & \text{for } x < -2, \\ |x|, & \text{for } -2 \le x \le 2, \\ \sqrt{x - 1}, & \text{for } x > 2 \end{cases}$

19. $f(x) = \begin{cases} \dfrac{x^2 - 1}{x + 1}, & \text{for } x \ne -1, \\ 3, & \text{for } x = -1 \end{cases}$

20. $f(x) = \text{INT}(x)$

21. $f(x) = \text{INT}(x - 3)$

Use a grapher. For each function in Exercises 22–24, estimate:

a) *the roots, or zeros;*

b) *where the function is increasing or decreasing;*

c) *any relative maxima or minima.*

22. $f(x) = 1.1x^3 - 9.04x^2 - 18x + 702$

23. $f(x) = \dfrac{20}{(x + 4)^2 + 1} - 5$

24. $f(x) = 0.8x\sqrt{16 - (x + 1)^2}$

25. *Inscribed Rectangle.*   A rectangle is inscribed in a semicircle of radius 2, as shown. The variable $x = $ half the length of the rectangle. Express the area of the rectangle as a function of x.

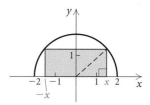

Each table of data contains input–output values for a function. Answer the following questions:

a) *Is the change in the inputs, x, the same?*

b) *Is the change in the outputs, y, the same?*

c) *Is the function linear?*

26.

x	y
-3	8
-2	11
-1	14
0	17
1	20
2	22
3	26

27.

x	y
20	11.8
30	24.2
40	36.6
50	49.0
60	61.4
70	73.8
80	86.2

28. Find the slope and the y-intercept of the graph of $-2x - y = 7$.

Write a point–slope equation for a line with the following characteristics.

29. $m = -\frac{2}{3}$, y-intercept $(0, -4)$

30. $m = 3$, passes through $(-2, -1)$

31. Passes through $(4, 1)$ and $(-2, -1)$

Determine whether the lines are parallel, perpendicular, or neither.

32. $3x - 2y = 8$,
$6x - 4y = 2$

33. $y - 2x = 4$,
$2y - 3x = -7$

34. $y = \frac{3}{2}x + 7$,
$y = -\frac{2}{3}x - 4$

Given the point $(1, -1)$ and the line $2x + 3y = 4$:

35. Find an equation of the line containing the given point and parallel to the given line.

36. Find an equation of the line containing the given point and perpendicular to the given line.

37. *Total Cost.* Clear County Cable Television charges a $25 installation fee and $20 per month for "basic" service. Write and graph an equation that can be used to determine the total cost, $C(t)$, of t months of basic cable television service. Find the total cost of 6 months of service.

38. *Temperature and Depth of the Earth.* The function T given by

$$T(d) = 10d + 20$$

can be used to determine the temperature T, in degrees Celsius, at a depth d, in kilometers, inside the earth.

a) Find $T(5)$, $T(20)$, and $T(1000)$.
b) Graph T.
c) The radius of the earth is about 5600 km. Use this fact to determine the domain of the function.

First, graph each equation and determine visually whether it is symmetric with respect to the x-axis, the y-axis, and the origin. Then verify your assertion algebraically.

39. $x^2 + y^2 = 4$

40. $y^2 = x^2 + 3$

41. $x + y = 3$

42. $y = x^2$

43. $y = x^3$

44. $y = x^4 - x^2$

Determine visually whether each function is even, odd, or neither even nor odd.

45.

46.

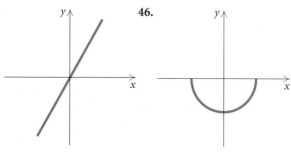

Test algebraically whether each function is even, odd, or neither even nor odd. Then check your work graphically, where possible, using a grapher.

47. $f(x) = 9 - x^2$

48. $f(x) = x^3 - 2x + 4$

49. $f(x) = x^7 - x^5$

50. $f(x) = |x|$

51. $f(x) = \sqrt{16 - x^2}$

52. $f(x) = \dfrac{10x}{x^2 + 1}$

Write an equation for a function that has a graph with the given characteristics. Check your answer on a grapher.

53. The shape of $y = x^2$, but shifted left 3 units

54. The shape of $y = \sqrt{x}$, but upside down and shifted right 3 units and up 4 units

55. The shape of $y = |x|$, but stretched vertically by a factor of 2 and shifted right 3 units

A graph of $y = f(x)$ is shown. No formula for f is given. Make a hand-drawn graph of each of the following.

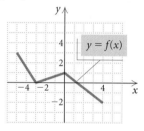

56. $y = f(x - 1)$

57. $y = f(2x)$

58. $y = -2f(x)$

59. $y = 3 + f(x)$

In Exercises 60 and 61, complete the square to:

a) *find the vertex;*
b) *find the line of symmetry;*
c) *determine whether there is a maximum or minimum value and find that value;*
d) *find the range.*

Then check your answers using a grapher.

60. $f(x) = -4x^2 + 3x - 1$

61. $f(x) = 5x^2 - 10x + 3$

In Exercises 62–65, match the equation with figures (a)–(d), which follow. Do not use a grapher except as a check.

a)

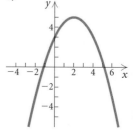

b)

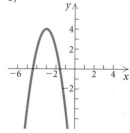

c)

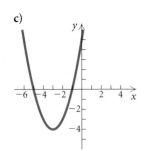

d)

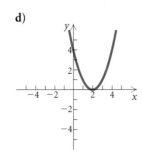

62. $y = (x - 2)^2$

63. $y = (x + 3)^2 - 4$

64. $y = -2(x + 3)^2 + 4$

65. $y = -\frac{1}{2}(x - 2)^2 + 5$

66. *Sidewalk Width.* A 60-ft by 80-ft parking lot is torn up to install a sidewalk of uniform width around its perimeter. The new area of the parking lot is two thirds of the old area. How wide is the sidewalk?

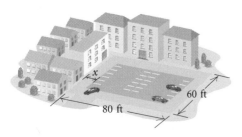

67. *Maximizing Volume.* The Berniers have 24 ft of flexible fencing with which to build a rectangular "toy corral" for a backyard. If the fencing is 2 ft high, what dimensions should the corral have in order to maximize its volume?

68. *Dimensions of a Box.* An open box is made from a 10-cm by 20-cm piece of aluminum by cutting a square from each corner and folding up the edges. The area of the resulting base is 90 cm². What is the length of the sides of the squares?

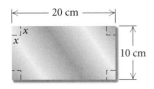

For each pair of functions:

a) *Find the domain of f, g, f + g, f − g, fg, f/g, f ∘ g, and g ∘ f.*

b) *Find (f + g)(x), (f − g)(x), fg(x), (f/g)(x), (f ∘ g)(x), and (g ∘ f)(x).*

69. $f(x) = \dfrac{4}{x^2}$; $g(x) = 3 - 2x$

70. $f(x) = 3x^2 + 4x$; $g(x) = 2x - 1$

71. Given the total-revenue and total-cost functions
$$R(x) = 120x - 0.5x^2 \quad \text{and} \quad C(x) = 15x + 6,$$
find total profit $P(x)$.

Find f(x) and g(x) such that h(x) = (f ∘ g)x:

72. $h(x) = \sqrt{5x + 2}$ **73.** $h(x) = 4(5x - 1)^2 + 9$

74. Find the inverse of the relation
$$\{(1.3, -2.7), (8, -3), (-5, 3), (6, -3), (7, -5)\}.$$

75. Find an equation of the inverse relation.
   a) $y = 3x^2 + 2x - 1$
   b) $0.8x^3 - 5.4y^2 = 3x$

Given each function:

a) *Determine whether it is one-to-one, using a grapher if desired.*

b) *If it is one-to-one, find a formula for the inverse.*

76. $f(x) = \sqrt{x - 6}$ **77.** $f(x) = x^3 - 8$

78. $f(x) = 3x^2 + 2x - 1$

79. Find $f(f^{-1}(657))$: $f(x) = \dfrac{4x^5 - 16x^{37}}{119x}$, $x > 0$.

Synthesis

80. ◆ Given that $f(x) = 4x^3 - 2x + 7$, find
   a) $f(x) + 2$;   **b)** $f(x + 2)$;   **c)** $f(x) + f(2)$.
   Then discuss how each expression differs from the other.

81. ◆ Given the graph of $y = f(x)$, explain and contrast the effect of the constant c on the graphs of $y = f(cx)$ and $y = cf(x)$.

82. ◆ Describe the difference between $f^{-1}(x)$ and $[f(x)]^{-1}$.

83. Use only a grapher. Determine whether the following functions are inverses of each other:
$$f(x) = \frac{4 + 3x}{x - 2}, \qquad g(x) = \frac{x + 4}{x - 3}.$$

Find the domain.

84. $f(x) = \dfrac{\sqrt{1 - x}}{x - |x|}$ **85.** $f(x) = (x - 9x^{-1})^{-1}$

86. Prove that the sum of two odd functions is odd.

87. Describe how the graph of $y = -f(-x)$ is obtained from the graph of $y = f(x)$.

The Trigonometric Functions 2

APPLICATION

Musical instruments can generate complex sine waves. Consider two tones whose graphs are $y = 2 \sin x$ and $y = \sin 2x$. The combination of the two tones produces a new sound whose graph is $y = 2 \sin x + \sin 2x$.

X	Y₁	Y₂
0	0	0
.7854	1.4142	1
1.5708	2	0
2.3562	1.4142	−1
3.1416	0	0
3.927	−1.414	1
4.7124	−2	0

X = 3.1416

$y_1 = 2 \sin x, \quad y_2 = \sin 2x,$
$y_3 = 2 \sin x + \sin 2x$

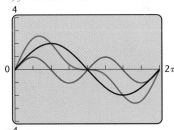

We now consider an important class of functions called *trigonometric*, or *circular*, *functions*. Historically, these functions arose from a study of triangles; hence the name trigonometric. We will begin our study with right triangles and degree measure and solve applications involving right triangles. Then we will consider trigonometric functions of angles or rotations of any size with both degree and radian measure. A circle of radius 1 (a unit circle) is then used to define the six basic trigonometric functions; hence the name circular functions. The domains and ranges of these functions consist of real numbers.

115

2.1

Trigonometric Functions of Acute Angles

- *Determine the six trigonometric ratios for a given acute angle of a right triangle.*
- *Determine the trigonometric function values of 30°, 45°, and 60°.*
- *Using a grapher, find function values for any acute angle, and given a function value of an acute angle, find the angle.*
- *Given the function values of an acute angle, find the function values of its complement.*

The Trigonometric Ratios

We begin our study of trigonometry by considering right triangles and acute angles measured in degrees. Recall that an **acute angle** is an angle that measures between 0° and 90°. Greek letters such as α (alpha), β (beta), γ (gamma), θ (theta), and ϕ (phi) are used to denote an angle. Consider a right triangle with one of its acute angles labeled θ. The side opposite the right angle is called the **hypotenuse**. The other sides of the triangle are referenced by their position relative to the acute angle θ. One side is opposite θ and one is adjacent to θ.

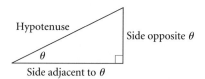

The *lengths* of the sides of the triangle are used to define the six trigonometric ratios:

sine (sin),	cosecant (csc),
cosine (cos),	secant (sec),
tangent (tan),	cotangent (cot).

The **sine of θ** is the *length* of the side opposite θ divided by the *length* of the hypotenuse:

$$\sin \theta = \frac{\text{length of side opposite } \theta}{\text{length of hypotenuse}}.$$

The ratio depends on angle θ and thus is a function of θ. The notation $\sin \theta$ actually means $\sin (\theta)$, where sin, or sine, is the name of the function.

The **cosine of θ** is the *length* of the side adjacent to θ divided by the *length* of the hypotenuse:

$$\cos \theta = \frac{\text{length of side adjacent to } \theta}{\text{length of hypotenuse}}.$$

The six trigonometric ratios, or trigonometric functions, are defined as follows.

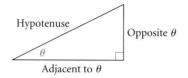

Trigonometric Function Values of an Acute Angle θ

Let θ be an acute angle of a right triangle. Then the six trigonometric functions of θ are as follows:

$$\sin \theta = \frac{\text{side opposite } \theta}{\text{hypotenuse}}, \qquad \csc \theta = \frac{\text{hypotenuse}}{\text{side opposite } \theta},$$

$$\cos \theta = \frac{\text{side adjacent to } \theta}{\text{hypotenuse}}, \qquad \sec \theta = \frac{\text{hypotenuse}}{\text{side adjacent to } \theta},$$

$$\tan \theta = \frac{\text{side opposite } \theta}{\text{side adjacent to } \theta}, \qquad \cot \theta = \frac{\text{side adjacent to } \theta}{\text{side opposite } \theta}.$$

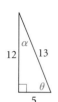

Example 1 In the right triangle shown at left, find the six trigonometric function values of (a) θ and (b) α.

SOLUTION We use the definitions.

a) $\sin \theta = \dfrac{\text{opp}}{\text{hyp}} = \dfrac{12}{13}, \qquad \csc \theta = \dfrac{\text{hyp}}{\text{opp}} = \dfrac{13}{12},$

$\cos \theta = \dfrac{\text{adj}}{\text{hyp}} = \dfrac{5}{13}, \qquad \sec \theta = \dfrac{\text{hyp}}{\text{adj}} = \dfrac{13}{5},$

$\tan \theta = \dfrac{\text{opp}}{\text{adj}} = \dfrac{12}{5}, \qquad \cot \theta = \dfrac{\text{adj}}{\text{opp}} = \dfrac{5}{12}$

The references to opposite, adjacent, and hypotenuse are relative to θ.

b) $\sin \alpha = \dfrac{\text{opp}}{\text{hyp}} = \dfrac{5}{13}, \qquad \csc \alpha = \dfrac{\text{hyp}}{\text{opp}} = \dfrac{13}{5},$

$\cos \alpha = \dfrac{\text{adj}}{\text{hyp}} = \dfrac{12}{13}, \qquad \sec \alpha = \dfrac{\text{hyp}}{\text{adj}} = \dfrac{13}{12},$

$\tan \alpha = \dfrac{\text{opp}}{\text{adj}} = \dfrac{5}{12}, \qquad \cot \alpha = \dfrac{\text{adj}}{\text{opp}} = \dfrac{12}{5}$

The references to opposite, adjacent, and hypotenuse are relative to α.

In Example 1(a), we note that the value of $\sin \theta$, $\frac{12}{13}$, is the reciprocal of $\frac{13}{12}$, the value of $\csc \theta$. Likewise, we see the same reciprocal relationship between the values of $\cos \theta$ and $\sec \theta$ and between the values of $\tan \theta$ and $\cot \theta$. For any angle, the cosecant, secant, and cotangent values are the reciprocals of the sine, cosine, and tangent function values, respectively.

Reciprocal Functions

$$\csc \theta = \frac{1}{\sin \theta}, \qquad \sec \theta = \frac{1}{\cos \theta}, \qquad \cot \theta = \frac{1}{\tan \theta}$$

If we know the values of the sine, cosine, and tangent functions of an angle, we can use these reciprocal relationships to find the values of the cosecant, secant, and cotangent functions of that angle.

Example 2 Given that $\sin \phi = \frac{4}{5}$, $\cos \phi = \frac{3}{5}$, and $\tan \phi = \frac{4}{3}$, find $\csc \phi$, $\sec \phi$, and $\tan \phi$.

SOLUTION   Using the reciprocal relationships, we have

$$\csc \phi = \frac{1}{\sin \phi} = \frac{1}{\frac{4}{5}} = \frac{5}{4}, \qquad \sec \phi = \frac{1}{\cos \phi} = \frac{1}{\frac{3}{5}} = \frac{5}{3},$$

and

$$\cot \phi = \frac{1}{\tan \phi} = \frac{1}{\frac{4}{3}} = \frac{3}{4}.$$

Triangles are said to be **similar** if their corresponding angles have the *same* measure. In similar triangles, the lengths of corresponding sides are in the same ratio. The following right triangles are similar—note that the corresponding angles are equal and the length of each side of the second triangle is four times the length of the corresponding side of the first triangle.

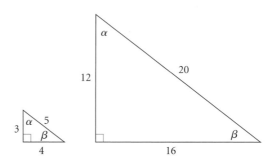

Let's observe the sine, cosine, and tangent values of β in each triangle. Can we expect these values to be the same?

FIRST TRIANGLE	SECOND TRIANGLE
$\sin \beta = \dfrac{3}{5}$	$\sin \beta = \dfrac{12}{20} = \dfrac{3}{5}$
$\cos \beta = \dfrac{4}{5}$	$\cos \beta = \dfrac{16}{20} = \dfrac{4}{5}$
$\tan \beta = \dfrac{3}{4}$	$\tan \beta = \dfrac{12}{16} = \dfrac{3}{4}$

For the two triangles, the corresponding values of $\sin \beta$, $\cos \beta$, and $\tan \beta$ are the same. The lengths of the sides are proportional—thus the *ratios* are the same. This must be the case because in order for the sine, cosine, and tangent to be functions, there must be only one output (the ratio) for each input (the angle β).

> The trigonometric function values of θ depend only on the size of the angle, not on the size of the triangle.

The Six Functions Related

We can find all six trigonometric function values of an acute angle when one of the function-value ratios is known.

Example 3 If $\sin \beta = \frac{6}{7}$ and β is an acute angle, find the other five trigonometric function values of β.

SOLUTION We know from the definition of the sine function that the ratio

$$\frac{6}{7} \quad \text{is} \quad \frac{\text{opp}}{\text{hyp}}.$$

Using this information, let's consider a right triangle in which the hypotenuse has length 7 and the side opposite β has length 6. To find the length of the side adjacent to β, we recall the *Pythagorean theorem*:

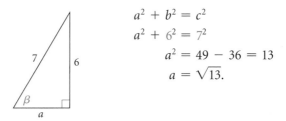

$$a^2 + b^2 = c^2$$
$$a^2 + 6^2 = 7^2$$
$$a^2 = 49 - 36 = 13$$
$$a = \sqrt{13}.$$

We now use the lengths of the three sides to find the other five ratios:

$$\sin \beta = \frac{6}{7}, \qquad\qquad\qquad \csc \beta = \frac{7}{6},$$

$$\cos \beta = \frac{\sqrt{13}}{7}, \qquad\qquad \sec \beta = \frac{7}{\sqrt{13}}, \quad \text{or} \quad \frac{7\sqrt{13}}{13},$$

$$\tan \beta = \frac{6}{\sqrt{13}}, \quad \text{or} \quad \frac{6\sqrt{13}}{13}, \qquad \cot \beta = \frac{\sqrt{13}}{6}.$$

Function Values of 30°, 45°, and 60°

In Examples 1 and 3, we found the trigonometric function values of an acute angle of a right triangle when the lengths of the three sides were known. In most situations, we are asked to find the function values when the measure of the acute angle is given. For certain special angles such as 30°, 45°, and 60°, which are frequently seen in applications, we can use geometry to determine the function values.

A right triangle with a 45° angle actually has two 45° angles. Thus the triangle is *isosceles*, and the legs are the same length. Let's consider such a triangle whose legs have length 1. Then if its hypotenuse has length c, we can find that length using the Pythagorean theorem as follows:

$$1^2 + 1^2 = c^2, \quad \text{or} \quad c^2 = 2, \quad \text{or} \quad c = \sqrt{2}.$$

Such a triangle is shown below. From this diagram, we can easily determine the trigonometric function values of 45°.

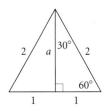

$$\sin 45° = \frac{\text{opp}}{\text{hyp}} = \frac{1}{\sqrt{2}} = \frac{\sqrt{2}}{2} \approx 0.7071,$$

$$\cos 45° = \frac{\text{adj}}{\text{hyp}} = \frac{1}{\sqrt{2}} = \frac{\sqrt{2}}{2} \approx 0.7071,$$

$$\tan 45° = \frac{\text{opp}}{\text{adj}} = \frac{1}{1} = 1$$

It is sufficient to find only the function values of the sine, cosine, and tangent, since the others are their reciprocals.

It is also possible to determine the function values of 30° and 60°. A right triangle with 30° and 60° acute angles is half of an equilateral triangle, as shown in the following figure. Thus if we choose an equilateral triangle whose sides have length 2 and take half of it, we obtain a right triangle that has a hypotenuse of length 2 and a leg of length 1. The other leg has length a, which can be found as follows:

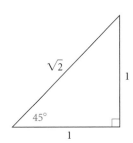

$$a^2 + 1^2 = 2^2$$
$$a^2 = 3$$
$$a = \sqrt{3}.$$

We can now determine the function values of 30° and 60°:

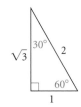

$$\sin 30° = \frac{1}{2} = 0.5, \qquad\qquad \sin 60° = \frac{\sqrt{3}}{2} \approx 0.8660,$$

$$\cos 30° = \frac{\sqrt{3}}{2} \approx 0.8660, \qquad\qquad \cos 60° = \frac{1}{2} = 0.5,$$

$$\tan 30° = \frac{1}{\sqrt{3}} = \frac{\sqrt{3}}{3} \approx 0.5774, \qquad\qquad \tan 60° = \frac{\sqrt{3}}{1} = \sqrt{3} \approx 1.7321$$

Since we will often use the function values of 30°, 45°, and 60°, either the triangles that yield them or the values themselves should be memorized.

Let's now use what we have learned about trigonometric functions of special angles to solve problems. We will consider such applications in greater detail in Section 2.2.

Example 4 *Height of a Hot-air Balloon.* A hot-air balloon ground crew drove 1.2 mi to an observation station as soon as the balloon began to rise. The initial observation estimated the angle between the ground and the line of sight to be 30°. Approximately how high was the balloon at that point? (We are assuming that the wind velocity was low and that the balloon rose vertically for the first few minutes.)

SOLUTION We begin with a sketch of the situation.

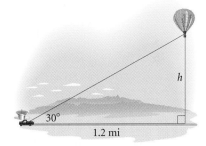

We know the measure of an acute angle and the length of its adjacent side. Since we want to determine the length of the opposite side, we can use the tangent ratio, or the cotangent ratio. Here we use the tangent ratio.

$$\tan 30° = \frac{\text{opp}}{\text{adj}} = \frac{h}{1.2}$$

$$1.2 \tan 30° = h$$

$$1.2\left(\frac{\sqrt{3}}{3}\right) = h \qquad \text{Substituting}$$

$$0.4\sqrt{3} = h$$

$$0.7 \approx h$$

The balloon is approximately 0.7 mi, or 3696 ft, high.

Function Values of Any Acute Angle

Historically, the measure of an angle has been expressed in degrees, minutes, and seconds. One minute, denoted 1′, is such that 60′ = 1°, or 1′ = $\frac{1}{60} \cdot (1°)$. One second, denoted 1″, is such that 60″ = 1′, or 1″ = $\frac{1}{60} \cdot (1')$. Then 61 degrees, 27 minutes, 4 seconds could be written as 61°27′4″. This D°M′S″ form was common before the widespread use of graphers. Now the preferred notation is to express fractional parts of degrees in *decimal degree form*. Although the D°M′S″ notation is still widely used in navigation, we will most often use the decimal form in this text.

Most graphers can convert D°M′S″ notation to decimal degree notation and vice versa. This is often accomplished using a feature in the ANGLE menu. Procedures among graphers vary.

Example 5   Convert 5°42′30″ to decimal degree notation.

SOLUTION   Using a feature in the ANGLE menu on a grapher (TI-82), we enter 5°42′30″ as 5′42′30′. Pressing ENTER gives us

$$5°42′30″ \approx 5.71°,$$

rounded to the nearest hundredth of a degree.   ▬

Example 6   Convert 72.18° to D°M′S″ notation.

SOLUTION   Using a grapher, we enter 72.18 and access the ▶DMS feature in the ANGLE menu. The result is

$$72.18° = 72°10′48″.$$ ▬

So far we have measured angles using degrees. Another useful unit for angle measure is the radian, which we will study in Section 2.4. Graphing calculators work with either degrees or radians. Be sure to use whichever mode is appropriate. In this section, we use the degree mode.

Keep in mind the difference between an exact answer and an approximation. For example,

$$\sin 60° = \frac{\sqrt{3}}{2}.$$ **This is exact!**

But on a grapher, you might get an answer like

$$\sin 60° \approx 0.8660254038.$$ **This is approximate!**

Graphers usually provide values only of the sine, cosine, and tangent functions. You can find values of the cosecant, secant, and cotangent by taking reciprocals of the sine, cosine, and tangent functions, respectively.

Example 7   Find the trigonometric function value, rounded to four decimal places, of each of the following.

a) tan 29.7°   **b)** sec 48°   **c)** sin 84°10′39″

SOLUTION

a) We check to be sure that the grapher is in degree mode. Some graphers require you to first enter 29.7 and then press the TAN key. Others require you to precede the number with the function key. The function value is

$$\tan 29.7° \approx 0.5703899297$$

$$\approx 0.5704.$$ Rounded to four decimal places

b) The secant function value can be found by taking the reciprocal of the cosine function value:

$$\sec 48° = \frac{1}{\cos 48°} \approx 1.49447655 \approx 1.4945.$$

c) Using a feature in the ANGLE menu on a grapher, we enter sin 84°10′39″ as sin 84′10′39′. Pressing ENTER gives us

$$\sin 84°10′39″ \approx 0.9948409474 \approx 0.9948.$$

We can use a TABLE feature on a grapher to find an angle for which we know a trigonometric function value.

Example 8 Find the acute angle, to the nearest tenth of a degree, whose sine value is approximately 0.20113.

SOLUTION With a grapher set in degree mode, we first enter the equation $y = \sin x$. With a minimum value of 0 and a step-value of 0.1, we scroll through the table of values looking for the y_1 value closest to 0.20113.

X	Y₁		
11.1	.19252		
11.2	.19423		
11.3	.19595		
11.4	.19766		
11.5	.19937		
11.6	**.20108**	←	— sin 11.6° ≈ 0.20108
11.7	.20279		
X = 11.6			

We find that 11.6° is the angle whose sine value is about 0.20113.

The quickest way to find the angle with a grapher is to use an inverse function key. (We first studied inverse functions in Section 1.8 and will consider inverse *trigonometric* functions in Section 3.4.) Usually two keys must be pressed in sequence. For this example, if we press

2nd   SIN .20113 ENTER ,

we find that the acute angle whose sine is 0.20113 is approximately 11.60304613°, or 11.6°.

Example 9 *Ladder Safety.* A paint crew has purchased new 30-ft extension ladders. The manufacturer states that the safest placement on a wall is to extend the ladder to 25 ft and to position the base $6\frac{1}{2}$ ft from the wall. (*Source:* R. D. Werner Co., Inc., Corporate Headquarters: P.O. Box 580, Greenville, PA 16125) What angle does this safest position determine with the ground?

SOLUTION We draw a diagram and then use the most convenient trigonometric function.

From the definition of the cosine function, we have

$$\cos \theta = \frac{6.5 \text{ ft}}{25 \text{ ft}} = 0.26.$$

Using a grapher, we find that

$$\theta \approx 74.92993786°.$$

Thus the ladder is at its safest position when it makes an angle of about 75° with the ground.   ▬

Cofunctions and Complements

We recall that two angles are **complementary** whenever the sum of their measures is 90°. Each is the complement of the other. In a right triangle, the acute angles are complementary, since the sum of all three angle measures is 180° and the right angle accounts for 90° of this total. Thus if one acute angle of a right triangle is θ, the other is $90° - \theta$.

Interactive Discovery

Find the six trigonometric function values of the acute angles given in this triangle. Note that 53° and 37° are complementary since $53° + 37° = 90°$.

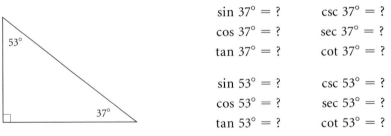

sin 37° = ?	csc 37° = ?
cos 37° = ?	sec 37° = ?
tan 37° = ?	cot 37° = ?
sin 53° = ?	csc 53° = ?
cos 53° = ?	sec 53° = ?
tan 53° = ?	cot 53° = ?

Try this with acute angles 20.3° and 69.7° as well. What pattern do you observe? Look for this same pattern in Example 1 earlier in this section.

Note that the sine of an angle is also the cosine of the angle's complement. Similarly, the tangent of an angle is the cotangent of the angle's complement, and the secant of an angle is the cosecant of the angle's complement. These pairs of functions are called **cofunctions**. A list of cofunction identities is as follows.

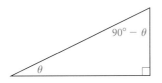

> **Cofunction Identities**
>
> $$\sin \theta = \cos (90° - \theta), \qquad \cos \theta = \sin (90° - \theta),$$
> $$\tan \theta = \cot (90° - \theta), \qquad \cot \theta = \tan (90° - \theta),$$
> $$\sec \theta = \csc (90° - \theta), \qquad \csc \theta = \sec (90° - \theta)$$

Example 10 Given that $\sin 18° \approx 0.3090$, $\cos 18° \approx 0.9511$, and $\tan 18° \approx 0.3249$, find the six trigonometric function values of 72°.

SOLUTION Using reciprocal relationships, we know that

$$\csc 18° = \frac{1}{\sin 18°} \approx 3.2361, \qquad \sec 18° = \frac{1}{\cos 18°} \approx 1.0515,$$

and $\quad \cot 18° = \dfrac{1}{\tan 18°} \approx 3.0777.$

Since 72° and 18° are complementary, we have

$$\sin 72° = \cos 18° \approx 0.9511, \qquad \cos 72° = \sin 18° \approx 0.3090,$$
$$\tan 72° = \cot 18° \approx 3.0777, \qquad \cot 72° = \tan 18° \approx 0.3249,$$
$$\sec 72° = \csc 18° \approx 3.2361, \qquad \csc 72° = \sec 18° \approx 1.0515.$$

2.1 Exercise Set

Find the six trigonometric function values of the specified angle.

1.
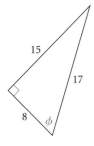

2.

3.

4.
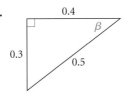

5.

6.

Given a function value of an acute angle, find the other five trigonometric function values.

7. $\sin \theta = \frac{24}{25}$ **8.** $\cos \theta = 0.7$

9. $\tan \phi = 2$ **10.** $\cot \theta = \frac{1}{3}$

11. $\csc \theta = 1.5$ **12.** $\sec \beta = \sqrt{17}$

Find the exact function value.

13. cos 45°

14. tan 30°

15. sec 60°

16. sin 45°

17. cot 60°

18. csc 45°

19. sin 30°

20. cos 60°

21. *Distance Between Bases.*   A baseball diamond is really a square 90 ft on a side. If a line is drawn from third base to first base, then a right triangle *QPH* is formed, where ∠*QPH* is 45°. Using a trigonometric function, find the distance from third base to first base.

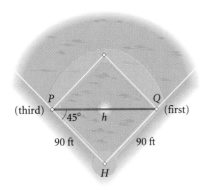

22. *Distance Across a River.*   Find the distance *a* across the river.

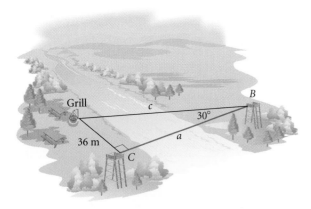

Convert to decimal degree notation. Round to two decimal places.

23. 9°43′

24. 52°15′

25. 35°50′

26. 64°53′

27. 3°2′

28. 19°47′23″

29. 49°38′46″

30. 76°11′34″

Convert to degrees, minutes, and seconds. Round to the nearest second.

31. 17.6°

32. 20.14°

33. 83.025°

34. 67.84°

35. 11.75°

36. 29.8°

37. 47.8268°

38. 0.253°

Find each function value. Round to four decimal places.

39. cos 51°

40. cot 17°

41. tan 4°13′

42. sin 26.1°

43. sec 38.43°

44. cos 74°10′40″

45. cos 40.35°

46. csc 45.2°

47. sin 69°

48. tan 63°48′

49. tan 85.4°

50. cos 4°

51. csc 89.5°

52. sec 35.28°

53. cot 30°25′6″

54. sin 59.2°

Using a grapher, find the acute angle θ, to the nearest tenth of a degree, for the given function value.

55. sin θ = 0.5125

56. tan θ = 2.032

57. tan θ = 0.2226

58. cos θ = 0.3842

59. sin θ = 0.9022

60. tan θ = 3.056

61. cos θ = 0.6879

62. sin θ = 0.4005

63. cot θ = 2.127

64. csc θ = 1.147

$$\left(Hint: \tan \theta = \frac{1}{\cot \theta}. \right)$$

65. sec θ = 1.279

66. cot θ = 1.351

Find the exact acute angle θ for the given function value.

67. $\sin \theta = \dfrac{\sqrt{2}}{2}$

68. $\cot \theta = \dfrac{\sqrt{3}}{3}$

69. $\cos \theta = \dfrac{1}{2}$

70. $\sin \theta = \dfrac{1}{2}$

71. tan θ = 1

72. $\cos \theta = \dfrac{\sqrt{3}}{2}$

Complete.

73. $\cos 20° = \underline{\hspace{1cm}} 70° = \dfrac{1}{\underline{\hspace{0.7cm}} 20°}$

74. $\sin 64° = \underline{\hspace{1cm}} 26° = \dfrac{1}{\underline{\hspace{0.7cm}} 64°}$

75. $\tan 52° = \cot \underline{\hspace{1cm}} = \dfrac{1}{\underline{\hspace{0.7cm}} 52°}$

76. $\sec 13° = \csc \underline{\hspace{1cm}} = \dfrac{1}{\underline{\hspace{0.7cm}} 13°}$

77. Given that

sin 65° ≈ 0.9063,   cos 65° ≈ 0.4226,
tan 65° ≈ 2.145,   cot 65° ≈ 0.4663,
sec 65° ≈ 2.366,   csc 65° ≈ 1.103,

find the six function values of 25°.

78. Given that sin 38.7° ≈ 0.6252, cos 38.7° ≈ 0.7804, and tan 38.7° ≈ 0.8012, find the six function values of 51.3°.

Skill Maintenance

Solve. Round to two decimal places.

79. $0.1284 = \dfrac{d}{11.8}$

80. $0.6045 = \dfrac{30}{h}$

81. $5(x + 20.5) = 1.8x$

82. $\dfrac{3}{4} = \dfrac{t}{0.42}$

83. Use the Pythagorean theorem to find the missing length. Round the answer to the nearest tenth.

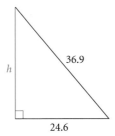

Synthesis

84. ◆ Explain the difference between two functions being reciprocal functions and two functions being cofunctions.

85. ◆ Explain why it is not necessary to memorize the function values for both 30° and 60°.

86. Given that cos 15.17° = 0.9651, find csc 74.83°.

87. Given that sec 49.2° = 1.5304, find sin 40.8°.

88. Find the six trigonometric function values of α.

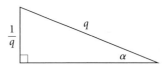

89. Show that the area of this right triangle is $\frac{1}{2}bc \sin A$.

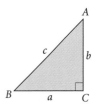

90. Show that the area of this triangle is $\frac{1}{2}ab \sin \theta$.

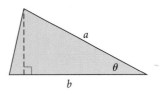

2.2
Applications of Right Triangles

• *Solve right triangles.*
• *Solve applied problems involving right triangles and trigonometric functions.*

Solving Right Triangles

Now that we can find function values for any acute angle, it is possible to *solve* right triangles. To **solve** a triangle means to find the lengths of all sides and the measures of all angles not already known. Triangles that are not right triangles will be studied later.

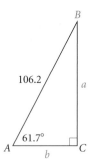

Example 1 In △ABC (shown at left), find a, b, and B, where a and b represent lengths of sides and B represents the measure of $\angle B$.

SOLUTION   In △ABC, we know three of the measures:

$$A = 61.7°, \qquad a = ?,$$
$$B = ?, \qquad b = ?,$$
$$C = 90°, \qquad c = 106.2.$$

Since the sum of the angle measures of any triangle is 180° and $C = 90°$, the sum of A and B is 90°. Thus,

$$B = 90° - A = 90° - 61.7° = 28.3°.$$

We are given an acute angle and the hypotenuse. This suggests that we can use the sine and cosine ratios to find a and b, respectively:

$$\sin 61.7° = \frac{\text{opp}}{\text{hyp}} = \frac{a}{106.2} \quad \text{and} \quad \cos 61.7° = \frac{\text{adj}}{\text{hyp}} = \frac{b}{106.2}.$$

Solving for a and b, we get

$$a = 106.2 \sin 61.7° \quad \text{and} \quad b = 106.2 \cos 61.7°$$
$$a \approx 93.5 \qquad\qquad\qquad b \approx 50.3.$$

Thus,

$$A = 61.7°, \qquad a \approx 93.5,$$
$$B = 28.3°, \qquad b \approx 50.3,$$
$$C = 90°, \qquad c = 106.2.$$

—

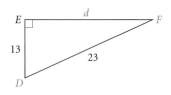

Example 2 In △DEF (shown at left), find D and F. Then find d.

SOLUTION   In △DEF, we know three of the measures:

$$D = ?, \qquad d = ?,$$
$$E = 90°, \qquad e = 23,$$
$$F = ?, \qquad f = 13.$$

We know the side adjacent to D and the hypotenuse. This suggests the use of the cosine ratio:

$$\cos D = \frac{\text{adj}}{\text{hyp}} = \frac{13}{23}.$$

We now use either the TABLE feature or an inverse function key to find the angle whose cosine is $\frac{13}{23}$. To the nearest hundredth of a degree,

$$D \approx 55.58°. \qquad \text{Pressing} \boxed{\text{2nd}} \boxed{\text{cos}} (13/23) \boxed{\text{ENTER}}$$

Since the sum of D and F is 90°, we can find F by subtracting:

$$F = 90° - D \approx 90° - 55.58° \approx 34.42°.$$

To find d, we can use either the Pythagorean theorem or a trigonometric function. We could use $\cos F$, $\sin D$, or the tangent or cotangent

Program

TRISOL90: This program solves right triangles.

ratios for either D or F. Let's use tan D:

$$\tan D = \frac{\text{opp}}{\text{adj}} = \frac{d}{13}, \quad \text{or} \quad \tan 55.58° \approx \frac{d}{13}.$$

Then

$$d \approx 13 \tan 55.58° \approx 19.$$

The six measures are

$$D \approx 55.58°, \qquad d \approx 19,$$
$$E = 90°, \qquad e = 23,$$
$$F \approx 34.42°, \qquad f = 13.$$

Applications

Right triangles can be used to model many applications in the real world. To solve such problems, we solve, or at least partially solve, the triangle.

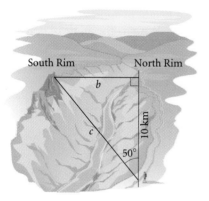

Example 3 *Hiking at the Grand Canyon.* A backpacker hiking east along the north rim of the Grand Canyon notices an unusual rock formation directly across the canyon. She decides to continue watching the landmark while hiking along the rim. In 2 hr, she has gone 10 km due east and the landmark is still visible but at approximately a 50° angle to the north rim. (See the figure at left.)

a) How many kilometers is she from the rock formation?

b) How far is it across the canyon at her starting point?

SOLUTION

a) We know the side adjacent to the 50° angle and want to find the hypotenuse. We can use the cosine ratio:

$$\cos 50° = \frac{10 \text{ km}}{c}$$

$$c = \frac{10 \text{ km}}{\cos 50°} \approx 15.6 \text{ km}.$$

After hiking for 10 km, she is approximately 15.6 km from the rock formation.

b) We know the side adjacent to the 50° angle and want to find the opposite side. We can use the tangent ratio:

$$\tan 50° = \frac{b}{10 \text{ km}}$$

$$b = 10 \text{ km} \cdot \tan 50° \approx 11.9 \text{ km}.$$

Thus it is approximately 11.9 km across the canyon at her starting point.

Many applications with right triangles involve an *angle of elevation* or an *angle of depression*. The angle between the horizontal and a line of sight above the horizontal is called an **angle of elevation**. The angle be-tween the horizontal and a line of sight below the horizontal is called an **angle of depression**. For example, suppose that you are looking straight ahead and then you move your eyes up to look at an approaching air-plane. The angle that your eyes pass through is an angle of elevation. If the pilot of the plane is looking forward and then looks down, the pilot's eyes pass through an angle of depression.

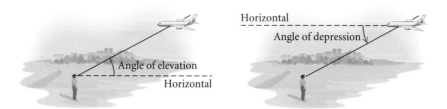

Example 4   *Aerial Photography.*   An aerial photographer who photo-graphs farm properties for a real estate company has determined from experience that the best photo is taken at a height of approximately 475 ft and a distance of 850 ft from the farmhouse. What is the angle of depres-sion from the plane to the house?

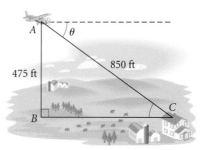

SOLUTION   When parallel lines are cut by a transversal, alternate interior angles are equal. Thus the angle of depression from the plane to the house is equal to the angle of elevation from the house to the plane, so we can use the right triangle shown in the figure. Since we know the side opposite $\angle C$ and the hypotenuse, we can find θ by using the sine function:

$$\sin \theta = \sin C = \frac{475 \text{ ft}}{850 \text{ ft}}.$$

Using the inverse sine key, we find that

$\theta \approx 34°.$ **Pressing** ⎡2nd⎤ ⎡sin⎤ (475/850) ⎡ENTER⎤

Thus the angle of depression is approximately 34°.   ▬

Example 5   *Cloud Height.*   To measure cloud height at night, a vertical beam of light is directed on a spot on the cloud. From a point 135 ft away from the light source, the angle of elevation to the spot is found to be 67.35°. Find the height of the cloud.

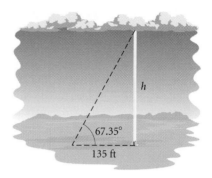

SOLUTION   From the figure, we have

$$\tan 67.35° = \frac{h}{135 \text{ ft}}$$

$$h = 135 \text{ ft} \cdot \tan 67.35° \approx 324 \text{ ft}.$$

The height of the cloud is about 324 ft.

Some applications of trigonometry involve the concept of direction, or bearing. In this text we present two ways of giving direction, the first in Example 6 below and the second in Exercise Set 2.3.

BEARING: FIRST-TYPE   One method of giving direction, or bearing, involves reference to a north–south line using an acute angle. For example, N55°W means 55° west of north and S67°E means 67° east of south.

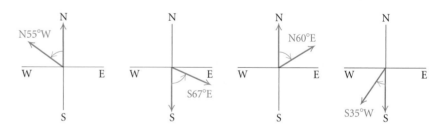

Example 6   *Distance to a Forest Fire.* A forest ranger at point *A* sights a fire directly south. A second ranger at point *B*, 7.5 mi east, sights the same fire at a bearing of S27°23′W. How far from *A* is the fire?

$$B = 90° - 27°23'$$
$$= 89°60' - 27°23'$$
$$= 62°37'$$

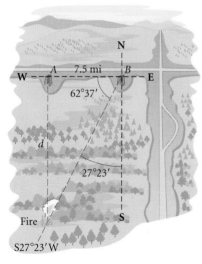

SOLUTION   From the figure at left, we see that the desired distance *d* is part of a right triangle, as shown. We have

$$\frac{d}{7.5 \text{ mi}} = \tan 62°37'$$

$$d = 7.5 \text{ mi} \tan 62°37' \approx 14.5 \text{ mi}.$$

The forest ranger at *A* is about 14.5 mi from the fire.

Example 7   *Comiskey Park.* In the new Comiskey Park, the home of the Chicago White Sox baseball team, the first row of seats in the upper deck is farther away from home plate than the last row of seats in the old Comiskey Park. Although there is no obstructed view in the new park, some of the fans still complain about the present distance from home plate to the upper deck of seats. From a seat in the last row of the upper deck directly behind the batter, the angle of depression to home plate is 29.9° and the angle of depression to the pitcher's mound is 24.2°. Find (a) the viewing distance to home plate and (b) the viewing distance to the pitcher's mound.

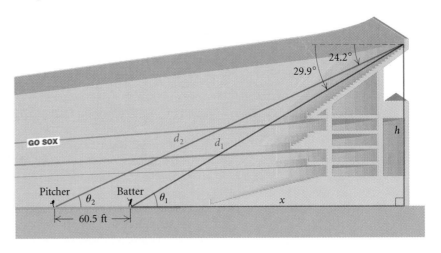

SOLUTION   From geometry we know that $\theta_1 = 29.9°$ and $\theta_2 = 24.2°$. The standard distance from home plate to the pitcher's mound is 60.5 ft. In the drawing, we let d_1 be the viewing distance to home plate, d_2 the viewing distance to the pitcher, h the elevation of the last row, and x the horizontal distance from the batter to a point directly below the seat in the last row of the upper deck.

We begin by determining the distance x. We use the tangent function with $\theta_1 = 29.9°$ and $\theta_2 = 24.2°$:

$$\tan 29.9° = \frac{h}{x} \qquad \text{and} \quad \tan 24.2° = \frac{h}{x + 60.5}$$

or

$$h = x \tan 29.9° \quad \text{and} \qquad h = (x + 60.5) \tan 24.2°.$$

Then substituting $x \tan 29.9°$ for h in the second equation, we obtain

$$x \tan 29.9° = (x + 60.5) \tan 24.2°.$$

Solving for x, we get

$$x \tan 29.9° = x \tan 24.2° + 60.5 \tan 24.2°$$

$$x(\tan 29.9° - \tan 24.2°) = 60.5 \tan 24.2°$$

$$x = \frac{60.5 \tan 24.2°}{\tan 29.9° - \tan 24.2°}$$

$$x \approx 216.5.$$

We can then find d_1 and d_2 using the cosine function:

$$\cos 29.9° = \frac{216.5}{d_1} \quad \text{and} \quad \cos 24.2° = \frac{216.5 + 60.5}{d_2}$$

or

$$d_1 = \frac{216.5}{\cos 29.9°} \quad \text{and} \quad d_2 = \frac{277}{\cos 24.2°}$$

$$d_1 \approx 249.7 \qquad\qquad d_2 \approx 303.7.$$

The distance to home plate is about 250 ft,* and the distance to the pitcher's mound is about 304 ft. ▬

2.2 | Exercise Set

Solve each of the following right triangles.

1.

2.

5.

6.

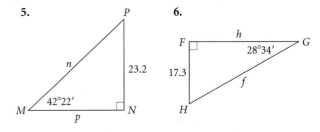

Solve each right triangle. (Standard lettering has been used.)

3.

4.

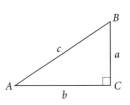

7. $A = 87°43'$, $a = 9.73$ 8. $a = 12.5$, $b = 18.3$

9. $b = 100$, $c = 450$ 10. $B = 56.5°$, $c = 0.0447$

11. $A = 47.58°$, $c = 48.3$ 12. $B = 20.6°$, $a = 7.5$

13. $A = 35°$, $b = 40$ 14. $B = 69.3°$, $b = 93.4$

15. $b = 1.86$, $c = 4.02$ 16. $a = 10.2$, $c = 20.4$

*In the old Comiskey Park, the distance to home plate was only 150 ft (*Chicago Tribune*, September 19, 1993).

17. *Safety Line to Raft.*   Each spring Madison uses his vacation time to ready his lake property for the summer. He wants to run a new safety line from point *B* on the shore to the corner of the anchored diving raft. The current safety line, which runs perpendicular to the shore line to point *A*, is 40 ft long. He estimates the angle from *B* to the corner of the raft to be 50°. Approximately how much rope does he need for the new safety line if he allows 5 ft of rope at each end to fasten the rope?

18. *Enclosing an Area.*   Glynis is enclosing a triangular area in a corner of her fenced rectangular backyard for her Labrador retriever. In order for a certain tree to be included in this pen, one side needs to be 14.5 ft and make a 53° angle with the new side. How long is the new side?

19. *Easel Display.*   A marketing group is designing an easel to display posters advertising their newest products. They want the easel to be 6 ft tall and fit flush against a wall. For optimal eye contact, the best angle between the front and back legs of the easel is 23°. How far from the wall should the front legs be placed in order to obtain this angle?

20. *Height of a Tree.*   A supervisor must train a new team of loggers to estimate the heights of trees. As an example, she walks off 40 ft from the base of a tree and estimates the angle of elevation to the

tree's peak to be 70°. Approximately how tall is the tree?

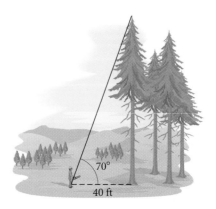

21. *Sand Dunes National Park.*   While visiting the Sand Dunes National Park in Colorado, Ian approximated the angle of elevation to the top of a sand dune to be 20°. After walking 800 ft closer, he guessed that the angle of elevation had increased by 15°. Approximately how tall is the dune he was observing?

22. *Tee-Shirt Design.*   A new tee-shirt design is to have a regular pentagon inscribed in a circle, as shown in the figure. Each side of the pentagon is to be 3.5 in. long. Find the radius of the circumscribed circle.

23. *Inscribed Octagon.*   A regular octagon is inscribed in a circle of radius 15.8 cm. Find the perimeter of the octagon.

24. *Height of a Weather Balloon.* A weather balloon is directly west of two observing stations that are 10 mi apart. The angles of elevation of the balloon from the two stations are 17.6° and 78.2°. How high is the balloon?

25. *Height of a Kite.* For a science fair project, a group of students tested different materials used to construct kites. Their instructor provided an instrument that accurately measures the angle of elevation. In one of the tests, the angle of elevation was 63.4° with 670 ft of string out. Assuming the string was taut, how high was the kite?

26. *Height of a Building.* A window washer on a ladder looks at a nearby building 100 ft away, noting that the angle of elevation of the top of the building is 18.7° and the angle of depression of the bottom of the building is 6.5°. How tall is the nearby building?

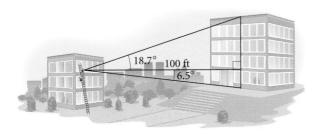

27. *Distance Between Towns.* From a hot-air balloon 2 km high, the angles of depression to two towns, in line with the balloon, are 81.2° and 13.5°. How far apart are the towns?

28. *Angle of Elevation.* What is the angle of elevation of the sun when a 35-ft mast casts a 20-ft shadow?

29. *Distance from a Lighthouse.* From the top of a lighthouse 55 ft above sea level, the angle of depression to a small boat is 11.3°. How far from the foot of the lighthouse is the boat?

30. *Lightning Detection.* In extremely large forests, it is not cost-effective to position forest rangers in towers or to use small aircraft to continually watch for fires. Since lightning is a frequent cause of fire, lightning detectors are now commonly used instead. These devices not only give a bearing on the location but also measure the intensity of the lightning. A detector at point Q is situated 15 mi west of a central fire station at point R. The bearing from Q to where lightning hits due south of R is S37.6°E. How far is the hit from point R?

31. *Lobster Boat.* A lobster boat is situated due west of a lighthouse. A barge is 12 km south of the lobster boat. From the barge, the bearing to the lighthouse is N63°20′E. How far is the lobster boat from the lighthouse?

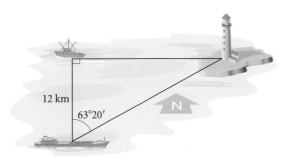

32. *Length of an Antenna.* A vertical antenna is mounted atop a 50-ft pole. From a point on the level ground 75 ft from the base of the pole, the antenna subtends an angle of 10.5°. Find the length of the antenna.

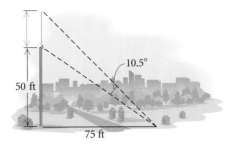

Skill Maintenance

Find the distance between the points.

33. $(8, -2)$ and $(-6, -4)$ **34.** $(-9, 3)$ and $(0, 0)$

Evaluate the function value exactly.

35. $\cos 30°$ **36.** $\sin 60°$

37. $\tan 30°$ **38.** $\sin 45°$

Synthesis

39. ◈ In this section, the trigonometric functions have been defined as functions of acute angles. Thus (0°, 90°) is the domain for each function. What appears to be the range for the sine, the cosine, and the tangent functions considering this limited domain?

40. ◈ Explain in your own words five ways in which length c can be determined in this triangle. Which way seems the most efficient?

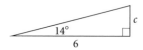

41. Find h, to the nearest tenth.

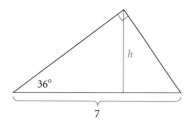

42. Find a, to the nearest tenth.

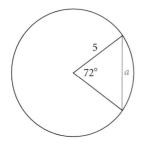

43. *Construction of Picnic Pavilions.* A construction company is mass-producing picnic pavilions for national parks, as shown in the figure. The rafter

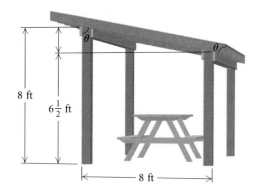

ends are to be sawed in such a way that they will be vertical when in place. The front wall is 8 ft high, the back wall is $6\frac{1}{2}$ ft high, and the distance between walls is 8 ft. At what angle should the rafters be cut?

44. *Diameter of a Pipe.* A V-gauge is used to find the diameter of a pipe. The advantage of such a device is that it is rugged, it is accurate, and it has no moving parts to break down. In the figure, the measure of angle *AVB* is 54°. A pipe is placed in the V-shaped slot and the distance *VP* is used to predict the diameter.

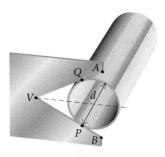

a) Suppose that the diameter of a pipe is 2 cm. What is the distance *VP*?
b) Suppose that the distance *VP* is 3.93 cm. What is the diameter of the pipe?
c) Find a formula for d in terms of *VP*.
d) Find a formula for *VP* in terms of d.

The line *VP* is calibrated by listing as its units the corresponding diameters. This, in effect, establishes a function between *VP* and d.

45. *Sound of an Airplane.* It is common experience to hear the sound of a low-flying airplane, and look at the wrong place in the sky to see the plane. Suppose that a plane is traveling directly at you at a speed of 200 mph and an altitude of 3000 ft, and you hear the sound at what seems to be an angle of inclination of 20°. At what angle θ should you actually look in order to see the plane? Consider the speed of sound to be 1100 ft/sec.

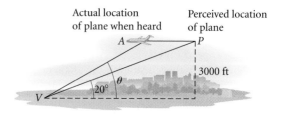

Actual location of plane when heard — Perceived location of plane

46. *Measuring the Radius of the Earth.* One way to measure the radius of the earth is to climb to the top of a mountain whose height above sea level is known and measure the angle between a vertical line to the center of the earth from the top of the mountain and a line drawn from the top of the mountain to the horizon, as shown in the figure. The height of Mt. Shasta in California is 14,162 ft. From the top of Mt. Shasta, one can see the horizon on the Pacific Ocean. The angle formed between a line to the horizon and the vertical is found to be 87°53'. Use this information to estimate the radius of the earth, in miles.

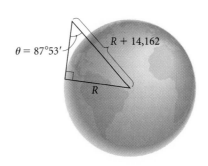

$\theta = 87°53'$

$R + 14,162$

R

2.3
Trigonometric Functions of Any Angle

- *Find angles that are coterminal with a given angle and find the complement and the supplement of a given angle.*
- *Determine the six trigonometric function values for any angle in standard position when the coordinates of a point on the terminal side are given.*
- *Find the function values for any angle whose terminal side lies on an axis.*
- *Find the function values for an angle whose terminal side makes an angle of 30°, 45°, or 60° with the x-axis.*
- *Use a grapher to find function values and angles.*

Angles, Rotations, and Degree Measure

An *angle* is a familiar figure in the world around us.

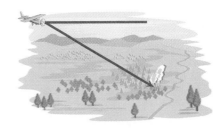

An **angle** is the union of two rays with a common endpoint called the **vertex**. In trigonometry, we often think of an angle as a **rotation**. To do so, think of locating a ray along the positive *x*-axis with its endpoint at the origin. This ray is called the **initial side** of the angle. Though we leave that ray fixed, think of making a copy of it and rotating it. A rotation *counterclockwise* is a **positive rotation**, and a rotation *clockwise* is a **negative rotation**. The ray at the end of the rotation is called the

terminal side of the angle. The angle formed is said to be in **standard position**.

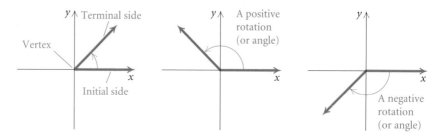

The measure of an angle or rotation may be given in degrees. The Babylonians developed the idea of dividing the circumference of a circle into 360 equal parts, or degrees. If we let the measure of one of these parts be 1°, then one complete positive revolution or rotation has a measure of 360°. One half of a revolution has a measure of 180°, one fourth of a revolution has a measure of 90°, and so on. We can also speak of an angle of measure 60°, 135°, 330°, or 420°. The terminal sides of these angles lie in quadrants I, II, IV, and I, respectively. The negative rotations −30°, −110°, and −225° represent angles with terminal sides in quadrants IV, III, and II, respectively.

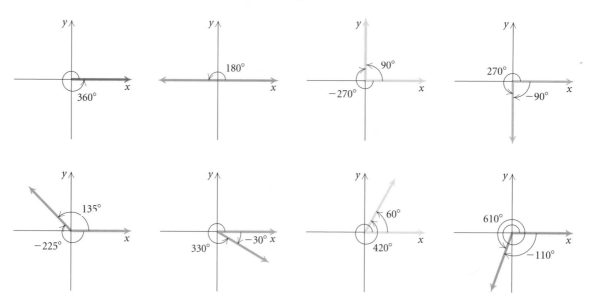

When the measure of an angle is greater than 360° or less than −360°, the rotating ray has gone through more than one complete rotation. If two or more angles have the same terminal side, the angles are said to be **coterminal**. To find angles coterminal with a given angle, we add or subtract multiples of 360°. For example, 420°, shown above, has the same terminal side as 60°, since 420° = 360° + 60°. Thus we say that angles of measure 60° and 420° are coterminal. The negative rotation that measures −300° is also coterminal with 60° because

$60° − 360° = −300°$. Other examples of coterminal angles shown above are

$$90° \text{ and } −270°,$$
$$−90° \text{ and } 270°,$$
$$135° \text{ and } −225°,$$
$$−30° \text{ and } 330°,$$
$$−110° \text{ and } 610°.$$

Example 1   Find two positive and two negative angles that are coterminal with (a) $51°$ and (b) $−7°$.

SOLUTION

a) We add and subtract multiples of $360°$. Many answers are possible.

$$51° + 360° = 411° \qquad 51° + 3(360°) = 1131°$$
$$51° − 360° = −309° \qquad 51° − 2(360°) = −669°$$

Thus angles of measure $411°$, $1131°$, $−309°$, and $−669°$ are coterminal with $51°$.

b) We have the following.

$$−7° + 360° = 353° \qquad −7° + 2(360°) = 713°$$
$$−7° − 360° = −367° \qquad −7° − 10(360°) = −3607°$$

Thus angles of measure $353°$, $713°$, $−367°$, and $−3607°$ are coterminal with $−7°$. ▬

Angles can also be classified by their measures, as seen in the following figure.

Right: $\theta = 90°$

Acute: $0° < \theta < 90°$

Obtuse: $90° < \theta < 180°$

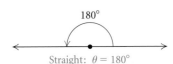

Straight: $\theta = 180°$

Two acute angles are **complementary** if their sum is $90°$. For example, angles that measure $10°$ and $80°$ are complementary because $10° + 80° = 90°$. Two positive angles are **supplementary** if their sum is $180°$. For example, angles that measure $45°$ and $135°$ are supplementary because $45° + 135° = 180°$.

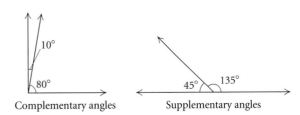

Complementary angles   Supplementary angles

Example 2   Find the complement and the supplement of 71.46°.

SOLUTION   We have

$$90° - 71.46° = 18.54°,$$
$$180° - 71.46° = 108.54°.$$

Thus the complement of 71.46° is 18.54° and the supplement is 108.54°.

Trigonometric Functions of Angles or Rotations

Many applied problems in trigonometry involve the use of angles that are not acute. Thus we need to extend the domains of the trigonometric functions defined in Section 2.1 to angles, or rotations, of any size. To do this, we first consider a right triangle with one vertex at the origin of a coordinate system and one vertex on the positive *x*-axis. The other vertex is at *P*, a point on the circle whose center is at the origin and whose radius *r* is the length of the hypotenuse of the triangle. This triangle is a **reference triangle** for angle *θ*, which is in standard position. Note that *y* is the length of the side opposite *θ* and *x* is the length of the side adjacent to *θ*.

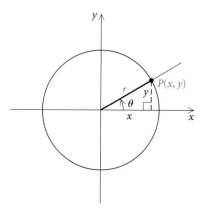

Recalling the definitions in Section 2.1, we note that three of the trigonometric functions of angle *θ* are defined as follows:

$$\sin \theta = \frac{\text{opp}}{\text{hyp}} = \frac{y}{r}, \qquad \cos \theta = \frac{\text{adj}}{\text{hyp}} = \frac{x}{r}, \qquad \tan \theta = \frac{\text{opp}}{\text{adj}} = \frac{y}{x}.$$

Since *x* and *y* are the coordinates of the point *P* and the length of the radius is the length of the hypotenuse, we can also define these functions as follows:

$$\sin \theta = \frac{y\text{-coordinate}}{\text{radius}},$$

$$\cos \theta = \frac{x\text{-coordinate}}{\text{radius}},$$

$$\tan \theta = \frac{y\text{-coordinate}}{x\text{-coordinate}}.$$

We will use these definitions for functions of angles of any measure. The following figures show angles whose terminal sides lie in quadrants II, III, and IV.

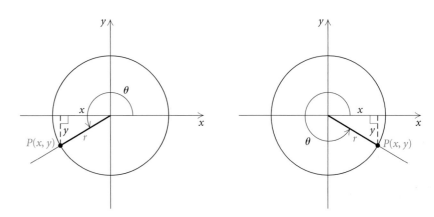

The point P, which is a point other than the vertex on the terminal side of an angle, can be anywhere on a circle of radius r. Each of its two coordinates may be positive, negative, or zero, depending on the location of the terminal side. A reference triangle can be drawn for any angle, as shown. The length of the radius, which is also the length of the hypotenuse of the reference triangle, is always considered positive. The angle is always measured from the positive half of the x-axis. Regardless of the location of P, we have the following definitions.

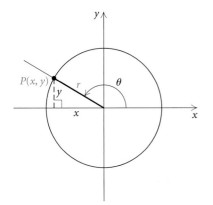

> ### Trigonometric Functions of Any Angle θ
>
> Suppose that $P(x, y)$ is any point on the terminal side of any angle θ in standard position, and r is the radius, or distance, from the origin of $P(x, y)$. Then the trigonometric functions are defined as follows.
>
> $$\sin \theta = \frac{y\text{-coordinate}}{\text{radius}} = \frac{y}{r} \qquad \csc \theta = \frac{\text{radius}}{y\text{-coordinate}} = \frac{r}{y}$$
>
> $$\cos \theta = \frac{x\text{-coordinate}}{\text{radius}} = \frac{x}{r} \qquad \sec \theta = \frac{\text{radius}}{x\text{-coordinate}} = \frac{r}{x}$$
>
> $$\tan \theta = \frac{y\text{-coordinate}}{x\text{-coordinate}} = \frac{y}{x} \qquad \cot \theta = \frac{x\text{-coordinate}}{y\text{-coordinate}} = \frac{x}{y}$$

Function values of the trigonometric functions can be positive, negative, or zero, depending on where the terminal side of the angle lies. The length of the radius is always positive. Thus the signs of the function values depend only on the coordinates of the point P on the terminal side. In the first quadrant, all function values are positive because both coordinates are positive. In the second quadrant, first coordinates are negative and second coordinates are positive; thus only the sine and the cosecant values are positive. Similarly, we can determine the signs, + or −, of the function values in the third and fourth quadrants. Because of the reciprocal relationships, we need to learn only the signs for the sine, cosine, and tangent functions.

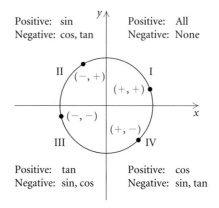

Positive: sin
Negative: cos, tan

Positive: All
Negative: None

Positive: tan
Negative: sin, cos

Positive: cos
Negative: sin, tan

Example 3   Find the six trigonometric function values for each angle shown.

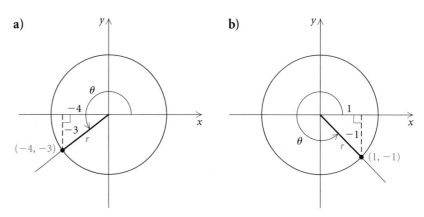

a)

b)

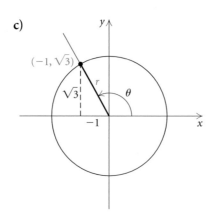

c)

SOLUTION

a) We first determine r, the distance from the origin $(0, 0)$ to the point $(-4, -3)$. The distance between $(0, 0)$ and any point (x, y) on the terminal side is

$$r = \sqrt{(x - 0)^2 + (y - 0)^2}$$
$$= \sqrt{x^2 + y^2}.$$

Substituting -4 for x and -3 for y, we find

$$r = \sqrt{(-4)^2 + (-3)^2}$$
$$= \sqrt{16 + 9} = \sqrt{25} = 5.$$

Using the definitions of the trigonometric functions, we can now find the function values of θ. We substitute -4 for x, -3 for y, and 5 for r:

$$\sin \theta = \frac{y}{r} = \frac{-3}{5} = -0.6, \qquad \csc \theta = \frac{r}{y} = \frac{5}{-3} \approx -1.67,$$

$$\cos \theta = \frac{x}{r} = \frac{-4}{5} = -0.8, \qquad \sec \theta = \frac{r}{x} = \frac{5}{-4} = -1.25,$$

$$\tan \theta = \frac{y}{x} = \frac{-3}{-4} = 0.75, \qquad \cot \theta = \frac{x}{y} = \frac{-4}{-3} \approx 1.33.$$

As expected, the tangent and the cotangent values are positive and the other four are negative. This is true for all angles in quadrant III.

b) We first determine r, the distance from the origin to the point $(1, -1)$:

$$r = \sqrt{1^2 + (-1)^2} = \sqrt{1 + 1} = \sqrt{2}.$$

Substituting 1 for x, -1 for y, and $\sqrt{2}$ for r, we find

$$\sin \theta = \frac{y}{r} = \frac{-1}{\sqrt{2}} = -\frac{\sqrt{2}}{2}, \qquad \csc \theta = \frac{r}{y} = \frac{\sqrt{2}}{-1} = -\sqrt{2},$$

$$\cos \theta = \frac{x}{r} = \frac{1}{\sqrt{2}} = \frac{\sqrt{2}}{2}, \qquad \sec \theta = \frac{r}{x} = \frac{\sqrt{2}}{1} = \sqrt{2},$$

$$\tan \theta = \frac{y}{x} = \frac{-1}{1} = -1, \qquad \cot \theta = \frac{x}{y} = \frac{1}{-1} = -1.$$

c) We determine r, the distance from the origin to the point $(-1, \sqrt{3})$:

$$r = \sqrt{(-1)^2 + (\sqrt{3})^2} = \sqrt{1 + 3} = \sqrt{4} = 2.$$

Substituting -1 for x, $\sqrt{3}$ for y, and 2 for r, we find the trigonometric function values of θ are

$$\sin \theta = \frac{\sqrt{3}}{2}, \qquad \csc \theta = \frac{2}{\sqrt{3}} = \frac{2\sqrt{3}}{3},$$

$$\cos \theta = \frac{-1}{2} = -\frac{1}{2}, \qquad \sec \theta = \frac{2}{-1} = -2,$$

$$\tan \theta = \frac{\sqrt{3}}{-1} = -\sqrt{3}, \qquad \cot \theta = \frac{-1}{\sqrt{3}} = -\frac{\sqrt{3}}{3}.$$

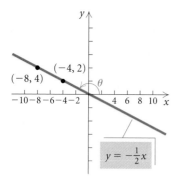

Any point other than the origin on the terminal side can be used to determine the trigonometric function values of that angle. The function values are the same regardless of which point is used. To illustrate this, let's consider an angle θ in standard position whose terminal side lies on the line $y = -\frac{1}{2}x$. We can determine two second-quadrant solutions of the equation, find the length r for each point, and then compare the sine, cosine, and tangent function values using each point.

If $x = -4$, then $y = -\frac{1}{2}(-4) = 2$.

If $x = -8$, then $y = -\frac{1}{2}(-8) = 4$.

For $(-4, 2)$, $r = \sqrt{(-4)^2 + 2^2} = \sqrt{20} = 2\sqrt{5}$.

For $(-8, 4)$, $r = \sqrt{(-8)^2 + 4^2} = \sqrt{80} = 4\sqrt{5}$.

Using $(-4, 2)$ and $r = 2\sqrt{5}$, we find that

$$\sin \theta = \frac{2}{2\sqrt{5}} = \frac{\sqrt{5}}{5}, \qquad \cos \theta = \frac{-4}{2\sqrt{5}} = -\frac{2\sqrt{5}}{5},$$

and $\tan \theta = \dfrac{2}{-4} = -\dfrac{1}{2}$.

Using $(-8, 4)$ and $r = 4\sqrt{5}$, we find that

$$\sin \theta = \frac{4}{4\sqrt{5}} = \frac{\sqrt{5}}{5}, \qquad \cos \theta = \frac{-8}{4\sqrt{5}} = -\frac{2\sqrt{5}}{5},$$

and $\tan \theta = \dfrac{4}{-8} = -\dfrac{1}{2}$.

We see that the function values are the same using either point. Any point other than the origin on the terminal side of an angle can be used to determine the trigonometric function values.

> The trigonometric function values of θ depend only on the angle, not on the choice of the point on the terminal side.

The Six Functions Related

When we know one of the function values of an angle, we can find the other five if we know the quadrant in which the terminal side lies. The idea is to sketch a reference triangle in the appropriate quadrant, use the Pythagorean theorem as needed to find the lengths of its sides, and then read off the ratios of the sides.

Example 4   Given that $\tan \theta = -\frac{2}{3}$ and θ is in the second quadrant, find the other function values.

SOLUTION   We first sketch a second-quadrant angle. Since

$$\tan \theta = \frac{y}{x} = -\frac{2}{3} = \frac{2}{-3}, \qquad \text{Expressing } -\frac{2}{3} \text{ as } \frac{2}{-3} \text{ since } \theta \text{ is in}$$

quadrant II

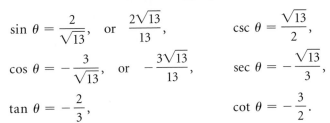

we make the legs of lengths 2 and 3. The hypotenuse must then have length $\sqrt{13}$. Now we can read off the appropriate ratios:

$$\sin \theta = \frac{2}{\sqrt{13}}, \quad \text{or} \quad \frac{2\sqrt{13}}{13}, \qquad\qquad \csc \theta = \frac{\sqrt{13}}{2},$$

$$\cos \theta = -\frac{3}{\sqrt{13}}, \quad \text{or} \quad -\frac{3\sqrt{13}}{13}, \qquad\qquad \sec \theta = -\frac{\sqrt{13}}{3},$$

$$\tan \theta = -\frac{2}{3}, \qquad\qquad \cot \theta = -\frac{3}{2}.$$

Terminal Side on an Axis

Suppose that the terminal side of an angle falls on one of the axes. Then one of the coordinates of any point on that side is 0. The definitions of the functions still apply, but in some cases, functions will not be defined because a denominator will be 0.

Example 5   Find the sine, cosine, and tangent values for 90°, 180°, 270°, and 360°.

SOLUTION   We first draw a sketch of each angle in standard position and label a point on the terminal side. Since the function values are the same for all points on the terminal side, we choose $(0, 1)$, $(-1, 0)$, $(0, -1)$, and $(1, 0)$ for convenience. Note also that $r = 1$ for each choice.

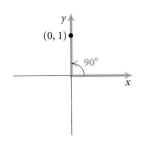   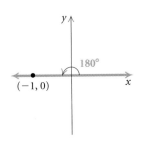   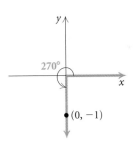   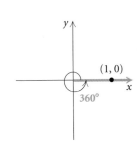

Then by the definitions we get

$$\sin 90° = \frac{1}{1} = 1, \qquad \sin 180° = \frac{0}{1} = 0, \qquad \sin 270° = \frac{-1}{1} = -1, \qquad \sin 360° = \frac{0}{1} = 0,$$

$$\cos 90° = \frac{0}{1} = 0, \qquad \cos 180° = \frac{-1}{1} = -1, \qquad \cos 270° = \frac{0}{1} = 0, \qquad \cos 360° = \frac{1}{1} = 1,$$

$$\tan 90° = \frac{1}{0}, \quad \text{Undefined} \qquad \tan 180° = \frac{0}{-1} = 0, \qquad \tan 270° = \frac{-1}{0}, \quad \text{Undefined} \qquad \tan 360° = \frac{0}{1} = 0.$$

In Example 5, all the values could be found using a grapher, but you will find that it is convenient to be able to compute them mentally. It is also helpful to note that coterminal angles have the same function values.

Example 6 Find each of the following.

a) sin (−90°) **b)** csc 540°

SOLUTION

a) We observe that −90° is coterminal with 270°. Thus,

$$\sin (-90°) = \sin 270° = \frac{-1}{1} = -1.$$

b) Since 540° = 180° + 360°, 540° and 180° are coterminal. Thus,

$$\csc 540° = \csc 180° = \frac{1}{\sin 180°} = \frac{1}{0}, \quad \text{which is undefined.}$$

Trigonometric values can always be checked with a grapher. When the value is undefined, the display will read

| ERR: DOMAIN | or | ERR: DIVIDE BY 0 |

Reference Angles: 30°, 45°, and 60°

We can also mentally determine trigonometric function values whenever the terminal side makes a 30°, 45°, or 60° angle with the *x*-axis. Consider, for example, an angle of 150°. The terminal side makes a 30° angle with the *x*-axis, since 180° − 150° = 30°. As the figure shows, △*ONP* is congruent to △*ON′P′*; then the ratios of the sides of the two triangles are the same.

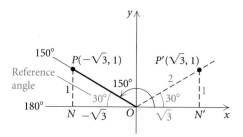

Thus the trigonometric function values are the same except perhaps for the sign. We could determine the function values directly from △*ONP*, but this is not necessary. If we remember that in quadrant II, the sine is positive and the cosine and the tangent are negative, we can simply use the function values of 30° that we already know and prefix the appropriate sign. Thus,

$$\sin 150° = \sin 30° = \frac{1}{2}, \qquad \cos 150° = -\cos 30° = -\frac{\sqrt{3}}{2},$$

and $\tan 150° = -\tan 30° = -\dfrac{\sqrt{3}}{3}.$

Triangle *ONP* is called a *reference triangle* and its acute angle, ∠*NOP*, is called a *reference angle*.

Reference Angle

The *reference angle* for an angle is the acute angle formed by the terminal side and the *x*-axis.

Example 7 Find the sine, cosine, and tangent function values for each of the following.

a) 225° **b)** −780°

SOLUTION

a) We draw a figure showing the terminal side of a 225° angle. The reference angle is 225° − 180°, or 45°.

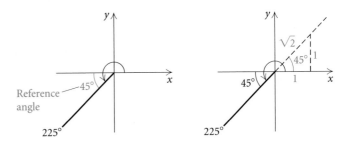

Recall from Section 2.1 that sin 45° = √2̄/2, cos 45° = √2̄/2, and tan 45° = 1. Also note that in the third quadrant, the sine and the cosine are negative and the tangent is positive. Thus we have

$$\sin 225° = -\frac{\sqrt{2}}{2}, \quad \cos 225° = -\frac{\sqrt{2}}{2}, \quad \text{and} \quad \tan 225° = 1.$$

b) We draw a figure showing the terminal side of a −780° angle. Since −780° + 2(360°) = −60°, we know that −780° and −60° are coterminal.

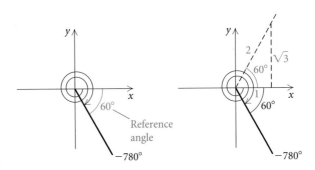

The reference angle for −60° is the acute angle formed by the terminal side and the *x*-axis. Thus the reference angle for −60° is 60°. We know that since −780° is a fourth-quadrant angle, the cosine is positive and the sine and the tangent are negative. Recalling that

$\sin 60° = \sqrt{3}/2$, $\cos 60° = 1/2$, and $\tan 60° = \sqrt{3}$, we have

$$\sin(-780°) = -\frac{\sqrt{3}}{2}, \qquad \cos(-780°) = \frac{1}{2},$$

and $\tan(-780°) = -\sqrt{3}$.

Function Values for Any Angle

When the terminal side of an angle falls on one of the axes or makes a 30°, 45°, or 60° angle with the x-axis, we can find exact function values without the use of a grapher. We can also find approximations of these values with a grapher. But this group is only a small subset of *all* angles. With a grapher, we can approximate the trigonometric function values of *any* angle. In fact, we can approximate function values of all angles without using a reference angle.

Example 8   Find each of the following function values.

a) $\cos 112°$ **b)** $\sec 500°$ **c)** $\tan(-83.4°)$

d) $\csc 351.75°$ **e)** $\cos 2400°$ **f)** $\tan 495°15'$

g) $\sin 175°40'9''$ **h)** $\cot(-135°)$

SOLUTION

FUNCTION VALUE	READOUT	ROUNDED
a) $\cos 112°$	$-.3746065934$	-0.3746
b) $\sec 500° = \dfrac{1}{\cos 500°}$	-1.305407289	-1.3054
c) $\tan(-83.4°)$	-8.642747461	-8.6427
d) $\csc 351.75° = \dfrac{1}{\sin 351.75°}$	-6.968999424	-6.9690
e) $\cos 2400°$	$-.5$	-0.5
f) $\tan 495°15' = \tan 495.25°$	$-.9913112106$	-0.9913
g) $\sin 175°40'9'' \approx \sin 175.67°$	$.0755008414$	0.0755
h) $\cot(-135°) = \dfrac{1}{\tan(-135°)}$	1	1

In many applications, we have a trigonometric function value and want to find the measure of an angle. To do so, we use either the TABLE feature or an inverse key on a grapher. When only acute angles are considered, there is only one output (angle) for each input (the trigonometric function value). This is not the case when we extend the domain of the trigonometric functions to the set of *all* angles. For a given function value, there is an infinite number of angles that have that function value. But there are only two such angles for each value in the range from 0° to 360°. To determine a unique answer, the quadrant in which the terminal

side lies must be specified. The grapher deals with this situation by giving the reference angle as an output for each function value that is entered as an input. With this reference angle and knowing the quadrant in
which the terminal side lies, we can find the specified angle.

Example 9 Given the function value and the quadrant restriction,
find θ.

a) $\sin \theta = 0.2812$, $90° < \theta < 180°$

b) $\cot \theta = -0.1611$, $270° < \theta < 360°$

SOLUTION

a) We first sketch the angle in the second quadrant. We then enter 0.2812
and find the reference angle to be approximately 16.33°.

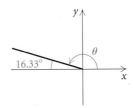

We find the angle θ by subtracting 16.33° from 180°:

$$180° - 16.33° = 163.67°.$$

Thus, $\theta \approx 163.67°$.

b) We begin by sketching the angle in the fourth quadrant. Because the
cotangent value is the reciprocal of the tangent value, we know that

$$\tan \theta \approx \frac{1}{-0.1611}$$

$$\approx -6.2073.$$

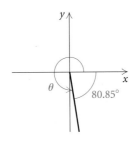

We then enter 6.2073, ignoring the fact that $\tan \theta$ is negative, and find
that the reference angle is approximately 80.85°. We find angle θ by
subtracting 80.85° from 360°:

$$360° - 80.85° = 279.15°.$$

Thus, $\theta \approx 279.15°$.

2.3 Exercise Set

For angles of the following measures, state in which quadrant the terminal side lies.

1. 187°   **2.** −14.3°   **3.** 245°15′

4. −120°   **5.** 800°   **6.** 1075°

7. −460.5°   **8.** 315°   **9.** −912°

10. 13°15′60″   **11.** 537°   **12.** −345.14°

Find two positive angles and two negative angles that are coterminal with the given angle. Answers may vary.

13. 74°   **14.** −81°   **15.** 115.3°

16. 275°10′   **17.** −180°   **18.** −310°

Find the complement and the supplement.

19. 17.11°   **20.** 47°38′   **21.** 12°3′14″

22. 9.038°   **23.** 45.2°   **24.** 67.31°

Find the six trigonometric function values for the angle shown.

25.

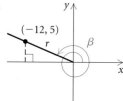

26.

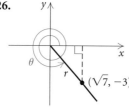

27.

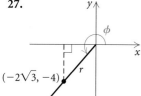

28.

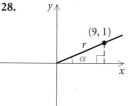

The terminal side of angle θ in standard position lies on the given line in the given quadrant. Find sin θ, cos θ, and tan θ.

29. $2x + 3y = 0$; quadrant IV

30. $4x + y = 0$; quadrant II

31. $5x - 4y = 0$; quadrant I

32. $y = 0.8x$; quadrant III

A function value and a quadrant are given. Find the other five function values.

33. $\sin \theta = -\dfrac{1}{3}$, quadrant III

34. $\tan \beta = 5$, quadrant I

35. $\cot \theta = -2$, quadrant IV

36. $\cos \alpha = -\dfrac{4}{5}$, quadrant II

37. $\cos \phi = \dfrac{3}{5}$, quadrant IV

38. $\sin \theta = -\dfrac{5}{13}$, quadrant III

Find the exact function value if it exists.

39. cos 150°   **40.** sec (−225°)

41. tan (−135°)   **42.** sin (−45°)

43. sin 7560°   **44.** tan 270°

45. cos 495°   **46.** tan 675°

47. csc (−210°)   **48.** sin 300°

49. cot 570°   **50.** cos (−120°)

51. tan 330°   **52.** cot 855°

53. sec (−90°)   **54.** sin 90°

55. cos (−180°)   **56.** csc 90°

57. tan 240°   **58.** cot (−180°)

59. sin 495°   **60.** sin 1050°

61. csc 225°   **62.** sin (−450°)

63. cos 0°   **64.** tan 480°

65. cot (−90°)   **66.** sec 315°

67. cos 90°   **68.** sin (−135°)

69. cos 270°   **70.** tan 0°

Find the signs of the six trigonometric function values for the given angles.

71. 319°   **72.** −57°

73. 194°   **74.** −620°

75. −215°   **76.** 290°

77. −272°   **78.** 91°

Use a calculator in Exercises 79–82, but do not use the trigonometric function keys.

79. Given that

$$\sin 41° = 0.6561,$$
$$\cos 41° = 0.7547,$$
$$\tan 41° = 0.8693,$$

find the trigonometric function values for 319°.

80. Given that

$$\sin 27° = 0.4540,$$
$$\cos 27° = 0.8910,$$
$$\tan 27° = 0.5095,$$

find the trigonometric function values for 333°.

81. Given that

$$\sin 65° = 0.9063,$$
$$\cos 65° = 0.4226,$$
$$\tan 65° = 2.1445,$$

find the trigonometric function values for 115°.

82. Given that

$$\sin 35° = 0.5736,$$
$$\cos 35° = 0.8192,$$
$$\tan 35° = 0.7002,$$

find the trigonometric function values for 215°.

Aerial Navigation.   In aerial navigation, directions are given in degrees clockwise from north. Thus, east is 90°, south is 180°, and west is 270°. Several aerial directions or bearings are given below.

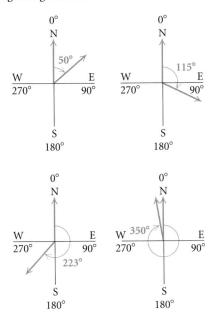

83. An airplane flies 150 km from an airport in a direction of 120°. How far east of the airport is the plane then? How far south?

84. An airplane leaves an airport and travels for 100 mi in a direction of 300°. How far north of the airport is the plane then? How far west?

85. An airplane travels at 150 km/h for 2 hr in a direction of 138° from Omaha. At the end of this time, how far south of Omaha is the plane?

86. An airplane travels at 120 km/h for 2 hr in a direction of 319° from Chicago. At the end of this time, how far north of Chicago is the plane?

Find the function value. Round to four decimal places.

87. tan 310.8°

88. cos 205.5°

89. cot 146.15°

90. sin (−16.4°)

91. sin 118°42′

92. cos 273°45′

93. cos (−295.8°)

94. tan 1086.2°

95. cos 5417°

96. sec 240°55′

97. csc 520°

98. sin 3824°

Given the function value and the quadrant restriction, find θ.

	FUNCTION VALUE	INTERVAL	θ
99.	$\sin \theta = -0.9956$	(270°, 360°)	
100.	$\tan \theta = 0.2460$	(180°, 270°)	
101.	$\cos \theta = -0.9388$	(180°, 270°)	
102.	$\sec \theta = -1.0485$	(90°, 180°)	
103.	$\tan \theta = -3.0545$	(270°, 360°)	
104.	$\sin \theta = -0.4313$	(180°, 270°)	
105.	$\csc \theta = 1.0480$	(0°, 90°)	
106.	$\cos \theta = -0.0990$	(90°, 180°)	

Skill Maintenance

Find decimal degree notation.

107. 44°10′35″

108. 382°20′16″

109. Convert 45 mph to ft/sec.

110. Convert 8.5 in./sec to ft/hr.

Synthesis

111. ◆ Why do the function values of θ depend only on the angle and not on the choice of a point on the terminal side?

112. ◆ Why is the domain of the tangent function different from the domains of the sine and the cosine functions?

113. *Valve Cap on a Bicycle.* The valve cap on a bicycle wheel is 12.5 in. from the center of the wheel. From the position shown, the wheel starts to roll. After the wheel has turned 390°, how far above the ground is the valve cap? Assume that the outer radius of the tire is 13.375 in.

114. *Seats of a Ferris Wheel.* The seats of a ferris wheel are 35 ft from the center of the wheel. When you board the wheel, you are 5 ft above the ground. After you have rotated through an angle of 765°, how far above the ground are you?

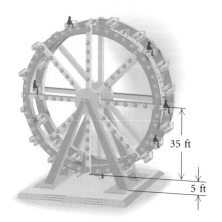

35 ft

5 ft

2.4
Radians, Arc Length, and Angular Speed

- *Find points on the unit circle determined by real numbers.*
- *Convert between radian and degree measure; find coterminal, complementary, and supplementary angles.*
- *Find the length of an arc of a circle; find the measure of a central angle of a circle.*
- *Convert between linear speed and angular speed.*

Another useful unit of angle measure is called a *radian*. To introduce radian measure, we use a circle centered at the origin with a radius of length 1. Such a circle is called a **unit circle**. Its equation is $x^2 + y^2 = 1$ (see Appendix A.2).

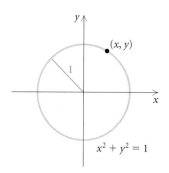

Distances on the Unit Circle

The circumference of a circle of radius r is $2\pi r$. Thus for the unit circle, where $r = 1$, the circumference is 2π. If a point starts at A and travels around the circle (Fig. 1), it will travel a distance of 2π. If it travels half-way around the circle (Fig. 2), it will travel a distance of $\frac{1}{2} \cdot 2\pi$, or π.

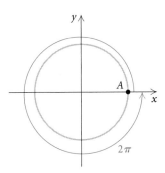

FIGURE 1

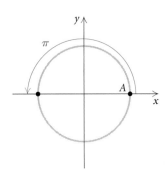

FIGURE 2

If a point C travels $\frac{1}{8}$ of the way around the circle (Fig. 3), it will travel a distance of $\frac{1}{8} \cdot 2\pi$, or $\pi/4$. Note that C is $\frac{1}{4}$ of the way from A to B. If a point D travels $\frac{1}{6}$ of the way around the circle (Fig. 4), it will travel a distance of $\frac{1}{6} \cdot 2\pi$, or $\pi/3$. Note that D is $\frac{1}{3}$ of the way from A to B.

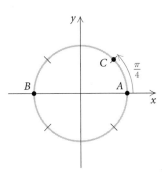

FIGURE 3

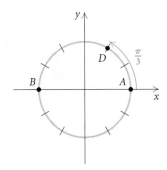

FIGURE 4

Example 1   How far will a point travel if it goes (a) $\frac{1}{4}$, (b) $\frac{1}{12}$, (c) $\frac{3}{8}$, and (d) $\frac{5}{6}$ of the way around the unit circle?

SOLUTION

a) $\frac{1}{4}$ of the total distance around the circle is $\frac{1}{4} \cdot 2\pi$, which is $\frac{1}{2} \cdot \pi$, or $\pi/2$.

b) The distance will be $\frac{1}{12} \cdot 2\pi$, which is $\frac{1}{6}\pi$, or $\pi/6$.

c) The distance will be $\frac{3}{8} \cdot 2\pi$, which is $\frac{3}{4}\pi$, or $3\pi/4$.

d) The distance will be $\frac{5}{6} \cdot 2\pi$, which is $\frac{5}{3}\pi$, or $5\pi/3$. Think of $5\pi/3$ as $\pi + \frac{2}{3}\pi$.

These distances are illustrated in the following figures.

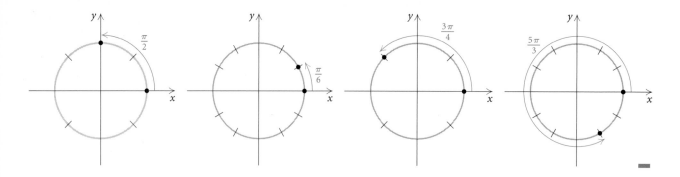

A point may travel completely around the circle and then continue. For example, if it goes around once and then continues $\frac{1}{4}$ of the way around, it will have traveled a distance of $2\pi + \frac{1}{4} \cdot 2\pi$, or $5\pi/2$ (Fig. 5). *Every* real number determines a point on the unit circle. For the positive number 10, for example, we start at A and travel counterclockwise a distance of 10. The point at which we stop is the point "determined" by the number 10 (Fig. 6).

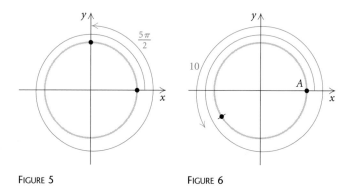

FIGURE 5   FIGURE 6

For a negative number, we move clockwise around the circle. Points for $-\pi/4$ and $-3\pi/2$ are shown in the figure below. The number 0 determines the point A.

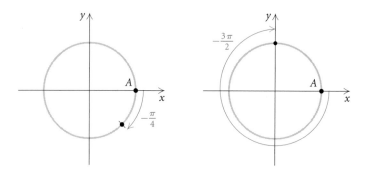

Example 2 On the unit circle, mark the point determined by each of the following real numbers.

a) $\dfrac{9\pi}{4}$ **b)** $-\dfrac{7\pi}{6}$

SOLUTION

a) Think of $9\pi/4$ as $2\pi + \frac{1}{4}\pi$ (see the figure on the left below). Since $9\pi/4 > 0$, the point moves counterclockwise. The point goes completely around once and then continues $\frac{1}{4}$ of the way from A to B.

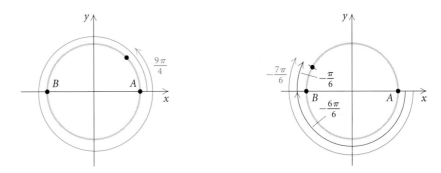

b) The number is negative, so the point moves clockwise. From A to B, the distance is π, or $\frac{6}{6}\pi$, so we need to go beyond B another distance of $\pi/6$, clockwise. (See the figure on the right above.) ▬

Radian Measure

Degree measure is a common unit of angle measure in many everyday applications. But in many scientific fields and in mathematics (calculus, in particular), there is another commonly used unit of measure called the *radian*.

Consider the unit circle whose radius has length 1. Suppose we measure, moving counterclockwise, an arc of length 1, and mark a point T on the circle.

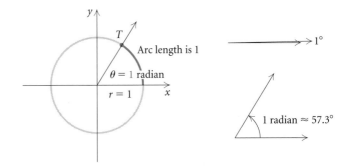

If we draw a ray from the origin through T, we have formed an angle. The measure of that angle is 1 **radian**. The word radian comes from the word *radius*. Thus measuring 1 "radius" along the circumference of the circle determines an angle whose measure is 1 *radian*. One radian is about 57.3°. Angles that measure 2 radians, 3 radians, and 6 radians are shown below.

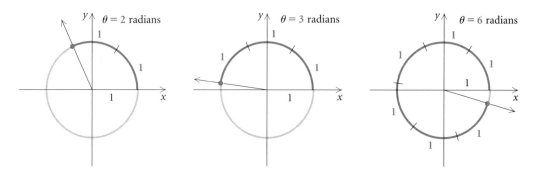

When we make a complete (counterclockwise) revolution, the terminal side coincides with the initial side on the positive *x*-axis. We then have an angle whose measure is 2π radians, or about 6.28 radians, which is the circumference of the circle:

$$2\pi r = 2\pi(1) = 2\pi.$$

Thus a rotation of 360° (1 revolution) has a measure of 2π radians. A half revolution is a rotation of 180°, or π radians. A quarter revolution is a rotation of 90°, or $\pi/2$ radians, and so on.

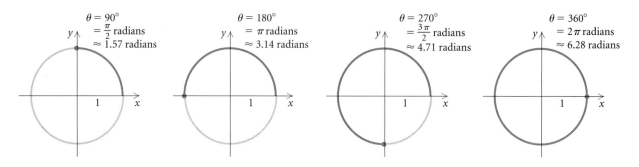

To convert between degrees and radians, we first note that

$$360° = 2\pi \text{ radians.}$$

It follows that

$$180° = \pi \text{ radians.}$$

To make conversions, we multiply by one, noting that:

$$\frac{\pi \text{ radians}}{180°} = \frac{180°}{\pi \text{ radians}} = 1.$$

Some graphers can convert directly from one measure to the other.

Example 3 Convert each of the following to radians.

a) 120°   **b)** −297.25°

Solution

a) $120° = 120° \cdot \dfrac{\pi \text{ radians}}{180°}$ Multiplying by 1

$ = \dfrac{120°}{180°}\pi \text{ radians}$

$ = \dfrac{2\pi}{3}\text{ radians,}\quad \text{or about} \quad 2.09 \text{ radians}$

b) $-297.25° = -297.25° \cdot \dfrac{\pi \text{ radians}}{180°}$

$ = -\dfrac{297.25°}{180°}\pi \text{ radians}$

$ = -\dfrac{297.25\pi}{180}\text{ radians}$

$ \approx -5.19 \text{ radians}$

Example 4 Convert each of the following to degrees.

a) $\dfrac{3\pi}{4}$ radians   **b)** 8.5 radians

Solution

a) $\dfrac{3\pi}{4}\text{ radians} = \dfrac{3\pi}{4}\text{ radians} \cdot \dfrac{180°}{\pi \text{ radians}}$ Multiplying by 1

$\phantom{\dfrac{3\pi}{4}\text{ radians}} = \dfrac{3\pi}{4\pi} \cdot 180° = \dfrac{3}{4} \cdot 180° = 135°$

b) $8.5 \text{ radians} = 8.5 \text{ radians} \cdot \dfrac{180°}{\pi \text{ radians}}$

$\phantom{8.5 \text{ radians}} = \dfrac{8.5(180°)}{\pi} \approx 487.01°$

The radian–degree equivalents of the most commonly used angle measures are illustrated in the following figures.

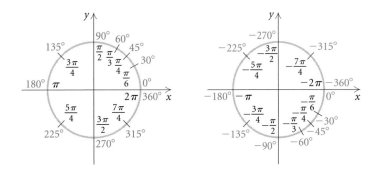

When a rotation is given in radians, the word "radians" is optional and is most often omitted. Thus if no unit is given for a rotation, the rotation is understood to be in radians.

We can also find coterminal, complementary, and supplementary angles in radian measure just as we did for degree measure in Section 2.3.

Example 5   Find a positive angle and a negative angle that are coterminal with $2\pi/3$. Many answers are possible.

SOLUTION   To find angles coterminal with a given angle, we add or subtract multiples of 2π:

$$\frac{2\pi}{3} + 2\pi = \frac{2\pi}{3} + \frac{6\pi}{3} = \frac{8\pi}{3},$$

$$\frac{2\pi}{3} - 3(2\pi) = \frac{2\pi}{3} - \frac{18\pi}{3} = -\frac{16\pi}{3}.$$

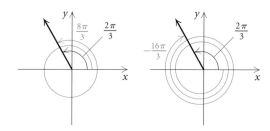

Thus, $8\pi/3$ and $-16\pi/3$ are coterminal with $2\pi/3$.

Example 6   Find the complement and the supplement of $\pi/6$.

SOLUTION   Since $90°$ equals $\pi/2$ radians, the complement of $\pi/6$ is

$$\frac{\pi}{2} - \frac{\pi}{6} = \frac{3\pi}{6} - \frac{\pi}{6} = \frac{2\pi}{6}, \quad \text{or} \quad \frac{\pi}{3}.$$

Since $180°$ equals π radians, the supplement of $\pi/6$ is

$$\pi - \frac{\pi}{6} = \frac{6\pi}{6} - \frac{\pi}{6} = \frac{5\pi}{6}.$$

Thus the complement of $\pi/6$ is $\pi/3$ and the supplement is $5\pi/6$.

Arc Length and Central Angles

Radian measure can be determined using a circle other than a unit circle. In the figure at left, a unit circle (with radius 1) is shown along with another circle (with radius r). The angle shown is a central angle of both circles.

From geometry, we know that the arcs that the angle subtends have their lengths in the same ratio as the radii of the circles. The radii of the

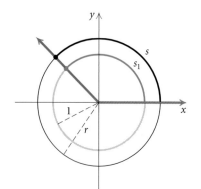

circles are r and 1. The corresponding arc lengths are s and s_1. Thus we have the proportion

$$\frac{s}{s_1} = \frac{r}{1},$$

which can also be written as

$$\frac{s_1}{1} = \frac{s}{r}.$$

Now s_1 is the *radian measure* of the rotation in question. It is common to use a Greek letter, such as θ, for the measure of an angle or rotation and the letter s for arc length. Adopting this convention, we rewrite the proportion above as

$$\theta = \frac{s}{r}.$$

In any circle, the measure (in radians) of a central angle, the arc length the angle subtends, and the length of the radius are related in this fashion. Or, in general, the following is true.

Radian Measure

The *radian measure* θ of a rotation is the ratio of the distance s traveled by a point at a radius r from the center of rotation, to the length of the radius r:

$$\theta = \frac{s}{r}.$$

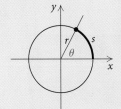

In using the above formula $\theta = s/r$, you must make sure that θ is in radians and that s and r are expressed in the same unit.

Example 7 Find the measure of a rotation in radians where a point 2 m from the center of rotation travels 4 m.

SOLUTION

$$\theta = \frac{s}{r} = \frac{4 \text{ m}}{2 \text{ m}} = 2 \qquad \text{The unit is understood to be radians.}$$

Example 8 Find the length of an arc of a circle of radius 5 cm associated with an angle of $\pi/3$ radians.

SOLUTION We have

$$\theta = \frac{s}{r}, \quad \text{or} \quad s = r\theta.$$

Therefore, $s = 5 \cdot \pi/3$ cm, or about 5.24 cm. ▬

A look at Examples 7 and 8 will show why the word radian is most often omitted. In Example 7, we have the division 4 m/2 m, which simplifies to the number 2, since m/m = 1. From this point of view, it would seem preferable to omit the word radians. In Example 8, had we used the word radians all the way through, our answer would have come out to be 5.23 cm-radians. It is a distance we seek; hence we know the unit should be centimeters. Thus we must omit the word radians. Since a measure in radians is simply a real number, it is usually preferable to omit the word radians.

Linear Speed and Angular Speed

Linear speed is defined to be distance traveled per unit of time. If we use v for linear speed, s for distance, and t for time, then

$$v = \frac{s}{t}.$$

Similarly, angular speed is defined to be amount of rotation per unit of time. For example, we might speak of the angular speed of a bicycle wheel as 150 revolutions per minute or the angular speed of the earth as 2π radians per day. The Greek letter ω (omega) is generally used for angular speed. Thus angular speed is defined as

$$\omega = \frac{\theta}{t}.$$

As an example of how these definitions can be applied, let's consider the refurbished carousel at the Children's Museum in Indianapolis, Indiana. It consists of three circular rows of animals. All animals, regardless of the row, travel at the same angular speed. But the animals in the outer row travel at a greater linear speed than those in the inner rows. What is the relationship between the linear speed v and the angular speed ω?

To develop the relationship we seek, recall that $\theta = s/r$. This is equivalent to

$$s = r\theta.$$

We divide by time, t, to obtain

$$\frac{s}{t} = r\,\frac{\theta}{t}.$$
$$\quad\downarrow\qquad\quad\downarrow$$
$$\quad v\qquad\quad \omega$$

Now s/t is linear speed v and θ/t is angular speed ω. Thus we have the relationship we seek, $v = r\omega$.

Linear Speed in Terms of Angular Speed

The *linear speed* v of a point a distance r from the center of rotation is given by

$$v = r\omega,$$

where ω is the *angular speed* in radians per unit of time.

For our new formula $v = r\omega$, the units of distance for v and r must be the same, ω must be in radians per unit of time, and the units of time for v and ω must be the same.

Example 9 *Linear Speed of an Earth Satellite.* An earth satellite in circular orbit 1200 km high makes one complete revolution every 90 min. What is its linear speed? Use 6400 km for the length of a radius of the earth.

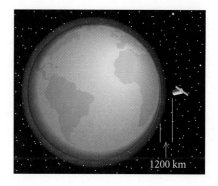

1200 km

SOLUTION   To use the formula $v = r\omega$, we will need to know r and ω:

$$r = 6400 \text{ km} + 1200 \text{ km} \qquad \text{Radius of earth plus height of satellite}$$

$$= 7600 \text{ km},$$

$$\omega = \frac{\theta}{t} = \frac{2\pi}{90 \text{ min}} = \frac{\pi}{45 \text{ min}}. \qquad \text{We have, as usual, omitted the word radians.}$$

Now, using $v = r\omega$, we have

$$v = 7600 \text{ km} \cdot \frac{\pi}{45 \text{ min}}$$

$$= \frac{7600\pi}{45} \cdot \frac{\text{km}}{\text{min}} \approx 531 \frac{\text{km}}{\text{min}}.$$

Thus the linear speed of the satellite is approximately 531 km/min.

Example 10 *Angular Speed of a Capstan.* An anchor is hoisted at a rate of 2 ft/sec as the chain is wound around a capstan with a 1.8-yd diameter. What is the angular speed of the capstan?

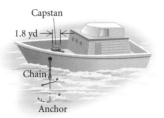

Capstan

1.8 yd

Chain

Anchor

SOLUTION We will use the formula $v = r\omega$ in the form $\omega = v/r$, taking care to use the proper units. Since v is given in feet per second, we need r in feet:

$$r = \frac{d}{2} = \frac{1.8}{2} \text{ yd} \cdot \frac{3 \text{ ft}}{1 \text{ yd}} = 2.7 \text{ ft}.$$

Then ω will be in radians per second:

$$\omega = \frac{v}{r} = \frac{2 \text{ ft/sec}}{2.7 \text{ ft}}$$

$$= \frac{2 \text{ ft}}{\text{sec}} \cdot \frac{1}{2.7 \text{ ft}} \approx 0.741/\text{sec}.$$

Thus the angular speed is approximately 0.741 radian/sec. ▬

The formulas $\theta = \omega t$ and $v = r\omega$ can be used in combination to find distances and angles in various situations involving rotational motion.

Example 11 *Angle of Revolution.* A Porsche 911 is traveling at a speed of 65 mph. Its tires have an outside diameter of 25.086 in. Find the angle through which a tire turns in 10 sec.

SOLUTION Recall that $\omega = \theta/t$, or $\theta = \omega t$. Thus we can find θ if we know ω and t. To find ω, we use the formula $v = r\omega$. For convenience, we first convert 65 mph to feet per second:

$$v = 65 \frac{\text{mi}}{\text{hr}} \cdot \frac{1 \text{ hr}}{60 \text{ min}} \cdot \frac{1 \text{ min}}{60 \text{ sec}} \cdot \frac{5280 \text{ ft}}{1 \text{ mi}}$$

$$\approx 95.333 \frac{\text{ft}}{\text{sec}}.$$

Now $r = d/2 = 25.086$ in./2 $= 12.543$ in. We will convert to feet, since v is in feet per second:

$$r = 12.543 \text{ in.} \cdot \frac{1 \text{ ft}}{12 \text{ in.}}$$

$$= \frac{12.543}{12} \text{ ft} \approx 1.045 \text{ ft.}$$

Using $v = r\omega$, we have

$$95.333 \frac{\text{ft}}{\text{sec}} = 1.045 \text{ ft} \cdot \omega,$$

so

$$\omega = \frac{95.333 \text{ ft/sec}}{1.045 \text{ ft}} \approx \frac{91.228}{\text{sec}}.$$

Then in 10 sec,

$$\theta = \omega t = \frac{91.228}{\text{sec}} \cdot 10 \text{ sec} \approx 912.$$

Thus the angle, in radians, through which the tire turns in 10 sec is 912.

2.4 | Exercise Set

For each of Exercises 1–4, sketch a unit circle and mark the points determined by the given real numbers.

1. a) $\dfrac{\pi}{4}$ **b)** $\dfrac{3\pi}{2}$ **c)** $\dfrac{3\pi}{4}$

d) π **e)** $\dfrac{11\pi}{4}$ **f)** $\dfrac{17\pi}{4}$

2. a) $\dfrac{\pi}{2}$ **b)** $\dfrac{5\pi}{4}$ **c)** 2π

d) $\dfrac{9\pi}{4}$ **e)** $\dfrac{13\pi}{4}$ **f)** $\dfrac{23\pi}{4}$

3. a) $\dfrac{\pi}{6}$ **b)** $\dfrac{2\pi}{3}$ **c)** $\dfrac{7\pi}{6}$

d) $\dfrac{10\pi}{6}$ **e)** $\dfrac{14\pi}{6}$ **f)** $\dfrac{23\pi}{4}$

4. a) $-\dfrac{\pi}{2}$ **b)** $-\dfrac{3\pi}{4}$ **c)** $-\dfrac{5\pi}{6}$

d) $-\dfrac{5\pi}{2}$ **e)** $-\dfrac{17\pi}{6}$ **f)** $-\dfrac{9\pi}{4}$

Find two real numbers between -2π and 2π that determine each of the points on the unit circle.

5.

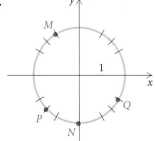

6.

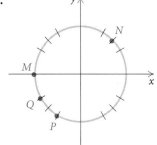

For Exercises 7 and 8, sketch a unit circle and mark the approximate location of the point determined by the given real number.

7. a) 2.4   **b)** 7.5   **c)** 32   **d)** 320

8. a) 0.25   **b)** 1.8   **c)** 47   **d)** 500

Find a positive angle and a negative angle that are coterminal with the given angle. Answers may vary.

9. $\dfrac{\pi}{4}$ **10.** $\dfrac{5\pi}{3}$ **11.** $\dfrac{7\pi}{6}$ **12.** π

Find the complement and the supplement.

13. $\dfrac{\pi}{3}$ **14.** $\dfrac{5\pi}{12}$ **15.** $\dfrac{3\pi}{8}$ **16.** $\dfrac{\pi}{4}$

Convert to radian measure. Leave your answer in terms of π.

17. 75°   **18.** 30°

19. 200°   **20.** −135°

21. −214.6°   **22.** 37.71°

23. −180°   **24.** 90°

Convert to radian measure. Round your answer to two decimal places.

25. 240°   **26.** 15°

27. −60°   **28.** 145°

29. 117.8°   **30.** −231.2°

31. 1.354°   **32.** 584°

Convert to degree measure. Round your answer to two decimal places.

33. $-\dfrac{3\pi}{4}$ **34.** $\dfrac{7\pi}{6}$

35. 8π **36.** $-\dfrac{\pi}{3}$

37. 1   **38.** −17.6

39. 2.347   **40.** 25

41. Certain positive angles are marked here in degrees. Find the corresponding radian measures.

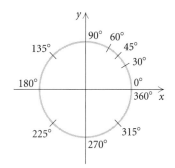

42. Certain negative angles are marked here in degrees. Find the corresponding radian measures.

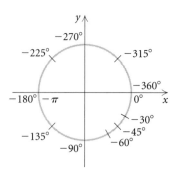

43. In a circle with a 120-cm radius, an arc 132 cm long subtends an angle of how many radians? how many degrees, to the nearest degree?

44. In a circle with a 5-m radius, how long is an arc associated with an angle of 2.1 radians?

45. *Angle of Revolution.* Through how many radians does the minute hand of a clock rotate from 12:40 P.M. to 1:30 P.M.?

46. *Angle of Revolution.* A tire on a Dodge Neon has an outside diameter of 23.468 in. Through what angle (in radians) does the tire turn while traveling 1 mi?

23.468 in.

47. *Linear Speed.* A flywheel with a 15-cm diameter is rotating at a rate of 7 radians/sec. What is the linear speed of a point on its rim, in centimeters per minute?

48. *Linear Speed.* A wheel with a 30-cm radius is rotating at a rate of 3 radians/sec. What is the linear speed of a point on its rim, in meters per minute?

49. *Angular Speed on a Printing Press.* This text was printed on a four-color Webb heatset offset press. A cylinder on this press has a 13.37-in. diameter. The linear speed of a point on the cylinder's surface is 16.53 feet per second. What is the angular speed of the cylinder in revolutions per hour? Printers often refer to the angular speed as impressions per hour (IPH). (*Source:* Mark Krahforst, Rand Printing Company, Taunton, MA)

50. *Linear Speeds on a Carousel.* When Alicia and Zoe ride the carousel described earlier in this section, Alicia always selects a horse on the outside row, whereas Zoe prefers the row closest to the center. These rows are 19 ft 3 in. and 13 ft 11 in. from the center, respectively. The angular speed of the carousel is 2.4 revolutions per minute. What is the difference, in miles per hour, in the linear speeds of Alicia and Zoe? (*Source:* The Children's Museum, Indianapolis, IN)

51. *Linear Speed at the Equator.* The earth has a 4000-mi radius and rotates one revolution every 24 hr. What is the linear speed of a point on the equator, in miles per hour?

52. *Linear Speed of the Earth.* The earth is 93,000,000 mi from the sun and traverses its orbit, which is nearly circular, every 365.25 days. What is the linear velocity of the earth in its orbit, in miles per hour?

53. *Determining the Speed of a River.* A water wheel has a 10-ft radius. To get a good approximation of

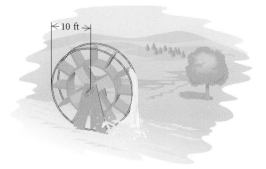

the speed of the river, you count the revolutions of the wheel and find that it makes 14 revolutions per minute (rpm). What is the speed of the river in miles per hour?

54. *The Tour de France.* Miguel Indurain won the 1995 Tour de France bicycle race. The wheel of his bicycle had a 67-cm diameter. His average linear speed during the 19th stage of the race was 48.461 km/h. What was the angular speed of the wheel in revolutions per hour? (*Source: Cycling Weekly,* July 29, 1995, p. 19, London)

55. *John Deere Tractor.* The center of a rear wheel on a John Deere 8300 farm tractor is 35 in. from the ground. Find the angle (in radians) through which a wheel rotates in 12 sec if the tractor is traveling at a speed of 22 mph.

Skill Maintenance

Graph each function. Sketch and label any vertical asymptote.

56. $f(x) = \dfrac{1}{x^2 - 25}$

57. $g(x) = x^3 - 2x + 1$

58. $f(x) = \dfrac{4}{x - 3}$

Determine the domain and the range of each function.

59. $f(x) = \dfrac{x - 4}{x + 2}$

60. $g(x) = \dfrac{x^2 - 9}{2x^2 - 7x - 15}$

Synthesis

61. ◈ Explain in your own words why it is preferable to omit the word, or unit, *radians* in radian measures.

62. ◈ In circular motion with a fixed angular speed, the length of the radius is directly proportional to the linear speed. Explain why with an example.

63. ◈ Two new cars are each driven at an average speed of 60 mph for an extended highway test drive of 2000 mi. The diameter of the wheels of the two cars are 15 in. and 16 in., respectively. If the cars use tires of equal durability and profile, differing only by the diameter, which car will probably need new tires first? Explain your result.

64. A point on the unit circle has y-coordinate $-\sqrt{21}/5$. What is its x-coordinate? Check with a grapher.

65. On the earth, one degree of latitude is how many kilometers? how many miles? (Assume that the radius of the earth is 6400 km, or 4000 mi, approximately.)

66. The **grad** is a unit of angle measure similar to a degree. A right angle has a measure of 100 grads. Convert each of the following to grads.

a) 48°   **b)** $\dfrac{5\pi}{7}$

67. A **mil** is a unit of angle measure. A right angle has a measure of 1600 mils. Convert each of the following to degrees, minutes, and seconds.

a) 100 mils   **b)** 350 mils

68. *Angular Speed of a Pulley.* Two pulleys, 50 cm and 30 cm in diameter, respectively, are connected by a belt. The larger pulley makes 12 revolutions per minute. Find the angular speed of the smaller pulley, in radians per second.

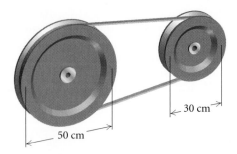

50 cm   30 cm

69. *Angular Speed of a Gear Wheel.* One gear wheel turns another, the teeth being on the rims. The wheels have 40-cm and 50-cm radii, and the smaller wheel rotates at 20 rpm. Find the angular speed of the larger wheel, in radians per second.

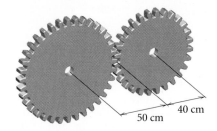

50 cm   40 cm

70. *Hands of a Clock.* At what times between noon and 1:00 P.M. are the hands of a clock perpendicular?

71. *Distance between Points on the Earth.* To find the distance between two points on the earth when their latitude and longitude are known, we can use a right triangle for an excellent approximation if the points are not too far apart. Point A is at latitude 38°27′30″ N, longitude 82°57′15″ W; and point B is at latitude 38°28′45″ N, longitude 82°56′30″ W. Find the distance from A to B in nautical miles (one minute of latitude is one nautical mile).

2.5
Circular Functions: Graphs and Properties

• *Given the coordinates of a point on the unit circle, find its reflections across the x-axis, the y-axis, and the origin.*
• *Determine the six trigonometric function values for a real number when the coordinates of the point on the unit circle determined by that real number are given.*
• *Find function values with a grapher for any real number.*
• *Graph the six circular functions and state their properties.*

The domains of the trigonometric functions, defined in Sections 2.1 and 2.3, have been sets of angles or rotations measured in a real number of degree units. We can also consider the domains to be sets of real numbers, or radians, introduced in Section 2.4. Many applications in calculus use the trigonometric functions referring only to radians.

Let's again consider radian measure and the unit circle. We defined radian measure for θ as

$$\theta = \frac{s}{r}.$$

When $r = 1$,

$$\theta = \frac{s}{1}, \quad \text{or} \quad \theta = s.$$

Thus the arc length s on the unit circle is the same as the radian measure of the angle θ.

In the figure above, the point (x, y) is the point where the terminal side of the angle with radian measure s intersects the unit circle. We can now extend our definitions of the trigonometric functions using domains of real numbers, or radians.

In the definitions, s can be considered the radian measure of an angle or the measure of an arc length on the unit circle. Either way, s is a real number. To each real number s, there corresponds an arc length s on the unit circle. Trigonometric functions with a domain of real numbers are called **circular functions**.

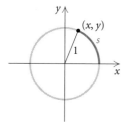

Basic Circular Functions

For a real number s that determines a point (x, y) on the unit circle:

$$\sin s = \text{second coordinate} = y,$$

$$\cos s = \text{first coordinate} = x,$$

$$\tan s = \frac{\text{second coordinate}}{\text{first coordinate}} = \frac{y}{x} \quad (x \neq 0),$$

$$\csc s = \frac{1}{\text{second coordinate}} = \frac{1}{y} \quad (y \neq 0),$$

$$\sec s = \frac{1}{\text{first coordinate}} = \frac{1}{x} \quad (x \neq 0),$$

$$\cot s = \frac{\text{first coordinate}}{\text{second coordinate}} = \frac{x}{y} \quad (y \neq 0).$$

Because of these definitions, we can consider the trigonometric functions with a domain of real numbers rather than angles. We can determine these values for a specific real number if we know the coordinates of the point on the unit circle determined by that number. As with degree measure, we can also find these function values directly with a grapher.

Reflections on the Unit Circle

Let's consider the unit circle and a few of its points. For any point (x, y) on the unit circle, with equation $x^2 + y^2 = 1$, we know that $-1 \leq x \leq 1$ and $-1 \leq y \leq 1$. If we know the x- or y-coordinate of a point on the unit circle, we can find the other coordinate. If $x = \frac{3}{5}$, then

$$\left(\tfrac{3}{5}\right)^2 + y^2 = 1$$
$$y^2 = 1 - \tfrac{9}{25} = \tfrac{16}{25}$$
$$y = \pm\tfrac{4}{5}.$$

Thus, $\left(\frac{3}{5}, \frac{4}{5}\right)$ and $\left(\frac{3}{5}, -\frac{4}{5}\right)$ are points on the unit circle. There are two points with an x-coordinate of $\frac{3}{5}$.

Now let's consider the radian measure $\pi/3$ and determine the coordinates of the point on the unit circle determined by $\pi/3$. We construct a right triangle by dropping a perpendicular segment from the point to the x-axis.

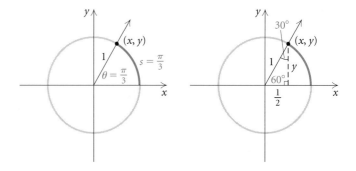

Since $\pi/3 = 60°$, we have a $30°-60°$ right triangle in which the side opposite the $30°$ angle is one half of the hypotenuse. The hypotenuse, or radius, is 1, so the side opposite the $30°$ angle is $\frac{1}{2}$. Using the Pythagorean theorem, we can find the other side:

$$\left(\frac{1}{2}\right)^2 + y^2 = 1$$
$$y^2 = 1 - \frac{1}{4} = \frac{3}{4}$$
$$y = \sqrt{\frac{3}{4}} = \frac{\sqrt{3}}{2}.$$

We know y is positive since the point is in the first quadrant. Thus the coordinates of the point determined by $\pi/3$ are $x = 1/2$ and $y = \sqrt{3}/2$, or $(1/2, \sqrt{3}/2)$. We can always check to see if a point is on the unit circle by substituting into the equation $x^2 + y^2 = 1$.

Because a unit circle is symmetric with respect to the x-axis, the y-axis, and the origin, we can use the coordinates of one point on the unit circle to find coordinates of its reflections.

Example 1 Each of the following points lies on the unit circle. Find their reflections across the x-axis, the y-axis, and the origin.

a) $\left(\dfrac{3}{5}, \dfrac{4}{5}\right)$ b) $\left(\dfrac{\sqrt{2}}{2}, \dfrac{\sqrt{2}}{2}\right)$ c) $\left(\dfrac{1}{2}, \dfrac{\sqrt{3}}{2}\right)$

SOLUTION

a)
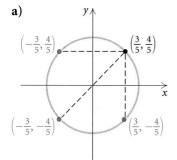

b)
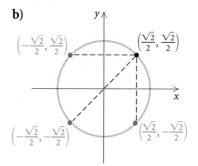

c)
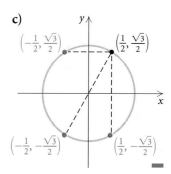

Finding Function Values

Knowing the coordinates of only a few points on the unit circle along with their reflections allows us to find trigonometric function values of the most frequently used real numbers, or radians. The coordinates on the unit circle below should be memorized.

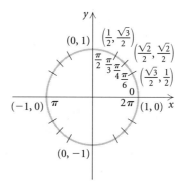

Example 2 Find each of the following function values.

a) $\tan \dfrac{\pi}{3}$ b) $\cos \dfrac{3\pi}{4}$ c) $\sin \left(-\dfrac{\pi}{6}\right)$

d) $\cos \dfrac{4\pi}{3}$ **e)** $\cot \pi$ **f)** $\csc \left(-\dfrac{7\pi}{2} \right)$

SOLUTION

a) The coordinates of the point determined by $\pi/3$ are $(1/2, \sqrt{3}/2)$. Thus,

$$\tan \frac{\pi}{3} = \frac{y}{x} = \frac{\sqrt{3}/2}{1/2} = \sqrt{3}.$$

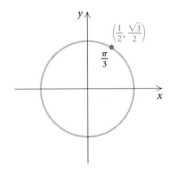

b) The reflection of $(\sqrt{2}/2, \sqrt{2}/2)$ across the y-axis is $(-\sqrt{2}/2, \sqrt{2}/2)$. Thus,

$$\cos \frac{3\pi}{4} = x = -\frac{\sqrt{2}}{2}.$$

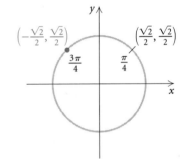

c) The reflection of $(\sqrt{3}/2, 1/2)$ across the x-axis is $(\sqrt{3}/2, -1/2)$. Thus,

$$\sin \left(-\frac{\pi}{6} \right) = y = -\frac{1}{2}.$$

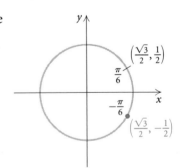

d) The reflection of $(1/2, \sqrt{3}/2)$ across the origin is $(-1/2, -\sqrt{3}/2)$. Thus,

$$\cos \frac{4\pi}{3} = x = -\frac{1}{2}.$$

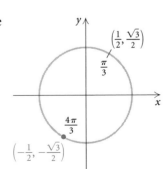

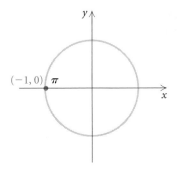

e) The coordinates of the point determined by π are $(-1, 0)$. Thus,

$$\cot \pi = \frac{x}{y} = \frac{-1}{0}, \quad \text{which is undefined.}$$

We can also think of $\cot \pi$ as the reciprocal of $\tan \pi$. Since $\tan \pi = y/x = 0/-1 = 0$ and the reciprocal of 0 is not defined, we know that $\cot \pi$ is undefined.

f) The coordinates of the point determined by $-7\pi/2$ are $(0, 1)$. Thus,

$$\csc \left(-\frac{7\pi}{2} \right) = \frac{1}{y} = \frac{1}{1} = 1.$$

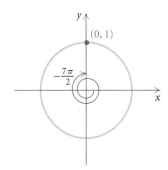

Using a grapher, we can find trigonometric function values of any real number without knowing the coordinates of the point it determines on the unit circle. Most graphers have both degree and radian modes. When finding function values of radian measures, or real numbers, we *must* set the grapher in radian mode.

Example 3 Find each of the following function values of radian measures using a grapher.

a) $\cos \dfrac{2\pi}{5}$ **b)** $\tan (-3)$ **c)** $\sin 24.9$ **d)** $\sec \dfrac{\pi}{7}$

SOLUTION

FUNCTION VALUE	READOUT	ROUNDED
a) $\cos \dfrac{2\pi}{5}$	.309016994	0.3090
b) $\tan (-3)$	.142546543	0.1425
c) $\sin 24.9$	−.230645706	−0.2306
d) $\sec \dfrac{\pi}{7}$	1.109916264	1.1099

Note in part (d) that the secant function value can be found by taking the reciprocal of the cosine value. Thus we enter $\cos \pi/7$ and use the reciprocal key.

Interactive Discovery

We can graph the unit circle on a grapher. We use parametric mode with the following window and let $X_{1T} = \cos T$ and $Y_{1T} = \sin T$.

WINDOW
  Tmin = 0
  Tmax = 2π, or 6.2831853
  Tstep = $\pi/12$, or 0.2617993
  Xmin = -1.5
  Xmax = 1.5
  Xscl = 1
  Ymin = -1
  Ymax = 1
  Yscl = 1

$T = 1.0471976$
$x = .5 \quad y = .8660254$

Using the trace key and an arrow key to move the cursor around the unit circle, we see the T, X, and Y values appear on the screen. What do they represent? (For more on parametric equations, see Section 5.8.)

From the definitions on page 167 and from the preceding exploration, we can relabel any point (x, y) on the unit circle as $(\cos s, \sin s)$, where s is any real number.

$(x, y) = (\cos s, \sin s)$

s

$(1, 0)$ x

Graphs of the Sine and Cosine Functions

Properties of functions can be easily observed from their graphs. We begin by graphing the sine and cosine functions. We make a table of values, plot the points, and then connect those points with a smooth curve. It is helpful to first draw a unit circle and label a few points with coordinates.

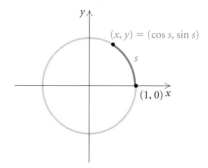

We can either use the coordinates as the function values or find approximate sine and cosine values directly with a grapher.

s	$\sin s$	$\cos s$
0	0	1
$\pi/6$	0.5	0.8660
$\pi/4$	0.7071	0.7071
$\pi/3$	0.8660	0.5
$\pi/2$	1	0
$3\pi/4$	0.7071	-0.7071
π	0	-1
$5\pi/4$	-0.7071	-0.7071
$3\pi/2$	-1	0
$7\pi/4$	-0.7071	0.7071
2π	0	1

s	$\sin s$	$\cos s$
0	0	1
$-\pi/6$	-0.5	0.8660
$-\pi/4$	-0.7071	0.7071
$-\pi/3$	-0.8660	0.5
$-\pi/2$	-1	0
$-3\pi/4$	-0.7071	-0.7071
$-\pi$	0	-1
$-5\pi/4$	0.7071	-0.7071
$-3\pi/2$	1	0
$-7\pi/4$	0.7071	0.7071
-2π	0	1

The graphs are as follows.

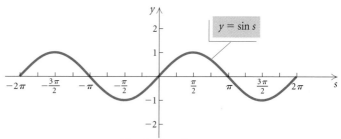

The sine function

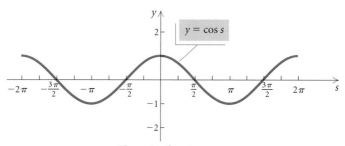

The cosine function

We can check our graphs with a grapher. When graphing trigonometric functions, the grapher can be in radian or degree mode. Here we use radian mode.

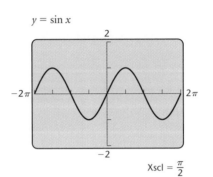

$y = \sin x$

Xscl $= \dfrac{\pi}{2}$

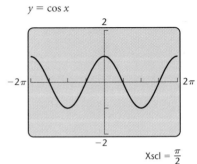

$y = \cos x$

Xscl $= \dfrac{\pi}{2}$

Interactive Discovery

Another way to construct the sine and cosine graphs is by considering the unit circle and transferring vertical distances for the sine function and horizontal distances for the cosine function. Using the parametric graphing feature of a grapher, we can visualize the transfer of these distances. We use PARAMETRIC mode with the following window and let $X_{1T} = \cos T - 1$ and $Y_{1T} = \sin T$ for the unit circle centered at $(-1, 0)$ and $X_{2T} = T$ and $Y_{2T} = \sin T$ for the sine curve.

$$
\begin{array}{lll}
\text{Tmin} = 0 & \text{Xmin} = -2 & \text{Ymin} = -3 \\
\text{Tmax} = 2\pi & \text{Xmax} = 2\pi & \text{Ymax} = 3 \\
\text{Tstep} = .1 & \text{Xscl} = \pi/2 & \text{Yscl} = 1
\end{array}
$$

With the grapher set to SIMUL mode, we can actually watch the sine function "unwind" from the unit circle. In the two screens below, we partially illustrate this animated procedure for the sine function.

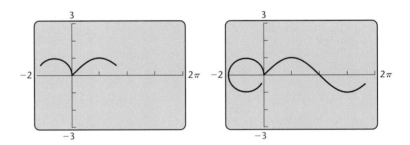

Consult your grapher's instruction manual for specific keystrokes and graph both the sine curve and the cosine curve in this manner. (For more on parametric equations, see Section 5.8.)

The sine and cosine functions are continuous functions. Note in the graph of the sine function that function values increase from 0 at $s = 0$ to 1 at $s = \pi/2$, then decrease to 0 at $s = \pi$, decrease further to -1 at $s = 3\pi/2$, and increase to 0 at 2π. The reverse pattern follows when s decreases from 0 to -2π. Note in the graph of the cosine function that function values start at 1 when $s = 0$, and decrease to 0 at $s = \pi/2$. They decrease further to -1 at $s = \pi$, then increase to 0 at $s = 3\pi/2$, and increase further to 1 at $s = 2\pi$. An identical pattern follows when s decreases from 0 to -2π.

Interactive Discovery

To use the TABLE feature on a grapher to verify the y-value patterns for $y = \sin x$ and $y = \cos x$, enter $\sin x$ for Y_1 and $\cos x$ for Y_2. We then set $TblMin = 0$ and $\triangle Tbl = \pi/12$ and begin scrolling.

X	Y₁	Y₂
0	0	1
.2618	.25882	.96593
.5236	.5	.86603
.7854	.70711	.70711
1.0472	.86603	.5
1.309	.96593	.25882
1.5708	1	0

X = 0

Can you determine the domain and the range for the sine function and for the cosine function?

From the unit circle, the graphs of the functions, and the TABLE feature on the grapher, we know that the domain of both the sine and cosine functions is the entire set of real numbers, $(-\infty, \infty)$. The range of each function is the set of all real numbers from -1 to 1, $[-1, 1]$.

A function with a repeating pattern is called **periodic**. The sine and cosine functions are examples of periodic functions. The function values of each function repeat themselves every 2π units. In other words, for any s, we have

$$\sin (s + 2\pi) = \sin s \quad \text{and} \quad \cos (s + 2\pi) = \cos s.$$

To see this another way, think of the part of the graph between 0 and 2π and note that the rest of the graph consists of copies of it. If we translate the graph of $y = \sin x$ or $y = \cos x$ to the left or right 2π units, we will obtain the original graph. We say that each of these functions has a period of 2π.

Periodic Function

A function f is said to be *periodic* if there exists a positive constant p such that

$$f(s + p) = f(s)$$

for all s in the domain of f. The smallest such positive number p is called the period of the function.

The period p can be thought of as the length of the shortest recurring interval.

We can also use the unit circle to verify that the period of the sine and cosine functions is 2π. Consider any real number s and the point T that it determines on a unit circle. If we increase s by 2π, the point determined by $s + 2\pi$ is again the point T. Hence for any real number s,

$$\sin (s + 2\pi) = \sin s \quad \text{and} \quad \cos (s + 2\pi) = \cos s.$$

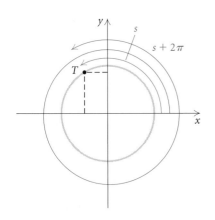

Thus we have shown that for any integer k, the following equations are identities:

$$\sin [s + k(2\pi)] = \sin s \quad \text{and} \quad \cos [s + k(2\pi)] = \cos s,$$

or

$$\sin s = \sin (s + 2k\pi) \quad \text{and} \quad \cos s = \cos (s + 2k\pi).$$

The **amplitude** of a periodic function is defined as one half of the distance between its maximum and minimum function values. It is always positive. Both the graphs and the unit circle verify that the maximum value of the sine and cosine functions is 1, whereas the minimum value of each is -1. Thus,

$$\text{the amplitude of the sine function} = \tfrac{1}{2}[1 - (-1)] = 1$$

and

$$\text{the amplitude of the cosine function is } \tfrac{1}{2}[1 - (-1)] = 1.$$

Interactive Discovery

Using the TABLE feature on a grapher, compare the y-values for $y_1 = \sin x$ and $y_2 = \sin (-x)$ and for $y_3 = \cos x$ and $y_4 = \cos (-x)$. We set TblMin $= 0$ and $\triangle$Tbl $= \pi/12$.

X	Y₁	Y₂
0	0	0
.2618	.25882	−.2588
.5236	.5	−.5
.7854	.70711	−.7071
1.0472	.86603	−.866
1.309	.96593	−.9659
1.5708	1	−1
X = 0		

X	Y₃	Y₄
0	1	1
.2618	.96593	.96593
.5236	.86603	.86603
.7854	.70711	.70711
1.0472	.5	.5
1.309	.25882	.25882
1.5708	0	0
X = 0		

What appears to be the relationship between $\sin x$ and $\sin (-x)$ and between $\cos x$ and $\cos (-x)$?

Consider any real number s and its opposite, $-s$. These numbers determine points T and T_1 on a unit circle that are symmetric with respect to the x-axis.

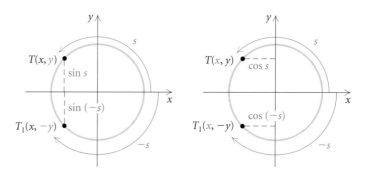

Because their second coordinates are opposites of each other, we know that for any number s,

$$\sin (-s) = -\sin s.$$

Because their first coordinates are the same, we know that for any number s,

$$\cos (-s) = \cos s.$$

Thus we have shown that the sine function is odd and the cosine function is even.

The following table is a summary of the properties of the sine and cosine functions.

SINE FUNCTION	COSINE FUNCTION
1. Continuous	**1.** Continuous
2. Period: 2π	**2.** Period: 2π
3. Domain: All real numbers	**3.** Domain: All real numbers
4. Range: $[-1, 1]$	**4.** Range: $[-1, 1]$
5. Amplitude: 1	**5.** Amplitude: 1
6. Odd: $\sin (-s) = -\sin s$	**6.** Even: $\cos (-s) = \cos s$

Graphs of the Tangent, Cotangent, Cosecant, and Secant Functions

To graph the tangent function, we could make a table of values with a grapher, but in this case it is easier to begin with the definition of tangent and the coordinates of a few points on the unit circle. We recall that

$$\tan s = \frac{y}{x} = \frac{\sin s}{\cos s}.$$

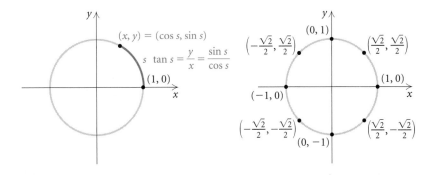

The tangent function is undefined when x, the first coordinate, is 0. That is, it is undefined for any number s whose cosine is 0:

$$s = \pm\frac{\pi}{2}, \pm\frac{3\pi}{2}, \pm\frac{5\pi}{2}, \dots.$$

We draw vertical asymptotes at these locations (see Fig. 1 below).

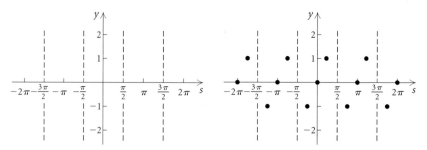

FIGURE 1   FIGURE 2

We also note that

$$\tan s = 0 \text{ at } s = 0, \pm\pi, \pm 2\pi, \pm 3\pi, \ldots,$$

$$\tan s = 1 \text{ at } s = \pm\frac{\pi}{4}, \pm\frac{5\pi}{4}, \pm\frac{9\pi}{4}, \ldots, \quad \text{and}$$

$$\tan s = -1 \text{ at } s = \pm\frac{3\pi}{4}, \pm\frac{7\pi}{4}, \pm\frac{11\pi}{4}, \ldots.$$

We can add these ordered pairs to the graph (see Fig. 2 above) and investigate the values in $(-\pi/2, \pi/2)$ with a grapher. Note that the function value is 0 when $s = 0$, and the values increase without bound as s increases toward $\pi/2$. The graph gets closer and closer to an asymptote as s gets closer to $\pi/2$, but it never touches the line. As s decreases from 0 to $-\pi/2$, the values decrease without bound. Again the graph gets closer and closer to an asymptote, but it never touches it. We now complete the graph.

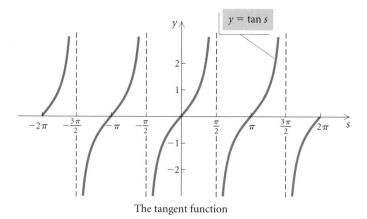

The tangent function

From the graph, we see that the tangent function is continuous except where it is not defined. Unlike the sine and cosine functions with a period of 2π, the tangent function has a period of π. Tan s is not defined where $\cos s = 0$. Thus the domain of the tangent function is the set of all real numbers except $(\pi/2) + k\pi$, where k is an integer. The range of the function is the set of all real numbers.

The cotangent function is undefined when y, the second coordinate, is 0—that is, it is undefined for any number s whose sine is 0. Thus the cotangent is undefined for $s = 0, \pm\pi, \pm 2\pi, \pm 3\pi, \ldots$. The graph of the function is shown below.

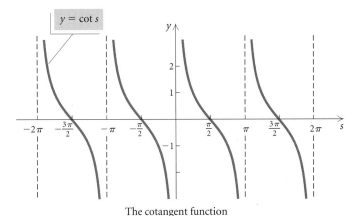

The cotangent function

The cosecant and sine functions are reciprocal functions, as are the secant and cosine functions. The graphs of the cosecant and secant functions can be constructed by finding the reciprocal of the values of the sine and cosine functions, respectively. Thus the functions will be positive together and negative together. The cosecant function is not defined for those numbers s whose sine is 0. The secant function is not defined for those numbers s whose cosine is 0. In the graphs below, the sine and cosine functions are shown for reference by the gray curves.

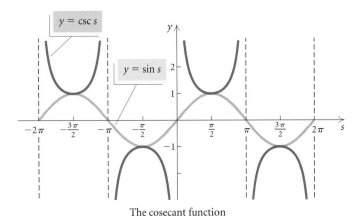

The cosecant function

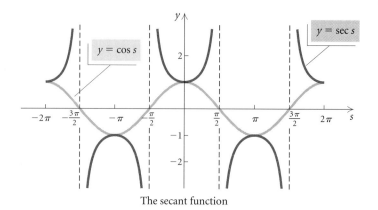

The secant function

The following is a summary of the basic properties of the tangent, cotangent, cosecant, and secant functions. These functions are continuous except where not defined.

Tangent Function	**Cotangent Function**
1. Period: π	**1.** Period: π
2. Domain: All real numbers except $(\pi/2) + k\pi$, where k is an integer	**2.** Domain: All real numbers except $k\pi$, where k is an integer
3. Range: All real numbers	**3.** Range: All real numbers
Cosecant Function	**Secant Function**
1. Period: 2π	**1.** Period: 2π
2. Domain: All real numbers except $k\pi$, where k is an integer	**2.** Domain: All real numbers except $(\pi/2) + k\pi$, where k is an integer
3. Range: $(-\infty, -1] \cup [1, \infty)$	**3.** Range: $(-\infty, -1] \cup [1, \infty)$

In this chapter, we have been using the letter s for arc length and have avoided the letters x and y, which generally represent first and second coordinates. In fact, we can represent the arc length on a unit circle by any variable, such as s, t, x, or θ. Each arc length determines a point that can be labeled with an ordered pair. The first coordinate of that ordered pair is the cosine of the arc length, and the second coordinate is the sine of the arc length. The identities we have developed hold no matter what symbols are used for variables—for example, $\cos(-s) = \cos s$, $\cos(-x) = \cos x$, $\cos(-\theta) = \cos \theta$, and $\cos(-t) = \cos t$.

2.5 | Exercise Set

The following points are on the unit circle. Find the coordinates of their reflections across (a) the x-axis, (b) the y-axis, and (c) the origin.

1. $\left(-\dfrac{3}{4}, \dfrac{\sqrt{7}}{4}\right)$

2. $\left(\dfrac{2}{3}, \dfrac{\sqrt{5}}{3}\right)$

3. $\left(\dfrac{2}{5}, -\dfrac{\sqrt{21}}{5}\right)$

4. $\left(-\dfrac{\sqrt{3}}{2}, -\dfrac{1}{2}\right)$

5. The number $\pi/4$ determines a point on the unit circle with coordinates $(\sqrt{2}/2, \sqrt{2}/2)$. What are the coordinates of the point determined by $-\pi/4$?

6. A number β determines a point on the unit circle with coordinates $(-2/3, \sqrt{5}/3)$. What are the coordinates of the point determined by $-\beta$?

Find the function value using coordinates of points on the unit circle.

7. $\sin \pi$

8. $\cos\left(-\dfrac{\pi}{3}\right)$

9. $\cot \dfrac{7\pi}{6}$

10. $\tan \dfrac{11\pi}{4}$

11. $\sin(-3\pi)$

12. $\csc \dfrac{3\pi}{4}$

13. $\cos \dfrac{5\pi}{6}$

14. $\tan\left(-\dfrac{\pi}{4}\right)$

15. $\cos 10\pi$

16. $\sec \dfrac{\pi}{2}$

17. $\cos \dfrac{\pi}{6}$

18. $\sin \dfrac{2\pi}{3}$

19. $\sin \dfrac{5\pi}{4}$

20. $\cos \dfrac{11\pi}{6}$

21. $\sin(-5\pi)$

22. $\tan \dfrac{3\pi}{2}$

23. $\cot \dfrac{5\pi}{2}$

24. $\tan \dfrac{5\pi}{3}$

Find the function value using a grapher. Round the answer to four decimal places.

25. $\tan \dfrac{\pi}{7}$

26. $\cos\left(-\dfrac{2\pi}{5}\right)$

27. $\sec 37$

28. $\sin 11.7$

29. $\cot 342$

30. $\tan 1.3$

31. $\cos 6\pi$

32. $\sin \dfrac{\pi}{10}$

33. $\csc 4.16$

34. $\sec \dfrac{10\pi}{7}$

35. $\tan \dfrac{7\pi}{4}$

36. $\cos 2000$

37. $\sin\left(-\dfrac{\pi}{4}\right)$

38. $\cot 7\pi$

39. $\sin 0$

40. $\cos(-29)$

41. $\tan \dfrac{2\pi}{9}$

42. $\sin \dfrac{8\pi}{3}$

In Exercises 43–48, make hand-drawn graphs.

43. a) Sketch a graph of $y = \sin x$.
  b) By reflecting the graph in part (a), sketch a graph of $y = \sin(-x)$.
  c) By reflecting the graph in part (a), sketch a graph of $y = -\sin x$.
  d) How do the graphs in parts (b) and (c) compare?

44. a) Sketch a graph of $y = \cos x$.
  b) By reflecting the graph in part (a), sketch a graph of $y = \cos(-x)$.
  c) By reflecting the graph in part (a), sketch a graph of $y = -\cos x$.
  d) How do the graphs in parts (a) and (b) compare?

45. a) Sketch a graph of $y = \sin x$.
  b) By translating, sketch a graph of $y = \sin(x + \pi)$.
  c) By reflecting the graph of part (a), sketch a graph of $y = -\sin x$.
  d) How do the graphs of parts (b) and (c) compare?

46. a) Sketch a graph of $y = \sin x$.
  b) By translating, sketch a graph of $y = \sin(x - \pi)$.
  c) By reflecting the graph of part (a), sketch a graph of $y = -\sin x$.
  d) How do the graphs of parts (b) and (c) compare?

47. a) Sketch a graph of $y = \cos x$.
  b) By translating, sketch a graph of $y = \cos(x + \pi)$.
  c) By reflecting the graph of part (a), sketch a graph of $y = -\cos x$.
  d) How do the graphs of parts (b) and (c) compare?

48. a) Sketch a graph of $y = \cos x$.
  b) By translating, sketch a graph of $y = \cos(x - \pi)$.
  c) By reflecting the graph of part (a), sketch a graph of $y = -\cos x$.
  d) How do the graphs of parts (b) and (c) compare?

49. Of the six circular functions, which are even? Which are odd?

50. Of the six circular functions, which have period π? Which have period 2π?

Consider the coordinates on the unit circle for Exercises 51–54.

51. In which quadrants is the tangent function positive? negative?

52. In which quadrants is the sine function positive? negative?

53. In which quadrants is the cosine function positive? negative?

54. In which quadrants is the cosecant function positive? negative?

Skill Maintenance

Graph both functions in the same viewing window and describe how g is a transformation of f.

55. $f(x) = x^2$, $g(x) = 2x^2 - 3$

56. $f(x) = x^2$, $g(x) = (x - 2)^2$

57. $f(x) = |x|$, $g(x) = \frac{1}{2}|x - 4| + 1$

58. $f(x) = x^3$, $g(x) = -x^3$

Write an equation for a function that has a graph with the given characteristics. Check with a grapher.

59. The shape of $y = x^3$, but reflected across the x-axis, shifted right 2 units, and shifted down 1 unit

60. The shape of $y = 1/x$, but shrunk vertically by a factor of $\frac{1}{4}$ and shifted up 3 units

Synthesis

61. ◆ Describe how the graphs of the sine and cosine functions are related.

62. ◆ Explain why both the sine and cosine functions are continuous, but the tangent function, defined as sine/cosine, is not continuous.

Complete. (For example, sin (x + 2π) = sin x.)

63. $\cos(-x) = $ _____

64. $\sin(-x) = $ _____

65. $\sin(x + 2k\pi)$, $k \in \mathbb{Z} = $ _____

66. $\cos(x + 2k\pi)$, $k \in \mathbb{Z} = $ _____

67. $\sin(\pi - x) = $ _____

68. $\cos(\pi - x) = $ _____

69. $\cos(x - \pi) = $ _____

70. $\cos(x + \pi) = $ _____

71. $\sin(x + \pi) = $ _____

72. $\sin(x - \pi) = $ _____

73. Find all numbers x that satisfy the following. Check with a grapher.

a) $\sin x = 1$ **b)** $\cos x = -1$
c) $\sin x = 0$ **d)** $\sin x < \cos x$

74. Find $f \circ g$ and $g \circ f$, where $f(x) = x^2 + 2x$ and $g(x) = \cos x$.

Use a grapher to determine the domain, the range, the period, and the amplitude of each of these functions.

75. $y = (\sin x)^2$ **76.** $y = |\cos x| + 1$

Determine the domain of each function.

77. $f(x) = \sqrt{\cos x}$ **78.** $g(x) = \dfrac{1}{\sin x}$

79. $f(x) = \dfrac{\sin x}{\cos x}$ **80.** $g(x) = \log(\sin x)$

81. Consider $(\sin x)/x$, where x is between 0 and $\pi/2$. As x approaches 0, this function approaches a limiting value. What is it?

82. One of the motivations for developing trigonometry with a unit circle is that you can actually "see" $\sin \theta$ and $\cos \theta$ on the circle. Note in the figure that $AP = \sin \theta$ and $OA = \cos \theta$. It turns out that you can also "see" the other four trigonometric functions. Prove each of the following.

a) $BD = \tan \theta$ **b)** $OD = \sec \theta$
c) $OE = \csc \theta$ **d)** $CE = \cot \theta$

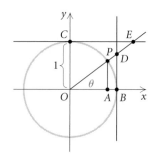

Graph.

83. $y = 3 \sin x$ **84.** $y = \sin |x|$

85. $y = \sin x + \cos x$ **86.** $y = |\cos x|$

2.6
Graphs of Transformed Sine and Cosine Functions

- *Graph transformations of $y = \sin x$ and $y = \cos x$ in the form*

$$y = A \sin (Cx - D) + B$$

and

$$y = A \cos (Cx - D) + B$$

and determine the amplitude, the period, and the phase shift.
- *Graph sums of functions.*

Variations of Basic Graphs

In Section 2.5, we graphed all six trigonometric functions. In this section, we will consider variations of the sine and cosine graphs. For example, we will graph equations like the following:

$$y = 5 \sin \tfrac{1}{2}x, \quad y = \cos (2x - \pi), \quad \text{and} \quad y = \tfrac{1}{2} \sin x - 3.$$

In particular, we are interested in graphs in the form

$$y = A \sin (Cx - D) + B$$

and

$$y = A \cos (Cx - D) + B,$$

where A, B, C, and D are constants. These constants have the effect of translating, reflecting, stretching, and shrinking the basic graphs. (It might be helpful to review Section 1.5.) Let's first examine the effect of each constant individually. Then we will consider the combined effects of more than one constant.

We first consider the effect of the constant B.

Interactive Discovery

Graph the first two functions in each group and then predict what the graph of the third function looks like. Then graph that function on a grapher.

$$y_1 = \sin x, \quad y_2 = \sin x + 3.5, \quad y_3 = \sin x - 4;$$
$$y_1 = \cos x, \quad y_2 = \cos x - 2, \quad y_3 = \cos x + 0.5.$$

What is the effect of the constant B on the graph of the basic function?

Example 1 Sketch a graph of $y = \sin x + 3$.

SOLUTION   The graph of $y = \sin x + 3$ is a vertical translation of the graph of $y = \sin x$ up 3 units. One way to sketch the graph is to first consider $y = \sin x$ on an interval of length 2π, say, $[0, 2\pi]$. The zeros of the function and the maximum and minimum values can be considered key points. These are

$$(0, 0), \quad \left(\frac{\pi}{2}, 1\right), \quad (\pi, 0), \quad \left(\frac{3\pi}{2}, -1\right), \quad (2\pi, 0).$$

These key points are transformed up 3 units to obtain the key points of the graph of $y = \sin x + 3$. These are

$$(0, 3), \quad \left(\frac{\pi}{2}, 4\right), \quad (\pi, 3), \quad \left(\frac{3\pi}{2}, 2\right), \quad (2\pi, 3).$$

The graph of $y = \sin x + 3$ can be sketched over the interval $[0, 2\pi]$ and extended to obtain the rest of the graph by repeating the graph over intervals of length 2π.

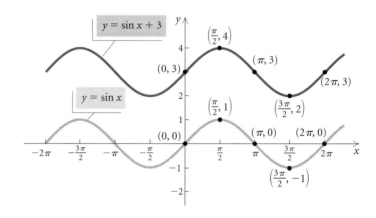

The constant B in

$$y = A \sin (Cx - D) + B \quad \text{and} \quad y = A \cos (Cx - D) + B$$

translates the graphs up B units if $B > 0$ or down $|B|$ units if $B < 0$.

Next we consider the effect of the constant A.

Interactive Discovery

Graph the first two functions in each group and then predict what the graph of the third function looks like. Then graph that function on a grapher.

$$y_1 = \sin x, \quad y_2 = 2 \sin x, \quad y_3 = \tfrac{1}{2} \sin x;$$
$$y_1 = \cos x, \quad y_2 = 0.8 \cos x, \quad y_3 = -0.8 \cos x.$$

What is the effect of the constant A on the graph of the basic function (a) when $0 < A < 1$? (b) when $A > 1$? (c) when $-1 < A < 0$? (d) when $A < -1$?

Example 2   Sketch a graph of $y = 2 \cos x$. What is the amplitude?

SOLUTION   The constant 2 in $y = 2 \cos x$ has the effect of stretching the graph of $y = \cos x$ vertically by a factor of 2 units. The function values of $y = \cos x$ are such that $-1 \le \cos x \le 1$. The function values of $y = 2 \cos x$ are such that $-2 \le 2 \cos x \le 2$. The maximum value of $y = 2 \cos x$ is 2, and the minimum value is -2. Thus the amplitude, A, is $\frac{1}{2}[2 - (-2)]$, or 2.

We draw the graph of $y = \cos x$ and consider its key points,

$$(0, 1), \quad \left(\frac{\pi}{2}, 0\right), \quad (\pi, -1), \quad \left(\frac{3\pi}{2}, 0\right), \quad (2\pi, 1),$$

over the interval $[0, 2\pi]$.

We then multiply the second coordinates by 2 to obtain the key points of $y = 2 \cos x$. These are

$$(0, 2), \quad \left(\frac{\pi}{2}, 0\right), \quad (\pi, -2), \quad \left(\frac{3\pi}{2}, 0\right), \quad (2\pi, 2).$$

We plot these points and sketch the graph over the interval $[0, 2\pi]$. Then we repeat this part of the graph over adjacent intervals of length 2π.

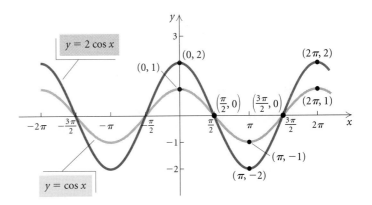

If $|A| > 1$, then there will be a vertical stretching. If $|A| < 1$, then there will be a vertical shrinking.

Amplitude

The *amplitude* of the graphs of $y = A \sin (Cx - D) + B$ and $y = A \cos (Cx - D) + B$ is $|A|$.

If $A < 0$, the graph is also reflected across the x-axis.

Example 3 Sketch a graph of $y = -\frac{1}{2} \sin x$.

SOLUTION The amplitude of the graph is $\left|-\frac{1}{2}\right|$, or $\frac{1}{2}$. The graph of $y = -\frac{1}{2} \sin x$ is a vertical shrinking and a reflection of the graph of $y = \sin x$ across the x-axis. In graphing, the key points of $y = \sin x$,

$$(0, 0), \quad \left(\frac{\pi}{2}, 1\right), \quad (\pi, 0), \quad \left(\frac{3\pi}{2}, -1\right), \quad (2\pi, 0),$$

are transformed to

$$(0, 0), \quad \left(\frac{\pi}{2}, -\frac{1}{2}\right), \quad (\pi, 0), \quad \left(\frac{3\pi}{2}, \frac{1}{2}\right), \quad (2\pi, 0).$$

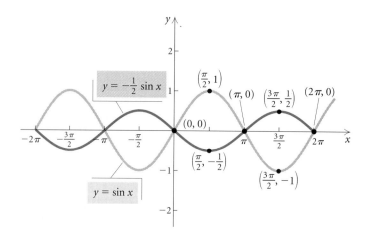

Now we consider the effect of the constant C.

**Interactive
Discovery**

The period of the sine and cosine functions is 2π. Each of the graphs in Examples 1–3 also has a period of 2π. In general, changes in the constants A and B do not change the period. To see what effect a change in the constant C will have on the basic functions, graph

$$y_1 = \cos x,$$
$$y_2 = \cos(-2x),$$
$$y_3 = \cos \frac{1}{2} x.$$

Then visualize what the graphs of $y_4 = \cos 4x$ and $y_5 = \cos\left(-\frac{1}{2}x\right)$ will look like and check your guesses with a grapher. What is the period of each of these graphs?

Example 4   Sketch a graph of $y = \sin 4x$. What is the period?

SOLUTION   The constant C has the effect of changing the period. Recall from Section 1.5 that the graph of $y = f(4x)$ is obtained from the graph of $y = f(x)$ by shrinking the graph horizontally. The new graph is obtained by dividing the first coordinate of each ordered-pair solution of $y = f(x)$ by 4. The key points of $y = \sin x$ are

$$(0, 0), \quad \left(\frac{\pi}{2}, 1\right), \quad (\pi, 0), \quad \left(\frac{3\pi}{2}, -1\right), \quad (2\pi, 0).$$

These are transformed to the key points of $y = \sin 4x$, which are

$$(0, 0), \quad \left(\frac{\pi}{8}, 1\right), \quad \left(\frac{\pi}{4}, 0\right), \quad \left(\frac{3\pi}{8}, -1\right), \quad \left(\frac{\pi}{2}, 0\right).$$

We plot these key points and sketch in the graph over the shortened interval $[0, \pi/2]$, which is of length $\pi/2$. Then we repeat the graph over other intervals of length $\pi/2$.

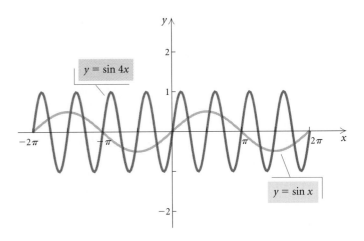

If $|C| < 1$, then there will be a horizontal stretching. If $|C| > 1$, then there will be a horizontal shrinking. If $C < 0$, the graph is also reflected across the y-axis.

Period

The *period* of the graphs of $y = A \sin (Cx - D) + B$ and $y = A \cos (Cx - D) + B$ is $\left| \dfrac{2\pi}{C} \right|$.

Now we examine the effect of the constant D.

Interactive Discovery

Graph

$$y_1 = \cos x,$$

$$y_2 = \cos (x - \pi),$$

$$y_3 = \cos \left(x + \frac{\pi}{2} \right),$$

$$y_4 = \cos \left(x - \frac{\pi}{4} \right).$$

Each curve has an amplitude of 1 and a period of 2π, but there are four distinct graphs. What is the effect of the constant D?

Example 5   Sketch a graph of $y = \sin \left(x - \dfrac{\pi}{2} \right)$.

SOLUTION   The amplitude is 1, and the period is 2π. Recall from Section 1.5 that the graph of $y = f(x - d)$ is obtained from the graph of $y = f(x)$ by translating the graph horizontally—to the right d units if $d > 0$ and to the left $|d|$ units if $d < 0$. The graph of $y = \sin (x - \pi/2)$ is a translation of the graph of $y = \sin x$ to the right $\pi/2$ units. The value

$\pi/2$ is called the *phase shift*. The key points of $y = \sin x$,

$$(0, 0), \quad \left(\frac{\pi}{2}, 1\right), \quad (\pi, 0), \quad \left(\frac{3\pi}{2}, -1\right), \quad (2\pi, 0),$$

are transformed by adding $\pi/2$ to each of the first coordinates to obtain the following key points of $y = \sin (x - \pi/2)$:

$$\left(\frac{\pi}{2}, 0\right), \quad (\pi, 1), \quad \left(\frac{3\pi}{2}, 0\right), \quad (2\pi, -1), \quad \left(\frac{5\pi}{2}, 0\right).$$

We plot these key points and sketch the curve over the interval $[\pi/2, 5\pi/2]$. Then we repeat the graph over other intervals of length 2π.

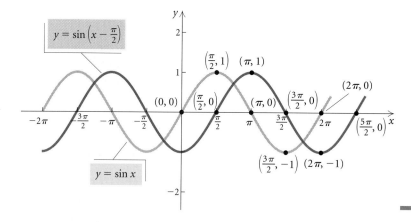

Now we consider combined transformations of graphs. It is helpful to rewrite

$$y = A \sin (Cx - D) + B \qquad \text{and} \quad y = A \cos (Cx - D) + B$$

as

$$y = A \sin \left[C\left(x - \frac{D}{C}\right)\right] + B \quad \text{and} \quad y = A \cos \left[C\left(x - \frac{D}{C}\right)\right] + B.$$

Example 6   Sketch a graph of $y = \cos (2x - \pi)$. What is the period?

SOLUTION   The graph of

$$y = \cos (2x - \pi)$$

is the same as the graph of

$$y = 1 \cdot \cos \left[2\left(x - \frac{\pi}{2}\right)\right].$$

The amplitude is 1. The factor 2 shrinks the period by half, making the period $|2\pi/2|$, or π. The $\pi/2$ translates the graph of $y = \cos 2x$ to the right $\pi/2$ units. Thus to form the graph, we first graph $y = \cos x$, followed by $y = \cos 2x$ and then $y = \cos [2(x - \pi/2)]$.

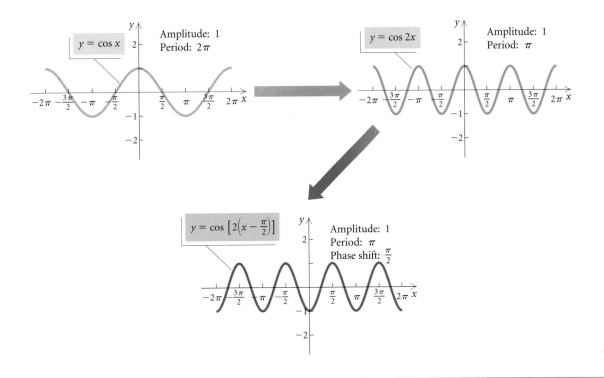

Phase Shift

The *phase shift* of the graphs

$$y = A \sin (Cx - D) + B = A \sin \left[C \left(x - \frac{D}{C} \right) \right] + B$$

and

$$y = A \cos (Cx - D) + B = A \cos \left[C \left(x - \frac{D}{C} \right) \right] + B$$

is the quantity $\dfrac{D}{C}$.

If $D/C > 0$, the graph is translated to the right D/C units. If $D/C < 0$, the graph is translated to the left $|D/C|$ units. Be sure that the horizontal stretching or shrinking based on the constant C is done before the translation based on the phase shift D/C.

Let's now summarize the effect of the constants. We carry out the procedures in the order listed.

Transformations of Sine and Cosine Functions

To graph

$$y = A \sin (Cx - D) + B = A \sin \left[C\left(x - \frac{D}{C} \right) \right] + B$$

and

$$y = A \cos (Cx - D) + B = A \cos \left[C\left(x - \frac{D}{C} \right) \right] + B,$$

follow the steps listed below in the order in which they are listed.

1. Stretch or shrink the graph horizontally according to C.

$|C| < 1$	Stretch horizontally
$|C| > 1$	Shrink horizontally
$C < 0$	Reflect across the y-axis

The *period* is $\left| \dfrac{2\pi}{C} \right|$.

2. Stretch or shrink the graph vertically according to A.

$|A| < 1$	Shrink vertically
$|A| > 1$	Stretch vertically
$A < 0$	Reflect across the x-axis

The *amplitude* is $|A|$.

3. Translate the graph horizontally according to D/C.

$\dfrac{D}{C} < 0$	$\left| \dfrac{D}{C} \right|$ units to the left
$\dfrac{D}{C} > 0$	$\dfrac{D}{C}$ units to the right

The *phase shift* is $\dfrac{D}{C}$.

4. Translate the graph vertically according to B.

$B < 0$	$|B|$ units down
$B > 0$	B units up

Example 7 Sketch a graph of $y = 3 \sin (2x + \pi/2) + 1$. Find the amplitude, the period, and the phase shift.

SOLUTION We first note that

$$y = 3 \sin \left(2x + \frac{\pi}{2} \right) + 1 = 3 \sin \left[2\left(x - \left(-\frac{\pi}{4} \right) \right) \right] + 1.$$

Then we have the following:

$$\text{Amplitude} = |A| = |3| = 3,$$

$$\text{Period} = \left|\frac{2\pi}{C}\right| = \left|\frac{2\pi}{2}\right| = \pi,$$

$$\text{Phase shift} = \frac{D}{C} = -\frac{\pi}{4}.$$

To create the final graph, we begin with the basic sine curve, $y = \sin x$. Then we sketch graphs of each of the following equations in sequence.

1. $y = \sin 2x$

2. $y = 3 \sin 2x$

3. $y = 3 \sin \left[2\left(x - \left(-\frac{\pi}{4}\right)\right)\right]$

4. $y = 3 \sin \left[2\left(x - \left(-\frac{\pi}{4}\right)\right)\right] + 1$

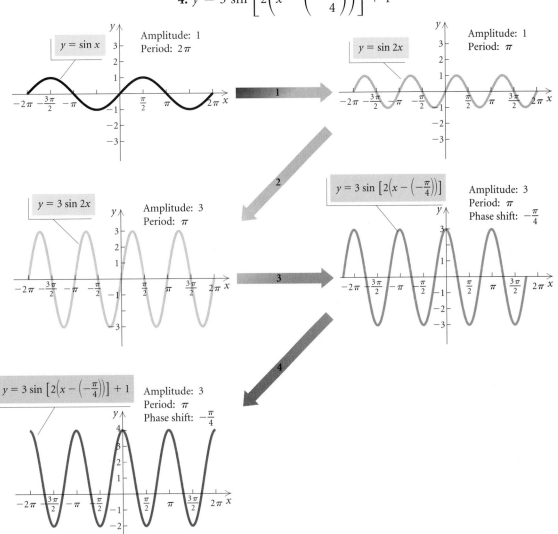

All the graphs in Examples 1–7 can be checked with a grapher. Even though it is faster and more accurate to graph with a grapher, graphing by hand gives us a greater understanding of the effect of changing the constants A, B, C, and D.

Graphers are especially convenient when a period or a phase shift is not a multiple of $\pi/4$, such as 4.3, 1, or $2\pi/3$.

Example 8 Graph $y = 3 \cos (2\pi x) - 1$. Find the amplitude, the period, and the phase shift.

SOLUTION First we note the following:

$$\text{Amplitude} = |A| = |3| = 3,$$

$$\text{Period} = \left|\frac{2\pi}{C}\right| = \left|\frac{2\pi}{2\pi}\right| = |1| = 1,$$

$$\text{Phase shift} = \frac{D}{C} = \frac{0}{2\pi} = 0.$$

There is no phase shift in this case because the constant $D = 0$. The graph has a vertical translation of the graph of the cosine function down 1 unit, an amplitude of 3, and a period of 1, so we can use $[-4, 4, -5, 5]$ as the viewing window. ▬

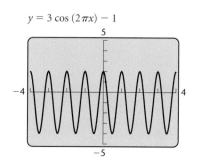

$y = 3 \cos (2\pi x) - 1$

The transformational techniques that we learned in this section for graphing the sine and cosine functions can also be applied in the same manner to the other trigonometric functions. A few of these transformations are addressed in the synthesis exercises in Exercise Set 2.6.

An **oscilloscope** is an electronic device that converts electrical signals into graphs like those in the preceding examples. These graphs are often called sine waves. By manipulating the controls, we can change the amplitude, the period, and the phase of sine waves. The oscilloscope has many applications, and the trigonometric functions play a major role in many of them.

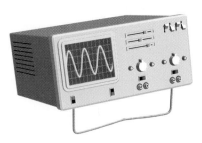

Graphs of Sums: Addition of Ordinates

The output of an electronic synthesizer used in the recording and playing of music can be converted into sine waves by an oscilloscope. The following graphs illustrate simple tones of different frequencies. The frequency of a simple tone is the number of vibrations in the signal of the tone per second. The loudness or intensity of the tone is reflected in the height of the graph (its amplitude). The three tones in the diagrams below all have the same intensity.

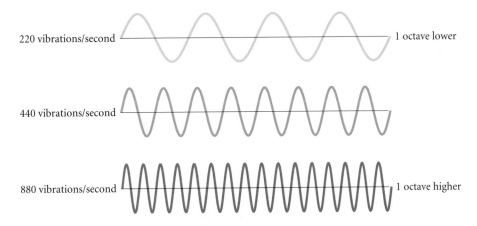

220 vibrations/second — 1 octave lower

440 vibrations/second

880 vibrations/second — 1 octave higher

Musical instruments can generate extremely complex sine waves. On a single instrument, overtones can become superimposed on a simple tone. When multiple notes are played simultaneously, graphs become very complicated. This can happen when multiple notes are played on a single instrument or a group of instruments, or even when the same simple note is played on different instruments.

Combinations of simple tones produce interesting curves. Consider two tones whose graphs are $y_1 = 2 \sin x$ and $y_2 = \sin 2x$. The combination of the two tones produces a new sound whose graph is $y = 2 \sin x + \sin 2x$, shown in the following example.

Example 9 Graph: $y = 2 \sin x + \sin 2x$.

SOLUTION We graph $y = 2 \sin x$ and $y = \sin 2x$ using the same set of axes.

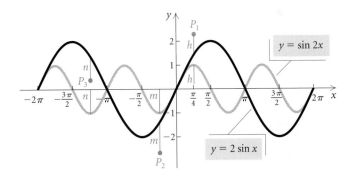

Now we graphically add some y-coordinates, or ordinates, to obtain points on the graph that we seek. At $x = \pi/4$, we transfer the distance h, which is the value of $\sin 2x$, up to add it to the value of $2 \sin x$. Point P_1 is on the graph that we seek. At $x = -\pi/4$, we use a similar procedure, but this time both ordinates are negative. Point P_2 is on the graph. At $x = -5\pi/4$, we add the negative ordinate of $\sin 2x$ to the positive ordinate of $2 \sin x$. Point P_3 is also on the graph. We continue to plot points in this fashion and then connect them to get the desired graph, shown at the top of the following page.

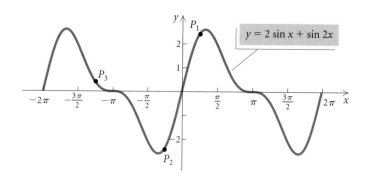

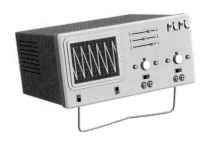

A "sawtooth function," such as the one shown on the oscilloscope at left, has numerous applications. This curve, which is actually not a function, can be approximated extremely well by adding, electronically, several sine and cosine functions.

With a grapher, we can quickly determine the period of a trigonometric function that is a combination of sine and cosine functions.

Example 10   Graph $y = 2 \cos x - \sin 3x$ and determine its period.

SOLUTION   We graph $y = 2 \cos x - \sin 3x$ with appropriate dimensions. The period appears to be 2π.

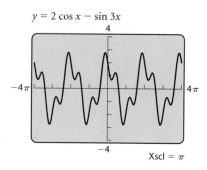

$y = 2 \cos x - \sin 3x$

$\text{Xscl} = \pi$

2.6 | *Exercise Set*

Determine the amplitude, the period, and the phase shift of each function, and, without a grapher, sketch the graph of the function. Then check the graph with a grapher.

1. $y = \sin x + 1$

2. $y = \frac{1}{4} \cos x$

3. $y = -3 \cos x$

4. $y = \sin (-2x)$

5. $y = 2 \sin \left(\frac{1}{2} x \right)$

6. $y = \cos \left(x - \frac{\pi}{2} \right)$

7. $y = \frac{1}{2} \sin \left(x + \frac{\pi}{2} \right)$

8. $y = \cos x - \frac{1}{2}$

9. $y = 3 \cos (x - \pi)$

10. $y = -\sin \left(\frac{1}{4} x \right) + 1$

11. $y = \frac{1}{3} \sin x - 4$

12. $y = \cos \left(\frac{1}{2} x + \frac{\pi}{2} \right)$

13. $y = -\cos (-x) + 2$

14. $y = \dfrac{1}{2} \sin \left(2x - \dfrac{\pi}{4} \right)$

Determine the amplitude, the period, and the phase shift of each function. Then graph the function with a grapher. Try to visualize the graph before creating it.

15. $y = 2 \cos \left(\dfrac{1}{2} x - \dfrac{\pi}{2} \right)$

16. $y = 4 \sin \left(\dfrac{1}{4} x + \dfrac{\pi}{8} \right)$

17. $y = -\dfrac{1}{2} \sin \left(2x + \dfrac{\pi}{2} \right)$

18. $y = -3 \cos (4x - \pi) + 2$

19. $y = 2 + 3 \cos (\pi x - 3)$

20. $y = 5 - 2 \cos \left(\dfrac{\pi}{2} x + \dfrac{\pi}{2} \right)$

21. $y = -\dfrac{1}{2} \cos (2\pi x) + 2$

22. $y = -2 \sin (-2x + \pi) - 2$

23. $y = -\sin \left(\dfrac{1}{2} x - \dfrac{\pi}{2} \right) + \dfrac{1}{2}$

24. $y = \dfrac{1}{3} \cos (-3x) + 1$

25. $y = \cos (-2\pi x) + 2$

26. $y = \dfrac{1}{2} \sin (2\pi x + \pi)$

27. $y = -\dfrac{1}{4} \cos (\pi x - 4)$

28. $y = 2 \sin (2\pi x + 1)$

In Exercises 29–36, without a grapher, match each function with graphs (a)–(h), which follow. Then check your work with a grapher.

a)
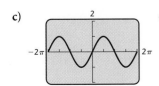

b)
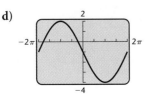

c)
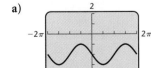

d)
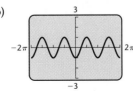

e)
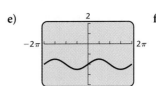

f)
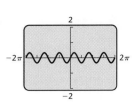

g)
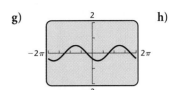

h)
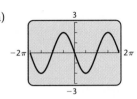

29. $y = -\cos 2x$

30. $y = \dfrac{1}{2} \sin x - 2$

31. $y = 2 \cos \left(x + \dfrac{\pi}{2} \right)$

32. $y = -3 \sin \dfrac{1}{2} x - 1$

33. $y = \sin (x - \pi) - 2$

34. $y = -\dfrac{1}{2} \cos \left(x - \dfrac{\pi}{4} \right)$

35. $y = \dfrac{1}{3} \sin 3x$

36. $y = \cos \left(x - \dfrac{\pi}{2} \right)$

Graph using addition of ordinates. Then check your graph with a grapher.

37. $y = 2 \cos x + \cos 2x$

38. $y = 3 \cos x + \cos 3x$

39. $y = \sin x + \cos 2x$

40. $y = 2 \sin x + \cos 2x$

41. $y = \sin x - \cos x$

42. $y = 3 \cos x - \sin x$

43. $y = 3 \cos x + \sin 2x$

44. $y = 3 \sin x - \cos 2x$

Use a grapher to graph each function.

45. $y = x + \sin x$

46. $y = \cos x - x$

47. $y = \cos 2x + 2x$

48. $y = \cos 3x + \sin 3x$

49. $y = 4 \cos 2x - 2 \sin x$

50. $y = 7.5 \cos x + \sin 2x$

Skill Maintenance

51. Add: $\dfrac{2x}{x^2 - 1} + \dfrac{4}{x - 1}$.

52. Multiply: $(x^2 - 4)(3x + 7)$.

53. Simplify: $\dfrac{2x^2 - 3x - 20}{2x^2 + 3x - 5}$.

54. Subtract: $\dfrac{1}{x - 6} - 4$.

Synthesis

55. ◆ In the equations $y = A \sin (Cx - D) + B$ and $y = A \cos (Cx - D) + B$, which constants translate the graphs and which constants stretch and shrink the graphs? Describe in your own words the effect of each constant.

56. ◆ In the transformation steps listed in this section, why must step (1) precede step (3)? Give an example that illustrates this.

Some graphers can use regression to fit a trigonometric function to a set of data.

57. *Sales.* Sales of certain products fluctuate in cycles. The data in the following table show the total sales of skis per month for a business in a northern climate.

x, MONTH	TOTAL SALES, y (IN THOUSANDS OF DOLLARS)
8. August	0
11. November	7
2. February	14
5. May	7
8. August	0

a) Using the SINE REGRESSION feature on a grapher, fit a sine function of the form $y = A \sin (Cx - D) + B$ to this set of data.

b) Approximate the total sales for December and for July.

58. *Daylight Hours.* The data in the following table give the number of daylight hours for certain days in Kajaani, Finland.

x, DAY	NUMBER OF DAYLIGHT HOURS, y
10. January 10	5.0
50. February 19	9.1
62. March 3	10.4
118. April 28	16.4
134. May 14	18.2
162. June 11	20.7
198. July 17	19.5
234. August 22	15.7
262. September 19	12.7
274. October 1	11.4
318. November 14	6.7
362. December 28	4.3

Source: The Astronomical Almanac, 1995, Washington: U.S. Government Printing Office.

a) Using the SINE REGRESSION feature on a grapher, model these data with an equation of the form $y = A \sin (Cx - D) + B$.

b) Approximate the number of daylight hours in Kajaani for April 22, July 4, and December 15.

The transformational techniques that we learned in this section for graphing the sine and cosine functions can also be applied to the other trigonometric functions. Sketch a graph of each of the following. Then check your work with a grapher.

59. $y = -\tan x$

60. $y = \tan (-x)$

61. $y = -2 + \cot x$

62. $y = -\dfrac{3}{2} \csc x$

63. $y = 2 \tan \dfrac{1}{2} x$

64. $y = \cot 2x$

65. $y = 2 \sec (x - \pi)$

66. $y = 4 \tan \left(\dfrac{1}{4} x + \dfrac{\pi}{8} \right)$

67. $y = 2 \csc \left(\dfrac{1}{2} x - \dfrac{3\pi}{4} \right)$

68. $y = 4 \sec (2x - \pi)$

Use a grapher to graph each function.

69. $f(x) = e^{-x/2} \cos x$

70. $f(x) = x \sin x$

71. $f(x) = 0.6x^2 \cos x$

72. $f(x) = 2^{-x} \cos x$

73. $f(x) = |\tan x|$

74. $f(x) = (\tan x)(\csc x)$

Use a grapher to graph each of the following over the given interval and approximate the zeros.

75. $f(x) = \dfrac{\sin x}{x}$; $[-12, 12]$

76. $f(x) = \dfrac{\cos x - 1}{x}$; $[-12, 12]$

77. $f(x) = x^3 \sin x$; $[-5, 5]$

78. $f(x) = \dfrac{(\sin x)^2}{x}$; $[-4, 4]$

79. *Temperature During an Illness.* The temperature T of a patient during a 12-day illness is given by

$$T(t) = 101.6° + 3° \sin\left(\frac{\pi}{8}t\right).$$

a) Graph the function over the interval $[0, 12]$.
b) What are the maximum and the minimum temperatures during the illness?

80. *Periodic Sales.* A company in a northern climate has sales of skis as given by

$$S(t) = 10\left(1 - \cos\frac{\pi}{6}t\right),$$

where t = time, in months ($t = 0$ corresponds to July 1), and $S(t)$ is in thousands of dollars.

a) Graph the function over a 12-month interval $[0, 12]$.
b) What is the period of the function?
c) What is the minimum amount of sales and when does it occur?
d) What is the maximum amount of sales and when does it occur?

81. *Satellite Location.* A satellite circles the earth in such a way that it is y miles from the equator (north or south, height not considered) t minutes after its launch, where

$$y(t) = 3000\left[\cos\frac{\pi}{45}(t - 10)\right].$$

a) Graph the function.
b) What are the amplitude, the period, and the phase shift?

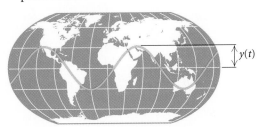

82. *Water Wave.* The cross-section of a water wave is given by

$$y = 3 \sin\left(\frac{\pi}{4}x + \frac{\pi}{4}\right),$$

where y is the vertical height of the water wave and x is the distance from the origin to the wave.

a) Graph the function.
b) What are the amplitude, the period, and the phase shift?

83. *Damped Oscillations.* Suppose that the motion of a spring is given by

$$d(t) = 6e^{-0.8t} \cos(6\pi t) + 4,$$

where d is the distance, in inches, of a weight from the point at which the spring is attached to a ceiling, after t seconds.

a) Graph the function over the interval $[0, 10]$.
b) How far do you think the spring is from the ceiling when the spring stops bobbing?

84. *Rotating Beacon.* A police car is parked 10 ft from a wall. On top of the car is a beacon rotating in such a way that the light is at a distance $d(t)$ from point Q after t seconds, where

$$d(t) = 10 \tan(2\pi t).$$

When d is positive, the light is pointing north of Q, and when d is negative, the light is pointing south of Q.

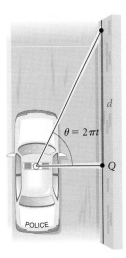

a) Graph the function over the interval $[0, 2]$.
b) Explain the meaning of the values of t for which the function is undefined.

CHAPTER 2

Summary and Review

Important Properties and Formulas

Trigonometric Function Values of an Acute Angle θ

Let θ be an acute angle of a right triangle. The six trigonometric functions of θ are as follows:

$$\sin \theta = \frac{\text{opp}}{\text{hyp}}, \quad \cos \theta = \frac{\text{adj}}{\text{hyp}}, \quad \tan \theta = \frac{\text{opp}}{\text{adj}}.$$

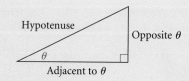

Reciprocal Functions

$$\csc \theta = \frac{1}{\sin \theta}, \quad \sec \theta = \frac{1}{\cos \theta}, \quad \cot \theta = \frac{1}{\tan \theta}$$

Function Values of Special Angles

	0°	30°	45°	60°	90°
sin	0	1/2	$\sqrt{2}/2$	$\sqrt{3}/2$	1
cos	1	$\sqrt{3}/2$	$\sqrt{2}/2$	1/2	0
tan	0	$\sqrt{3}/3$	1	$\sqrt{3}$	Undefined

Cofunction Identities

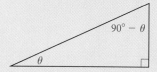

$$\sin \theta = \cos (90° - \theta) \quad \cos \theta = \sin (90° - \theta)$$
$$\tan \theta = \cot (90° - \theta) \quad \cot \theta = \tan (90° - \theta)$$
$$\sec \theta = \csc (90° - \theta) \quad \csc \theta = \sec (90° - \theta)$$

Trigonometric Functions of Any Angle θ

$P(x, y)$ is any point on the terminal side of any angle θ in standard position, and r is the distance from the origin to $P(x, y)$.

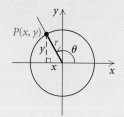

$$\sin \theta = \frac{y}{r}, \quad \cos \theta = \frac{x}{r}, \quad \tan \theta = \frac{y}{x}$$

(continued)

Signs of Function Values

The signs of the function values depend only on the coordinates of the point P on the terminal side.

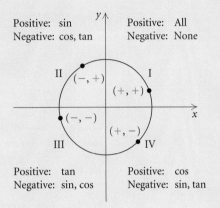

Positive: sin
Negative: cos, tan

Positive: All
Negative: None

Positive: tan
Negative: sin, cos

Positive: cos
Negative: sin, tan

Radian–Degree Equivalents

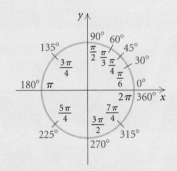

Linear Speed in Terms of Angular Speed

$$v = r\omega$$

Basic Circular Functions

For a real number s that determines a point (x, y) on the *unit circle*:

$$\sin s = y, \qquad \cos s = x, \qquad \tan s = \frac{y}{x}.$$

Sine (Odd Function): $\sin (-s) = -\sin s$

Cosine (Even Function): $\cos (-s) = \cos s$

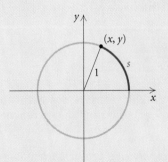

Transformations of Sine and Cosine Functions

To graph $y = A \sin (Cx - D) + B$ and $y = A \cos (Cx - D) + B$:

1. Stretch or shrink the graph horizontally according to C.

2. Stretch or shrink the graph vertically according to A.

3. Translate the graph horizontally according to D/C.

4. Translate the graph vertically according to B.

REVIEW EXERCISES

1. Find the six trigonometric function values of the specified angle.

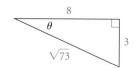

Find the exact function value, if it exists.

2. cos 45°

3. cot 60°

4. cos 495°

5. sin 150°

6. sec (−270°)

7. tan (−600°)

8. Convert 22.27° to degrees, minutes, and seconds. Round to the nearest second.

9. Convert 47°33′27″ to decimal degree notation. Round to two decimal places.

Find the function value. Round to four decimal places.

10. tan 2184°

11. sec 27.9°

12. cos 18°13′42″

13. sin 245°24′

14. cot (−33.2°)

15. sin 556.13°

Find θ in the interval indicated. Round the answer to the nearest tenth of a degree.

16. cos θ = −0.9041, (180°, 270°)

17. tan θ = 1.0799, (0°, 90°)

Find the exact acute angle θ, in degrees, given the function value.

18. $\sin \theta = \dfrac{\sqrt{3}}{2}$

19. $\tan \theta = \sqrt{3}$

20. Given that sin 59.1° ≈ 0.8581, cos 59.1° ≈ 0.5135, and tan 59.1° ≈ 1.6709, find the six function values for 30.9°.

Solve each of the following right triangles. Standard lettering has been used.

21. a = 7.3, c = 8.6

22. a = 30.5, B = 51.17°

23. One leg of a right triangle bears east. The hypotenuse is 734 m long and bears N57°23′E. Find the perimeter of the triangle.

24. An observer's eye is 6 ft above the floor. A mural is being viewed. The bottom of the mural is at floor level. The observer looks down 13° to see the bottom and up 17° to see the top. How tall is the mural?

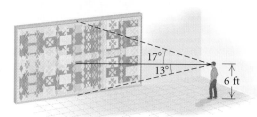

Find a positive angle and a negative angle that are coterminal with the given angle. Answers may vary.

25. 65°

26. $\dfrac{7\pi}{3}$

Find the complement and the supplement.

27. 13.4°

28. $\dfrac{\pi}{6}$

29. Find the six trigonometric function values for the angle θ shown.

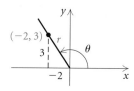

30. Given that tan θ = 2/√5 and that the terminal side is in quadrant III, find the other five function values.

31. An airplane travels at 530 mph for $3\frac{1}{2}$ hr in a direction of 160° from Minneapolis, Minnesota. At the end of that time, how far south of Minneapolis is the airplane?

32. On a unit circle, mark and label the points determined by 7π/6, −3π/4, −π/3, and 9π/4.

For angles of the following measures, state in which quadrant the terminal side lies, convert to radian measure in terms of π, and convert to radian measure not in terms of π.

33. 145.2°

34. −30°

Convert to degree measure. Round the answer to two decimal places.

35. $\dfrac{3\pi}{2}$

36. 3

37. Find the length of an arc of a circle, given a central angle of $\pi/4$ and a radius of 7 cm.

38. An arc 18 m long on a circle of radius 8 m subtends an angle of how many radians? how many degrees, to the nearest degree?

39. Inside La Madeleine French Bakery and Cafe in Houston, Texas, there is one of the few remaining working watermills in the world. The 300-yr-old French-built waterwheel has a radius of 7 ft and makes one complete revolution in 70 sec. What is the linear speed in feet per minute of a point on the rim? (*Source:* La Madeleine French Bakery and Cafe, Houston, TX)

40. An automobile wheel has a diameter of 14 in. If the car travels at a speed of 55 mph, what is the angular velocity, in radians per hour, of a point on the edge of the wheel?

41. The point $\left(\frac{3}{5}, -\frac{4}{5}\right)$ is on a unit circle. Find the coordinates of its reflections across the x-axis, the y-axis, and the origin.

Find the exact function value, if it exists.

42. $\cos \pi$ **43.** $\tan \dfrac{5\pi}{4}$ **44.** $\sin \dfrac{5\pi}{3}$

45. $\sin \left(-\dfrac{7\pi}{6}\right)$ **46.** $\tan \dfrac{\pi}{6}$ **47.** $\cos (-13\pi)$

Find the function value. Round to four decimal places.

48. $\sin 24$ **49.** $\cos (-75)$

50. $\cot 16\pi$ **51.** $\tan \dfrac{3\pi}{7}$

52. $\sec 14.3$ **53.** $\cos \left(-\dfrac{\pi}{5}\right)$

54. Make a hand-drawn graph from -2π to 2π for each of the six trigonometric functions.

55. What is the period of each of the six trigonometric functions?

56. Complete the following table.

FUNCTION	DOMAIN	RANGE
sine		
cosine		
tangent		

57. Complete the following table with the sign of the specified trigonometric function value in each of the four quadrants.

FUNCTION	I	II	III	IV
sine				
cosine				
tangent				

Determine the amplitude, the period, and the phase shift of each function; and without a grapher, sketch the graph of the function. Then check the graph with a grapher.

58. $y = \sin \left(x + \dfrac{\pi}{2}\right)$

59. $y = 3 + \dfrac{1}{2} \cos \left(2x - \dfrac{\pi}{2}\right)$

In Exercises 60–63, without a grapher, match each function with graphs (a)–(d), which follow. Then check your work with a grapher.

a)

b)

c)

d)

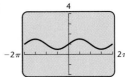

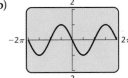

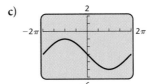

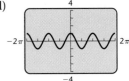

60. $y = \cos 2x$

61. $y = \dfrac{1}{2} \sin x + 1$

62. $y = -2 \sin \dfrac{1}{2} x - 3$

63. $y = -\cos \left(x - \dfrac{\pi}{2}\right)$

64. Sketch a graph of $y = 3 \cos x + \sin x$ for values of x between 0 and 2π.

Synthesis

65. ◆ Compare the terms radian and degree.

66. ◆ Describe the shape of the graph of the cosine function. How many maximum values are there of the cosine function? Where do they occur?

67. ◆ Explain the disadvantage of a grapher when graphing a function like

$$f(x) = \frac{\sin x}{x}.$$

68. ◆ Does $5 \sin x = 7$ have a solution for x? Why or why not?

69. For what values of x in $(0, \pi/2]$ is $\sin x < x$ true?

70. Graph $y = 3 \sin (x/2)$ and determine the domain, the range, and the period.

71. In the window below, $y_1 = \sin x$ is shown and y_2 is shown in red. Express y_2 as a transformation of the graph of y_1.

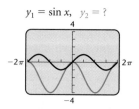

72. Find the domain of $y = \log (\cos x)$.

73. Given that $\sin x = 0.6144$ and that the terminal side is in quadrant II, find the other basic circular function values.

Trigonometric Identities, Inverse Functions, and Equations

3

A satellite circles the earth in such a manner that it is y miles from the equator (north or south) t minutes after its launch, where

$$y = 5000\left[\cos\frac{\pi}{45}(t - 10)\right].$$

This trigonometric equation can be used to determine the times t in the first 4 hr that the satellite is 3000 mi north of the equator.

X	Y1	
23.26	3006.3	
23.27	3003.5	
23.28	**3000.7**	
23.29	2997.9	
23.3	2995.1	
23.31	2992.3	
23.32	2989.5	

X = 23.28

$y_1 = 5000[\cos\frac{\pi}{45}(x - 10)], \quad y_2 = 3000$

6000

0   240

−6000

There are a number of relationships among the trigonometric functions, given by identities, that are important in algebraic and trigonometric manipulations. A large part of this chapter is devoted to those identities and their use in solving trigonometric equations. We also provide a detailed examination of inverses of the trigonometric functions. This chapter provides not only a basis for applications but also a foundation for further work in mathematics.

3.1

Identities: Pythagorean and Sum and Difference

- *State the Pythagorean identities.*
- *Simplify and manipulate expressions containing trigonometric expressions.*
- *Use the sum and difference identities to find function values.*

Recall that an identity is an equation that is true for all *possible* replacements of the variables. The following is a list of the identities studied in Chapter 2.

Basic Identities

$$\sin x = \frac{1}{\csc x}, \qquad \csc x = \frac{1}{\sin x}, \qquad \begin{aligned} \sin(-x) &= -\sin x, \\ \cos(-x) &= \cos x, \end{aligned}$$

$$\cos x = \frac{1}{\sec x}, \qquad \sec x = \frac{1}{\cos x}, \qquad \tan(-x) = -\tan x,$$

$$\tan x = \frac{1}{\cot x}, \qquad \cot x = \frac{1}{\tan x}, \qquad \tan x = \frac{\sin x}{\cos x},$$

$$\cot x = \frac{\cos x}{\sin x}$$

In this section, we will develop some other important identities.

Pythagorean Identities

We now consider three other identities that are fundamental to a study of trigonometry. They are called the *Pythagorean identities*. Recall that the equation of a unit circle in the xy-plane is $x^2 + y^2 = 1$.

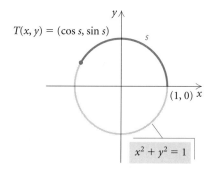

For any point on the unit circle, the coordinates x and y satisfy this equation. Suppose that a real number s determines a point T on the unit circle, with coordinates (x, y), or $(\cos s, \sin s)$. Then $x = \cos s$ and $y = \sin s$. Substituting $\cos s$ for x and $\sin s$ for y in the equation of the

unit circle gives us the identity

$$(\cos s)^2 + (\sin s)^2 = 1,$$

which can be expressed as

$$\mathbf{sin^2\ s + cos^2\ s = 1.}$$

It is conventional in trigonometry to use the notation $\sin^2 s$ rather than $(\sin s)^2$, which uses parentheses. Note that $\sin^2 s \neq \sin s^2$.

$y = (\sin x)^2$

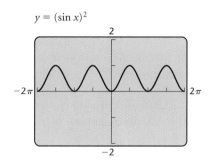

$y = \sin x^2$

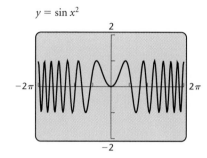

The identity $\sin^2 s + \cos^2 s = 1$ relates the sine and the cosine of any real number s. It is an important **Pythagorean identity**.

Interactive Discovery

Addition of y-values provides a unique way of developing the identity $\sin^2 x + \cos^2 x = 1$. First graph $y_1 = \sin^2 x$ and $y_2 = \cos^2 x$. By adding the y-values, visually graph the sum, $y_3 = \sin^2 x + \cos^2 x$. Then graph y_3 with a grapher and check your prediction. The resulting graph appears to be the line $y_4 = 1$, which is the graph of the expression on the right of the equals sign. These graphs do not prove the identity, but they do provide a check in the interval $[-2\pi, 2\pi]$.

$y_1 = \sin^2 x$

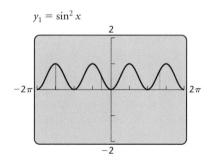

$y_2 = \cos^2 x$

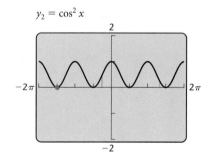

$y_3 = \sin^2 x + \cos^2 x,\ \ y_4 = 1$

We can divide on both sides of the preceding identity by $\sin^2 s$:

$$\frac{\sin^2 s}{\sin^2 s} + \frac{\cos^2 s}{\sin^2 s} = \frac{1}{\sin^2 s}.$$

Simplifying gives us a second identity:

$$1 + \cot^2 s = \csc^2 s.$$

This equation is true for any replacement of s by a real number for which $\sin^2 s \neq 0$, since we divided by $\sin^2 s$. But the numbers for which $\sin^2 s = 0$ (or $\sin s = 0$) are exactly the ones for which the cotangent and cosecant functions are undefined. Hence our new equation holds for all real numbers s for which $\cot s$ and $\csc s$ are defined and is thus an identity.

The third Pythagorean identity, obtained by dividing on both sides of the first Pythagorean identity by $\cos^2 s$, is

$$\tan^2 s + 1 = \sec^2 s.$$

The identities we have developed hold no matter what symbols are used for the variables. For example, $\sin^2 s + \cos^2 s = 1$, $\sin^2 \theta + \cos^2 \theta = 1$, and $\sin^2 x + \cos^2 x = 1$.

Pythagorean Identities

$\sin^2 x + \cos^2 x = 1,$

$1 + \cot^2 x = \csc^2 x,$

$1 + \tan^2 x = \sec^2 x$

It is important to remember the Pythagorean identities and how they are developed.

Simplifying Trigonometric Expressions

We can factor, simplify, and manipulate trigonometric expressions in the same way that we manipulate strictly algebraic expressions.

Example 1 Multiply and simplify: $\cos x (\tan x - \sec x)$.

SOLUTION

$\cos x (\tan x - \sec x)$

$\quad = \cos x \tan x - \cos x \sec x$ Multiplying

$\quad = \cos x \dfrac{\sin x}{\cos x} - \cos x \dfrac{1}{\cos x}$ Recalling the identities $\tan x = \dfrac{\sin x}{\cos x}$ and $\sec x = \dfrac{1}{\cos x}$ and substituting

$\quad = \sin x - 1$ Simplifying

There is no general procedure for creating an identity as in Example 1, but it is often helpful to write everything in terms of sines and cosines.

Example 2 Factor and simplify: $\sin^2 x \cos^2 x + \cos^4 x$.

SOLUTION

$$\sin^2 x \cos^2 x + \cos^4 x$$
$$= \cos^2 x \, (\sin^2 x + \cos^2 x) \qquad \text{Removing a common factor}$$
$$= \cos^2 x \qquad \qquad \text{Using } \sin^2 x + \cos^2 x = 1 \quad \blacksquare$$

Identities can be checked with a grapher. First graph the expression on the left side of the equals sign. Then graph the expression on the right side using the same screen. If the two graphs are indistinguishable, then we have a partial verification that the equation is an identity. Of course, we can never see the entire graph, so there can always be some doubt. Also, the graphs may not overlap precisely, but you may not be able to tell because the difference between the graphs may be less than the width of a pixel.

For example, consider the identity in Example 1:

$$\cos x \, (\tan x - \sec x) = \sin x - 1.$$

Graph each side, recalling that we may need to enter $\sec x$ as $1/\cos x$:

$$y_1 = \cos x \, (\tan x - \sec x) = \cos x \, [\tan x - (1/\cos x)]$$

and

$$y_2 = \sin x - 1.$$

As you can see in the screen below, the graphs appear to be identical. Thus, $\cos x \, (\tan x - \sec x) = \sin x - 1$ is most likely an identity.

$y_1 = \cos x \, (\tan x - \sec x), \ y_2 = \sin x - 1$

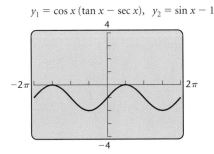

The TABLE feature can also be used to check identities. Note in the table above that the function values are the same except for those values of x for which $\cos x = 0$. The domain of y_1 excludes these values. The domain of y_2 is the set of all real numbers. Thus all real numbers except $\pm\pi/2, \pm 3\pi/2, \pm 5\pi/2, \ldots$ are possible replacements for x in the identity. Recall that an identity is an equation that is true for all *possible* replacements. Note that when the grapher graphs y_1, it simplifies the expression to $\sin x - 1$ before graphing. This is why the graph is continuous and does not show the domain restriction for y_1.

$y_1 = \cos x\,(\tan x - \sec x),$
$y_2 = \cos x - 1$

If we had simplified incorrectly in Example 1 and had gotten $\cos x - 1$ on the right instead of $\sin x - 1$, two different graphs would have appeared in the window. Thus we would have known that we did not have an identity and that $\cos x\,(\tan x - \sec x) \neq \cos x - 1$.

Example 3 Simplify each of the following trigonometric expressions.

a) $\dfrac{\cot(-\theta)}{\csc(-\theta)}$

b) $\dfrac{2\sin^2 t + \sin t - 3}{1 - \cos^2 t - \sin t}$

SOLUTION

a) $\dfrac{\cot(-\theta)}{\csc(-\theta)} = \dfrac{\dfrac{\cos(-\theta)}{\sin(-\theta)}}{\dfrac{1}{\sin(-\theta)}} = \dfrac{\cos(-\theta)}{\sin(-\theta)} \cdot \sin(-\theta) = \cos(-\theta) = \cos\theta$

b) $\dfrac{2\sin^2 t + \sin t - 3}{1 - \cos^2 t - \sin t}$

$= \dfrac{2\sin^2 t + \sin t - 3}{\sin^2 t - \sin t}$ Using $\sin^2 x + \cos^2 x = 1$, or $\sin^2 x = 1 - \cos^2 x$, and substituting $\sin^2 t$ for $1 - \cos^2 t$

$= \dfrac{(2\sin t + 3)(\sin t - 1)}{\sin t\,(\sin t - 1)}$ Factoring in both numerator and denominator

$= \dfrac{2\sin t + 3}{\sin t}$ Simplifying

Check the results on a grapher.   ▬

We can add and subtract trigonometric fractional expressions in the same way that we did algebraic expressions.

Example 4 Add and simplify: $\dfrac{\cos x}{1 + \sin x} + \tan x.$

SOLUTION

$\dfrac{\cos x}{1 + \sin x} + \tan x = \dfrac{\cos x}{1 + \sin x} + \dfrac{\sin x}{\cos x}$ Using $\tan x = \dfrac{\sin x}{\cos x}$

$= \dfrac{\cos x}{1 + \sin x} \cdot \dfrac{\cos x}{\cos x} + \dfrac{\sin x}{\cos x} \cdot \dfrac{1 + \sin x}{1 + \sin x}$

    Multiplying by forms of 1

$= \dfrac{\cos^2 x + \sin x + \sin^2 x}{\cos x\,(1 + \sin x)}$ Adding

$= \dfrac{1 + \sin x}{\cos x\,(1 + \sin x)}$ Using $\sin^2 x + \cos^2 x = 1$

$= \dfrac{1}{\cos x},$ or $\sec x$ Simplifying

Check the result on a grapher.   ▬

When radicals occur, the use of absolute value is sometimes necessary, but it can be difficult to determine when to use it. In Examples 5 and 6, we will assume that all radicands are nonnegative. This means that the identities are meant to be confined to certain quadrants.

Example 5 Multiply and simplify: $\sqrt{\sin^3 x \cos x} \cdot \sqrt{\cos x}$.

SOLUTION

$$\sqrt{\sin^3 x \cos x} \cdot \sqrt{\cos x} = \sqrt{\sin^3 x \cos^2 x}$$
$$= \sqrt{\sin^2 x \cos^2 x \sin x}$$
$$= \sin x \cos x \sqrt{\sin x}$$

Example 6 Rationalize the denominator: $\sqrt{\dfrac{2}{\tan x}}$.

SOLUTION

$$\sqrt{\frac{2}{\tan x}} = \sqrt{\frac{2}{\tan x} \cdot \frac{\tan x}{\tan x}} = \sqrt{\frac{2 \tan x}{\tan^2 x}} = \frac{\sqrt{2 \tan x}}{\tan x}$$

Often in calculus, a substitution is a useful manipulation, as we show in the following example.

Example 7 Express $\sqrt{9 + x^2}$ as a trigonometric function of θ without using radicals by letting $x = 3 \tan \theta$. Assume that $0 < \theta < \pi/2$. Then find $\sin \theta$ and $\cos \theta$.

SOLUTION   We have

$$\sqrt{9 + x^2} = \sqrt{9 + (3 \tan \theta)^2} \qquad \text{Substituting } 3 \tan \theta \text{ for } x$$
$$= \sqrt{9 + 9 \tan^2 \theta}$$
$$= \sqrt{9(1 + \tan^2 \theta)} \qquad \text{Factoring}$$
$$= \sqrt{9 \sec^2 \theta} \qquad \text{Using } 1 + \tan^2 x = \sec^2 x$$
$$= 3|\sec \theta| = 3 \sec \theta. \qquad \text{For } 0 < \theta < \pi/2, \sec \theta > 0.$$

We can express $\sqrt{9 + x^2} = 3 \sec \theta$ as

$$\sec \theta = \frac{\sqrt{9 + x^2}}{3}.$$

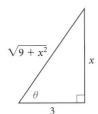

In a right triangle, we know that $\sec \theta$ is hypotenuse/adjacent, when θ is one of the acute angles. Using the Pythagorean theorem, we can determine that the side opposite θ is x. Then from the right triangle, we see that

$$\sin \theta = \frac{x}{\sqrt{9 + x^2}} \quad \text{and} \quad \cos \theta = \frac{3}{\sqrt{9 + x^2}}.$$

Sum and Difference Identities

We now develop some important identities involving sums or differences of two numbers (or angles), beginning with an identity for the cosine of the difference of two numbers. We use the letters u and v for these numbers.

Let's consider a real number u in the interval $[\pi/2, \pi]$ and a real number v in the interval $[0, \pi/2]$. These determine points A and B on the unit circle as shown. The arc length s is $u - v$, and we know that $0 \leq s \leq \pi$. Recall that the coordinates of A are $(\cos u, \sin u)$, and the coordinates of B are $(\cos v, \sin v)$.

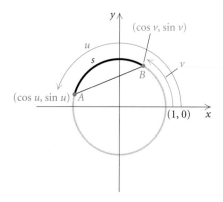

Using the distance formula, we can write an expression for the distance AB:

$$AB = \sqrt{(\cos u - \cos v)^2 + (\sin u - \sin v)^2}.$$

This can be simplified as follows:

$$
\begin{aligned}
AB &= \sqrt{\cos^2 u - 2\cos u \cos v + \cos^2 v + \sin^2 u - 2\sin u \sin v + \sin^2 v} \\
&= \sqrt{(\sin^2 u + \cos^2 u) + (\sin^2 v + \cos^2 v) - 2(\cos u \cos v + \sin u \sin v)} \\
&= \sqrt{2 - 2(\cos u \cos v + \sin u \sin v)}.
\end{aligned}
$$

Now let's imagine rotating the circle above so that point B is at $(1, 0)$. Although the coordinates of point A are now $(\cos s, \sin s)$, the distance AB has not changed.

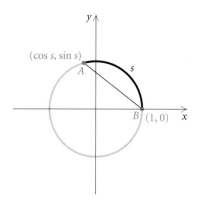

Again we use the distance formula to write an expression for the distance AB:

$$AB = \sqrt{(\cos s - 1)^2 + (\sin s - 0)^2}.$$

This simplifies as follows:

$$AB = \sqrt{\cos^2 s - 2 \cos s + 1 + \sin^2 s}$$
$$= \sqrt{(\sin^2 s + \cos^2 s) + 1 - 2 \cos s}$$
$$= \sqrt{2 - 2 \cos s}.$$

Equating our two expressions for AB, we obtain

$$\sqrt{2 - 2(\cos u \cos v + \sin u \sin v)} = \sqrt{2 - 2 \cos s}.$$

Solving this equation for $\cos s$ gives

$$\cos s = \cos u \cos v + \sin u \sin v. \tag{1}$$

But $s = u - v$, so we have the equation

$$\cos (u - v) = \cos u \cos v + \sin u \sin v. \tag{2}$$

Formula (1) above holds when s is the length of the shortest arc from A to B. Given any real numbers u and v, the length of the shortest arc from A to B is not always $u - v$. In fact, it could be $v - u$. However, since $\cos (-x) = \cos x$, we know that $\cos (v - u) = \cos (u - v)$. Thus, $\cos s$ is always equal to $\cos (u - v)$. Formula (2) holds for all real numbers u and v. That formula is thus the identity we sought:

$$\mathbf{\cos (u - v) = \cos u \cos v + \sin u \sin v.}$$

On a grapher, graph

$$y_1 = \cos (x - 3)$$

and $$y_2 = \cos x \cos 3 + \sin x \sin 3$$

to illustrate this result.

The cosine sum formula follows easily from the one we have just derived. Let's consider $\cos (u + v)$. This is equal to $\cos [u - (-v)]$, and by the identity above, we have

$$\cos (u + v) = \cos [u - (-v)] = \cos u \cos (-v) + \sin u \sin (-v).$$

But $\cos (-v) = \cos v$ and $\sin (-v) = -\sin v$, so the identity we seek is the following:

$$\mathbf{\cos (u + v) = \cos u \cos v - \sin u \sin v.}$$

On a grapher, graph

$$y_1 = \cos (x + 2)$$

and $$y_2 = \cos x \cos 2 - \sin x \sin 2$$

to illustrate this result.

Example 8   Find $\cos (5\pi/12)$ exactly.

SOLUTION   We can express $5\pi/12$ as a difference of two numbers whose sine and cosine values are known:

$$\frac{5\pi}{12} = \frac{9\pi}{12} - \frac{4\pi}{12}, \quad \text{or} \quad \frac{3\pi}{4} - \frac{\pi}{3}.$$

Then, using $\cos (u - v) = \cos u \cos v + \sin u \sin v$, we have

$$\cos \frac{5\pi}{12} = \cos \left(\frac{3\pi}{4} - \frac{\pi}{3} \right) = \cos \frac{3\pi}{4} \cos \frac{\pi}{3} + \sin \frac{3\pi}{4} \sin \frac{\pi}{3}$$

$$= -\frac{\sqrt{2}}{2} \cdot \frac{1}{2} + \frac{\sqrt{2}}{2} \cdot \frac{\sqrt{3}}{2}$$

$$= -\frac{\sqrt{2}}{4} + \frac{\sqrt{6}}{4} = \frac{\sqrt{6} - \sqrt{2}}{4}.$$

We can check with a grapher:

$$\cos \frac{5\pi}{12} \approx 0.2588 \quad \text{and} \quad \frac{\sqrt{6} - \sqrt{2}}{4} \approx 0.2588.$$

Consider $\cos (\pi/2 - \theta)$. We can use the identity for the cosine of a difference to simplify as follows:

$$\cos \left(\frac{\pi}{2} - \theta \right) = \cos \frac{\pi}{2} \cos \theta + \sin \frac{\pi}{2} \sin \theta \qquad \begin{array}{l}\text{This identity}\\\text{appeared first}\\\text{in Section 2.1.}\end{array}$$

$$= 0 \cdot \cos \theta + 1 \cdot \sin \theta = \sin \theta.$$

Thus we have developed the identity

$$\sin \theta = \cos \left(\frac{\pi}{2} - \theta \right). \tag{3}$$

This identity holds for any real number θ. From it we can obtain an identity for the sine function. We first let α be any real number. Then we replace θ in $\sin \theta = \cos (\pi/2 - \theta)$ by $\pi/2 - \alpha$. This gives us

$$\sin \left(\frac{\pi}{2} - \alpha \right) = \cos \left[\frac{\pi}{2} - \left(\frac{\pi}{2} - \alpha \right) \right] = \cos \alpha,$$

which yields the identity

$$\cos \alpha = \sin \left(\frac{\pi}{2} - \alpha \right). \tag{4}$$

Using identities (3) and (4) and the identity for the cosine of a difference, we can obtain an identity for the sine of a sum. We start with identity (3) and substitute $u + v$ for θ:

$$\sin \theta = \cos \left(\frac{\pi}{2} - \theta \right) \qquad \text{Identity (3)}$$

$$\sin (u + v) = \cos \left[\frac{\pi}{2} - (u + v) \right] \qquad \text{Substituting } u + v \text{ for } \theta$$

$$= \cos \left[\left(\frac{\pi}{2} - u \right) - v \right]$$

$$= \cos \left(\frac{\pi}{2} - u \right) \cos v + \sin \left(\frac{\pi}{2} - u \right) \sin v$$

$$\begin{array}{l}\text{Using the identity for the}\\\text{cosine of a difference}\end{array}$$

$$= \sin u \cos v + \cos u \sin v. \qquad \text{Using identities (3) and (4)}$$

Thus the identity we seek is

$$\sin (u + v) = \sin u \cos v + \cos u \sin v.$$

To find a formula for the sine of a difference, we can use the identity just derived, substituting $-v$ for v:

$$\sin (u + (-v)) = \sin u \cos (-v) + \cos u \sin (-v).$$

Simplifying gives us

$$\sin (u - v) = \sin u \cos v - \cos u \sin v.$$

Example 9 Find $\sin 105°$ exactly.

SOLUTION   We can express $105°$ as the sum of two measures:

$$105° = 45° + 60°.$$

Then

$$\begin{aligned} \sin 105° &= \sin (45° + 60°) \\ &= \sin 45° \cos 60° + \cos 45° \sin 60° \\ &\qquad \text{Using } \sin (u + v) = \sin u \cos v + \cos u \sin v \\ &= \frac{\sqrt{2}}{2} \cdot \frac{1}{2} + \frac{\sqrt{2}}{2} \cdot \frac{\sqrt{3}}{2} \\ &= \frac{\sqrt{2} + \sqrt{6}}{4}. \end{aligned}$$

We can easily check this result on a grapher:

$$\sin 105° \approx 0.9659 \quad \text{and} \quad \frac{\sqrt{2} + \sqrt{6}}{4} \approx 0.9659.$$

Formulas for the tangent of a sum or a difference can be derived using identities already established. A summary of the sum and difference identities follows.*

Sum and Difference Identities

$$\sin (u \pm v) = \sin u \cos v \pm \cos u \sin v,$$

$$\cos (u \pm v) = \cos u \cos v \mp \sin u \sin v,$$

$$\tan (u \pm v) = \frac{\tan u \pm \tan v}{1 \mp \tan u \tan v}$$

Example 10 Find $\tan 15°$ exactly.

SOLUTION   We rewrite $15°$ as $45° - 30°$ and use the identity for the tan-

---

*There are six identities here, half of them obtained by using the signs shown in color.

gent of a difference:

$$\tan 15° = \tan(45° - 30°)$$

$$= \frac{\tan 45° - \tan 30°}{1 + \tan 45° \tan 30°}$$

$$= \frac{1 - \sqrt{3}/3}{1 + 1 \cdot \sqrt{3}/3}$$

$$= \frac{3 - \sqrt{3}}{3 + \sqrt{3}}.$$

Example 11   Assume that $\sin \alpha = \frac{2}{3}$ and $\sin \beta = \frac{1}{3}$ and that α and β are between 0 and $\pi/2$. Then evaluate $\sin (\alpha + \beta)$.

SOLUTION   Using the identity for the sine of a sum, we have

$$\sin (\alpha + \beta) = \sin \alpha \cos \beta + \cos \alpha \sin \beta$$

$$= \tfrac{2}{3} \cos \beta + \tfrac{1}{3} \cos \alpha.$$

To finish, we need to know the values of $\cos \beta$ and $\cos \alpha$. Using reference triangles and the Pythagorean theorem, we can determine these values from the diagrams:

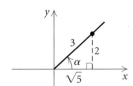

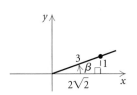

$$\cos \alpha = \frac{\sqrt{5}}{3} \quad \text{and} \quad \cos \beta = \frac{2\sqrt{2}}{3}.$$

Substituting these values gives us

$$\sin (\alpha + \beta) = \frac{2}{3} \cdot \frac{2\sqrt{2}}{3} + \frac{1}{3} \cdot \frac{\sqrt{5}}{3}$$

$$= \frac{4}{9} \sqrt{2} + \frac{1}{9} \sqrt{5}, \quad \text{or} \quad \frac{4\sqrt{2} + \sqrt{5}}{9}.$$

3.1 | *Exercise Set*

Multiply and simplify. Check your result with a grapher.

1. $(\sin x - \cos x)(\sin x + \cos x)$

2. $\tan x (\cos x - \csc x)$

3. $\cos y \sin y (\sec y + \csc y)$

4. $(\sin x + \cos x)(\sec x + \csc x)$

5. $(\sin \phi - \cos \phi)^2$

6. $(1 + \tan x)^2$

7. $(\sin x + \csc x)(\sin^2 x + \csc^2 x - 1)$

8. $(1 - \sin t)(1 + \sin t)$

Factor and simplify. Check your result with a grapher.

9. $\sin x \cos x + \cos^2 x$

10. $\tan^2 \theta - \cot^2 \theta$

11. $\sin^4 x - \cos^4 x$

12. $4 \sin^2 y + 8 \sin y + 4$

13. $2 \cos^2 x + \cos x - 3$

14. $3 \cot^2 \beta + 6 \cot \beta + 3$

15. $\sin^3 x + 27$

16. $1 - 125 \tan^3 s$

Simplify and check with a grapher.

17. $\dfrac{\sin^2 x \cos x}{\cos^2 x \sin x}$

18. $\dfrac{30 \sin^3 x \cos x}{6 \cos^2 x \sin x}$

19. $\dfrac{\sin^2 x + 2 \sin x + 1}{\sin x + 1}$

20. $\dfrac{\cos^2 \alpha - 1}{\cos \alpha + 1}$

21. $\dfrac{4 \tan t \sec t + 2 \sec t}{6 \tan t \sec t + 2 \sec t}$

22. $\dfrac{\csc (-x)}{\cot (-x)}$

23. $\dfrac{\sin^4 x - \cos^4 x}{\sin^2 x - \cos^2 x}$

24. $\dfrac{4 \cos^3 x}{\sin^2 x} \cdot \left(\dfrac{\sin x}{4 \cos x}\right)^2$

25. $\dfrac{5 \cos \phi}{\sin^2 \phi} \cdot \dfrac{\sin^2 \phi - \sin \phi \cos \phi}{\sin^2 \phi - \cos^2 \phi}$

26. $\dfrac{\tan^2 y}{\sec y} \div \dfrac{3 \tan^3 y}{\sec y}$

27. $\dfrac{1}{\sin^2 s - \cos^2 s} - \dfrac{2}{\cos s - \sin s}$

28. $\left(\dfrac{\sin x}{\cos x}\right)^2 - \dfrac{1}{\cos^2 x}$

29. $\dfrac{\sin^2 \theta - 9}{2 \cos \theta + 1} \cdot \dfrac{10 \cos \theta + 5}{3 \sin \theta + 9}$

30. $\dfrac{9 \cos^2 \alpha - 25}{2 \cos \alpha - 2} \cdot \dfrac{\cos^2 \alpha - 1}{6 \cos \alpha - 10}$

Simplify and check with a grapher. Assume that all radicands are nonnegative.

31. $\sqrt{\sin^2 x \cos x} \cdot \sqrt{\cos x}$

32. $\sqrt{\cos^2 x \sin x} \cdot \sqrt{\sin x}$

33. $\sqrt{\cos \alpha \sin^2 \alpha} - \sqrt{\cos^3 \alpha}$

34. $\sqrt{\tan^2 x - 2 \tan x \sin x + \sin^2 x}$

35. $(1 - \sqrt{\sin y})(\sqrt{\sin y} + 1)$

36. $\sqrt{\cos \theta}(\sqrt{2 \cos \theta} + \sqrt{\sin \theta \cos \theta})$

Rationalize the denominator.

37. $\sqrt{\dfrac{\sin x}{\cos x}}$

38. $\sqrt{\dfrac{\cos x}{\tan x}}$

39. $\sqrt{\dfrac{\cos^2 y}{2 \sin^2 y}}$

40. $\sqrt{\dfrac{1 - \cos \beta}{1 + \cos \beta}}$

Rationalize the numerator.

41. $\sqrt{\dfrac{\cos x}{\sin x}}$

42. $\sqrt{\dfrac{\sin x}{\cot x}}$

43. $\sqrt{\dfrac{1 + \sin y}{1 - \sin y}}$

44. $\sqrt{\dfrac{\cos^2 x}{2 \sin^2 x}}$

Use the given substitution to express the given radical expression as a trigonometric function without radicals.

Assume that $a > 0$ and $0 < \theta < \pi/2$. Then find expressions for the indicated trigonometric functions.

45. Let $x = a \sin \theta$ in $\sqrt{a^2 - x^2}$. Then find $\cos \theta$ and $\tan \theta$.

46. Let $x = 2 \tan \theta$ in $\sqrt{4 + x^2}$. Then find $\sin \theta$ and $\cos \theta$.

47. Let $x = 3 \sec \theta$ in $\sqrt{x^2 - 9}$. Then find $\sin \theta$ and $\cos \theta$.

48. Let $x = a \sec \theta$ in $\sqrt{x^2 - a^2}$. Then find $\sin \theta$ and $\cos \theta$.

Use the given substitution to express the given radical expression as a trigonometric function without radicals. Assume that $0 < \theta < \pi/2$.

49. Let $x = \sin \theta$ in $\dfrac{x^2}{\sqrt{1 - x^2}}$.

50. Let $x = 4 \sec \theta$ in $\dfrac{\sqrt{x^2 - 16}}{x^2}$.

Use the sum and difference identities to evaluate exactly. Then check with a grapher.

51. $\sin \dfrac{\pi}{12}$ **52.** $\cos 75°$ **53.** $\tan 105°$

54. $\tan \dfrac{5\pi}{12}$ **55.** $\cos 15°$ **56.** $\sin \dfrac{7\pi}{12}$

First write each of the following as a trigonometric function of a single angle; then evaluate.

57. $\sin 37° \cos 22° + \cos 37° \sin 22°$

58. $\cos 83° \cos 53° + \sin 83° \sin 53°$

59. $\dfrac{\tan 20° + \tan 32°}{1 - \tan 20° \tan 32°}$

60. $\dfrac{\tan 35° - \tan 12°}{1 + \tan 35° \tan 12°}$

Assuming that $\sin u = \frac{3}{5}$ and $\sin v = \frac{4}{5}$ and that u and v are between 0 and $\pi/2$, evaluate each of the following exactly.

61. $\cos (u + v)$ **62.** $\tan (u - v)$ **63.** $\sin (u - v)$

Assuming that $\sin \theta = 0.6249$ and $\cos \phi = 0.1102$ and that both θ and ϕ are first-quadrant angles, evaluate each of the following.

64. $\sin (\theta - \phi)$ **65.** $\tan (\theta + \phi)$ **66.** $\cos (\theta + \phi)$

Simplify.

67. $\sin (\alpha + \beta) + \sin (\alpha - \beta)$

68. $\cos (\alpha + \beta) - \cos (\alpha - \beta)$

69. $\cos (u + v) \cos v + \sin (u + v) \sin v$

70. $\sin (u - v) \cos v + \cos (u - v) \sin v$

Skill Maintenance

Solve.

71. $2x - 3 = 2\left(x - \frac{3}{2}\right)$

72. $x - 7 = x + 3.4$

Given that $\sin 31° = 0.5150$ and $\cos 31° = 0.8572$, find the specified function value.

73. $\sec 59°$

74. $\tan 59°$

Synthesis

75. ◆ What is the difference between a trigonometric equation that is an identity and a trigonometric equation that is not an identity? Give an example of each.

76. ◆ Why is it possible with a grapher to *disprove* that an equation is an identity but not to *prove* that one is?

Angles Between Lines. *One of the identities gives an easy way to find an angle formed by two lines. Consider two lines with equations l_1: $y = m_1 x + b_1$ and l_2: $y = m_2 x + b_2$.*

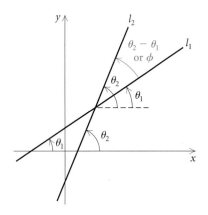

The slopes m_1 and m_2 are the tangents of the angles θ_1 and θ_2 that the lines form with the positive direction of the x-axis. Thus we have $m_1 = \tan \theta_1$ and $m_2 = \tan \theta_2$. To find the measure of $\theta_2 - \theta_1$, or ϕ, we proceed as follows:

$$\tan \phi = \tan (\theta_2 - \theta_1)$$
$$= \frac{\tan \theta_2 - \tan \theta_1}{1 + \tan \theta_2 \tan \theta_1}$$
$$= \frac{m_2 - m_1}{1 + m_2 m_1}.$$

This formula also holds when the lines are taken in the reverse order. When ϕ is acute, $\tan \phi$ will be positive. When ϕ is obtuse, $\tan \phi$ will be negative.

Find the measure of the angle from l_1 to l_2.

77. l_1: $2x = 3 - 2y$,
l_2: $x + y = 5$

78. l_1: $3y = \sqrt{3}x + 3$,
l_2: $y = \sqrt{3}x + 2$

79. l_1: $y = 3$,
l_2: $x + y = 5$

80. l_1: $2x + y - 4 = 0$,
l_2: $y - 2x + 5 = 0$

81. *Circus Guy Wire.* In a circus, a guy wire A is attached to the top of a 30-ft pole. Wire B is used for performers to walk up to the tight wire, 10 ft above the ground. Find the angle ϕ between the wires if they are attached to the ground 40 ft from the pole.

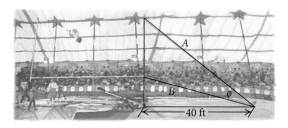

82. Given that $f(x) = \sin x$, show that
$$\frac{f(x + h) - f(x)}{h} = \sin x \left(\frac{\cos h - 1}{h}\right) + \cos x \left(\frac{\sin h}{h}\right).$$

83. Given that $f(x) = \cos x$, show that
$$\frac{f(x + h) - f(x)}{h} = \cos x \left(\frac{\cos h - 1}{h}\right) - \sin x \left(\frac{\sin h}{h}\right).$$

Show that each of the following is not an identity by first finding a replacement or replacements for which the sides of the equation do not name the same number. Then use a grapher to show that the equation is not an identity.

84. $\sqrt{\sin^2 \theta} = \sin \theta$

85. $\dfrac{\sin 5x}{x} = \sin 5$

86. $\sin (-x) = \sin x$

87. $\cos (2\alpha) = 2 \cos \alpha$

88. $\tan^2 \theta + \cot^2 \theta = 1$

89. $\dfrac{\cos 6x}{\cos x} = 6$

Find the slope of line l_1, where m_2 is the slope of line l_2 and ϕ is the smallest positive angle from l_1 to l_2.

90. $m_2 = \frac{4}{3}$, $\phi = 45°$

91. $m_2 = \frac{2}{3}$, $\phi = 30°$

92. Line l_1 contains the points $(-2, 4)$ and $(5, -1)$. Find the slope of line l_2 such that the angle from l_1 to l_2 is $45°$.

93. Line l_1 contains the points $(-3, 7)$ and $(-3, -2)$. Line l_2 contains $(0, -4)$ and $(2, 6)$. Find the smallest positive angle from l_1 to l_2.

94. Find an identity for $\sin 2\theta$. (*Hint:* $2\theta = \theta + \theta$.)

95. Find an identity for $\cos 2\theta$. (*Hint:* $2\theta = \theta + \theta$.)

Derive each identity. Check with a grapher.

96. $\sin\left(x - \dfrac{3\pi}{2}\right) = \cos x$

97. $\tan\left(x + \dfrac{\pi}{4}\right) = \dfrac{1 + \tan x}{1 - \tan x}$

98. $\dfrac{\sin(\alpha + \beta)}{\cos(\alpha - \beta)} = \dfrac{\tan \alpha + \tan \beta}{1 + \tan \alpha \tan \beta}$

99. $\sin(\alpha + \beta) + \sin(\alpha - \beta) = 2 \sin \alpha \cos \beta$

3.2

Identities: Cofunction, Double-Angle, and Half-Angle

- *Use cofunction identities to derive other identities.*
- *Use the double-angle identities to find function values of twice an angle when one function value is known for that angle.*
- *Use the half-angle identities to find function values of half an angle when one function value is known for that angle.*
- *Simplify trigonometric expressions using the double-angle and half-angle identities.*

Cofunction Identities

Each of the identities listed below yields a conversion to a *cofunction*. For this reason, we call them cofunction identities.

Cofunction Identities

$$\sin\left(\frac{\pi}{2} - x\right) = \cos x, \qquad \cos\left(\frac{\pi}{2} - x\right) = \sin x,$$

$$\tan\left(\frac{\pi}{2} - x\right) = \cot x, \qquad \cot\left(\frac{\pi}{2} - x\right) = \tan x,$$

$$\sec\left(\frac{\pi}{2} - x\right) = \csc x, \qquad \csc\left(\frac{\pi}{2} - x\right) = \sec x$$

We verified the first two of these identities in Section 3.1. The other four can be proved using the first two and the definitions of the trigonometric functions. These identities hold for all real numbers, and thus, for all degree measures, but if we restrict θ to values such that $0° < \theta < 90°$, then we have a special application to the acute angles of a right triangle.

Interactive Discovery

Graph $y_1 = \sin x$ and $y_2 = \sin(x + \pi/2)$ on the same screen. Note that the graph of y_2 is a translation of the graph of y_1 to the left $\pi/2$ units. Now graph $y_3 = \cos x$. We observe that the graph of y_2 is the same as the graph of y_3. This leads to a possible identity:

$$\sin\left(x + \frac{\pi}{2}\right) = \cos x.$$

(continued)

Repeat this exploration for

$$y = \sin\left(x - \frac{\pi}{2}\right),$$

$$y = \cos\left(x + \frac{\pi}{2}\right),$$

and $$y = \cos\left(x - \frac{\pi}{2}\right).$$

What possible identities do you observe?

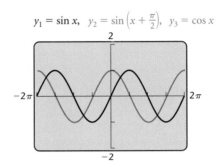

$$y_1 = \sin x, \quad y_2 = \sin\left(x + \frac{\pi}{2}\right), \quad y_3 = \cos x$$

The identity $\sin (x + \pi/2) = \cos x$ can be proved using the identity for the sine of a sum developed in Section 3.1.

Example 1   Prove the identity $\sin (\theta + \pi/2) = \cos \theta$.

SOLUTION

$$\sin\left(\theta + \frac{\pi}{2}\right) = \sin \theta \cos \frac{\pi}{2} + \cos \theta \sin \frac{\pi}{2} \qquad \text{Using } \sin (u + v) = \\ \sin u \cos v + \\ \cos u \sin v$$

$$= \sin \theta \cdot 0 + \cos \theta \cdot 1$$

$$= \cos \theta$$

We now state four more cofunction identities. These new identities that involve the sine and cosine functions can be verified using previously established identities as seen in Example 1.

Cofunction Identities for the Sine and Cosine

$$\sin\left(x \pm \frac{\pi}{2}\right) = \pm\cos x, \qquad \cos\left(x \pm \frac{\pi}{2}\right) = \mp\sin x$$

Example 2   Find an identity for each of the following.

a) $\tan\left(x + \dfrac{\pi}{2}\right)$ **b)** $\sec (x - 90°)$

SOLUTION

a) We have

$$\tan\left(x + \frac{\pi}{2}\right) = \frac{\sin\left(x + \dfrac{\pi}{2}\right)}{\cos\left(x + \dfrac{\pi}{2}\right)} \qquad \text{Using } \tan x = \frac{\sin x}{\cos x}$$

$$= \frac{\cos x}{-\sin x} \qquad \text{Using cofunction identities}$$

$$= -\cot x.$$

Thus the identity we seek is

$$\tan\left(x + \frac{\pi}{2}\right) = -\cot x.$$

b) We have

$$\sec(x - 90°) = \frac{1}{\cos(x - 90°)} = \frac{1}{\sin x} = \csc x.$$

Thus, $\sec(x - 90°) = \csc x.$

Double-Angle Identities

To develop these identities, we will use the sum formulas from the preceding section. We first develop a formula for $\sin 2x$. Recall that

$$\sin(u + v) = \sin u \cos v + \cos u \sin v.$$

We will consider a number x and substitute it for both u and v in this identity. Doing so gives us

$$\sin(x + x) = \sin 2x$$
$$= \sin x \cos x + \cos x \sin x$$
$$= 2 \sin x \cos x.$$

Our first double-angle identity is thus

$$\textbf{sin } 2x = \textbf{2 sin } x \textbf{ cos } x.$$

Graphers provide visual partial checks of identities. We can graph $y_1 = \sin 2x$ and $y_2 = 2 \sin x \cos x$ and see that they appear to have the same graph in $[-2\pi, 2\pi]$.

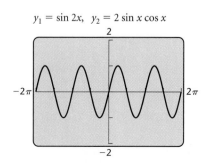

$y_1 = \sin 2x, \quad y_2 = 2 \sin x \cos x$

Double-angle identities for the cosine and tangent functions can be derived and checked with a grapher in much the same way as the identity above:

$$\cos 2x = \cos^2 x - \sin^2 x,$$

$$\tan 2x = \frac{2 \tan x}{1 - \tan^2 x}.$$

Example 3   Given that $\tan \theta = -\frac{3}{4}$ and θ is in quadrant II, find each of the following.

a) $\sin 2\theta$ **b)** $\cos 2\theta$

c) $\tan 2\theta$ **d)** The quadrant in which 2θ lies

SOLUTION   By drawing a diagram as shown, we find that

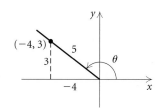

$$\sin \theta = \frac{3}{5} \quad \text{and} \quad \cos \theta = -\frac{4}{5}.$$

Thus we have the following.

a) $\sin 2\theta = 2 \sin \theta \cos \theta = 2 \cdot \frac{3}{5} \cdot \left(-\frac{4}{5}\right) = -\frac{24}{25}$

b) $\cos 2\theta = \cos^2 \theta - \sin^2 \theta = \left(-\frac{4}{5}\right)^2 - \left(\frac{3}{5}\right)^2 = \frac{16}{25} - \frac{9}{25} = \frac{7}{25}$

c) $\tan 2\theta = \dfrac{2 \tan \theta}{1 - \tan^2 \theta} = \dfrac{2 \cdot \left(-\frac{3}{4}\right)}{1 - \left(-\frac{3}{4}\right)^2} = \dfrac{-\frac{3}{2}}{1 - \frac{9}{16}} = -\frac{24}{7}$

Note that $\tan 2\theta$ could have been found more easily in this case by simply dividing:

$$\tan 2\theta = \frac{\sin 2\theta}{\cos 2\theta} = \frac{-\frac{24}{25}}{\frac{7}{25}} = -\frac{24}{7}.$$

d) Since $\sin 2\theta$ is negative and $\cos 2\theta$ is positive, we know that 2θ is in quadrant IV.   ▬

Two other useful identities for $\cos 2x$ can be derived easily, as follows.

$$\cos 2x = \cos^2 x - \sin^2 x \qquad\qquad \cos 2x = \cos^2 x - \sin^2 x$$
$$= (1 - \sin^2 x) - \sin^2 x \qquad\qquad = \cos^2 x - (1 - \cos^2 x)$$
$$= 1 - 2 \sin^2 x \qquad\qquad\qquad = 2 \cos^2 x - 1$$

Double-Angle Identities

$$\sin 2x = 2 \sin x \cos x, \qquad\qquad \cos 2x = \cos^2 x - \sin^2 x$$

$$\tan 2x = \frac{2 \tan x}{1 - \tan^2 x}, \qquad\qquad\quad = 1 - 2 \sin^2 x$$

$$= 2 \cos^2 x - 1$$

Solving the last two cosine double-angle identities for $\sin^2 x$ and $\cos^2 x$, respectively, we can obtain two more identities:

$$\sin^2 x = \frac{1 - \cos 2x}{2}$$

and

$$\cos^2 x = \frac{1 + \cos 2x}{2}.$$

Using division and the last two identities gives us the following useful identity:

$$\tan^2 x = \frac{1 - \cos 2x}{1 + \cos 2x}.$$

Example 4   Find an equivalent expression for each of the following.

a) $\sin 3\theta$ in terms of function values of θ

b) $\cos^3 x$ in terms of function values of x or $2x$, raised only to the first power

SOLUTION

a) $\sin 3\theta = \sin(2\theta + \theta)$

$\qquad = \sin 2\theta \cos \theta + \cos 2\theta \sin \theta$

$\qquad = (2 \sin \theta \cos \theta) \cos \theta + (2 \cos^2 \theta - 1) \sin \theta$

$\qquad\qquad$ Using $\sin 2\theta = 2 \sin \theta \cos \theta$ and $\cos 2\theta = 2 \cos^2 \theta - 1$

$\qquad = 2 \sin \theta \cos^2 \theta + 2 \sin \theta \cos^2 \theta - \sin \theta$

$\qquad = 4 \sin \theta \cos^2 \theta - \sin \theta$

We could also substitute $\cos^2 \theta - \sin^2 \theta$ or $1 - 2 \sin^2 \theta$ for $\cos 2\theta$. Each substitution leads to a different result, but all results are equivalent.

b) $\cos^3 x = \cos^2 x \cos x$

$\qquad = \dfrac{1 + \cos 2x}{2} \cos x$ ▬

Half-Angle Identities

To develop more identities, we take square roots and replace x by $x/2$. For example,

$$\sin^2 x = \frac{1 - \cos 2x}{2} \longrightarrow \left| \sin \frac{x}{2} \right| = \sqrt{\frac{1 - \cos x}{2}}.$$

The half-angle formula is the one on the right above. We can eliminate the absolute-value sign by using $\pm$ signs, with the understanding that our use of $+$ and $-$ depends on the quadrant in which the angle $x/2$ lies. Half-angle identities for the cosine and tangent functions can be derived in a similar manner. Two additional formulas for the half-angle tangent identity are listed on the following page.

Half-Angle Identities

$$\sin \frac{x}{2} = \pm \sqrt{\frac{1 - \cos x}{2}}, \qquad \tan \frac{x}{2} = \pm \sqrt{\frac{1 - \cos x}{1 + \cos x}}$$

$$\cos \frac{x}{2} = \pm \sqrt{\frac{1 + \cos x}{2}}, \qquad\qquad = \frac{\sin x}{1 + \cos x}$$

$$= \frac{1 - \cos x}{\sin x}$$

Example 5   Find $\tan (\pi/8)$ exactly. Then check the answer with a grapher.

SOLUTION   We have

$$\tan \frac{\pi}{8} = \tan \frac{\frac{\pi}{4}}{2} = \frac{\sin \frac{\pi}{4}}{1 + \cos \frac{\pi}{4}} = \frac{\frac{\sqrt{2}}{2}}{1 + \frac{\sqrt{2}}{2}}$$

$$= \frac{\sqrt{2}}{2 + \sqrt{2}} = \sqrt{2} - 1.$$

This value checks:

$$\tan \frac{\pi}{8} \approx 0.4142 \quad \text{and} \quad \sqrt{2} - 1 \approx 0.4142.$$

     The identities that we have developed are also useful for simplifying trigonometric expressions.

Example 6   Simplify each of the following.

a) $\dfrac{\sin x \cos x}{\frac{1}{2} \cos 2x}$

b) $2 \sin^2 \dfrac{x}{2} + \cos x$

SOLUTION

a) We can obtain $2 \sin x \cos x$ in the numerator by multiplying the expression by $\frac{2}{2}$:

$$\frac{\sin x \cos x}{\frac{1}{2} \cos 2x} = \frac{2}{2} \cdot \frac{\sin x \cos x}{\frac{1}{2} \cos 2x}$$

$$= \frac{2 \sin x \cos x}{\cos 2x}$$

$$= \frac{\sin 2x}{\cos 2x} \qquad \text{Using } \sin 2x = 2 \sin x \cos x$$

$$= \tan 2x.$$

b) We have

$$2 \sin^2 \frac{x}{2} + \cos x = 2\left(\frac{1 - \cos x}{2}\right) + \cos x$$

$$\text{Using } \sin \frac{x}{2} = \pm\sqrt{\frac{1 - \cos x}{2}}, \text{ or } \sin^2 \frac{x}{2} = \frac{1 - \cos x}{2}$$

$$= 1 - \cos x + \cos x$$

$$= 1.$$

To check this on $[-2\pi, 2\pi]$, we graph

$$y_1 = 2 \sin^2 \frac{x}{2} + \cos x \quad \text{and} \quad y_2 = 1.$$

Note that the graphs appear to be the same.   ▬

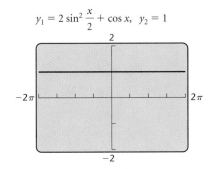

$y_1 = 2 \sin^2 \dfrac{x}{2} + \cos x, \ y_2 = 1$

3.2 | Exercise Set

1. Given that sin $(3\pi/10) \approx 0.8090$ and cos $(3\pi/10) \approx 0.5878$, find each of the following.

   a) The other four function values for $3\pi/10$
   b) The six function values for $\pi/5$

2. Given that

$$\sin \frac{\pi}{12} = \frac{\sqrt{2 - \sqrt{3}}}{2} \quad \text{and} \quad \cos \frac{\pi}{12} = \frac{\sqrt{2 + \sqrt{3}}}{2},$$

find each of the following.

   a) The other four function values for $\pi/12$
   b) The six function values for $5\pi/12$

3. Given that sin $\theta = \frac{1}{3}$ and that the terminal side is in quadrant II, find each of the following.

   a) The other function values for θ
   b) The six function values for $\pi/2 - \theta$
   c) The six function values for $\theta - \pi/2$

4. Given that cos $\phi = \frac{4}{5}$ and that the terminal side is in quadrant IV, find each of the following.

   a) The other function values for ϕ
   b) The six function values for $\pi/2 - \phi$
   c) The six function values for $\phi + \pi/2$

Find an equivalent expression for each of the following.

5. $\sec\left(x + \dfrac{\pi}{2}\right)$ **6.** $\cot\left(x - \dfrac{\pi}{2}\right)$

7. $\tan\left(x - \dfrac{\pi}{2}\right)$ **8.** $\csc\left(x + \dfrac{\pi}{2}\right)$

Find sin 2θ, cos 2θ, tan 2θ, and the quadrant in which 2θ lies.

9. sin $\theta = \dfrac{4}{5}$, θ in quadrant I

10. cos $\theta = \dfrac{5}{13}$, θ in quadrant I

11. cos $\theta = -\dfrac{3}{5}$, θ in quadrant III

12. tan $\theta = -\dfrac{5}{8}$, θ in quadrant II

13. tan $\theta = -\dfrac{5}{12}$, θ in quadrant II

14. sin $\theta = -\dfrac{\sqrt{10}}{10}$, θ in quadrant IV

15. Find an equivalent expression for cos $4x$ in terms of function values of x.

16. Find an equivalent expression for $\sin^4 \theta$ in terms of function values of θ, 2θ, or 4θ, raised only to the first power.

Use the half-angle identities to evaluate exactly.

17. cos $15°$ **18.** tan $67.5°$ **19.** sin $112.5°$

20. $\cos \dfrac{\pi}{8}$ **21.** tan $75°$ **22.** $\sin \dfrac{5\pi}{12}$

Given that sin $\theta = 0.3416$ and θ is in quadrant I, find each of the following.

23. sin 2θ **24.** $\cos \dfrac{\theta}{2}$

25. $\sin \dfrac{\theta}{2}$ **26.** sin 4θ

Use a grapher to determine which of the following expressions asserts an identity. Then prove the identity algebraically.

27. $\dfrac{\cos 2x}{\cos x - \sin x} = \cdots$

   a) $1 + \cos x$ **b)** $\cos x - \sin x$

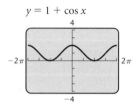 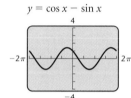

   c) $-\cot x$ **d)** $\sin x \, (\cot x + 1)$

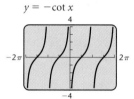 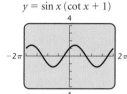

28. $2 \cos^2 \dfrac{x}{2} = \cdots$

   a) $\sin x \, (\csc x + \tan x)$ **b)** $\sin x - 2 \cos x$
   c) $2(\cos^2 x - \sin^2 x)$ **d)** $1 + \cos x$

29. $\dfrac{\sin 2x}{2 \cos x} = \cdots$

   a) $\cos x$ **b)** $\tan x$
   c) $\cos x + \sin x$ **d)** $\sin x$

30. $2 \sin \dfrac{\theta}{2} \cos \dfrac{\theta}{2} = \cdots$

   a) $\cos^2 \theta$ **b)** $\sin \dfrac{\theta}{2}$

   c) $\sin \theta$ **d)** $\sin \theta - \cos \theta$

Simplify. Check your results with a grapher.

31. $2 \cos^2 \dfrac{x}{2} - 1$

32. $\cos^4 x - \sin^4 x$

33. $(\sin x - \cos x)^2 + \sin 2x$

34. $(\sin x + \cos x)^2$

35. $\dfrac{2 - \sec^2 x}{\sec^2 x}$

36. $\dfrac{1 + \sin 2x + \cos 2x}{1 + \sin 2x - \cos 2x}$

37. $(-4 \cos x \sin x + 2 \cos 2x)^2 +$
   $(2 \cos 2x + 4 \sin x \cos x)^2$

38. $2 \sin x \cos^3 x - 2 \sin^3 x \cos x$

Skill Maintenance

Complete the identity.

39. $1 - \cos^2 x =$ **40.** $\sec^2 x - \tan^2 x =$

41. $\sin^2 x - 1 =$ **42.** $1 + \cot^2 x =$

Synthesis

43. ◆ Discuss and compare the graphs of $y_1 = \sin x$, $y_2 = \sin 2x$, and $y_3 = \sin (x/2)$.

44. ◆ Find all errors in the following:

$$2 \sin^2 2x + \cos 4x$$
$$= 2(2 \sin x \cos x)^2 + 2 \cos 2x$$
$$= 8 \sin^2 x \cos^2 x + 2(\cos^2 x + \sin^2 x)$$
$$= 8 \sin^2 x \cos^2 x + 2.$$

Then verify with a grapher that

$$2 \sin^2 2x + \cos 4x = 8 \sin^2 x \cos^2 x + 2$$

is not an identity.

45. Given that $\cos 51° \approx 0.6293$, find the six function values for $141°$.

Simplify. Check your results with a grapher.

46. $\sin \left(\dfrac{\pi}{2} - x \right) [\sec x - \cos x]$

47. $\cos (\pi - x) + \cot x \sin \left(x - \dfrac{\pi}{2} \right)$

48. $\dfrac{\cos x - \sin \left(\dfrac{\pi}{2} - x \right) \sin x}{\cos x - \cos (\pi - x) \tan x}$

49. $\dfrac{\cos^2 y \sin \left(y + \dfrac{\pi}{2} \right)}{\sin^2 y \sin \left(\dfrac{\pi}{2} - y \right)}$

Find $\sin \theta$, $\cos \theta$, and $\tan \theta$ under the given conditions.

50. $\cos 2\theta = \dfrac{7}{12}, \dfrac{3\pi}{2} \leq 2\theta \leq 2\pi$

51. $\tan \dfrac{\theta}{2} = -\dfrac{5}{3}, \pi < \theta \leq \dfrac{3\pi}{2}$

52. *Nautical Mile.* *Latitude* is used to measure north–south location on the earth between the equator and the poles. For example, Chicago has latitude $42°$N. (See the figure.) In Great Britain, the

nautical mile is defined as the length of a minute of arc of the earth's radius. Since the earth is flattened slightly at the poles, a British nautical mile varies with latitude. In fact, it is given, in feet, by the function

$$N(\phi) = 6066 - 31 \cos 2\phi,$$

where ϕ is the latitude in degrees.

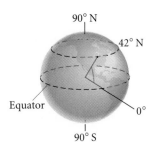

Equator

90° N

42° N

0°

90° S

a) What is the length of a British nautical mile at Chicago?
b) What is the length of a British nautical mile at the North Pole?
c) Express $N(\phi)$ in terms of $\cos \phi$ only. That is, do not use the double angle.

53. *Acceleration Due to Gravity.*   The acceleration due to gravity is often denoted by g in a formula such as $S = \frac{1}{2}gt^2$, where S is the distance that an object falls in time t. The number g relates to motion near the earth's surface and is usually considered constant. In fact, however, g is not constant, but varies slightly with latitude. If ϕ stands for latitude, in degrees, g is given with good approximation by the formula

$$g = 9.78049(1 + 0.005288 \sin^2 \phi - 0.000006 \sin^2 2\phi),$$

where g is measured in meters per second per second at sea level.

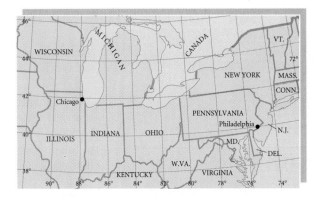

a) Chicago has latitude 42°N. Find g.
b) Philadelphia has latitude 40°N. Find g.
c) Express g in terms of $\sin \phi$ only. That is, eliminate the double angle.
d) Where on earth is g greatest? least? Use the MAX–MIN feature on a grapher.

3.3

Proving Trigonometric Identities

• *Prove identities using other identities.*

The Logic of Proving Identities

We outline two algebraic methods for proving identities.

Method 1. Start with either the left or the right side of the equation and deduce the other side. For example, suppose you are trying to prove that the equation $P = Q$ is an identity. You might try to produce a string of statements like the following, which start at P and end with Q:

$$P = S_1$$
$$= S_2$$
$$\vdots$$
$$= Q.$$

Method 2. Work with each side separately until you deduce the same expression. For example, suppose you are trying to prove that $P = Q$ is an identity. You might be able to produce two strings of statements like the following, each ending with the same statement S.

$$
\begin{array}{ll}
P = S_1 & Q = R_1 \\
 = S_2 & = R_2 \\
 \ \vdots & \ \vdots \\
 = S. & = S.
\end{array}
$$

The number of steps in each string might be different, but in each case the result is S.

A first step in learning to prove identities is to have at hand a list of the identities that you have already learned. Such a list is on the inside back cover of this text. Ask your instructor which ones you are expected to memorize. The more identities you prove, the easier it will be to prove new ones. A list of helpful hints follows.

Hints for Proving Identities

1. Use methods 1 or 2 previously outlined.

2. Work with the more complex side first.

3. Carry out any algebraic manipulations, such as adding, subtracting, multiplying, or factoring.

4. Multiplying by 1 can be helpful when rational expressions are involved.

5. Converting all expressions to sines and cosines is often helpful.

6. Try something! Put your pencil to work and get involved. You will be amazed at how often this leads to success.

7. Use a grapher for a partial check of your final answer.

Proving Identities

In the work that follows, method 1 is used in Examples 1–3 and method 2 is used in Examples 4 and 5.

Example 1   Prove the identity $1 + \sin 2\theta = (\sin \theta + \cos \theta)^2$.

Solution   Let's use method 1. We begin with the right side and deduce the left side:

$$
\begin{array}{ll}
(\sin \theta + \cos \theta)^2 = \sin^2 \theta + 2 \sin \theta \cos \theta + \cos^2 \theta & \text{Squaring} \\[4pt]
 = 1 + 2 \sin \theta \cos \theta & \text{Recalling the identity} \\
& \sin^2 x + \cos^2 x = 1 \text{ and} \\
& \text{substituting} \\[4pt]
 = 1 + \sin 2\theta. & \text{Using } \sin 2x = \\
& 2 \sin x \cos x
\end{array}
$$

We could also begin with the left side and deduce the right side:

$$1 + \sin 2\theta = 1 + 2 \sin \theta \cos \theta \qquad \text{Using } \sin 2x = 2 \sin x \cos x$$
$$= \sin^2 \theta + 2 \sin \theta \cos \theta + \cos^2 \theta \qquad \begin{array}{l}\text{Replacing 1 with}\\ \sin^2 \theta + \cos^2 \theta\end{array}$$
$$= (\sin \theta + \cos \theta)^2. \qquad \text{Factoring} \qquad \rule[0.3ex]{1.5em}{0.7ex}$$

Technology allows us to do partial checks. We graph each side of the equation and look to see if the graphs appear to be the same. For instance, in Example 1 we can graph

$$y_1 = 1 + \sin 2x \quad \text{and} \quad y_2 = (\sin x + \cos x)^2$$

in the same screen and observe that the graphs appear to be identical in $[-2\pi, 2\pi]$.

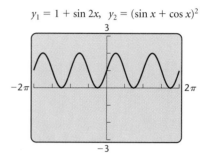

$$y_1 = 1 + \sin 2x, \quad y_2 = (\sin x + \cos x)^2$$

Keep in mind that this check does not *prove* an identity. But it is a useful tool for checking over specific intervals.

Example 2 Prove the identity

$$\frac{\sec t - 1}{t \sec t} = \frac{1 - \cos t}{t}.$$

SOLUTION   We use method 1, starting with the left side. Note that the left side involves sec t, whereas the right side involves cos t, so it might be wise to make use of an identity that involves these two expressions. That basic identity is sec $t = 1/\cos t$.

$$\frac{\sec t - 1}{t \sec t} = \frac{\dfrac{1}{\cos t} - 1}{t \, \dfrac{1}{\cos t}} \qquad \text{Substituting } 1/\cos t \text{ for sec } t$$

$$= \left(\frac{1}{\cos t} - 1\right) \cdot \frac{\cos t}{t}$$

$$= \frac{1}{t} - \frac{\cos t}{t} \qquad \text{Multiplying}$$

$$= \frac{1 - \cos t}{t} \qquad \begin{array}{l}\text{Note that all these steps}\\ \text{are reversible.}\end{array}$$

We started with the left side and deduced the right side, so the proof is complete. Check with a grapher. $\rule[0.3ex]{1.5em}{0.7ex}$

Example 3   Prove the identity

$$\frac{\sin 2x}{\sin x} - \frac{\cos 2x}{\cos x} = \sec x.$$

SOLUTION

$$\frac{\sin 2x}{\sin x} - \frac{\cos 2x}{\cos x} = \frac{2 \sin x \cos x}{\sin x} - \frac{\cos^2 x - \sin^2 x}{\cos x} \qquad \text{Using double-angle identities}$$

$$= 2 \cos x - \frac{\cos^2 x - \sin^2 x}{\cos x} \qquad \text{Simplifying}$$

$$= \frac{2 \cos^2 x}{\cos x} - \frac{\cos^2 x - \sin^2 x}{\cos x} \qquad \text{Multiplying } 2 \cos x \text{ by 1, or } \cos x / \cos x$$

$$= \frac{2 \cos^2 x - \cos^2 x + \sin^2 x}{\cos x} \qquad \text{Subtracting}$$

$$= \frac{\cos^2 x + \sin^2 x}{\cos x}$$

$$= \frac{1}{\cos x} \qquad \text{Using a Pythagorean identity}$$

$$= \sec x \qquad \text{Recalling a basic identity}$$

Check with a grapher.   ▬

Example 4   Prove the identity

$$\sin^2 x \tan^2 x = \tan^2 x - \sin^2 x.$$

SOLUTION   For this proof, we are going to work with each side separately using method 2. We try to deduce the same expression on each side. In practice, you might work on one side for awhile, then work on the other side separately, and then go back to the other side. In other words, you work back and forth until you arrive at the same expression. Let's start with the right side:

$$\tan^2 x - \sin^2 x = \frac{\sin^2 x}{\cos^2 x} - \sin^2 x \qquad \text{Recalling the identity } \tan x = \frac{\sin x}{\cos x} \text{ and substituting}$$

$$= \frac{\sin^2 x}{\cos^2 x} - \sin^2 x \cdot \frac{\cos^2 x}{\cos^2 x} \qquad \text{Multiplying by 1 in order to subtract}$$

$$= \frac{\sin^2 x - \sin^2 x \cos^2 x}{\cos^2 x} \qquad \text{Carrying out the subtraction}$$

$$= \frac{\sin^2 x \, (1 - \cos^2 x)}{\cos^2 x} \qquad \text{Factoring}$$

$$= \frac{\sin^2 x \sin^2 x}{\cos^2 x} \qquad \text{Recalling the identity } \sin^2 x + \cos^2 x = 1 \text{ or } 1 - \cos^2 x = \sin^2 x \text{ and substituting}$$

$$= \frac{\sin^4 x}{\cos^2 x}.$$

At this point, we stop and work with the left side, $\sin^2 x \tan^2 x$, of the original identity and try to end with the same expression that we ended with on the right side:

$$\sin^2 x \tan^2 x = \sin^2 x \frac{\sin^2 x}{\cos^2 x}$$

Recalling the identity $\tan x = \dfrac{\sin x}{\cos x}$ and substituting

$$= \frac{\sin^4 x}{\cos^2 x}.$$

We have deduced the same expression from each side, so the proof is complete. Check with a grapher. ▬

Example 5 Prove the identity

$$\cot \phi + \csc \phi = \frac{\sin \phi}{1 - \cos \phi}.$$

SOLUTION   We are again using method 2, beginning with the left side:

$$\cot \phi + \csc \phi = \frac{\cos \phi}{\sin \phi} + \frac{1}{\sin \phi}$$

Using basic identities

$$= \frac{1 + \cos \phi}{\sin \phi}.$$

Adding

At this point, we stop and work with the right side of the original identity:

$$\frac{\sin \phi}{1 - \cos \phi} = \frac{\sin \phi}{1 - \cos \phi} \cdot \frac{1 + \cos \phi}{1 + \cos \phi}$$

Multiplying by 1

$$= \frac{\sin \phi \, (1 + \cos \phi)}{1 - \cos^2 \phi}$$

$$= \frac{\sin \phi \, (1 + \cos \phi)}{\sin^2 \phi}$$

Using $\sin^2 x = 1 - \cos^2 x$

$$= \frac{1 + \cos \phi}{\sin \phi}.$$

Simplifying

The proof is complete since we deduced the same expression from each side. Check with a grapher. ▬

3.3 Exercise Set

Prove each of the following identities. Check your work with a grapher.

1. $\sec x - \sin x \tan x = \cos x$

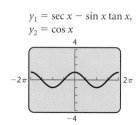

$$y_1 = \sec x - \sin x \tan x,$$
$$y_2 = \cos x$$

2. $\dfrac{1 + \cos \theta}{\sin \theta} + \dfrac{\sin \theta}{\cos \theta} = \dfrac{\cos \theta + 1}{\sin \theta \cos \theta}$

$$y_1 = \frac{1 + \cos x}{\sin x} + \frac{\sin x}{\cos x},$$
$$y_2 = \frac{\cos x + 1}{\sin x \cos x}$$

3. $\dfrac{1 - \cos x}{\sin x} = \dfrac{\sin x}{1 + \cos x}$

4. $\dfrac{1 + \tan y}{1 + \cot y} = \dfrac{\sec y}{\csc y}$

5. $\dfrac{1 + \tan \theta}{1 - \tan \theta} + \dfrac{1 + \cot \theta}{1 - \cot \theta} = 0$

6. $\dfrac{\sin x + \cos x}{\sec x + \csc x} = \dfrac{\sin x}{\sec x}$

7. $\dfrac{\cos^2 \alpha + \cot \alpha}{\cos^2 \alpha - \cot \alpha} = \dfrac{\cos^2 \alpha \tan \alpha + 1}{\cos^2 \alpha \tan \alpha - 1}$

8. $\sec 2\theta = \dfrac{\sec^2 \theta}{2 - \sec^2 \theta}$

9. $\dfrac{2 \tan \theta}{1 + \tan^2 \theta} = \sin 2\theta$

10. $\dfrac{\cos (u - v)}{\cos u \sin v} = \tan u + \cot v$

11. $1 - \cos 5\theta \cos 3\theta - \sin 5\theta \sin 3\theta = 2 \sin^2 \theta$

12. $\cos^4 x - \sin^4 x = \cos 2x$

13. $2 \sin \theta \cos^3 \theta + 2 \sin^3 \theta \cos \theta = \sin 2\theta$

14. $\dfrac{\tan 3t - \tan t}{1 + \tan 3t \tan t} = \dfrac{2 \tan t}{1 - \tan^2 t}$

15. $\dfrac{\tan x - \sin x}{2 \tan x} = \sin^2 \dfrac{x}{2}$

16. $\dfrac{\cos^3 \beta - \sin^3 \beta}{\cos \beta - \sin \beta} = \dfrac{2 + \sin 2\beta}{2}$

17. $\sin (\alpha + \beta) \sin (\alpha - \beta) = \sin^2 \alpha - \sin^2 \beta$

18. $\cos^2 x (1 - \sec^2 x) = -\sin^2 x$

19. $\tan \theta (\tan \theta + \cot \theta) = \sec^2 \theta$

20. $\dfrac{\cos \theta + \sin \theta}{\cos \theta} = 1 + \tan \theta$

21. $\dfrac{1 + \cos^2 x}{\sin^2 x} = 2 \csc^2 x - 1$

22. $\dfrac{\tan y + \cot y}{\csc y} = \sec y$

23. $\dfrac{1 + \sin x}{1 - \sin x} + \dfrac{\sin x - 1}{1 + \sin x} = 4 \sec x \tan x$

24. $\tan \theta - \cot \theta = (\sec \theta - \csc \theta)(\sin \theta + \cos \theta)$

25. $\cos^2 \alpha \cot^2 \alpha = \cot^2 \alpha - \cos^2 \alpha$

26. $\dfrac{\tan x + \cot x}{\sec x + \csc x} = \dfrac{1}{\cos x + \sin x}$

27. $2 \sin^2 \theta \cos^2 \theta + \cos^4 \theta = 1 - \sin^4 \theta$

28. $\dfrac{\cot \theta}{\csc \theta - 1} = \dfrac{\csc \theta + 1}{\cot \theta}$

29. $\dfrac{1 + \sin x}{1 - \sin x} = (\sec x + \tan x)^2$

30. $\sec^4 s - \tan^2 s = \tan^4 s + \sec^2 s$

In Exercises 31–36, use a grapher to determine which expression (A)–(F) on the right can be used to complete an identity. Then try to prove that identity algebraically.

31. $\dfrac{\cos x + \cot x}{1 + \csc x}$ **A.** $\dfrac{\sin^3 x - \cos^3 x}{\sin x - \cos x}$

32. $\cot x + \csc x$ **B.** $\cos x$

33. $\sin x \cos x + 1$ **C.** $\tan x + \cot x$

34. $2 \cos^2 x - 1$ **D.** $\cos^3 x + \sin^3 x$

35. $\dfrac{1}{\cot x \sin^2 x}$ **E.** $\dfrac{\sin x}{1 - \cos x}$

36. $(\cos x + \sin x)(1 - \sin x \cos x)$ **F.** $\cos^4 x - \sin^4 x$

Skill Maintenance

For each function:

a) *Graph the function.*
b) *Determine whether the function is one-to-one.*
c) *If the function is one-to-one, find an equation for its inverse.*
d) *Graph the inverse of the function.*

37. $f(x) = 3x - 2$ **38.** $f(x) = x^3 + 1$

39. $f(x) = x^2 - 4, \ x \geq 0$ **40.** $f(x) = \sqrt{x + 2}$

Synthesis

41. ◆ What restrictions must be placed on the variable in each of the following identities? Why?

a) $\sin 2x = \dfrac{2 \tan x}{1 + \tan^2 x}$

b) $\dfrac{1 - \cos x}{\sin x} = \dfrac{\sin x}{1 + \cos x}$

c) $2 \sin x \cos^3 x + 2 \sin^3 x \cos x = \sin 2x$

42. ◆ Consider each of the following functions:

$$y_1 = \frac{\pi}{2} - \frac{4}{\pi}\left(\cos x + \frac{\cos 3x}{9} + \frac{\cos 5x}{25}\right)$$

and

$$y_2 = \frac{\pi}{2} - \frac{4}{\pi}\left(\cos x + \frac{\cos 3x}{9} + \frac{\cos 5x}{25} + \frac{\cos 7x}{49}\right).$$

Use a grapher to graph both functions using the viewing window $[-10, 10, -1, 4]$, with Xscl $= 1$ and Yscl $= 1$. On the basis of your graphs, would you consider

$$\frac{\pi}{2} - \frac{4}{\pi}\left(\cos x + \frac{\cos 3x}{9} + \frac{\cos 5x}{25}\right)$$

$$= \frac{\pi}{2} - \frac{4}{\pi}\left(\cos x + \frac{\cos 3x}{9} + \frac{\cos 5x}{25} + \frac{\cos 7x}{49}\right)$$

to be an identity? Now change the viewing window and use ZOOM and TRACE and the TABLE feature to examine the graphs in more detail. What do you discover about the graphs? Do we have an identity? What caution must be used in determining whether the equation is an identity considering only the graphs?

Prove each identity and verify your results with a grapher.

43. $\ln |\tan x| = -\ln |\cot x|$

44. $\ln |\sec \theta + \tan \theta| = -\ln |\sec \theta - \tan \theta|$

45. Prove the identity

$\log (\cos x - \sin x) + \log (\cos x + \sin x) = \log \cos 2x.$

46. *Mechanics.* The following equation occurs in the study of mechanics:

$$\sin \theta = \frac{I_1 \cos \phi}{\sqrt{(I_1 \cos \phi)^2 + (I_2 \sin \phi)^2}}.$$

It can happen that $I_1 = I_2$. Assuming that this happens, simplify the equation.

47. *Alternating Current.* In the theory of alternating current, the following equation occurs:

$$R = \frac{1}{\omega C(\tan \theta + \tan \phi)}.$$

Show that this equation is equivalent to

$$R = \frac{\cos \theta \cos \phi}{\omega C \sin (\theta + \phi)}.$$

48. *Electrical Theory.* In electrical theory, the following equations occur:

$$E_1 = \sqrt{2}E_t \cos \left(\theta + \frac{\pi}{P}\right)$$

and

$$E_2 = \sqrt{2}E_t \cos \left(\theta - \frac{\pi}{P}\right).$$

Assuming that these equations hold, show that

$$\frac{E_1 + E_2}{2} = \sqrt{2}E_t \cos \theta \cos \frac{\pi}{P}$$

and

$$\frac{E_1 - E_2}{2} = -\sqrt{2}E_t \sin \theta \sin \frac{\pi}{P}.$$

3.4

Inverses of the Trigonometric Functions

- *Find values of the inverse trigonometric functions.*
- *Simplify expressions such as sin (sin⁻¹ x) and sin⁻¹ (sin x).*
- *Simplify expressions involving compositions such as* $\sin\left(\cos^{-1}\frac{1}{2}\right)$ *without using a calculator.*
- *Simplify expressions such as sin arctan (a/b) by making a drawing and reading off appropriate ratios.*

In this section, we develop inverse trigonometric functions. It may be helpful for you to review the material on inverse functions in Section 1.8.

The graphs of the sine, cosine, and tangent functions follow. Do these functions have inverses that are functions? They do if they are one-to-one, which means that they pass the horizontal-line test.

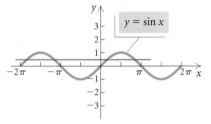
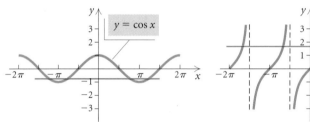

Note that each function has a horizontal line (shown in red) that crosses the graph more than once. Therefore, none of them has an inverse that is a function.

The graphs of an equation and its inverse are reflections of each other across the line $y = x$. Let's examine the inverses of each of the three functions graphed above.

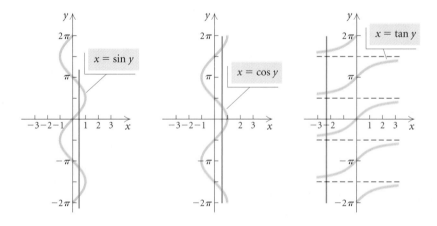

We can check again to see whether these are graphs of functions by using the vertical-line test. In each case, there is a vertical line (shown in red) that crosses the graph more than once, so each fails to be a function.

Let's look specifically at the graph of the inverse of $y = \sin x$, which is $x = \sin y$. Consider the input $x = \frac{1}{2}$. On the graph of the inverse of

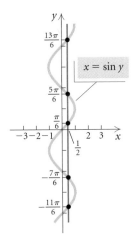

the sine functions, we draw a vertical line at $x = \frac{1}{2}$, as shown in the figure at left. The vertical line intersects the graph at points whose y-values are such that $\frac{1}{2} = \sin y$. Some numbers whose sine is $\frac{1}{2}$ are $\pi/6$, $5\pi/6$, $-7\pi/6$, and so on. The complete set of values is given by $\pi/6 + 2k\pi$ and $5\pi/6 + 2k\pi$, where k is an integer. Indeed, the vertical line crosses the curve at many points, verifying that the inverse of the sine function is not a function.

Restricting Ranges to Define Inverse Functions

Recall that a function like $f(x) = x^2$ does not have an inverse that is a function, but by restricting the domain of f to nonnegative numbers, we have a new squaring function, $f(x) = x^2$, $x \geq 0$, that has an inverse, $f^{-1}(x) = \sqrt{x}$. This is equivalent to restricting the range of the inverse relation to exclude ordered pairs that contain negative numbers.

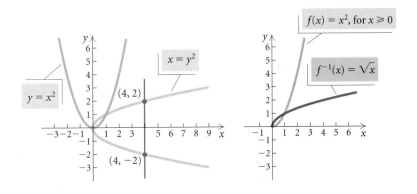

In a similar manner, we can define new trigonometric functions whose inverses are functions. We can do this by restricting either the domains of the basic trigonometric functions or the ranges of their inverse relations. This can be done in many ways, but the restrictions illustrated below with solid black curves are fairly standard in mathematics.

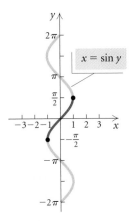

FIGURE 1

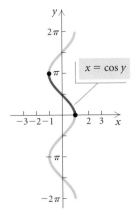

FIGURE 2

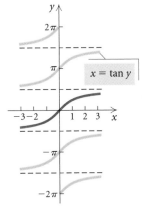

FIGURE 3

For the inverse sine function, we choose a range close to the origin that allows all inputs in the interval $[-1, 1]$ to have function values. Thus we choose the interval $[-\pi/2, \pi/2]$ for the range (Fig. 1). For the inverse cosine function, we choose a range close to the origin that allows all inputs in the interval $[-1, 1]$ to have function values. This is the interval $[0, \pi]$ (Fig. 2). For the inverse tangent function, we choose a range close to the origin that allows all real numbers to have function values. The interval $(-\pi/2, \pi/2)$ satisfies this requirement (Fig. 3).

Inverse Trigonometric Functions

FUNCTION	DOMAIN	RANGE
$y = \sin^{-1} x$ $= \arcsin x$, where $x = \sin y$	$[-1, 1]$	$[-\pi/2, \pi/2]$
$y = \cos^{-1} x$ $= \arccos x$, where $x = \cos y$	$[-1, 1]$	$[0, \pi]$
$y = \tan^{-1} x$ $= \arctan x$, where $x = \tan y$	$(-\infty, \infty)$	$(-\pi/2, \pi/2)$

The notation $\arcsin x$ arises because the function is the length of an arc on the unit circle for which the sine is x. **The notation $\sin^{-1} x$ is not exponential notation. It does not mean $1/\sin x$!** Either of the two kinds of notation above can be read "the inverse sine of x" or "the arc sine of x" or "the number (or angle) whose sine is x."

The graphs of the inverse trigonometric functions are as follows.

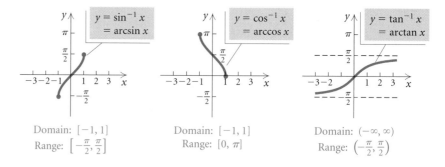

Domain: $[-1, 1]$
Range: $\left[-\frac{\pi}{2}, \frac{\pi}{2}\right]$

Domain: $[-1, 1]$
Range: $[0, \pi]$

Domain: $(-\infty, \infty)$
Range: $\left(-\frac{\pi}{2}, \frac{\pi}{2}\right)$

Interactive Discovery

Inverse trigonometric functions can be graphed on a grapher. Graph $y = \sin^{-1} x$ making sure the grapher is in RADIAN mode and changing the viewing window to $[-3, 3, -\pi, \pi]$, with Xscl = 1 and Yscl = $\pi/2$. Now try graphing $y = \cos^{-1} x$ and $y = \tan^{-1} x$. Then trace to confirm the domain and the range of each inverse with its graph.

The following diagrams show the restricted ranges for the inverse trigonometric functions on a unit circle. These ranges should be memorized. The missing endpoints indicate inputs that are not in the domain of the original function.

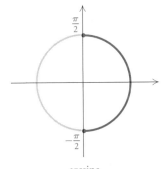

arcsine
Range $\left[-\frac{\pi}{2}, \frac{\pi}{2}\right]$

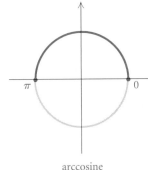

arccosine
Range $[0, \pi]$

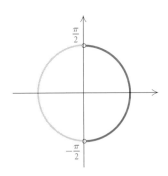

arctangent
Range $\left(-\frac{\pi}{2}, \frac{\pi}{2}\right)$

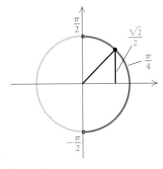

FIGURE 4

Example 1 Find each of the following function values.

a) $\arcsin \dfrac{\sqrt{2}}{2}$ **b)** $\cos^{-1}\left(-\dfrac{1}{2}\right)$ **c)** $\tan^{-1}\left(-\dfrac{\sqrt{3}}{3}\right)$

SOLUTION

a) In the restricted range $[-\pi/2, \pi/2]$, the only number with a sine of $\sqrt{2}/2$ is $\pi/4$. Thus, $\arcsin(\sqrt{2}/2) = \pi/4$, or $45°$. (See Fig. 4 at left.)

b) The only number with a cosine of $-\frac{1}{2}$ in the restricted range $[0, \pi]$ is $2\pi/3$. Thus, $\cos^{-1}\left(-\frac{1}{2}\right) = 2\pi/3$, or $120°$. (See Fig. 5 at left.)

c) The only number in the restricted range $(-\pi/2, \pi/2)$ with a tangent of $-\sqrt{3}/3$ is $-\pi/6$. Thus, $\tan^{-1}(-\sqrt{3}/3)$ is $-\pi/6$, or $-30°$. (See Fig. 6 at left.)

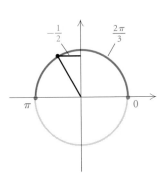

FIGURE 5

We can also use a grapher to find inverse trigonometric function values (see Section 2.1, p. 123). On most graphers, we can find inverse function values in both radians and degrees simply by selecting the appropriate mode. The key strokes involved in finding inverse function values vary with the grapher. Be sure to read the instructions for the grapher you are using.

Finding $\sin^{-1} 1$ provides a quick way to check the mode setting. If you get 1.570796327, which is about $\pi/2$, you know that the values are in radians. If you get 90, you know that the values are in degrees.

Example 2 Approximate each of the following function values in both radians and degrees.

a) $\cos^{-1}(-0.2689)$

b) $\arctan(-0.2623)$

c) $\sin^{-1} 0.20345$

d) $\arccos 1.318$

e) $\csc^{-1} 8.205$

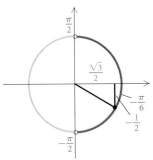

FIGURE 6

SOLUTION

FUNCTION VALUE	MODE	READOUT	ROUNDED
a) $\cos^{-1}(-0.2689)$	Radian	1.843047111	1.8430
	Degree	105.5988209	105.6°
b) $\arctan(-0.2623)$	Radian	−.256521214	−0.2565
	Degree	−14.69758292	−14.7°
c) $\sin^{-1} 0.20345$	Radian	.204880336	0.2049
	Degree	11.73877855	11.7°
d) $\arccos 1.318$	Radian	ERR:DOMAIN	
	Degree	ERR:DOMAIN	

The value 1.318 is not in $[-1, 1]$, the domain of the arccosine function.

e) The cosecant function is the reciprocal of the sine function:

$\csc^{-1} 8.205 =$
$\sin^{-1}(1/8.205)$

	MODE	READOUT	ROUNDED
	Radian	.122180665	0.1222
	Degree	7.000436462	7.0°

The following is a summary of the domains and ranges of the trigonometric functions together with a summary of the domains and ranges of the inverse trigonometric functions. For completeness, we have included the arccosecant, the arcsecant, and the arccotangent, though there is a lack of uniformity in their definitions in mathematical literature.

FUNCTION	DOMAIN	RANGE	INVERSE FUNCTION	DOMAIN	RANGE
sin	All reals, $(-\infty, \infty)$	$[-1, 1]$	$\sin^{-1}$	$[-1, 1]$	$\left[-\dfrac{\pi}{2}, \dfrac{\pi}{2}\right]$
cos	All reals, $(-\infty, \infty)$	$[-1, 1]$	$\cos^{-1}$	$[-1, 1]$	$[0, \pi]$
tan	All reals except $k\pi/2$, k odd	All reals, $(-\infty, \infty)$	$\tan^{-1}$	All reals, $(-\infty, \infty)$	$\left(-\dfrac{\pi}{2}, \dfrac{\pi}{2}\right)$
csc	All reals except $k\pi$	$(-\infty, -1] \cup [1, \infty)$	$\csc^{-1}$	$(-\infty, -1] \cup [1, \infty)$	$\left[-\dfrac{\pi}{2}, 0\right) \cup \left(0, \dfrac{\pi}{2}\right]$
sec	All reals except $k\pi/2$, k odd	$(-\infty, -1] \cup [1, \infty)$	$\sec^{-1}$	$(-\infty, -1] \cup [1, \infty)$	$\left[0, \dfrac{\pi}{2}\right) \cup \left(\dfrac{\pi}{2}, \pi\right]$
cot	All reals except $k\pi$	All reals, $(-\infty, \infty)$	$\cot^{-1}$	All reals, $(-\infty, \infty)$	$\left[-\dfrac{\pi}{2}, 0\right) \cup \left(0, \dfrac{\pi}{2}\right]$

Composition of Trigonometric Functions and Their Inverses

Various compositions of trigonometric functions and their inverses often occur in practice. For example, we might want to try to simplify an expression such as

$$\sin\left(\sin^{-1} x\right) \quad \text{or} \quad \sin\left(\text{arccot}\, \frac{x}{2}\right).$$

In the expression on the left, we are finding "the sine of a number whose sine is x." Recall from Section 1.8 that if a function f has an inverse that is also a function, then

$$f(f^{-1}(x)) = x, \quad \text{for all } x \text{ in the domain of } f^{-1},$$

and $\quad f^{-1}(f(x)) = x, \quad \text{for all } x \text{ in the domain of } f.$

Thus, if $f(x) = \sin x$ and $f^{-1}(x) = \sin^{-1} x$, then

$$\sin\left(\sin^{-1} x\right) = x, \quad \textbf{for all } x \textbf{ in the domain of } \sin^{-1},$$

which is any number in the interval $[-1, 1]$. Similar results hold for the other trigonometric functions.

Composition of Trigonometric Functions

$\sin\left(\sin^{-1} x\right) = x, \quad$ for all x in the domain of $\sin^{-1}$.
$\cos\left(\cos^{-1} x\right) = x, \quad$ for all x in the domain of $\cos^{-1}$.
$\tan\left(\tan^{-1} x\right) = x, \quad$ for all x in the domain of $\tan^{-1}$.

Example 3 Simplify each of the following.

a) $\cos\left(\cos^{-1} \dfrac{\sqrt{3}}{2}\right)$

b) $\sin\left(\sin^{-1} 1.8\right)$

SOLUTION

a) Since $\sqrt{3}/2$ is in the domain of $\cos^{-1}$, $[-1, 1]$, it follows that

$$\cos\left(\cos^{-1} \frac{\sqrt{3}}{2}\right) = \frac{\sqrt{3}}{2}.$$

b) Since 1.8 is not in $[-1, 1]$, the domain of $\sin^{-1}$, we cannot evaluate this expression. We know that there is no number with a sine of 1.8. Since we cannot find $\sin^{-1} 1.8$, we state that $\sin\left(\sin^{-1} 1.8\right)$ does not exist.　　　▬

Now let's consider an expression like $\sin^{-1}\left(\sin x\right)$. We might also suspect that this is equal to x for any x, but this is not true unless x is in the range of the $\sin^{-1}$ function. Note that in order to define $\sin^{-1}$, we had to restrict the domain of the sine function. In doing so, we restricted the range of the inverse sine function. Thus,

$$\sin^{-1}\left(\sin x\right) = x, \quad \textbf{for all } x \textbf{ in the range of } \sin^{-1}.$$

Similar results hold for the other trigonometric functions.

Special Cases

$\sin^{-1}(\sin x) = x,$ for all x in the range of $\sin^{-1}$.

$\cos^{-1}(\cos x) = x,$ for all x in the range of $\cos^{-1}$.

$\tan^{-1}(\tan x) = x,$ for all x in the range of $\tan^{-1}$.

Example 4   Simplify each of the following.

a) $\tan^{-1}\left(\tan \dfrac{\pi}{6}\right)$ **b)** $\sin^{-1}\left(\sin \dfrac{3\pi}{4}\right)$

SOLUTION

a) Since $\pi/6$ is in $(-\pi/2,\ \pi/2)$, the range of the $\tan^{-1}$ function, we can use $\tan^{-1}(\tan x) = x$. Thus,

$$\tan^{-1}\left(\tan \frac{\pi}{6}\right) = \frac{\pi}{6}.$$

b) Note that $3\pi/4$ is not in $[-\pi/2,\ \pi/2]$, the range of the $\sin^{-1}$ function. Thus we *cannot* apply $\sin^{-1}(\sin x) = x$. Instead we first find $\sin(3\pi/4)$, which is $\sqrt{2}/2$, and substitute:

$$\sin^{-1}\left(\sin \frac{3\pi}{4}\right) = \sin^{-1}\left(\frac{\sqrt{2}}{2}\right) = \frac{\pi}{4}.$$

Now we find some other function compositions.

Example 5   Simplify each of the following.

a) $\sin[\arctan(-1)]$ **b)** $\cos^{-1}\left(\sin \dfrac{\pi}{2}\right)$

SOLUTION

a) $\sin[\arctan(-1)] = \sin\left[-\dfrac{\pi}{4}\right] = -\dfrac{\sqrt{2}}{2}$

b) $\cos^{-1}\left(\sin \dfrac{\pi}{2}\right) = \cos^{-1}(1) = 0$

Next let's consider

$$\cos\left(\arcsin \frac{3}{5}\right).$$

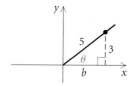

Without using a grapher, we cannot find $\arcsin \frac{3}{5}$. However, we can still evaluate the entire expression by sketching a reference triangle. We are looking for angle θ such that $\arcsin \frac{3}{5} = \theta$, or $\sin \theta = \frac{3}{5}$. Since $\arcsin$ is defined in $[-\pi/2,\ \pi/2]$ and $\frac{3}{5} > 0$, we know that θ is in quadrant I. We sketch a reference right triangle, as shown at left. The angle θ in this triangle is an angle whose sine is $\frac{3}{5}$. We wish to find the cosine of this angle.

Since the triangle is a right triangle, we can find the length of the base, b. It is 4. Thus we know that $\cos \theta = b/5$, or $\frac{4}{5}$. Therefore,

$$\cos \left(\arcsin \frac{3}{5} \right) = \frac{4}{5}.$$

Example 6   Find $\sin \left(\operatorname{arccot} \dfrac{x}{2} \right)$.

SOLUTION   We draw a right triangle whose legs have lengths $|x|$ and 2, so that $\cot \theta = x/2$. If x is negative, we get the triangle in standard position shown on the left below. If x is positive, we get the triangle shown on the right.

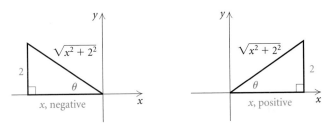

We find the length of the hypotenuse and then read off the sine ratio. In either case, we get

$$\sin \left(\operatorname{arccot} \frac{x}{2} \right) = \frac{2}{\sqrt{x^2 + 2^2}}, \quad \text{or} \quad \frac{2}{\sqrt{x^2 + 4}}.$$

In the following example, we use a sum identity to evaluate an expression.

Example 7   Evaluate: $\sin \left(\sin^{-1} \frac{1}{2} + \cos^{-1} \frac{5}{13} \right)$.

SOLUTION   The expression is a sine of a sum, so we use the identity

$$\sin (u + v) = \sin u \cos v + \cos u \sin v.$$

Thus,

$$\sin \left(\sin^{-1} \frac{1}{2} + \cos^{-1} \frac{5}{13} \right)$$

$$= \sin \left(\sin^{-1} \frac{1}{2} \right) \cdot \cos \left(\cos^{-1} \frac{5}{13} \right) + \cos \left(\sin^{-1} \frac{1}{2} \right) \cdot \sin \left(\cos^{-1} \frac{5}{13} \right)$$

$$= \frac{1}{2} \cdot \frac{5}{13} + \cos \left(\sin^{-1} \frac{1}{2} \right) \cdot \sin \left(\cos^{-1} \frac{5}{13} \right) \qquad \text{Using composition identities}$$

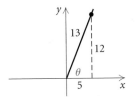

Now since $\sin^{-1} \frac{1}{2} = \pi/6$, $\cos \left(\sin^{-1} \frac{1}{2} \right)$ simplifies to $\sqrt{3}/2$. To find $\sin \left(\cos^{-1} \frac{5}{13} \right)$, we use a reference triangle in quadrant I and determine that the sine of the angle whose cosine is $\frac{5}{13}$ is $\frac{12}{13}$. Our expression now simplifies to

$$\frac{1}{2} \cdot \frac{5}{13} + \frac{\sqrt{3}}{2} \cdot \frac{12}{13}, \quad \text{or} \quad \frac{5 + 12\sqrt{3}}{26}.$$

Thus,

$$\sin\left(\sin^{-1}\frac{1}{2} + \cos^{-1}\frac{5}{13}\right) = \frac{5 + 12\sqrt{3}}{26}.$$

We can check with a grapher:

$$\sin\left(\sin^{-1}\frac{1}{2} + \cos^{-1}\frac{5}{13}\right) \approx 0.9917 \quad \text{and} \quad \frac{5 + 12\sqrt{3}}{26} \approx 0.9917.$$

3.4 Exercise Set

Find each of the following exactly in radians and degrees.

1. $\sin^{-1}\left(-\dfrac{\sqrt{3}}{2}\right)$

2. $\cos^{-1}\dfrac{1}{2}$

3. $\arctan 1$

4. $\arcsin 0$

5. $\arccos\dfrac{\sqrt{2}}{2}$

6. $\sec^{-1}\sqrt{2}$

7. $\tan^{-1} 0$

8. $\arctan\dfrac{\sqrt{3}}{3}$

9. $\cos^{-1}\dfrac{\sqrt{3}}{2}$

10. $\cot^{-1}\left(-\dfrac{\sqrt{3}}{3}\right)$

11. $\csc^{-1} 2$

12. $\sin^{-1}\dfrac{1}{2}$

13. $\text{arccot}\,(-\sqrt{3})$

14. $\tan^{-1}(-1)$

15. $\arcsin\left(-\dfrac{1}{2}\right)$

16. $\arccos\left(-\dfrac{\sqrt{2}}{2}\right)$

17. $\cos^{-1} 0$

18. $\sin^{-1}\dfrac{\sqrt{3}}{2}$

19. $\text{arcsec}\,2$

20. $\text{arccsc}\,(-1)$

Use a grapher to find each of the following in radians, rounded to three decimal places, and in degrees, rounded to the nearest tenth of a degree.

21. $\arctan 0.3673$

22. $\cos^{-1}(-0.2935)$

23. $\sin^{-1} 0.9613$

24. $\arcsin(-0.6199)$

25. $\cos^{-1}(-0.9810)$

26. $\tan^{-1} 158$

27. $\csc^{-1}(-6.2774)$

28. $\sec^{-1} 1.1677$

29. $\tan^{-1}(1.091)$

30. $\cot^{-1} 1.265$

31. $\arcsin(-0.8192)$

32. $\arccos(-0.2716)$

33. State the domains of the inverse sine, inverse cosine, and inverse tangent functions.

34. State the ranges of the inverse sine, inverse cosine, and inverse tangent functions.

35. *Angle of Depression.*   An airplane is flying at an altitude of 2000 ft toward an island. The straight-line distance from the airplane to the island is d feet. Express θ, the angle of depression, as an inverse sine function of d.

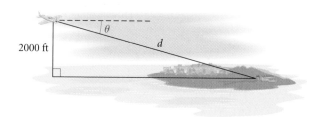

36. *Angle of Inclination.*   A guy wire is attached to the top of a 50-ft pole and stretched to a point that is d feet from the bottom of the pole. Express β, the angle of inclination, as a function of d.

Evaluate.

37. $\sin(\arcsin 0.3)$

38. $\tan[\tan^{-1}(-4.2)]$

39. $\cos^{-1}\left[\cos\left(-\dfrac{\pi}{4}\right)\right]$

40. $\arcsin\left(\sin\dfrac{2\pi}{3}\right)$

41. $\sin^{-1}\left(\sin\dfrac{\pi}{5}\right)$

42. $\cot^{-1}\left(\cot\dfrac{2\pi}{3}\right)$

43. $\tan^{-1}\left(\tan\dfrac{2\pi}{3}\right)$

44. $\cos^{-1}\left(\cos\dfrac{\pi}{7}\right)$

45. $\sin\left(\arctan\dfrac{\sqrt{3}}{3}\right)$

46. $\cos\left(\arcsin\dfrac{\sqrt{3}}{2}\right)$

47. $\tan\left(\cos^{-1}\dfrac{\sqrt{2}}{2}\right)$

48. $\cos^{-1}(\sin\pi)$

49. $\arcsin\left(\cos\dfrac{\pi}{6}\right)$

50. $\sin^{-1}\left[\tan\left(-\dfrac{\pi}{4}\right)\right]$

51. $\tan(\arcsin 0.1)$

52. $\cos\left(\tan^{-1}\dfrac{\sqrt{3}}{4}\right)$

Find.

53. $\sin\left(\arctan\dfrac{a}{3}\right)$

54. $\tan\left(\cos^{-1}\dfrac{3}{x}\right)$

55. $\cot\left(\sin^{-1}\dfrac{p}{q}\right)$

56. $\sin(\cos^{-1}x)$

57. $\tan\left(\arcsin\dfrac{p}{\sqrt{p^2+9}}\right)$

58. $\tan\left(\dfrac{1}{2}\arcsin\dfrac{1}{2}\right)$

59. $\cos\left(\dfrac{1}{2}\arcsin\dfrac{\sqrt{3}}{2}\right)$

60. $\sin\left(2\cos^{-1}\dfrac{3}{5}\right)$

Evaluate.

61. $\cos\left(\sin^{-1}\dfrac{\sqrt{2}}{2}+\cos^{-1}\dfrac{3}{5}\right)$

62. $\sin\left(\sin^{-1}\dfrac{1}{2}+\cos^{-1}\dfrac{3}{5}\right)$

63. $\sin(\sin^{-1}x+\cos^{-1}y)$

64. $\cos(\sin^{-1}x-\cos^{-1}y)$

65. $\sin(\sin^{-1}0.6032+\cos^{-1}0.4621)$

66. $\cos(\sin^{-1}0.7325-\cos^{-1}0.4838)$

Skill Maintenance

Solve.

67. $2x^2=5x$

68. $3x^2+5x-10=18$

69. $x^4+5x^2-36=0$

70. $x^2-10x+1=0$

71. $\sqrt{x-2}=5$

72. $x=\sqrt{x+7}+5$

Synthesis

73. ◆ Explain in your own words why the ranges of the inverse trigonometric functions are restricted.

74. ◆ How does the graph of $y=\sin^{-1}x$ differ from the graph of $y=\sin x$?

75. ◆ Why is it that

$$\sin\dfrac{5\pi}{6}=\dfrac{1}{2},$$

but

$$\sin^{-1}\left(\dfrac{1}{2}\right)\neq\dfrac{5\pi}{6}?$$

76. Use a calculator to approximate the following expression:

$$16\tan^{-1}\dfrac{1}{5}-4\tan^{-1}\dfrac{1}{239}.$$

What number does this expression seem to approximate?

Prove each identity.

77. $\sin^{-1}x+\cos^{-1}x=\dfrac{\pi}{2}$

78. $\tan^{-1}x+\cot^{-1}x=\dfrac{\pi}{2}$

79. $\sin^{-1}x=\tan^{-1}\dfrac{x}{\sqrt{1-x^2}}$

80. $\tan^{-1}x=\sin^{-1}\dfrac{x}{\sqrt{x^2+1}}$

81. $\arcsin x=\arccos\sqrt{1-x^2},\quad$ for $x\geq 0$

82. $\arccos x=\arctan\dfrac{\sqrt{1-x^2}}{x},\quad$ for $x>0$

83. *Height of a Mural.* An art student's eye is at a point A, looking at a mural of height h, with the bottom of the mural y feet above the eye. The eye is x feet from the wall. Write an expression for θ in terms of x, y, and h. Then evaluate the expression when $x=20$ ft, $y=7$ ft, and $h=25$ ft.

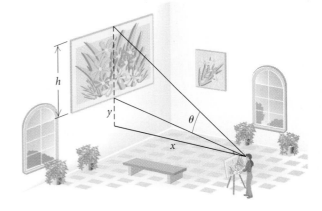

3.5
Solving Trigonometric Equations

• *Solve trigonometric equations.*

When an equation contains a trigonometric expression with a variable, such as cos x, it is called a trigonometric equation. Some trigonometric equations are identities, such as $\sin^2 x + \cos^2 x = 1$. Now we consider equations, such as $2 \cos x = -1$, that are usually not identities. As we have done for other types of equations, we will solve such equations by finding all values for x that make the equation true.

Example 1 Solve: $2 \cos x = -1$.

SOLUTION   We first solve for cos x:

$$2 \cos x = -1$$

$$\cos x = -\frac{1}{2}.$$

The solutions are numbers that have a cosine of $-\frac{1}{2}$. To find them, we use the unit circle (see Section 2.5).

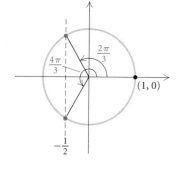

There are just two points on it for which the cosine is $-\frac{1}{2}$, as shown in the figure at left. They are the points for $2\pi/3$ and $4\pi/3$. These numbers, plus any multiple of 2π, are the solutions:

$$\frac{2\pi}{3} + 2k\pi \quad \text{or} \quad \frac{4\pi}{3} + 2k\pi,$$

where k is any integer. In degrees, the solutions are

$$120° + k \cdot 360° \quad \text{or} \quad 240° + k \cdot 360°,$$

where k is any integer.

To check the solution to $2 \cos x = -1$, or cos $x = -\frac{1}{2}$, we can graph $y_1 = \cos x$ and $y_2 = -\frac{1}{2}$ on the same set of axes. Using $\pi/3$ as the Xscl facilitates our reading of the solutions. First let's graph these equations in the interval from 0 to 2π as shown in the figure on the left below. The only solutions in $[0, 2\pi)$ are $2\pi/3$ and $4\pi/3$.

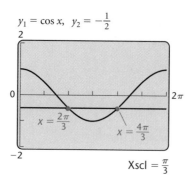

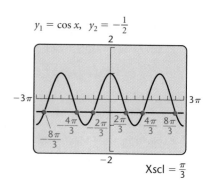

Now let's change the viewing window dimensions to $[-3\pi, 3\pi, -2, 2]$ and graph again. Since the cosine function is periodic, there is an infinite number of solutions. A few of these appear in the graph on the right above. From the graph, we see that the solutions are $2\pi/3 + 2k\pi$ and $4\pi/3 + 2k\pi$.

Example 2   Solve: $4 \sin^2 x = 1$.

SOLUTION   We begin by solving for $\sin x$:

$$4 \sin^2 x = 1$$

$$\sin^2 x = \frac{1}{4}$$

$$\sin x = \pm\frac{1}{2}.$$

Again we use the unit circle to find those numbers having a sine of $\frac{1}{2}$ or $-\frac{1}{2}$. The solutions are

$$\frac{\pi}{6} + 2k\pi, \quad \frac{5\pi}{6} + 2k\pi, \quad \frac{7\pi}{6} + 2k\pi, \quad \text{and} \quad \frac{11\pi}{6} + 2k\pi,$$

where k is any integer. In degrees, the solutions are

$$30° + k \cdot 360°, \quad 150° + k \cdot 360°, \quad 210° + k \cdot 360°, \quad \text{and}$$
$$330° + k \cdot 360°,$$

where k is any integer.

The general solutions listed above could be condensed using odd as well as even multiples of π:

$$\frac{\pi}{6} + k\pi \quad \text{and} \quad \frac{5\pi}{6} + k\pi,$$

or, in degrees,

$$30° + k \cdot 180° \quad \text{and} \quad 150° + k \cdot 180°,$$

where k is any integer.

Let's do a partial check with a grapher, checking only the solutions in $[0, 2\pi)$. We graph $y_1 = \sin x$, $y_2 = \frac{1}{2}$, and $y_3 = -\frac{1}{2}$ and observe that the solutions in $[0, 2\pi)$ are $\pi/6$, $5\pi/6$, $7\pi/6$, and $11\pi/6$.

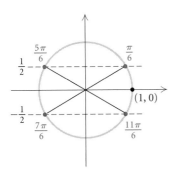

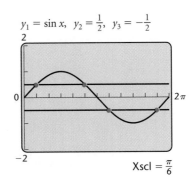

$y_1 = \sin x, \ y_2 = \frac{1}{2}, \ y_3 = -\frac{1}{2}$

$\text{Xscl} = \frac{\pi}{6}$

In most applications, it is sufficient to find just the solutions from 0 to 2π or from 0° to 360°. We then remember that any multiple of 2π, or 360°, can be added to obtain the rest of the solutions.

We must be careful to find all solutions in $[0, 2\pi)$ when solving trigonometric equations involving double angles.

Example 3 Solve $3 \tan 2x = -3$ in the interval $[0, 2\pi)$.

SOLUTION We first solve for $\tan 2x$:

$$3 \tan 2x = -3$$
$$\tan 2x = -1.$$

We are looking for solutions x to the equation for which

$$0 \le x < 2\pi.$$

Multiplying by 2, we get

$$0 \le 2x < 4\pi,$$

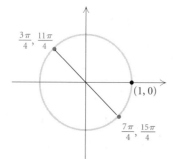

which is the interval we use when solving $\tan 2x = -1$.

Using the unit circle, we find points $2x$ in $[0, 4\pi)$ for which $\tan 2x = -1$. These values of $2x$ are as follows:

$$2x = \frac{3\pi}{4}, \quad \frac{7\pi}{4}, \quad \frac{11\pi}{4}, \quad \text{and} \quad \frac{15\pi}{4}.$$

Thus the desired values of x in $[0, 2\pi)$ are each of these values divided by 2. Therefore,

$$x = \frac{3\pi}{8}, \quad \frac{7\pi}{8}, \quad \frac{11\pi}{8}, \quad \text{and} \quad \frac{15\pi}{8}.$$

Graphers are needed for some trigonometric equations. Answers can be found in radians or degrees, depending on the mode setting.

Example 4 Solve $\frac{1}{2} \cos \phi + 1 = 1.2108$ in $[0, 360°)$.

SOLUTION We have

$$\frac{1}{2} \cos \phi + 1 = 1.2108$$

$$\frac{1}{2} \cos \phi = 0.2108$$

$$\cos \phi = 0.4216.$$

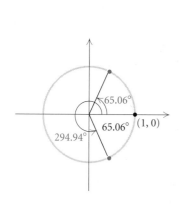

Using a grapher, we find that the reference angle, $\arccos 0.4216$, is

$$\phi \approx 65.06°.$$

Since $\cos \phi$ is positive, the solutions are in quadrants I and IV. The solutions in $[0, 360°)$ are

$$65.06° \quad \text{and} \quad 294.94°.$$

In some cases, we may need to apply some algebra before concerning ourselves with the trigonometry.

Example 5   Solve $2 \cos^2 u = 1 - \cos u$ in $[0, 2\pi)$.

SOLUTION   It is instructive to begin with a grapher. We can use either of two methods to determine the solutions graphically.

Method 1. We graph the left side of the equation as one function and then the right side as another function. Then we look for points of intersection. Here it is helpful to use $\text{Xscl} = \pi/3$.

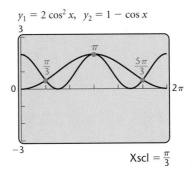

$y_1 = 2 \cos^2 x, \;\; y_2 = 1 - \cos x$

The solutions in $[0, 2\pi)$ appear to be $\pi/3$, π, and $5\pi/3$. We could use the INTERSECT feature and find approximate solutions, 1.05, 3.14, and 5.24.

Method 2. We rewrite the equation

$$2 \cos^2 u = 1 - \cos u$$

as

$$2 \cos^2 u + \cos u - 1 = 0.$$

Then we graph $y = 2 \cos^2 u + \cos u - 1$ and determine the zeros of the function.

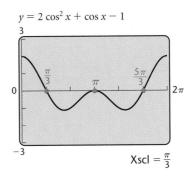

$y = 2 \cos^2 x + \cos x - 1$

Again the solutions appear to be $\pi/3$, π, and $5\pi/3$. We could use the SOLVE feature and find approximate solutions, 1.05, 3.14, and 5.24.

To confirm the solutions algebraically, we use the principle of zero products:

$$2 \cos^2 u = 1 - \cos u$$

$$2 \cos^2 u + \cos u - 1 = 0$$

$$(2 \cos u - 1)(\cos u + 1) = 0 \qquad \text{Factoring}$$

$$2 \cos u - 1 = 0 \quad or \quad \cos u + 1 = 0 \qquad \text{Principle of zero products}$$

$$2 \cos u = 1 \quad or \qquad \cos u = -1$$

$$\cos u = \frac{1}{2} \quad or \qquad \cos u = -1.$$

Thus,

$$u = \frac{\pi}{3}, \frac{5\pi}{3} \quad or \quad u = \pi.$$

The solutions in $[0, 2\pi)$ are $\pi/3$, π, and $5\pi/3$.

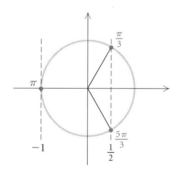

Example 6 Solve: $\sin^2 \beta - \sin \beta = 0$ in $[0, 2\pi)$.

SOLUTION   We have

$$\sin^2 \beta - \sin \beta = 0$$

$$\sin \beta (\sin \beta - 1) = 0 \qquad \text{Factoring}$$

$$\sin \beta = 0 \quad or \quad \sin \beta - 1 = 0$$

$$\sin \beta = 0 \quad or \quad \sin \beta = 1$$

$$\beta = 0, \pi \quad or \qquad \beta = \frac{\pi}{2}.$$

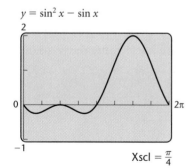

$y = \sin^2 x - \sin x$

$Xscl = \frac{\pi}{4}$

The solutions in $[0, 2\pi)$ are 0, $\pi/2$, and π.

If a trigonometric equation is quadratic but difficult or impossible to factor, we use the quadratic formula.

Example 7 Solve $10 \sin^2 x - 12 \sin x - 7 = 0$ in $[0°, 360°)$.

SOLUTION   This equation is quadratic in $\sin x$ with $a = 10$, $b = -12$, and $c = -7$. Substituting into the quadratic formula, we get

$$\sin x = \frac{12 \pm \sqrt{144 + 280}}{20} \qquad \textbf{Using the quadratic formula}$$

$$= \frac{12 \pm \sqrt{424}}{20}$$

$$\approx \frac{12 \pm 20.5913}{20}$$

$$\sin x \approx 1.6296 \quad \text{or} \quad \sin x \approx -0.4296.$$

Since sine values are never greater than 1, the first of the equations has no solution. Using the other equation, we find the reference angle to be 25.44°. Since $\sin x$ is negative, the solutions are in quadrants III and IV. Thus the solutions in $[0°, 360°)$ are

$$180° + 25.44° = 205.44° \quad \text{and} \quad 360° - 25.44° = 334.56°. \quad \rule[0.1em]{1.2em}{0.4em}$$

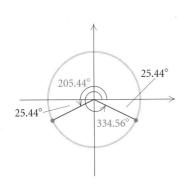

Trigonometric equations can involve more than one function.

Example 8   Solve $2 \cos^2 x \tan x = \tan x$ in $[0, 2\pi)$.

SOLUTION   With a grapher, we can determine that there are six solutions. If we let $\text{Xscl} = \pi/4$, the solutions are read more easily. In the figures at left, we show two different methods of solving graphically. Each illustrates that the solutions in $[0, 2\pi)$ are

$$0, \quad \frac{\pi}{4}, \quad \frac{3\pi}{4}, \quad \pi, \quad \frac{5\pi}{4}, \quad \text{and} \quad \frac{7\pi}{4}.$$

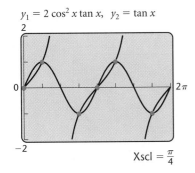

$y_1 = 2 \cos^2 x \tan x, \quad y_2 = \tan x$

$\text{Xscl} = \frac{\pi}{4}$

We can verify these solutions algebraically, as follows:

$$2 \cos^2 x \tan x = \tan x$$

$$2 \cos^2 x \tan x - \tan x = 0$$

$$\tan x \,(2 \cos^2 x - 1) = 0$$

$$\tan x = 0 \qquad \text{or} \qquad 2 \cos^2 x - 1 = 0$$

$$\cos^2 x = \frac{1}{2}$$

$$\cos x = \pm \frac{\sqrt{2}}{2}$$

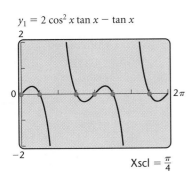

$y_1 = 2 \cos^2 x \tan x - \tan x$

$\text{Xscl} = \frac{\pi}{4}$

$$x = 0, \pi \quad \text{or} \qquad x = \frac{\pi}{4}, \frac{3\pi}{4}, \frac{5\pi}{4}, \frac{7\pi}{4}.$$

These solutions check with those found graphically. Thus, $x = 0, \pi/4,$ $3\pi/4, \pi, 5\pi/4,$ and $7\pi/4$. $\quad \rule[0.1em]{1.2em}{0.4em}$

When a trigonometric equation involves more than one function, it is sometimes helpful to use identities to rewrite the equation in terms of a single function.

Example 9   Solve $\sin x + \cos x = 1$ in $[0, 2\pi)$.

We have

$$\sin x + \cos x = 1$$

$$\sin^2 x + 2 \sin x \cos x + \cos^2 x = 1$$

Squaring both sides

$$2 \sin x \cos x + 1 = 1$$

Using $\sin^2 x + \cos^2 x = 1$

$$2 \sin x \cos x = 0$$

$$\sin 2x = 0.$$

Using $2 \sin x \cos x = \sin 2x$

We are looking for solutions x to the equation for which $0 \le x < 2\pi$. Multiplying by 2, we get $0 \le 2x < 4\pi$, which is the interval we consider to solve $\sin 2x = 0$. These values of $2x$ are 0, π, 2π, and 3π. Thus the desired values of x in $[0, 2\pi)$ satisfying this equation are 0, $\pi/2$, π, and $3\pi/2$. Now we check these in the original equation:

$$\sin 0 + \cos 0 = 0 + 1 = 1,$$

$$\sin \frac{\pi}{2} + \cos \frac{\pi}{2} = 1 + 0 = 1,$$

$$\sin \pi + \cos \pi = 0 + (-1) = -1,$$

$$\sin \frac{3\pi}{2} + \cos \frac{3\pi}{2} = (-1) + 0 = -1.$$

We find that π and $3\pi/2$ do not check, but the other values do. Thus the solutions are

$$0 \quad \text{and} \quad \frac{\pi}{2}.$$

When the solution process involves squaring both sides, values are often obtained that are not solutions of the original equation. As we saw in this example, it is important to check the possible solutions.

We can graph the left side and then the right side of the equation as seen in the first window below. Then we look for points of intersection. We could also rewrite the equation as $\sin x + \cos x - 1 = 0$, graph the left side, and look for the zeros of the function, as illustrated in the second window below. In each window, we see the solutions in $[0, 2\pi)$ as 0 and $\pi/2$.

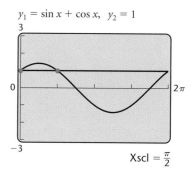

$y_1 = \sin x + \cos x, \quad y_2 = 1$

$\text{Xscl} = \frac{\pi}{2}$

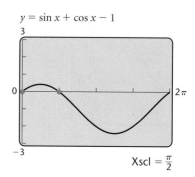

$y = \sin x + \cos x - 1$

$\text{Xscl} = \frac{\pi}{2}$

This example illustrates a valuable advantage of the grapher—that is, with a grapher, extraneous solutions do not appear.

Example 10 Solve $\cos 2x + \sin x = 1$ in $[0, 2\pi)$.

ALGEBRAIC SOLUTION

We have

$$\cos 2x + \sin x = 1$$
$$1 - 2\sin^2 x + \sin x = 1 \qquad \text{Using the identity } \cos 2x = 1 - 2\sin^2 x$$
$$-2\sin^2 x + \sin x = 0$$
$$\sin x\,(1 - 2\sin x) = 0 \qquad \text{Factoring}$$
$$\sin x = 0 \quad \text{or} \quad 1 - 2\sin x = 0 \qquad \text{Principle of zero products}$$
$$\sin x = 0 \quad \text{or} \quad \sin x = \frac{1}{2}$$
$$x = 0,\ \pi \quad \text{or} \quad x = \frac{\pi}{6},\ \frac{5\pi}{6}.$$

All values check. The solutions in $[0, 2\pi)$ are 0, $\pi/6$, $5\pi/6$, and π.

GRAPHICAL SOLUTION

We graph $y_1 = \cos 2x + \sin x - 1$ and look for the zeros of the function.

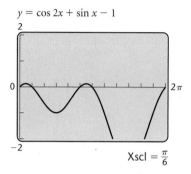

$y = \cos 2x + \sin x - 1$

$\text{Xscl} = \frac{\pi}{6}$

The solutions in $[0, 2\pi)$ are 0, $\pi/6$, $5\pi/6$, and π, or, approximately, 0, 0.524, 2.618, and 3.142.

Example 11 Solve $\tan^2 x + \sec x - 1 = 0$ in $[0, 2\pi)$.

ALGEBRAIC SOLUTION

We have

$$\tan^2 x + \sec x - 1 = 0$$
$$\sec^2 x - 1 + \sec x - 1 = 0 \qquad \text{Using the identity } 1 + \tan^2 x = \sec^2 x$$
$$\sec^2 x + \sec x - 2 = 0$$
$$(\sec x + 2)(\sec x - 1) = 0 \qquad \text{Factoring}$$
$$\sec x = -2 \quad \text{or} \quad \sec x = 1 \qquad \text{Principle of zero products}$$
$$x = \frac{2\pi}{3},\ \frac{4\pi}{3} \quad \text{or} \quad x = 0$$

All these values check. The solutions in $[0, 2\pi)$ are 0, $2\pi/3$, and $4\pi/3$.

GRAPHICAL SOLUTION

We graph $y = \tan^2 x + \sec x - 1$, but we must enter this equation in the form

$$y_1 = \left(\frac{\sin x}{\cos x}\right)^2 + \left(\frac{1}{\cos x}\right) - 1$$

in some graphers. We look for the zeros of the function.

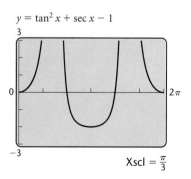

$y = \tan^2 x + \sec x - 1$

$\text{Xscl} = \frac{\pi}{3}$

The solutions in $[0, 2\pi)$ are 0, $2\pi/3$, and $4\pi/3$.

Sometimes the only plan of action is to approximate solutions with a grapher.

Example 12 Solve each of the following in $[0, 2\pi)$.

a) $x^2 - 1.5 = \cos x$

b) $\sin x - \cos x = \cot x$

SOLUTION

a) In the screen on the left, we graph $y_1 = x^2 - 1.5$ and $y_2 = \cos x$ and look for points of intersection. In the screen on the right, we graph $y_1 = x^2 - 1.5 - \cos x$ and look for the zeros of the function.

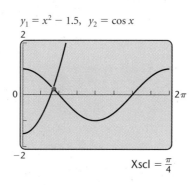
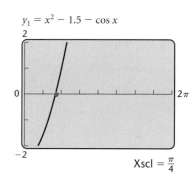

We determine the solution in $[0, 2\pi)$ to be approximately 1.32.

b) In the screen on the left, we graph $y_1 = \sin x - \cos x$ and $y_2 = \cot x$ and determine the points of intersection. In the screen on the right, we graph the function $y_1 = \sin x - \cos x - \cot x$ and determine the zeros.

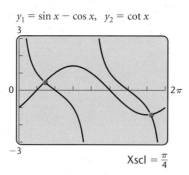
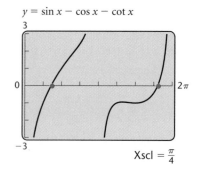

Each method leads to the approximate solutions 1.13 and 5.66 in $[0, 2\pi)$.

3.5 | Exercise Set

Solve, finding all solutions. Express the solutions in both radians and degrees.

1. $\cos x = \dfrac{\sqrt{3}}{2}$ **2.** $\sin x = -\dfrac{\sqrt{2}}{2}$

3. $\tan x = -\sqrt{3}$ **4.** $\cos x = \dfrac{1}{2}$

Solve, finding all solutions in $[0, 2\pi)$ or $[0°, 360°)$. Verify your answer with a grapher.

5. $2 \cos x - 1 = -1.2814$

6. $\sin x + 3 = 2.0816$

7. $2 \sin x + \sqrt{3} = 0$

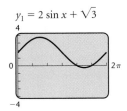

$y_1 = 2 \sin x + \sqrt{3}$

8. $2 \tan x - 4 = 1$

$y_1 = 2 \tan x - 4,$
$y_2 = 1$

9. $2 \cos^2 x = 1$

10. $\csc^2 x - 4 = 0$

11. $2 \sin^2 x + \sin x = 1$

12. $\cos^2 x + 2 \cos x = 3$

13. $2 \cos^2 x - \sqrt{3} \cos x = 0$

14. $2 \sin^2 \theta + 7 \sin \theta = 4$

15. $6 \cos^2 \phi + 5 \cos \phi + 1 = 0$

16. $2 \sin t \cos t + 2 \sin t - \cos t - 1 = 0$

17. $\sin 2x \cos x - \sin x = 0$

18. $5 \sin^2 x - 8 \sin x = 3$

19. $\cos^2 x + 6 \cos x + 4 = 0$

20. $2 \tan^2 x = 3 \tan x + 7$

21. $7 = \cot^2 x + 4 \cot x$

22. $3 \sin^2 x = 3 \sin x + 2$

Solve, finding all solutions in $[0, 2\pi)$. Then check each solution with a grapher.

23. $\cos 2x - \sin x = 1$

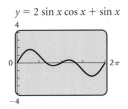

$y_1 = \cos 2x - \sin x,$
$y_2 = 1$

24. $2 \sin x \cos x + \sin x = 0$

$y = 2 \sin x \cos x + \sin x$

25. $\sin 4x - 2 \sin 2x = 0$

26. $\tan x \sin x - \tan x = 0$

27. $\sin 2x \cos x + \sin x = 0$

28. $\cos 2x \sin x + \sin x = 0$

29. $2 \sec x \tan x + 2 \sec x + \tan x + 1 = 0$

30. $\sin 2x \sin x - \cos 2x \cos x = -\cos x$

31. $\sin 2x + \sin x + 2 \cos x + 1 = 0$

32. $\tan^2 x + 4 = 2 \sec^2 x + \tan x$

33. $\sec^2 x - 2 \tan^2 x = 0$

34. $\cot x = \tan (2x - 3\pi)$

35. $2 \cos x + 2 \sin x = \sqrt{6}$

36. $\sqrt{3} \cos x - \sin x = 1$

37. $\sec^2 x + 2 \tan x = 6$

38. $5 \cos 2x + \sin x = 4$

39. $\cos (\pi - x) + \sin \left(x - \dfrac{\pi}{2} \right) = 1$

40. $\dfrac{\sin^2 x - 1}{\cos \left(\dfrac{\pi}{2} - x \right) + 1} = \dfrac{\sqrt{2}}{2} - 1$

Solve with a grapher, finding all solutions in $[0, 2\pi)$.

41. $x \sin x = 1$

42. $x^2 + 2 = \sin x$

43. $2 \cos^2 x = x + 1$

44. $x \cos x - 2 = 0$

45. $\cos x - 2 = x^2 - 3x$

46. $\sin x = \tan \dfrac{x}{2}$

Skill Maintenance

Solve the right triangle.

47.

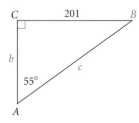

48.

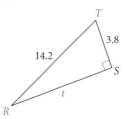

Solve.

49. $\dfrac{x}{27} = \dfrac{4}{3}$

50. $\dfrac{0.01}{0.7} = \dfrac{0.2}{h}$

Synthesis

51. ◆ Jan lists her answer to a problem as $\pi/6 + k\pi$, for any integer k, while Jacob lists his answer as $\pi/6 + 2k\pi$ and $7\pi/6 + 2\pi k$, for any integer k. Are their answers equivalent? Why or why not?

52. ◆ Under what circumstances will a grapher give exact solutions of a trigonometric equation?

Solve in $[0, 2\pi)$. *Check your results with a grapher.*

53. $|\sin x| = \dfrac{\sqrt{3}}{2}$

54. $|\cos x| = \dfrac{1}{2}$

55. $\sqrt{\tan x} = \sqrt[4]{3}$

56. $12 \sin x - 7\sqrt{\sin x} + 1 = 0$

57. $\ln (\cos x) = 0$

58. $e^{\sin x} = 1$

59. $\sin (\ln x) = -1$

60. $e^{\ln (\sin x)} = 1$

61. *Temperature During an Illness.* The temperature T, in degrees Fahrenheit, of a patient t days into a 12-day illness is given by

$$T(t) = 101.6° + 3° \sin \left(\dfrac{\pi}{8} t \right).$$

Find the times t during the illness at which the patient's temperature was $103°$.

62. *Satellite Location.* A satellite circles the earth in such a manner that it is y miles from the equator (north or south, height from the surface not considered) t minutes after its launch, where

$$y = 5000 \left[\cos \dfrac{\pi}{45} (t - 10) \right].$$

At what times t in the interval $[0, 240]$, the first 4 hr, is the satellite 3000 mi north of the equator?

63. *Nautical Mile.* (See Exercise 52 in Exercise Set 3.2.) In Great Britain, the *nautical mile* is defined as the length of a minute of arc of the earth's radius. Since the earth is flattened at the poles, a British nautical mile varies with latitude. In fact, it is given, in feet, by the function

$$N(\phi) = 6066 - 31 \cos 2\phi,$$

where ϕ is the latitude in degrees. At what latitude north is the length of a British nautical mile found to be 6040 ft?

64. *Acceleration Due to Gravity.* (See Exercise 53 in Exercise Set 3.2.) The acceleration due to gravity is often denoted by g in a formula such as $S = \frac{1}{2}gt^2$, where $S =$ the distance that an object falls in t seconds. The number g is generally considered constant, but in fact it varies slightly with latitude. If ϕ stands for latitude, in degrees, an excellent approximation of g is given by the formula

$$g = 9.78049(1 + 0.005288 \sin^2 \phi - 0.000006 \sin^2 2\phi),$$

where g is measured in meters per second per second at sea level. At what latitude north does $g = 9.8$?

Solve.

65. $\arccos x = \arccos \frac{3}{5} - \arcsin \frac{4}{5}$

66. $\sin^{-1} x = \tan^{-1} \frac{1}{3} + \tan^{-1} \frac{1}{2}$

67. Suppose that $\sin x = 5 \cos x$. Find $\sin x \cos x$.

3 *Summary and Review*

Important Properties and Formulas

Basic Identities

$$\sin x = \frac{1}{\csc x},$$

$$\cos x = \frac{1}{\sec x},$$

$$\tan x = \frac{1}{\cot x},$$

$$\tan x = \frac{\sin x}{\cos x},$$

$$\cot x = \frac{\cos x}{\sin x},$$

$$\sin (-x) = -\sin x,$$

$$\cos (-x) = \cos x,$$

$$\tan (-x) = -\tan x$$

Pythagorean Identities

$$\sin^2 x + \cos^2 x = 1,$$

$$1 + \cot^2 x = \csc^2 x,$$

$$1 + \tan^2 x = \sec^2 x$$

Sum and Difference Identities

$$\sin (u \pm v) = \sin u \cos v \pm \cos u \sin v,$$

$$\cos (u \pm v) = \cos u \cos v \mp \sin u \sin v,$$

$$\tan (u \pm v) = \frac{\tan u \pm \tan v}{1 \mp \tan u \tan v}$$

Cofunction Identities

$$\sin \left(\frac{\pi}{2} - x \right) = \cos x,$$

$$\tan \left(\frac{\pi}{2} - x \right) = \cot x,$$

$$\sec \left(\frac{\pi}{2} - x \right) = \csc x,$$

$$\sin \left(x \pm \frac{\pi}{2} \right) = \pm \cos x,$$

$$\cos \left(x \pm \frac{\pi}{2} \right) = \mp \sin x$$

Double-Angle Identities

$$\sin 2x = 2 \sin x \cos x,$$

$$\cos 2x = \cos^2 x - \sin^2 x$$

$$= 1 - 2 \sin^2 x$$

$$= 2 \cos^2 x - 1,$$

$$\tan 2x = \frac{2 \tan x}{1 - \tan^2 x}$$

Half-Angle Identities

$$\sin \frac{x}{2} = \pm \sqrt{\frac{1 - \cos x}{2}},$$

$$\cos \frac{x}{2} = \pm \sqrt{\frac{1 + \cos x}{2}},$$

$$\tan \frac{x}{2} = \pm \sqrt{\frac{1 - \cos x}{1 + \cos x}},$$

$$= \frac{\sin x}{1 + \cos x}$$

$$= \frac{1 - \cos x}{\sin x}$$

Inverse Trigonometric Functions

FUNCTION	DOMAIN	RANGE
$y = \sin^{-1} x$	$[-1, 1]$	$\left[-\dfrac{\pi}{2}, \dfrac{\pi}{2} \right]$
$y = \cos^{-1} x$	$[-1, 1]$	$[0, \pi]$
$y = \tan^{-1} x$	$(-\infty, \infty)$	$\left(-\dfrac{\pi}{2}, \dfrac{\pi}{2} \right)$

(continued)

Composition of Trigonometric Functions

The following are true for any x in the domain of the inverse function:

$$\sin (\sin^{-1} x) = x,$$
$$\cos (\cos^{-1} x) = x,$$
$$\tan (\tan^{-1} x) = x.$$

The following are true for any x in the range of the inverse function:

$$\sin^{-1} (\sin x) = x,$$
$$\cos^{-1} (\cos x) = x,$$
$$\tan^{-1} (\tan x) = x.$$

REVIEW EXERCISES

Complete the Pythagorean identity.

1. $1 + \cot^2 x =$

2. $\sin^2 x + \cos^2 x =$

Multiply and simplify. Check with a grapher.

3. $(\tan y - \cot y)(\tan y + \cot y)$

4. $(\cos x + \sec x)^2$

Factor and simplify. Check with a grapher.

5. $\sec x \csc x - \csc^2 x$

6. $3 \sin^2 y - 7 \sin y - 20$

7. $1000 - \cos^3 u$

Simplify and check with a grapher.

8. $\dfrac{\sec^4 x - \tan^4 x}{\sec^2 x + \tan^2 x}$

9. $\dfrac{2 \sin^2 x}{\cos^3 x} \cdot \left(\dfrac{\cos x}{2 \sin x} \right)^2$

10. $\dfrac{3 \sin x}{\cos^2 x} \cdot \dfrac{\cos^2 x + \cos x \sin x}{\sin^2 x - \cos^2 x}$

11. $\dfrac{3}{\cos y - \sin y} - \dfrac{2}{\sin^2 y - \cos^2 y}$

12. $\left(\dfrac{\cot x}{\csc x} \right)^2 + \dfrac{1}{\csc^2 x}$

13. $\dfrac{4 \sin x \cos^2 x}{16 \sin^2 x \cos x}$

14. Simplify. Assume the radicand is nonnegative.

$$\sqrt{\sin^2 x + 2 \cos x \sin x + \cos^2 x}$$

15. Rationalize the denominator: $\sqrt{\dfrac{1 + \sin x}{1 - \sin x}}$.

16. Rationalize the numerator: $\sqrt{\dfrac{\cos x}{\tan x}}$.

17. Given that $x = 3 \tan \theta$, express $\sqrt{9 + x^2}$ as a trigonometric function without radicals. Assume that $0 < \theta < \pi/2$.

Use the sum and difference formulas to write equivalent expressions. You need not simplify.

18. $\cos \left(x + \dfrac{3\pi}{2} \right)$ **19.** $\tan (45° - 30°)$

20. Simplify: $\cos 27° \cos 16° + \sin 27° \sin 16°$.

21. Find $\cos 165°$ exactly.

22. Given that $\tan \alpha = \sqrt{3}$ and $\sin \beta = \sqrt{2}/2$ and that α and β are between 0 and $\pi/2$, evaluate $\tan (\alpha - \beta)$ exactly.

23. Assume that $\sin \theta = 0.5812$ and $\cos \phi = 0.2341$ and that both θ and ϕ are first-quadrant angles. Evaluate $\cos (\theta + \phi)$.

Complete the cofunction identity.

24. $\cos \left(x + \dfrac{\pi}{2} \right) =$ **25.** $\cos \left(\dfrac{\pi}{2} - x \right) =$

26. $\sin \left(x - \dfrac{\pi}{2} \right) =$

27. Given that $\cos \alpha = -\dfrac{3}{5}$ and that the terminal side is in quadrant III:

a) Find the other function values for α.
b) Find the six function values for $\pi/2 - \alpha$.
c) Find the six function values for $\alpha + \pi/2$.

28. Find an equivalent expression for $\csc \left(x - \dfrac{\pi}{2} \right)$.

29. Find $\tan 2\theta$, $\cos 2\theta$, and $\sin 2\theta$ and the quadrant in which 2θ lies, where $\cos \theta = -\dfrac{4}{5}$ and θ is in quadrant III.

30. Find $\sin \dfrac{\pi}{8}$ exactly.

31. Given that $\sin \beta = 0.2183$ and β is in quadrant I, find $\sin 2\beta$, $\cos \dfrac{\beta}{2}$, and $\cos 4\beta$.

Simplify and check with a grapher.

32. $1 - 2 \sin^2 \dfrac{x}{2}$

33. $(\sin x + \cos x)^2 - \sin 2x$

34. $2 \sin x \cos^3 x + 2 \sin^3 x \cos x$

35. $\dfrac{2 \cot x}{\cot^2 x - 1}$

Prove each of the following identities. Check with a grapher.

36. $\dfrac{1 - \sin x}{\cos x} = \dfrac{\cos x}{1 + \sin x}$

37. $\dfrac{1 + \cos 2\theta}{\sin 2\theta} = \cot \theta$

38. $\dfrac{\tan y + \sin y}{2 \tan \theta} = \cos^2 \dfrac{y}{2}$

39. $\dfrac{\sin x - \cos x}{\cos^2 x} = \dfrac{\tan^2 x - 1}{\sin x + \cos x}$

In Exercises 40–43, use a grapher to determine which expression (A)–(D) on the right can be used to complete an identity. Then prove the identity algebraically.

40. $\csc x - \cos x \cot x$ **A.** $\dfrac{\csc x}{\sec x}$

41. $\dfrac{1}{\sin x \cos x} - \dfrac{\cos x}{\sin x}$ **B.** $\sin x$

42. $\dfrac{\cot x - 1}{1 - \tan x}$ **C.** $\dfrac{2}{\sin x}$

43. $\dfrac{\cos x + 1}{\sin x} + \dfrac{\sin x}{\cos x + 1}$ **D.** $\dfrac{\sin x \cos x}{1 - \sin^2 x}$

Find each of the following exactly in both radians and degrees.

44. $\sin^{-1} \left(-\dfrac{1}{2} \right)$ **45.** $\cos^{-1} \dfrac{\sqrt{3}}{2}$

46. $\arctan 1$ **47.** $\arcsin 0$

Use a grapher to find each of the following in radians, rounded to three decimal places, and in degrees, rounded to the nearest tenth of a degree.

48. $\arccos (-0.2194)$ **49.** $\cot^{-1} 2.381$

Evaluate.

50. $\cos \left(\cos^{-1} \dfrac{1}{2} \right)$ **51.** $\tan^{-1} \left(\tan \dfrac{\sqrt{3}}{3} \right)$

52. $\sin^{-1} \left(\sin \dfrac{\pi}{7} \right)$ **53.** $\cos \left(\arcsin \dfrac{\sqrt{2}}{2} \right)$

Find.

54. $\cos \left(\arctan \dfrac{b}{3} \right)$ **55.** $\cos \left(2 \sin^{-1} \dfrac{4}{5} \right)$

Solve, finding all solutions. Express the solutions in both radians and degrees.

56. $\cos x = -\dfrac{\sqrt{2}}{2}$

57. $\tan x = \sqrt{3}$

Solve, finding all solutions in $[0, 2\pi)$.

58. $4 \sin^2 x = 1$

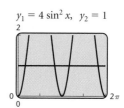
$y_1 = 4 \sin^2 x, \;\; y_2 = 1$

59. $\sin 2x \sin x - \cos x = 0$

$y = \sin 2x \sin x - \cos x$

60. $2 \cos^2 x + 3 \cos x = -1$

61. $\sin^2 x - 7 \sin x = 0$

62. $\csc^2 x - 2 \cot^2 x = 0$

63. $\sin 4x + 2 \sin 2x = 0$

64. $2 \cos x + 2 \sin x = \sqrt{2}$

65. $6 \tan^2 x = 5 \tan x + \sec^2 x$

Solve with a grapher, finding all solutions in $[-2\pi, 2\pi]$.

66. $x \cos x = 1$

67. $2 \sin^2 x = x + 1$

Synthesis

68. ◆ Prove the identity $2 \cos^2 x - 1 = \cos^4 x - \sin^4 x$ in three ways:

  a) Start with the left side and deduce the right (method 1).

  b) Start with the right side and deduce the left (method 1).

  c) Work with each side separately until you deduce the same expression (method 2).

  Then determine the most efficient method and explain why.

69. ◆ Why are the ranges of the inverse trigonometric functions restricted?

70. ◆ Explain why $\tan(x + 450°)$ cannot be simplified using the tangent sum formula, but can be simplified using the sine and cosine sum formulas.

71. Find the measure of the angle from l_1 to l_2:

$$l_1:\ x + y = 3 \qquad l_2:\ 2x - y = 5.$$

72. Find an identity for $\cos(u + v)$ involving only cosines.

73. Simplify: $\cos\left(\dfrac{\pi}{2} - x\right)[\csc x - \sin x]$.

74. Find $\sin \theta$, $\cos \theta$, and $\tan \theta$ under the given conditions:

$$\sin 2\theta = \frac{1}{5}, \quad \frac{\pi}{2} \le 2\theta < \pi.$$

75. Prove the following equation to be an identity and verify the results with a grapher:

$$\ln e^{\sin t} = \sin t.$$

76. Graph: $y = \sec^{-1} x$.

77. Show that

$$\tan^{-1} x = \frac{\sin^{-1} x}{\cos^{-1} x}$$

is *not* an identity.

78. Solve $e^{\cos x} = 1$ in $[0, 2\pi)$. Check with a grapher.

Applications of Trigonometry 4

Triangle trigonometry is important in applications such as large-scale construction, navigation, and surveying. In this chapter, we continue the study of triangle trigonometry that we began in Chapter 4. We will find that the trigonometric functions can also be used to solve triangles that are not right triangles.

The study of complex numbers begun in Chapter 2 is continued in this chapter. Complex numbers have applications in both electronics and engineering. We also introduce the polar coordinate system and graphs of polar equations.

This chapter also introduces the idea of a vector, which is related to the study of triangles. A vector is a quantity that has a direction. Vectors have many practical applications in the physical sciences.

APPLICATION

During a rescue mission, a Marine fighter pilot receives data on an unidentified aircraft from an AWACS plane and is instructed to intercept the enemy plane. The diagram shown below pops up on the screen. The law of sines,

$$\frac{x}{\sin X} = \frac{y}{\sin Y} = \frac{z}{\sin Z},$$

can be used to determine how far the pilot must fly.

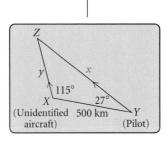

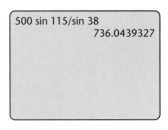

500 sin 115/sin 38
736.0439327

4.1 The Law of Sines
4.2 The Law of Cosines
4.3 Introduction to Complex Numbers
4.4 Complex Numbers: Trigonometric Form
4.5 Vectors
4.6 The Dot Product
SUMMARY AND REVIEW

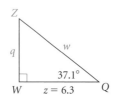

4.1
The Law of Sines

- *Use the law of sines to solve triangles.*
- *Find the area of any triangle given the lengths of two sides and the measure of the included angle.*

To **solve a triangle** means to find the lengths of all its sides and the measures of all its angles. We solved right triangles in Section 2.2. For review, let's solve the right triangle shown below. We begin by listing the known measures.

$$Q = 37.1° \qquad q = ?$$
$$W = 90° \qquad w = ?$$
$$Z = ? \qquad z = 6.3$$

Since the sum of the three angle measures of any triangle is 180°, we can immediately find the third angle:

$$Z = 180° - (90° + 37.1°) = 52.9°.$$

Then using the tangent and cosine ratios, respectively, we can find q and w:

$$\tan 37.1° = \frac{q}{6.3}, \quad \text{or} \quad q = 6.3 \tan 37.1° \approx 4.8,$$

and

$$\cos 37.1° = \frac{6.3}{w}, \quad \text{or} \quad w = \frac{6.3}{\cos 37.1°} \approx 7.9.$$

Now all six measures are known and we have solved triangle QWZ.

$$Q = 37.1° \qquad q \approx 4.8$$
$$W = 90° \qquad w \approx 7.9$$
$$Z = 52.9° \qquad z = 6.3$$

Solving Oblique Triangles

The trigonometric functions can also be used to solve triangles that are not right triangles. Such triangles are called **oblique**. Any triangle, right or oblique, can be solved *if at least one side and any other two measures are known*. The five possible situations are illustrated below.

> **1.** AAS: Two angles of a triangle and a side opposite one of them are known.

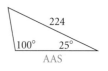

2. ASA: Two angles of a triangle and the included side are known.

37.5

ASA

3. SSA: Two sides of a triangle and an angle opposite one of them are known. (In this case, there may be no solution, one solution, or two solutions. The latter is known as the ambiguous case.)

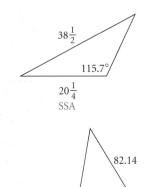

$38\frac{1}{2}$

115.7°

$20\frac{1}{4}$

SSA

4. SAS: Two sides of a triangle and the included angle are known.

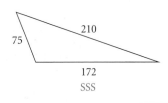

82.14

58°

19.05

SAS

5. SSS: All three sides of the triangle are known.

210

75

172

SSS

The list above does not include the situation in which only the three angle measures are given. The reason for this lies in the fact that the angle measures determine *only the shape* of the triangle and *not the size,* as shown with the following triangles.

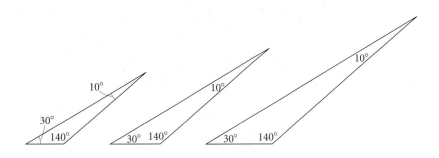

Thus we cannot solve a triangle when only the three angle measures are given.

In order to solve oblique triangles, we need to derive the *law of sines* and the *law of cosines*. The law of sines applies to the first three situations listed above. The law of cosines, which we develop in Section 4.2, applies to the last two situations.

The Law of Sines

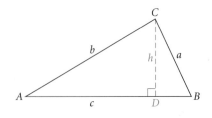

We consider any oblique triangle. It may or may not have an obtuse angle. Although we look at only the acute-triangle case, the derivation of the obtuse-triangle case is essentially the same.

In acute $\triangle ABC$ above, we have drawn an altitude from vertex C. It has length h. From $\triangle ADC$, we have

$$\sin A = \frac{h}{b}, \quad \text{or} \quad h = b \sin A.$$

From $\triangle BDC$, we have

$$\sin B = \frac{h}{a}, \quad \text{or} \quad h = a \sin B.$$

With $h = b \sin A$ and $h = a \sin B$, we now have

$$a \sin B = b \sin A$$

$$\frac{a \sin B}{\sin A \sin B} = \frac{b \sin A}{\sin A \sin B} \qquad \textbf{Dividing by sin } A \textbf{ sin } B$$

$$\frac{a}{\sin A} = \frac{b}{\sin B}. \qquad \textbf{Simplifying}$$

There is no danger of dividing by 0 here because we are dealing with triangles whose angles are never $0°$ or $180°$. Thus the sine value will never be 0.

If we were to consider altitudes from vertex A and vertex B in the triangle shown above, the same argument would give us

$$\frac{b}{\sin B} = \frac{c}{\sin C} \quad \text{and} \quad \frac{a}{\sin A} = \frac{c}{\sin C}.$$

We combine these results to obtain the law of sines.

The Law of Sines
In any triangle ABC,

$$\frac{a}{\sin A} = \frac{b}{\sin B} = \frac{c}{\sin C}.$$

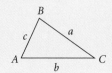

In any triangle, the sides are proportional to the sines of the opposite angles.

Solving Triangles (AAS and ASA)

When two angles and a side of any triangle are known, the law of sines can be used to solve the triangle.

Example 1 In $\triangle EFG$, $e = 4.56$, $E = 43°$, and $G = 57°$. Solve the triangle.

Solution We first draw a sketch. We know three of the six measures.

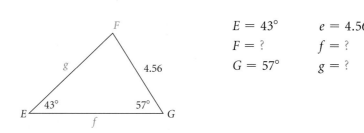

$$E = 43° \qquad e = 4.56$$
$$F = ? \qquad f = ?$$
$$G = 57° \qquad g = ?$$

From the figure, we see that we have the AAS situation. We begin by finding F:

$$F = 180° - (43° + 57°) = 80°.$$

We can now find the other two sides, using the law of sines:

$$\frac{f}{\sin F} = \frac{e}{\sin E}$$

$$\frac{f}{\sin 80°} = \frac{4.56}{\sin 43°} \qquad \text{Substituting}$$

$$f = \frac{4.56 \sin 80°}{\sin 43°} \qquad \text{Solving for } f$$

$$f \approx 6.58;$$

$$\frac{g}{\sin G} = \frac{e}{\sin E}$$

$$\frac{g}{\sin 57°} = \frac{4.56}{\sin 43°} \qquad \text{Substituting}$$

$$g = \frac{4.56 \sin 57°}{\sin 43°} \qquad \text{Solving for } g$$

$$g \approx 5.61.$$

Thus, we have found the following to solve the triangle:

$$E = 43°, \qquad e = 4.56,$$
$$F = 80°, \qquad f \approx 6.58,$$
$$G = 57°, \qquad g \approx 5.61.$$

The law of sines is frequently used in determining distances.

Example 2 *Rescue Mission.* During a rescue mission, a Marine fighter pilot receives data on an unidentified aircraft from an AWACS plane and is instructed to intercept the aircraft. The diagram shown below pops up on the screen, but before the distance to the point of interception appears on the screen, communications are jammed. Fortunately, the pilot remembers the law of sines. How far must the pilot fly?

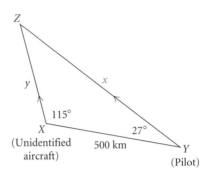

SOLUTION   We let x represent the distance that the pilot must fly in order to intercept the aircraft and Z represent the point of interception. We first find angle Z:

$$Z = 180° - (115° + 27°)$$
$$= 38°.$$

Because this application involves the ASA situation, we use the law of sines to determine x:

$$\frac{x}{\sin X} = \frac{z}{\sin Z}$$

$$\frac{x}{\sin 115°} = \frac{500}{\sin 38°} \qquad \text{Substituting}$$

$$x = \frac{500 \sin 115°}{\sin 38°} \qquad \text{Solving for } x$$

$$x \approx 736.$$

Thus the pilot must fly approximately 736 km in order to intercept the unidentified aircraft. ▬

Solving Triangles (SSA)

When two sides of a triangle and an angle opposite one of them are known, the law of sines can be used to solve the triangle.

Suppose for $\triangle ABC$ that b, c, and B are given. The various possibilities are as shown in the eight cases below: 5 cases when B is acute and 3 cases when B is obtuse. Note that $b < c$ in cases 1, 2, 3, and 6; $b = c$ in cases 4 and 7; and $b > c$ in cases 5 and 8.

Angle B Is Acute

Case 1: No solution
$b < c$; side b is too short to reach the base. No triangle is formed.

Case 2: One solution
$b < c$; side b just reaches the base and is perpendicular to it.

Case 3: Two solutions
$b < c$; an arc of radius b meets the base at two points. (This case is called the *ambiguous case.*)

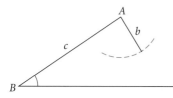

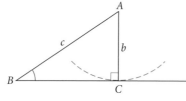

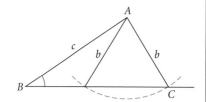

Case 4: One solution
$b = c$; an arc of radius b meets the base at just one point, other than B.

Case 5: One solution
$b > c$; an arc of radius b meets the base at just one point.

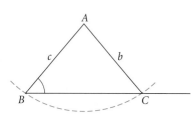

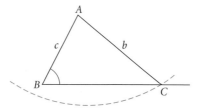

Angle B Is Obtuse

Case 6: No solution
$b < c$; side b is too short to reach the base. No triangle is formed.

Case 7: No solution
$b = c$; an arc of radius b meets the base only at point B. No triangle is formed.

Case 8: One solution
$b > c$; an arc of radius b meets the base at just one point.

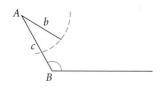

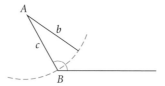

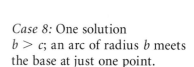

The eight cases above lead us to three possibilities in the SSA situation: *no* solution, *one* solution, or *two* solutions. Since there are two solutions in case 3, this possibility is referred to as the **ambiguous case**. Let's investigate these possibilities further, looking for ways to recognize the number of solutions.

Example 3   *No solution.*   In $\triangle QRS$, $q = 15$, $r = 28$, and $Q = 43.6°$. Solve the triangle.

SOLUTION   We make a drawing and list the known measures.

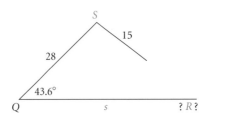

$$Q = 43.6° \qquad q = 15$$
$$R = ? \qquad r = 28$$
$$S = ? \qquad s = ?$$

We observe the SSA situation and use the law of sines to find R:

$$\frac{q}{\sin Q} = \frac{r}{\sin R}$$

$$\frac{15}{\sin 43.6°} = \frac{28}{\sin R} \qquad \text{Substituting}$$

$$\sin R = \frac{28 \sin 43.6°}{15} \qquad \text{Solving for } \sin R$$

$$\sin R \approx 1.2873.$$

Since there is no angle with a sine greater than 1, there is *no solution*.

Example 4   *One solution.*   In $\triangle XYZ$, $x = 23.5$, $y = 9.8$, and $X = 39.7°$. Solve the triangle.

SOLUTION   We make a drawing and organize the given information.

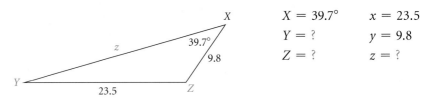

$$X = 39.7° \qquad x = 23.5$$
$$Y = ? \qquad y = 9.8$$
$$Z = ? \qquad z = ?$$

We see the SSA situation and begin by finding Y with the law of sines:

$$\frac{x}{\sin X} = \frac{y}{\sin Y}$$

$$\frac{23.5}{\sin 39.7°} = \frac{9.8}{\sin Y} \qquad \text{Substituting}$$

$$\sin Y = \frac{9.8 \sin 39.7°}{23.5} \qquad \text{Solving for } \sin Y$$

$$\sin Y \approx 0.2664.$$

Then $Y = 15.4°$ or $Y = 164.6°$, to the nearest tenth of a degree. An angle of $164.6°$ cannot be an angle of this triangle because it already has an angle of $39.7°$ and these two angles would total more than $180°$. Thus,

$15.4°$ is the only possibility for Y. Therefore,

$$Z \approx 180° - (39.7° + 15.4°) \approx 124.9°.$$

We now find z:

$$\frac{z}{\sin Z} = \frac{x}{\sin X}$$

$$\frac{z}{\sin 124.9°} = \frac{23.5}{\sin 39.7°} \qquad \text{Substituting}$$

$$z = \frac{23.5 \sin 124.9°}{\sin 39.7°} \qquad \text{Solving for } z$$

$$z \approx 30.2.$$

We now have solved the triangle:

$$X = 39.7°, \qquad x = 23.5,$$
$$Y \approx 15.4°, \qquad y = 9.8,$$
$$Z \approx 124.9°, \qquad z \approx 30.2.$$

In the ambiguous case, there are two possible solutions.

Example 5 *Two solutions.* In $\triangle ABC$, $b = 15$, $c = 20$, and $B = 29°$. Solve the triangle.

SOLUTION We make a drawing, list the known measures, and see that we again have the SSA situation.

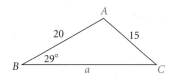

$$A = ? \qquad a = ?$$
$$B = 29° \qquad b = 15$$
$$C = ? \qquad c = 20$$

We first find C:

$$\frac{b}{\sin B} = \frac{c}{\sin C}$$

$$\frac{15}{\sin 29°} = \frac{20}{\sin C} \qquad \text{Substituting}$$

$$\sin C = \frac{20 \sin 29°}{15} \approx 0.6464. \qquad \text{Solving for } \sin C$$

Program

TRISOL: This program solves triangles when three parts are given. It covers unsolvable triangles as well as multiple solutions.

There are two angles less than $180°$ with a sine of 0.6464. They are $40°$ and $140°$, to the nearest degree. This gives us two possible solutions.

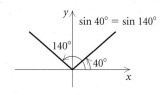

Possible Solution I.

If $C = 40°$, then

$$A = 180° - (29° + 40°) = 111°.$$

Then we find a:

$$\frac{a}{\sin A} = \frac{b}{\sin B}$$

$$\frac{a}{\sin 111°} = \frac{15}{\sin 29°}$$

$$a = \frac{15 \sin 111°}{\sin 29°} \approx 29.$$

These measures make a triangle as shown below; thus we have a solution.

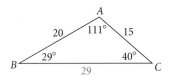

Possible Solution II.

If $C = 140°$, then

$$A = 180° - (29° + 140°) = 11°.$$

Then we find a:

$$\frac{a}{\sin A} = \frac{b}{\sin B}$$

$$\frac{a}{\sin 11°} = \frac{15}{\sin 29°}$$

$$a = \frac{15 \sin 11°}{\sin 29°} \approx 6.$$

These measures make a triangle as shown below; thus we have a second solution.

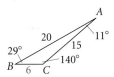

Examples 3–5 illustrate the SSA situation. Note that we need not memorize the eight cases or the procedures in finding no solution, one solution, or two solutions. When we are using the law of sines, the sine value leads us directly to the correct solution or solutions.

The Area of a Triangle

The familiar formula for the area of a triangle, $A = \frac{1}{2}bh$, can be used only when h is known. However, we can use the method used to derive the law of sines to derive an area formula that does not involve the height.

Consider a general triangle $\triangle ABC$, with area K, as shown below.

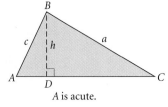

A is acute.

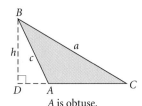

A is obtuse.

In each $\triangle ADB$,

$$\sin A = \frac{h}{c}, \quad \text{or} \quad h = c \sin A.$$

Substituting into the formula $K = \frac{1}{2}bh$, we get

$$K = \frac{1}{2}bc \sin A.$$

Any pair of sides and the included angle could have been used. Thus we also have

$$K = \frac{1}{2} ab \sin C \quad \text{and} \quad K = \frac{1}{2} ac \sin B.$$

The Area of a Triangle

The area K of any $\triangle ABC$ is one half the product of the lengths of two sides and the sine of the included angle:

$$K = \frac{1}{2} bc \sin A = \frac{1}{2} ab \sin C = \frac{1}{2} ac \sin B.$$

Example 6 *Area of the Peace Monument.* Through the Mentoring in the City Program sponsored by Marian College, Indianapolis, Indiana, children have turned a vacant downtown lot into a monument for peace.[*] This community project brought together neighborhood volunteers, businesses, and government in hopes of showing children how to develop positive nonviolent ways of dealing with conflict. A landscape architect[†] used the children's drawings and ideas to design a triangular-shaped peace garden. Two sides of the property, formed by Indiana Avenue and Senate Avenue, measure 182 ft and 230 ft, respectively, and together form a 44.7° angle. The third side of the garden, formed by an apartment building, measures 163 ft. What is the area of this property?

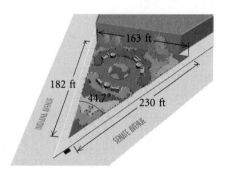

SOLUTION   Since we do not know a height of the triangle, we use the area formula:

$$K = \frac{1}{2} bc \sin A$$
$$K = \frac{1}{2} \cdot 182 \text{ ft} \cdot 230 \text{ ft} \cdot \sin 44.7°$$
$$K \approx 14{,}722 \text{ ft}^2.$$

The area of the property is approximately 14,722 ft².

---

[*]*The Indianapolis Star*, August 6, 1995, p. J8.
[†]Alan Day, a landscape architect with Browning Day Mullins Dierdorf, Inc., donated his time to this project.

4.1 Exercise Set

Solve the triangle, if possible.

1. $B = 38°, C = 21°, b = 24$
2. $A = 131°, C = 23°, b = 10$
3. $A = 36.5°, a = 24, b = 34$
4. $B = 118.3°, C = 45.6°, b = 42.1$
5. $C = 61°10', c = 30.3, b = 24.2$
6. $A = 126.5°, a = 17.2, c = 13.5$
7. $c = 3$ mi, $B = 37.48°, C = 32.16°$
8. $a = 2345$ mi, $b = 2345$ mi, $A = 124.67°$
9. $b = 56.78$ yd, $c = 56.78$ yd, $C = 83.78°$
10. $A = 129°32', C = 18°28', b = 1204$ in.
11. $a = 20.01$ cm, $b = 10.07$ cm, $A = 30.3°$
12. $b = 4.157$ km, $c = 3.446$ km, $C = 51°48'$
13. $A = 89°, a = 15.6$ in., $b = 18.4$ in.
14. $C = 46°32', a = 56.2$ m, $c = 22.1$ m
15. $a = 200$ m, $A = 32.76°, C = 21.97°$
16. $B = 115°, c = 45.6$ yd, $b = 23.8$ yd

Find the area of the triangle.

17. $B = 42°, a = 7.2$ ft, $c = 3.4$ ft
18. $A = 17°12', b = 10$ in., $c = 13$ in.
19. $C = 82°54', a = 4$ yd, $b = 6$ yd
20. $C = 75.16°, a = 1.5$ m, $b = 2.1$ m
21. $B = 135.2°, a = 46.12$ ft, $c = 36.74$ ft
22. $A = 113°, b = 18.2$ cm, $c = 23.7$ cm

Solve. Keep in mind the two types of bearing considered in Sections 2.2 and 2.3.

23. *Area of Back Yard.*   A new homeowner has a triangular-shaped back yard. Two of the three sides measure 53 ft and 42 ft and form an included angle of 135°. To determine the amount of fertilizer and grass seed to be purchased, the owner has to know, or at least approximate, the area of the yard. Find the area of the yard to the nearest square foot.

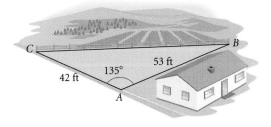

24. *Boarding Stable.*   A rancher operates a boarding stable and temporarily needs to make an extra pen. He has a piece of rope 38 ft long and plans to tie the rope to one end of the barn (S) and run the rope around a tree (T) and back to the barn (Q). The tree is 21 ft from where the rope is first tied, and the rope from the barn to the tree makes an angle of 35° with the barn. Does the rancher have enough rope if he allows $4\frac{1}{2}$ ft at each end to fasten the rope?

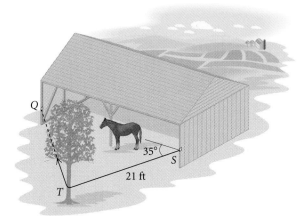

25. *Rock Concert.*   In preparation for an outdoor rock concert, a stage crew must determine how far apart to place the two large speaker columns on stage. What generally works best is to place them at 50° angles to the center of the front row. The distance from the center of the front row to each of the speakers is 10 ft. How far apart does the crew need to place the speakers on stage?

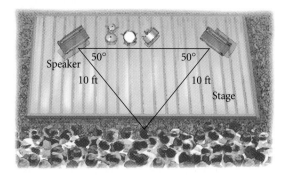

26. *Lunar Crater.*   Points A and B are on opposite sides of a lunar crater. Point C is 50 m from A. The measure of $\angle BAC$ is determined to be 112° and the measure of $\angle ACB$ is determined to be 42°. What is the width of the crater?

27. *Length of Pole.*   A pole leans away from the sun at an angle of 7° to the vertical. When the angle of

elevation of the sun is 51°, the pole casts a shadow 47 ft long on level ground. How long is the pole?

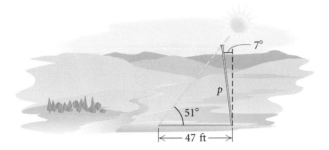

28. *Reconnaissance Airplane.* A reconnaissance airplane leaves its airport on the east coast of the United States and flies in a direction of 085°. Because of bad weather, it returns to another airport 230 km to the north of its home base. For the return trip, it flies in a direction of 283°. What is the total distance that the airplane flew?

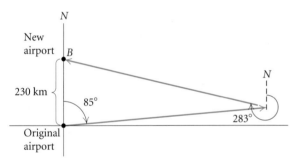

29. *Fire Tower.* A ranger in fire tower A spots a fire at a direction of 295°. A ranger in fire tower B, located 45 mi at a direction of 045° from tower A, spots the same fire at a direction of 255°. How far from tower A is the fire? from tower B?

30. *Lighthouse.* A boat leaves a lighthouse A and sails 5.1 km. At this time it is sighted from lighthouse B, 7.2 km west of A. The bearing of the boat from B is N65°10′E. How far is the boat from B?

31. *Mackinac Island.* Mackinac Island is located 35 mi N65°20′W of Cheboygan, Michigan, where the Coast Guard cutter Mackinaw is stationed. A freighter in distress radios the Coast Guard cutter for help. It radios its position as N25°40′E of Mackinac Island and N10°10′W of Cheboygan. How far is the freighter from Cheboygan?

32. *Gears.* Three gears are arranged as shown in the figure below. Find the angle φ.

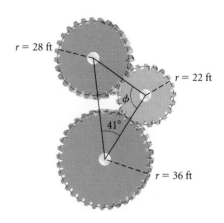

Skill Maintenance

Find the acute angle, A, in both radians and degrees, for the given function value.

33. $\cos A = 0.2213$

34. $\cos A = 1.5612$

Convert to decimal degree notation.

35. 18°14′20″

36. 125°3′42″

Synthesis

37. ◆ Explain why the law of sines cannot be used to find the first angle when solving a triangle given three sides.

38. ◆ We considered eight cases of solving triangles given two sides and an angle opposite one of them. Describe the relationship between side b and the height h in each.

39. Prove the following area formulas for a general triangle ABC with area represented by K.

$$K = \frac{a^2 \sin B \sin C}{2 \sin A}$$
$$= \frac{c^2 \sin A \sin B}{2 \sin C}$$
$$= \frac{b^2 \sin C \sin A}{2 \sin B}$$

40. *Area of a Parallelogram.* Prove that the area of a parallelogram is the product of two sides and the sine of the included angle.

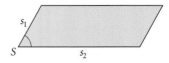

41. *Area of a Quadrilateral.* Prove that the area of a quadrilateral is one half the product of the lengths of its diagonals and the sine of the angle between the diagonals.

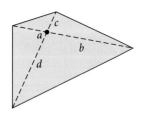

42. Find *d*.

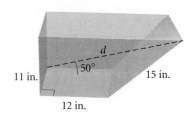

$$\dfrac{4.2}{}$$

The Law of Cosines

- *Use the law of cosines to solve triangles.*
- *Determine whether the law of sines or the law of cosines should be applied to solve a triangle, and then solve the triangle.*

The law of sines is used to solve triangles given a side and two angles (AAS and ASA) or given two sides and an angle opposite one of them (SSA). A second law, called the *law of cosines,* is needed to solve triangles given two sides and the included angle (SAS) or given three sides (SSS).

The Law of Cosines

To derive this property, we consider any $\triangle ABC$ placed on a coordinate system. We position the origin at one of the vertices—say, *C*—and the positive half of the *x*-axis along one of the sides—say, *CB*. Let (x, y) be the coordinates of vertex *A*. Point *B* has coordinates $(a, 0)$ and point *C* has coordinates $(0, 0)$.

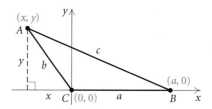

Then

$$\cos C = \frac{x}{b}, \quad \text{so} \quad x = b \cos C$$

and

$$\sin C = \frac{y}{b}, \quad \text{so} \quad y = b \sin C.$$

Thus point A has coordinates

$(b \cos C, b \sin C)$.

Next, we use the distance formula to determine c^2:

$c^2 = (x - a)^2 + (y - 0)^2$,   See Appendix A.2

or $c^2 = (b \cos C - a)^2 + (b \sin C - 0)^2$.

Now we multiply and simplify:

$$c^2 = b^2 \cos^2 C - 2ab \cos C + a^2 + b^2 \sin^2 C$$
$$= a^2 + b^2(\sin^2 C + \cos^2 C) - 2ab \cos C$$
$$= a^2 + b^2 - 2ab \cos C.$$

Had we placed the origin at one of the other vertices, we would have obtained

$$a^2 = b^2 + c^2 - 2bc \cos A$$

or $b^2 = a^2 + c^2 - 2ac \cos B.$

The Law of Cosines

In any triangle ABC,

$$a^2 = b^2 + c^2 - 2bc \cos A,$$
$$b^2 = a^2 + c^2 - 2ac \cos B,$$

or $c^2 = a^2 + b^2 - 2ab \cos C.$

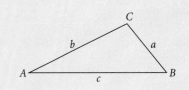

Thus, in any triangle, the square of a side is the sum of the squares of the other two sides, minus twice the product of those sides and the cosine of the included angle.

Solving Triangles (SAS)

When two sides of a triangle and the included angle are known, we can use the law of cosines to find the third side. The law of cosines or the law of sines can then be used to finish solving the triangle.

Example 1 Solve $\triangle ABC$ if $a = 32$, $c = 48$, and $B = 125.2°$.

SOLUTION We first label a triangle with the known and unknown measures.

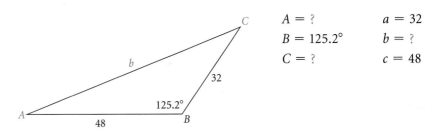

$A = ?$	$a = 32$
$B = 125.2°$	$b = ?$
$C = ?$	$c = 48$

We can find the third side using the law of cosines, as follows:

$$b^2 = a^2 + c^2 - 2ac \cos B$$
$$b^2 = 32^2 + 48^2 - 2 \cdot 32 \cdot 48 \cos 125.2° \qquad \text{Substituting}$$
$$b^2 \approx 5098.8$$
$$b \approx 71.$$

We now have $a = 32$, $b \approx 71$, and $c = 48$, and we need to find the other two angle measures. At this point, we can find them in two ways. One way uses the law of sines. The ambiguous case may arise, however, and we would have to be alert to this possibility. The advantage of using the law of cosines again is that if we solve for the cosine and find that its value is negative, then we know that the angle is obtuse. If the value of the cosine is positive, then the angle is acute. Thus we use the law of cosines to find a second angle.

Let's find angle A. We select the formula from the law of cosines that contains $\cos A$ and substitute:

$$a^2 = b^2 + c^2 - 2bc \cos A$$
$$32^2 = 71^2 + 48^2 - 2 \cdot 71 \cdot 48 \cos A \qquad \text{Substituting}$$
$$1024 = 5041 + 2304 - 6816 \cos A$$
$$-6321 = -6816 \cos A$$
$$\cos A \approx 0.9273768$$
$$A \approx 22.0°.$$

The third angle is now easy to find:

$$C \approx 180° - (125.2° + 22.0°) \approx 32.8°.$$

Thus,

$$A \approx 22.0°, \qquad a = 32,$$
$$B = 125.2°, \qquad b \approx 71,$$
$$C \approx 32.8°, \qquad c = 48.$$

Due to errors created by rounding, answers may vary depending on the order in which they are found. Had we found the measure of angle C first in Example 1, the angle measures would have been $C \approx 34.1°$ and $A \approx 20.7°$. The answers at the back of the book were generated by considering alphabetical order. Variances in rounding also change the answers. Had we used 71.4 for b in Example 1, the angle measures would have been $A \approx 21.5°$ and $C \approx 33.3°$.

Suppose we used the law of sines at the outset in Example 1 to find b. We were given only three measures: $a = 32$, $c = 48$, and $B = 125.2°$. When substituting these measures into the proportions, we see that there is not enough information to use the law of sines:

$$\frac{a}{\sin A} = \frac{b}{\sin B} \;\rightarrow\; \frac{32}{\sin A} = \frac{b}{\sin 125.2°},$$
$$\frac{b}{\sin B} = \frac{c}{\sin C} \;\rightarrow\; \frac{b}{\sin 125.2°} = \frac{48}{\sin C},$$

$$\frac{a}{\sin A} = \frac{c}{\sin C} \rightarrow \frac{32}{\sin A} = \frac{48}{\sin C}.$$

In all three situations, the resulting equation, after the substitutions, still has two unknowns. Thus we cannot use the law of sines to find *b*.

Solving Triangles (SSS)

When all three sides of a triangle are known, the law of cosines can be used to solve the triangle.

Example 2   Solve $\triangle RST$ if $r = 3.5$, $s = 4.7$, and $t = 2.8$.

SOLUTION   We sketch a triangle and label it with the given measures.

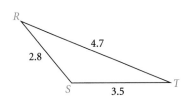

$R = ?$	$r = 3.5$
$S = ?$	$s = 4.7$
$T = ?$	$t = 2.8$

Since we do not know any of the angle measures, we cannot use the law of sines. We begin instead by finding angle *S* with the law of cosines. We choose the formula that contains cos *S*:

$$s^2 = r^2 + t^2 - 2rt \cos S$$
$$(4.7)^2 = (3.5)^2 + (2.8)^2 - 2(3.5)(2.8) \cos S \qquad \text{Substituting}$$
$$\cos S = \frac{(3.5)^2 + (2.8)^2 - (4.7)^2}{2(3.5)(2.8)}$$
$$\cos S \approx -0.1020408$$
$$S \approx 95.86°.$$

Similarly, we find angle *R*:

$$r^2 = s^2 + t^2 - 2st \cos R$$
$$(3.5)^2 = (4.7)^2 + (2.8)^2 - 2(4.7)(2.8) \cos R$$
$$\cos R = \frac{(4.7)^2 + (2.8)^2 - (3.5)^2}{2(4.7)(2.8)}$$
$$\cos R \approx 0.6717325$$
$$R \approx 47.80°.$$

Then

$$T \approx 180° - (95.86° + 47.80°) \approx 36.34°.$$

Thus,

$$R \approx 47.80°, \qquad r = 3.5,$$
$$S \approx 95.86°, \qquad s = 4.7,$$
$$T \approx 36.34°, \qquad t = 2.8.$$

Example 3   *Knife Bevel.* Knifemakers know that the *bevel* of the blade (the angle formed at the cutting edge of the blade) determines the cutting characteristics of the knife. A small bevel like that of a straight razor makes for a keen edge, but is impractical for heavy-duty cutting because the edge dulls quickly and is prone to chipping. A large bevel is suitable for heavy-duty work like chopping wood. Survival knives, being universal in application, are a compromise between small and large bevels. The diagram at left illustrates the blade of a hand-made Randall Model 18 survival knife. What is its bevel? (*Source:* Randall Made Knives, P.O. Box 1988, Orlando, FL 32802)

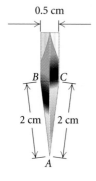

SOLUTION   We know three sides of a triangle. We can use the law of cosines to find the bevel, angle A.

$$a^2 = b^2 + c^2 - 2bc \cos A$$
$$(0.5)^2 = 2^2 + 2^2 - 2 \cdot 2 \cdot 2 \cdot \cos A$$
$$\cos A = \frac{4 + 4 - 0.25}{8}$$
$$\cos A = 0.96875$$
$$A \approx 14.36°.$$

Thus the bevel is approximately 14.36°.

Choosing the Appropriate Law

The following summarizes the situations in which to use the law of sines and the law of cosines.

To solve an oblique triangle:	
Use the *law of sines* for:	Use the *law of cosines* for:
AAS	SAS
ASA	SSS
SSA	

The law of cosines can also be used for the SSA situation, but since the process involves solving a quadratic equation, we do not include that option in the list above.

Example 4   In $\triangle ABC$, three measures are given. Determine which law to use when solving the triangle. You need not solve the triangle.

a) $a = 14$, $b = 23$, $c = 10$ **b)** $a = 207$, $B = 43.8°$, $C = 57.6°$
c) $A = 112°$, $C = 37°$, $a = 84.7$ **d)** $B = 101°$, $a = 960$, $c = 1042$
e) $b = 17.26$, $a = 27.29$, $A = 39°$ **f)** $A = 61°$, $B = 39°$, $C = 80°$

SOLUTION   It is helpful to make a sketch of a triangle with the given information. The triangle need not be drawn to scale. The given parts are shown in color.

FIGURE		SITUATION	LAW TO USE
a)	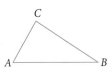	SSS	Law of Cosines
b)	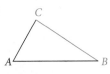	ASA	Law of Sines
c)	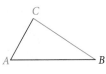	AAS	Law of Sines
d)	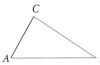	SAS	Law of Cosines
e)	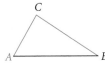	SSA	Law of Sines
f)	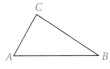	AAA	Cannot be solved

4.2 Exercise Set

Solve the triangle, if possible.

1. $A = 30°$, $b = 12$, $c = 24$
2. $B = 133°$, $a = 12$, $c = 15$
3. $a = 12$, $b = 14$, $c = 20$
4. $a = 22.3$, $b = 22.3$, $c = 36.1$
5. $B = 72°40'$, $c = 16$ m, $a = 78$ m
6. $C = 22.28°$, $a = 25.4$ cm, $b = 73.8$ cm
7. $a = 16$ m, $b = 20$ m, $c = 32$ m
8. $B = 72.66°$, $a = 23.78$ km, $c = 25.74$ km

9. $a = 2$ ft, $b = 3$ ft, $c = 8$ ft
10. $A = 96°13'$, $b = 15.8$ yd, $c = 18.4$ yd
11. $a = 26.12$ km, $b = 21.34$ km, $c = 19.25$ km
12. $C = 28°43'$, $a = 6$ mm, $b = 9$ mm
13. $a = 60.12$ mi, $b = 40.23$ mi, $C = 48.7°$
14. $a = 11.2$ cm, $b = 5.4$ cm, $c = 7$ cm
15. $b = 10.2$ in., $c = 17.3$ in., $A = 53.456°$
16. $a = 17$ yd, $b = 15.4$ yd, $c = 1.5$ yd

Determine which law applies. Then solve the triangle.

17. $A = 70°$, $B = 12°$, $b = 21.4$
18. $a = 15$, $c = 7$, $B = 62°$
19. $a = 3.3$, $b = 2.7$, $c = 2.8$
20. $a = 1.5$, $b = 2.5$, $A = 58°$

21. $A = 40.2°, B = 39.8°, C = 100°$

22. $a = 60, b = 40, C = 47°$

23. $a = 3.6, b = 6.2, c = 4.1$

24. $B = 110°30', C = 8°10', c = 0.912$

Solve.

25. *Poachers.* A park ranger establishes an observation post from which to watch for poachers. Despite losing her map, the ranger does have a compass and a rangefinder. She observes some poachers, and the rangefinder indicates that they are 500 ft from her position. They are headed toward big game that she knows to be 375 ft from her position. Using her compass, she finds that the poachers' azimuth (the direction measured as an angle from north) is 355° and that of the big game is 42°. What is the distance between the poachers and the game?

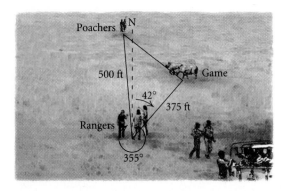

26. *Circus Highwire Act.* A circus highwire act walks up an approach wire to reach a highwire. The approach wire is 122 ft long and is currently anchored so that the angle it forms with the ground is at the maximum of 35°. A greater approach angle causes the aerialists to slip. However, the aerialists find that there is enough room to anchor the approach wire 30 ft back in order to make the approach angle less severe. When this is done, how much farther will they have to walk up the approach wire, and what will the new approach angle be?

27. *In-line Skater.* An in-line skater skates on a fitness trail along the Pacific Ocean from point A to point B. As shown below, two streets intersecting at point C also intersect the trail at A and B. In his car, the skater found the lengths of AC and BC to be approximately 0.5 mi and 1.3 mi, respectively. From a map, he estimates the included angle at C to be 110°. How far did he skate from A to B?

28. *Baseball Bunt.* A batter in a baseball game drops a bunt down the first-base line. It rolls 34 ft at an angle of 25° with the base path. The pitcher's mound is 60.5 ft from home plate. How far must the pitcher travel to pick up the ball? (*Hint:* A baseball diamond is a square.)

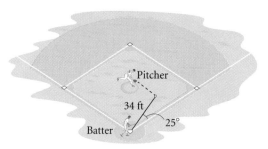

29. *Ships.* Two ships leave harbor at the same time. The first sails N15°W at 25 knots (a knot is one nautical mile per hour). The second sails N32°E at 20 knots. After 2 hr, how far apart are the ships?

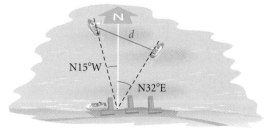

30. *Survival Trip.* A group of college students is learning to navigate for an upcoming survival trip.

On a map, they have been given three points at which they are to check in. The map also shows the distances between the points. However, to navigate they need to know the angle measurements. Calculate the angles for them.

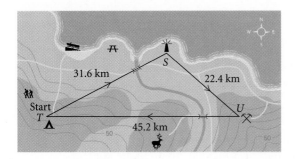

31. *Airplanes.* Two airplanes leave an airport at the same time. The first flies 150 km/h in a direction of 320°. The second flies 200 km/h in a direction of 200°. After 3 hr, how far apart are the planes?

32. *Slow-pitch Softball.* A slow-pitch softball diamond is a square 65 ft on a side. The pitcher's mound is 46 ft from home. How far is it from the pitcher's mound to first base?

33. *Isosceles Trapezoid.* The longer base of an isosceles trapezoid measures 14 ft. The nonparallel sides measure 10 ft, and the base angles measure 80°.

a) Find the length of a diagonal.
b) Find the area.

34. *Area of Sail.* A sail that is in the shape of an isosceles triangle has a vertex angle of 38°. The angle is included by two sides, each measuring 20 ft. Find the area of the sail.

35. Three circles are arranged as shown in the figure below. Find the length *PQ*.

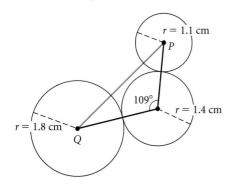

36. *Swimming Pool.* A triangular swimming pool measures 44 ft on one side and 32.8 ft on another side. These sides form an angle that measures 40.8°. How long is the other side?

Skill Maintenance

37. Find the absolute value: $|-5|$.

Find the values.

38. $\cos \dfrac{\pi}{6}$

39. $\sin 45°$

40. $\sin 300°$

41. $\cos \left(-\dfrac{2\pi}{3}\right)$

42. Multiply: $(1 - i)(1 + i)$.

Synthesis

43. ◈ Try to solve this triangle using the law of cosines. Then explain why it is easier to solve it using the law of sines.

44. ◈ Explain why we cannot solve a triangle given SAS with the law of sines.

45. *Canyon Depth.* A bridge is being built across a canyon. The length of the bridge is 5045 ft. From the deepest point in the canyon, the angles of elevation of the ends of the bridge are 78° and 72°. How deep is the canyon?

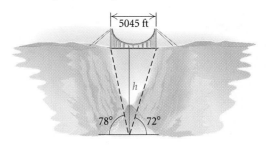

46. *Heron's Formula.* If *a*, *b*, and *c* are the lengths of the sides of a triangle, then the area *K* of the triangle is given by

$$K = \sqrt{s(s - a)(s - b)(s - c)},$$

where $s = \frac{1}{2}(a + b + c)$. The number *s* is called the **semiperimeter**. Prove Heron's formula. (*Hint:* Use the area formula $K = \frac{1}{2}bc \sin A$ developed in Section 4.1. Then use Heron's formula to find the area of the triangular swimming pool described in Exercise 36.)

47. *Area of Isosceles Triangle.*   Find a formula for the area of an isosceles triangle in terms of the congruent sides and their included angle. Under what conditions will the area of a triangle with fixed congruent sides be maximum?

48. *Reconnaissance Plane.*   A reconnaissance plane patrolling at 5000 ft sights a submarine at bearing 35° and at an angle of depression of 25°. A carrier is at bearing 105° and at an angle of depression of 60°. How far is the submarine from the carrier?

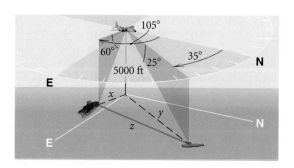

$\dfrac{4.3}{}$

Introduction to Complex Numbers

- *Perform computations involving complex numbers.*
- *Solve quadratic equations for which the solutions are nonreal.*

At this point, we need to expand the real-number system to one called the **complex-number system** in which certain polynomial functions have zeros that may not be real numbers.

The Complex Numbers

Let's begin by considering the polynomial function $f(x) = x^2 - 5$.
To find the zeros of the function, we solve the equation $x^2 - 5 = 0$.

Example 1   Solve: $x^2 - 5 = 0$.

ALGEBRAIC SOLUTION

We use the principle of square roots:

$$x^2 - 5 = 0$$
$$x^2 = 5 \qquad \text{Adding 5}$$
$$x = \pm\sqrt{5}. \qquad \begin{array}{l}\text{Using the}\\\text{principle of}\\\text{square roots}\end{array}$$

The solutions of the equation are $-\sqrt{5}$ and $\sqrt{5}$. The solution set is $\{-\sqrt{5}, \sqrt{5}\}$. Note that $\pm\sqrt{5} \approx \pm 2.236$.

GRAPHICAL SOLUTION

To solve $x^2 - 5 = 0$, we first graph the function $f(x) = x^2 - 5$ and estimate its zeros.

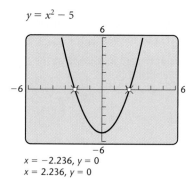

$y = x^2 - 5$

$x = -2.236, y = 0$
$x = 2.236, y = 0$

We see that there are zeros near -2 and 2. Using the ZERO, or ROOT, feature in the CALC menu to approximate the zeros, we get the approximations -2.236 and 2.236. The zeros of the function are the solutions of the equation $x^2 - 5 = 0$. The solution set is $\{-2.236, 2.236\}$.

Note that the real solutions found with the grapher were approximations. Now let's consider an equation that has no real-number solutions.

Example 2   Solve: $x^2 + 1 = 0$.

$$x^2 + 1 = 0$$

$$x^2 = -1 \qquad \text{Subtracting 1}$$

$$x = \pm\sqrt{-1} \qquad \text{Using the principle of square roots}$$

There are no real-number square roots of negative numbers. Thus the equation $x^2 + 1 = 0$ has no real-number solutions.

We first graph the function $f(x) = x^2 + 1$.

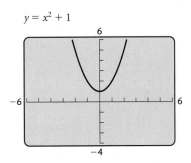

We see that since the graph does not cross the x-axis, it has no x-intercepts. This means that the function $f(x) = x^2 + 1$ has no real-number zeros. Thus there are no real-number solutions of the equation $x^2 + 1 = 0$. ▬

A new kind of number, called *imaginary*, was defined (invented) so that negative numbers would have square roots and equations like the one in Example 2 would have solutions. These new numbers make use of a number named i, defined so that $i = \sqrt{-1}$ and $i^2 = -1$.

The Number i

The number i is defined such that

$$i = \sqrt{-1} \quad \text{and} \quad i^2 = -1.$$

To express roots of negative numbers in terms of i, we can use the fact that $\sqrt{-p} = \sqrt{-1}\sqrt{p}$ when p is a positive real number.

Example 3   Express each number in terms of i.

a) $\sqrt{-7}$ b) $\sqrt{-16}$

c) $-\sqrt{-13}$ d) $-\sqrt{-64}$

e) $\sqrt{-48}$

SOLUTION

i is **not** under the radical.

a) $\sqrt{-7} = \sqrt{-1 \cdot 7} = \sqrt{-1} \cdot \sqrt{7} = i\sqrt{7}$, or $\sqrt{7}i$

b) $\sqrt{-16} = \sqrt{-1 \cdot 16} = \sqrt{-1} \cdot \sqrt{16} = i \cdot 4 = 4i$

c) $-\sqrt{-13} = -\sqrt{-1 \cdot 13} = -\sqrt{-1} \cdot \sqrt{13} = -i\sqrt{13}$, or $-\sqrt{13}i$

d) $-\sqrt{-64} = -\sqrt{-1 \cdot 64} = -\sqrt{-1} \cdot \sqrt{64} = -i \cdot 8 = -8i$

e) $\sqrt{-48} = \sqrt{-1 \cdot 48} = \sqrt{-1} \cdot \sqrt{48}$
$$= i\sqrt{48} = i \cdot 4\sqrt{3} = 4\sqrt{3}i, \quad \text{or} \quad 4i\sqrt{3}$$

Imaginary Number

An *imaginary number* is a number that can be written $a + bi$, where a and b are real numbers and $b \neq 0$.

Don't let the word "imaginary" mislead you. Imaginary numbers have very "real" uses in engineering and science. The following are examples of imaginary numbers:

$$\left.\begin{array}{l} -6 + 2i \\ \frac{3}{5} - \frac{1}{4}i \\ \pi - 3.7i \\ 3 + i\sqrt{2} \end{array}\right\} \quad \text{(here } a \neq 0, b \neq 0\text{);}$$

$$17i \qquad \text{(here } a = 0, b \neq 0\text{).}$$

The number i and the numbers in Example 3 are also examples of *imaginary numbers* with $a = 0$. In Example 2, the solutions of the equation $x^2 + 1 = 0$ are imaginary numbers: $x = \pm\sqrt{-1} = \pm i$.

When a and b are real numbers and b could be 0, the number $a + bi$ is said to be **complex**. The real number a is said to be the **real part** of $a + bi$, and the real number b is said to be the **imaginary part**.

Complex Number

A *complex number* is any number that can be written $a + bi$, where a and b are any real numbers. (Note that a and b both can be 0.)

The following are examples of complex numbers:

$$7 + 3i \qquad \text{(here } a \neq 0, b \neq 0\text{);}$$
$$8 \qquad \text{(here } a \neq 0, b = 0\text{);}$$
$$4i \qquad \text{(here } a = 0, b \neq 0\text{);}$$
$$0 \qquad \text{(here } a = 0, b = 0\text{).}$$

Note that when $b = 0$, $a + 0i = a$, so every real number is a complex number. Complex numbers like $17i$ or $4i$, in which $a = 0$ and $b \neq 0$, are imaginary numbers with no real part. Such numbers are called **pure imaginary numbers**. The relationships among various types of complex numbers are shown below.

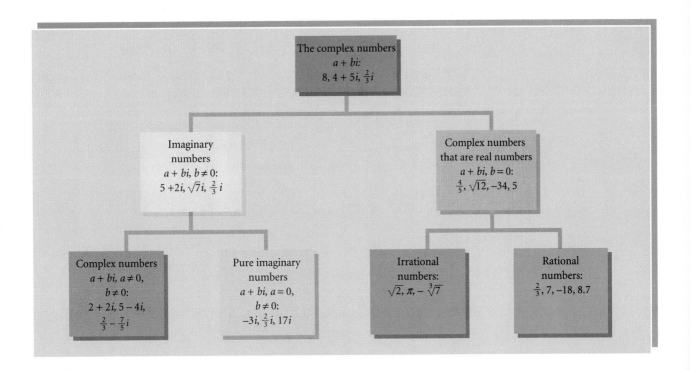

Addition and Subtraction

The complex numbers obey the commutative, associative, and distributive laws. Thus we can add and subtract them as we do binomials.

Example 4   Add or subtract and simplify each of the following.

a) $(8 + 6i) + (3 + 2i)$ 　　　　　　　　　**b)** $(4 + 5i) - (6 - 3i)$

SOLUTION

a) $(8 + 6i) + (3 + 2i) = (8 + 3) + (6i + 2i)$
　　　　　　　　　　Collecting the real parts and the imaginary parts
　　　　　　　　$= 11 + (6 + 2)i = 11 + 8i$

b) $(4 + 5i) - (6 - 3i) = (4 - 6) + [5i - (-3i)]$
　　　　　　　　　Note that the 6 and the $-3i$ are both being subtracted.
　　　　　　　　　　$= -2 + 8i$

Multiplication

For complex numbers, the property $\sqrt{a}\sqrt{b} = \sqrt{ab}$ does *not* hold in general, but it does hold when $a = -1$ and b is a nonnegative number. To multiply square roots of negative real numbers, we first express them in terms of i. For example,

$$\sqrt{-2} \cdot \sqrt{-5} = \sqrt{-1} \cdot \sqrt{2} \cdot \sqrt{-1} \cdot \sqrt{5} = i\sqrt{2} \cdot i\sqrt{5}$$
$$= i^2\sqrt{10} = -1\sqrt{10} = -\sqrt{10} \quad \text{is correct!}$$

But

$$\sqrt{-2} \cdot \sqrt{-5} = \sqrt{(-2)(-5)} = \sqrt{10} \quad \text{is wrong!}$$

Keeping this and the fact that $i^2 = -1$ in mind, we multiply in much the same way that we do with real numbers.

Example 5 Multiply and simplify each of the following.

a) $\sqrt{-16} \cdot \sqrt{-25}$ **b)** $\sqrt{-5} \cdot \sqrt{-7}$ **c)** $-4i(3 - 5i)$

d) $(1 + 2i)(1 + 3i)$ **e)** $(3 - 7i)^2$

SOLUTION

a) $\sqrt{-16} \cdot \sqrt{-25} = \sqrt{-1} \cdot \sqrt{16} \cdot \sqrt{-1} \cdot \sqrt{25}$

$$= i \cdot 4 \cdot i \cdot 5$$
$$= i^2 \cdot 20$$
$$= -1 \cdot 20 \qquad i^2 = -1$$
$$= -20$$

b) $\sqrt{-5} \cdot \sqrt{-7} = \sqrt{-1} \cdot \sqrt{5} \cdot \sqrt{-1} \cdot \sqrt{7}$

$$= i \cdot \sqrt{5} \cdot i \cdot \sqrt{7}$$
$$= i^2 \cdot \sqrt{35}$$
$$= -1 \cdot \sqrt{35} \qquad i^2 = -1$$
$$= -\sqrt{35}$$

c) $-4i(3 - 5i) = -4i \cdot 3 + (-4i)(-5i)$ Using the distributive law

$$= -12i + 20i^2$$
$$= -12i - 20 \qquad\qquad i^2 = -1$$
$$= -20 - 12i \qquad\qquad \text{Writing in the form } a + bi$$

d) $(1 + 2i)(1 + 3i) = 1 + 3i + 2i + 6i^2$ Multiplying each term of one number by every term of the other (FOIL)

$$= 1 + 3i + 2i - 6 \qquad i^2 = -1$$
$$= -5 + 5i \qquad\qquad \text{Collecting like terms}$$

e) $(3 - 7i)^2 = 3^2 - 2 \cdot 3 \cdot 7i + (7i)^2$ Recall that $(A - B)^2 = A^2 - 2AB + B^2$

$$= 9 - 42i + 49i^2$$
$$= 9 - 42i - 49 \qquad\qquad i^2 = -1$$
$$= -40 - 42i$$

Recall that -1 raised to an *even* power is 1, and -1 raised to an *odd* power is -1. Simplifying powers of i can then be done by using the fact that $i^2 = -1$ and expressing the given power of i in terms of i^2. Consider the following:

$$i = \sqrt{-1}, \qquad\qquad i^4 = (i^2)^2 = (-1)^2 = 1,$$
$$i^2 = -1, \qquad\qquad i^5 = i^4 \cdot i = (i^2)^2 \cdot i = (-1)^2 \cdot i = i,$$
$$i^3 = i^2 \cdot i = (-1)i = -i, \qquad i^6 = (i^2)^3 = (-1)^3 = -1.$$

Note that the powers of i cycle themselves through the values i, -1, $-i$, and 1.

Example 6   Simplify each of the following.

a) i^{37} **b)** i^{58} **c)** i^{75} **d)** i^{80}

SOLUTION

a) $i^{37} = i^{36} \cdot i = (i^2)^{18} \cdot i = (-1)^{18} \cdot i = 1 \cdot i = i$
b) $i^{58} = (i^2)^{29} = (-1)^{29} = -1$
c) $i^{75} = i^{74} \cdot i = (i^2)^{37} \cdot i = (-1)^{37} \cdot i = -1 \cdot i = -i$
d) $i^{80} = (i^2)^{40} = (-1)^{40} = 1$

Conjugates and Division

Conjugates of complex numbers are defined as follows.

Conjugate of a Complex Number
The *conjugate* of a complex number $a + bi$ is $a - bi$, and the *conjugate* of $a - bi$ is $a + bi$.

Example 7   Find the conjugate of each of the following.

a) $-3 + 7i$ **b)** $14 - 5i$ **c)** $4i$

SOLUTION

a) The conjugate of $-3 + 7i$ is $-3 - 7i$.
b) The conjugate of $14 - 5i$ is $14 + 5i$.
c) The conjugate of $4i$ is $-4i$.

The product of a complex number and its conjugate is a real number.

Example 8   Multiply each of the following.

a) $(5 + 7i)(5 - 7i)$ **b)** $(8i)(-8i)$

SOLUTION

a) $(5 + 7i)(5 - 7i) = 5^2 - (7i)^2$ Using $(A + B)(A - B) = A^2 - B^2$
$$= 25 - 49i^2$$
$$= 25 - 49(-1)$$
$$= 74$$

b) $(8i)(-8i) = -64i^2$
$$= -64(-1)$$
$$= 64$$

When we are dividing complex numbers, conjugates are used.

Example 9 Divide $2 - 5i$ by $1 - 6i$.

SOLUTION   We write fractional notation and then multiply by 1:

$$\frac{2 - 5i}{1 - 6i} = \frac{2 - 5i}{1 - 6i} \cdot \frac{1 + 6i}{1 + 6i}$$ Note that $1 + 6i$ is the conjugate of the divisor.

$$= \frac{(2 - 5i)(1 + 6i)}{(1 - 6i)(1 + 6i)} = \frac{2 + 7i - 30i^2}{1 - 36i^2}$$

$$= \frac{32 + 7i}{1 + 36} = \frac{32}{37} + \frac{7}{37}i.$$

Solving Quadratic Equations

If the graph of a quadratic function, $f(x) = ax^2 + bx + c$, has no x-intercepts, then the solutions of the quadratic equation, $ax^2 + bx + c = 0$, are imagainary and cannot be found graphically.

Example 10   Solve: $x^2 + 5x + 8 = 0$.

r-- ALGEBRAIC SOLUTION

To find the solutions, we use the quadratic formula. Here

$$a = 1, \quad b = 5, \quad c = 8;$$

$$x = \frac{-b \pm \sqrt{b^2 - 4ac}}{2a}$$

$$= \frac{-5 \pm \sqrt{5^2 - 4(1)(8)}}{2 \cdot 1}$$ Substituting

$$= \frac{-5 \pm \sqrt{-7}}{2}$$ Simplifying

$$= \frac{-5 \pm \sqrt{7}i}{2}.$$

The solutions are $-\dfrac{5}{2} - \dfrac{\sqrt{7}}{2}i$ and $-\dfrac{5}{2} + \dfrac{\sqrt{7}}{2}i$.

r-- GRAPHICAL SOLUTION

The graph of the function $f(x) = x^2 + 5x + 8$ shows no x-intercepts.

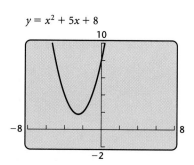

$y = x^2 + 5x + 8$

Thus there are no real-number solutions of the equation $x^2 + 5x + 8 = 0$. This is a quadratic equation that cannot be solved graphically.

The Discriminant

From the quadratic formula, we know that the solutions x_1 and x_2 of a quadratic equation are given by

$$x_1 = \frac{-b + \sqrt{b^2 - 4ac}}{2a} \quad \text{and} \quad x_2 = \frac{-b - \sqrt{b^2 - 4ac}}{2a}.$$

The expression $b^2 - 4ac$ shows the nature of the solutions. This expression is called the **discriminant**.

Discriminant

For $ax^2 + bx + c = 0$:

$b^2 - 4ac = 0$	One real-number solution;
$b^2 - 4ac > 0$	Two different real-number solutions;
$b^2 - 4ac < 0$	Two different imaginary-number solutions, complex conjugates.

4.3 Exercise Set

Express in terms of i.

1. $\sqrt{-17}$ **2.** $\sqrt{-25}$

3. $\sqrt{-49}$ **4.** $\sqrt{-13}$

5. $-\sqrt{-81}$ **6.** $-\sqrt{-20}$

7. $6 - \sqrt{-84}$ **8.** $7 - \sqrt{-60}$

9. $-\sqrt{76} + \sqrt{-125}$ **10.** $\sqrt{-4} + \sqrt{-12}$

11. $\sqrt{-5}\sqrt{-11}$ **12.** $-\sqrt{-9}\sqrt{-7}$

13. $\dfrac{-\sqrt{25}}{\sqrt{-16}}$ **14.** $\dfrac{-\sqrt{-36}}{\sqrt{-9}}$

Simplify. Write answers in the form $a + bi$, where a and b are real numbers.

15. $(-5 + 3i) + (7 + 8i)$

16. $(-6 - 5i) + (9 + 2i)$

17. $(4 - 9i) + (1 - 3i)$

18. $(7 - 2i) + (4 - 5i)$

19. $(3 + \sqrt{-16}) + (2 + \sqrt{-25})$

20. $(7 - \sqrt{-36}) + (2 + \sqrt{-9})$

21. $(10 + 7i) - (5 + 3i)$

22. $(-3 - 4i) - (8 - i)$

23. $(13 + 9i) - (8 + 2i)$

24. $(-7 + 12i) - (3 - 6i)$

25. $(1 + 3i)(1 - 4i)$ **26.** $(1 - 2i)(1 + 3i)$

27. $(2 + 3i)(2 + 5i)$ **28.** $(3 - 5i)(8 - 2i)$

29. $7i(2 - 5i)$ **30.** $3i(6 + 4i)$

31. $(3 + \sqrt{-16})(2 + \sqrt{-25})$

32. $(7 - \sqrt{-16})(2 + \sqrt{-9})$

33. $(5 - 4i)(5 + 4i)$ **34.** $(5 + 9i)(5 - 9i)$

35. $(4 + 2i)^2$ **36.** $(5 - 4i)^2$

37. $(-2 + 7i)^2$ **38.** $(-3 + 2i)^2$

39. $\dfrac{2 + \sqrt{3}i}{5 - 4i}$ **40.** $\dfrac{\sqrt{5} + 3i}{1 - i}$

41. $\dfrac{4 + i}{-3 - 2i}$ **42.** $\dfrac{5 - i}{-7 + 2i}$

43. $\dfrac{3}{5 - 11i}$ **44.** $\dfrac{i}{2 + i}$

45. $\dfrac{1 + i}{(1 - i)^2}$ **46.** $\dfrac{1 - i}{(1 + i)^2}$

47. $\dfrac{4 - 2i}{1 + i} + \dfrac{2 - 5i}{1 + i}$ **48.** $\dfrac{3 + 2i}{1 - i} + \dfrac{6 + 2i}{1 - i}$

Simplify.

49. i^{11} **50.** i^7 **51.** i^{35} **52.** i^{24}

53. i^{64} **54.** i^{42} **55.** $(-i)^{71}$ **56.** $(-i)^6$

57. $(5i)^4$ **58.** $(2i)^5$

Solve.

59. $x^2 + 10 = 0$ **60.** $x^2 = -14$

61. $x^2 + 8x + 25 = 0$ **62.** $x^2 + 6x + 13 = 0$

63. $x^2 + x + 2 = 0$ **64.** $x^2 + 1 = x$

65. $5x^2 + 2 = x$ **66.** $3x^2 + 4 = 5x$

For each of the following, consider only $b^2 - 4ac$, the discriminant of the quadratic formula, to determine whether imaginary solutions exist.

67. $4x^2 = 8x + 5$ **68.** $4x^2 - 12x + 9 = 0$

69. $x^2 + 3x + 4 = 0$ **70.** $x^2 - 2x + 4 = 0$

71. $5t^2 - 7t = 0$ **72.** $5t^2 - 4t = 11$

Skill Maintenance

Factor.

73. $x^2 + 8x + 16$ **74.** $y^2 + 20y + 100$

75. $x^2 - 10x + 25$ **76.** $x^2 - 18x + 81$

Synthesis

77. ◆ Is the sum of two imaginary numbers always an imaginary number? Why or why not? What about the product?

78. ◆ Is the domain of every polynomial function $\mathbb{R}$? Why or why not?

Solve.

79. $(2 + i)x - i = 5 + i$

80. $5i = (3 + i)x + i$

81. $5ix + 3 + 2i = (3 - 2i)x + 3i$

82. $(2 + 3i)x - 2i = 2ix + 5 - 4i$

Determine whether each of the following is true or false.

83. The sum of two numbers that are conjugates of each other is always a real number.

84. The conjugate of a sum is the sum of the conjugates of the individual complex numbers.

85. The conjugate of a product is the product of the conjugates of the individual complex numbers.

Let $z = a + bi$ and $\bar{z} = a - bi$.

86. Find a general expression for $1/z$.

87. Find a general expression for $z\bar{z}$.

88. Solve $z + 6\bar{z} = 7$ for z.

4.4
Complex Numbers: Trigonometric Form

- *Graph complex numbers.*
- *Given a complex number in standard form, find trigonometric, or polar, notation; and given a complex number in trigonometric form, find standard notation.*
- *Use trigonometric notation to multiply and divide complex numbers.*
- *Use DeMoivre's theorem to raise complex numbers to powers.*
- *Find the nth roots of a complex number.*

Graphical Representation

Just as real numbers can be graphed on a line, complex numbers can be graphed on a plane. We graph a complex number $a + bi$ in the same way that we graph an ordered pair of real numbers (a, b). However, in place of an x-axis, we have a real axis, and in place of a y-axis, we have an imaginary axis. Horizontal distances correspond to the real part of a number. Vertical distances correspond to the imaginary part.

Example 1   Graph each of the following complex numbers.

a) $3 + 2i$ b) $-4 - 5i$ c) $-3i$

d) $-1 + 3i$ e) 2

SOLUTION

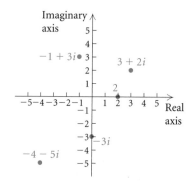

We recall that the absolute value of a real number is its distance from 0 on the number line. The absolute value of a complex number is its distance from the origin in the complex plane. For example, if $z = a + bi$, then using the distance formula, we have

$$|z| = |a + bi| = \sqrt{(a - 0)^2 + (b - 0)^2}$$
$$= \sqrt{a^2 + b^2}.$$

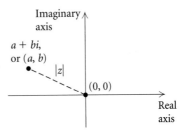

Absolute Value of a Complex Number

The absolute value of a complex number $a + bi$ is

$$|a + bi| = \sqrt{a^2 + b^2}.$$

Example 2   Find the absolute value of each of the following.

a) $3 + 4i$ **b)** $-2 - i$ **c)** $\dfrac{4}{5}i$

SOLUTION

a) $|3 + 4i| = \sqrt{3^2 + 4^2} = \sqrt{9 + 16} = \sqrt{25} = 5$

b) $|-2 - i| = \sqrt{(-2)^2 + (-1)^2} = \sqrt{5}$

c) $\left|\dfrac{4}{5}i\right| = \left|0 + \dfrac{4}{5}i\right| = \sqrt{\left(\dfrac{4}{5}\right)^2} = \dfrac{4}{5}$

Trigonometric Notation for Complex Numbers

Now let's consider a nonzero complex number $a + bi$. Suppose that its absolute value is r. If we let θ be an angle in standard position whose terminal side passes through the point (a, b), as shown in the figure, then

$$\cos \theta = \frac{a}{r}, \quad \text{or} \quad a = r \cos \theta$$

and

$$\sin \theta = \frac{b}{r}, \quad \text{or} \quad b = r \sin \theta.$$

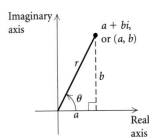

Substituting these values for a and b into the $(a + bi)$ notation, we get

$$a + bi = r \cos \theta + (r \sin \theta)i$$
$$= r(\cos \theta + i \sin \theta).$$

This is **trigonometric notation** for a complex number $a + bi$. The number r is called the **absolute value** of $a + bi$, and θ is called the **argument** of $a + bi$. Trigonometric notation for a complex number is also called **polar notation**.

Trigonometric Notation for Complex Numbers

$$a + bi = r(\cos \theta + i \sin \theta)$$

To find trigonometric notation for a complex number given in standard notation, we must find r and determine the angle θ for which $\sin \theta = b/r$ and $\cos \theta = a/r$.

Example 3   Find trigonometric notation for each of the following complex numbers.

a) $1 + i$

b) $\sqrt{3} - i$

SOLUTION

a) We note that $a = 1$ and $b = 1$. Then

$$r = \sqrt{a^2 + b^2} = \sqrt{1^2 + 1^2} = \sqrt{2},$$

$$\sin \theta = \frac{b}{r} = \frac{1}{\sqrt{2}}, \quad \text{or} \quad \frac{\sqrt{2}}{2},$$

and

$$\cos \theta = \frac{1}{\sqrt{2}}, \quad \text{or} \quad \frac{\sqrt{2}}{2}.$$

Thus, $\theta = \pi/4$, or $45°$, and we have

$$1 + i = \sqrt{2}\left(\cos \frac{\pi}{4} + i \sin \frac{\pi}{4} \right),$$

or $1 + i = \sqrt{2}(\cos 45° + i \sin 45°)$.

b) We see that $a = \sqrt{3}$ and $b = -1$. Then

$$r = \sqrt{(\sqrt{3})^2 + (-1)^2} = 2, \quad \sin \theta = -\frac{1}{2}, \quad \text{and} \quad \cos \theta = \frac{\sqrt{3}}{2}.$$

Thus, $\theta = 11\pi/6$, or $330°$, and we have

$$\sqrt{3} - i = 2\left(\cos \frac{11\pi}{6} + i \sin \frac{11\pi}{6} \right),$$

or $\sqrt{3} - i = 2(\cos 330° + i \sin 330°)$.

In changing to trigonometric notation, note that there are many angles satisfying the given conditions. We ordinarily choose the smallest positive angle.

To change from trigonometric notation to standard notation, $a + bi$, we recall that $a = r \cos \theta$ and $b = r \sin \theta$.

Example 4   Find standard notation, $a + bi$, for each of the following complex numbers.

a) $2(\cos 120° + i \sin 120°)$ **b)** $\sqrt{8}\left(\cos \dfrac{7\pi}{4} + i \sin \dfrac{7\pi}{4}\right)$

SOLUTION

a) Rewriting, we have

$$2(\cos 120° + i \sin 120°) = 2 \cos 120° + (2 \sin 120°)i.$$

Thus,

$$a = 2 \cos 120° = 2 \cdot \left(-\frac{1}{2}\right) = -1$$

and

$$b = 2 \sin 120° = 2 \cdot \frac{\sqrt{3}}{2} = \sqrt{3},$$

so

$$2(\cos 120° + i \sin 120°) = -1 + \sqrt{3}i.$$

b) Rewriting, we have

$$\sqrt{8}\left(\cos \frac{7\pi}{4} + i \sin \frac{7\pi}{4}\right) = \sqrt{8} \cos \frac{7\pi}{4} + \left(\sqrt{8} \sin \frac{7\pi}{4}\right)i.$$

Thus,

$$a = \sqrt{8} \cos \frac{7\pi}{4} = \sqrt{8} \cdot \frac{\sqrt{2}}{2} = 2$$

and

$$b = \sqrt{8} \sin \frac{7\pi}{4} = \sqrt{8} \cdot \left(-\frac{\sqrt{2}}{2}\right) = -2,$$

so

$$\sqrt{8}\left(\cos \frac{7\pi}{4} + i \sin \frac{7\pi}{4}\right) = 2 - 2i.$$

Multiplication and Division with Trigonometric Notation

Multiplication and division of complex numbers is easier to manage with trigonometric notation than with standard notation. We simply multiply the absolute values and add the arguments. Let's state this in a more formal manner.

Complex Numbers: Multiplication

For any complex numbers $r_1(\cos \theta_1 + i \sin \theta_1)$ and $r_2(\cos \theta_2 + i \sin \theta_2)$,

$$r_1(\cos \theta_1 + i \sin \theta_1) \cdot r_2(\cos \theta_2 + i \sin \theta_2)$$
$$= r_1 r_2[\cos (\theta_1 + \theta_2) + i \sin (\theta_1 + \theta_2)].$$

Proof: Let's first multiply $a_1 + b_1 i$ by $a_2 + b_2 i$:

$$(a_1 + b_1 i)(a_2 + b_2 i) = (a_1 a_2 - b_1 b_2) + (a_2 b_1 + a_1 b_2)i.$$

Recall that

$$a_1 = r_1 \cos \theta_1, \qquad b_1 = r_1 \sin \theta_1,$$

and

$$a_2 = r_2 \cos \theta_2, \qquad b_2 = r_2 \sin \theta_2.$$

We substitute these in the product above, to obtain

$$r_1(\cos \theta_1 + i \sin \theta_1) \cdot r_2(\cos \theta_2 + i \sin \theta_2)$$
$$= (r_1 r_2 \cos \theta_1 \cos \theta_2 - r_1 r_2 \sin \theta_1 \sin \theta_2)$$
$$+ (r_1 r_2 \sin \theta_1 \cos \theta_2 + r_1 r_2 \cos \theta_1 \sin \theta_2)i.$$

This simplifies to

$$r_1 r_2(\cos \theta_1 \cos \theta_2 - \sin \theta_1 \sin \theta_2) + r_1 r_2(\sin \theta_1 \cos \theta_2 + \cos \theta_1 \sin \theta_2)i.$$

Now, using identities for sums of angles, we simplify, obtaining

$$r_1 r_2 \cos (\theta_1 + \theta_2) + r_1 r_2 \sin (\theta_1 + \theta_2)i,$$

which was to be shown.

Example 5   Multiply and express the answer to each of the following in standard notation.

a) $3(\cos 40° + i \sin 40°)$ and $4(\cos 20° + i \sin 20°)$

b) $2(\cos \pi + i \sin \pi)$ and $3\left[\cos \left(-\dfrac{\pi}{2} \right) + i \sin \left(-\dfrac{\pi}{2} \right) \right]$

SOLUTION

a) $3(\cos 40° + i \sin 40°) \cdot 4(\cos 20° + i \sin 20°)$
$$= 3 \cdot 4 \cdot [\cos (40° + 20°) + i \sin (40° + 20°)]$$
$$= 12(\cos 60° + i \sin 60°)$$
$$= 12\left(\frac{1}{2} + \frac{\sqrt{3}}{2} i \right)$$
$$= 6 + 6\sqrt{3}i$$

b) $2(\cos \pi + i \sin \pi) \cdot 3\left[\cos\left(-\dfrac{\pi}{2}\right) + i \sin\left(-\dfrac{\pi}{2}\right)\right]$

$$= 2 \cdot 3 \cdot \left[\cos\left(\pi + \left(-\dfrac{\pi}{2}\right)\right) + i \sin\left(\pi + \left(-\dfrac{\pi}{2}\right)\right)\right]$$

$$= 6\left(\cos\dfrac{\pi}{2} + i \sin\dfrac{\pi}{2}\right)$$

$$= 6(0 + i \cdot 1)$$

$$= 6i$$

We can check by multiplying with standard notation. Let's check Example 5(b):

$$2(\cos \pi + i \sin \pi) = 2(-1 + 0) = -2$$

and

$$3\left[\cos\left(-\dfrac{\pi}{2}\right) + i \sin\left(-\dfrac{\pi}{2}\right)\right] = 3[0 + i(-1)] = -3i.$$

Thus,

$$-2 \cdot (-3i) = 6i.$$

Example 6 Convert to trigonometric notation and multiply:

$$(1 + i)(\sqrt{3} - i).$$

SOLUTION We first find trigonometric notation:

$$1 + i = \sqrt{2}(\cos 45° + i \sin 45°), \qquad \text{See Example 3(a).}$$
$$\sqrt{3} - i = 2(\cos 330° + i \sin 330°). \qquad \text{See Example 3(b).}$$

Then we multiply:

$$\sqrt{2}(\cos 45° + i \sin 45°) \cdot 2(\cos 330° + i \sin 330°)$$
$$= 2\sqrt{2}[\cos(45° + 330°) + i \sin(45° + 330°)]$$
$$= 2\sqrt{2}(\cos 375° + i \sin 375°).$$

To divide complex numbers, we divide the absolute values and subtract the arguments. We state this fact below, but omit the proof.

Complex Numbers: Division

For any complex numbers $r_1(\cos \theta_1 + i \sin \theta_1)$ and $r_2(\cos \theta_2 + i \sin \theta_2)$, $r_2 \neq 0$,

$$\dfrac{r_1(\cos \theta_1 + i \sin \theta_1)}{r_2(\cos \theta_2 + i \sin \theta_2)} = \dfrac{r_1}{r_2}[\cos(\theta_1 - \theta_2) + i \sin(\theta_1 - \theta_2)].$$

Example 7   Divide

$$2\left(\cos \frac{3\pi}{2} + i \sin \frac{3\pi}{2}\right) \quad \text{by} \quad 4\left(\cos \frac{\pi}{2} + i \sin \frac{\pi}{2}\right)$$

and express the solution in standard notation.

SOLUTION   We have

$$\frac{2\left(\cos \dfrac{3\pi}{2} + i \sin \dfrac{3\pi}{2}\right)}{4\left(\cos \dfrac{\pi}{2} + i \sin \dfrac{\pi}{2}\right)}$$

$$= \frac{2}{4}\left[\cos \left(\frac{3\pi}{2} - \frac{\pi}{2}\right) + i \sin \left(\frac{3\pi}{2} - \frac{\pi}{2}\right)\right]$$

$$= \frac{1}{2}(\cos \pi + i \sin \pi)$$

$$= \frac{1}{2}(-1 + i \cdot 0)$$

$$= -\frac{1}{2}.$$

Example 8   Convert to trigonometric notation and divide:

$$\frac{1 + i}{1 - i}.$$

SOLUTION   We first convert to trigonometric notation:

$$1 + i = \sqrt{2}(\cos 45° + i \sin 45°), \qquad \text{See Example 3(a).}$$
$$1 - i = \sqrt{2}(\cos 315° + i \sin 315°).$$

We now divide:

$$\frac{\sqrt{2}(\cos 45° + i \sin 45°)}{\sqrt{2}(\cos 315° + i \sin 315°)}$$

$$= 1[\cos (45° - 315°) + i \sin (45° - 315°)]$$
$$= \cos (-270°) + i \sin (-270°)$$
$$= 0 + i \cdot 1$$
$$= i.$$

Powers of Complex Numbers

An important theorem about powers and roots of complex numbers is named for the French mathematician Abraham DeMoivre (1667–1754). Let's consider the square of a complex number $r(\cos \theta + i \sin \theta)$:

$$[r(\cos \theta + i \sin \theta)]^2 = [r(\cos \theta + i \sin \theta)] \cdot [r(\cos \theta + i \sin \theta)]$$
$$= r \cdot r \cdot [\cos (\theta + \theta) + i \sin (\theta + \theta)]$$
$$= r^2(\cos 2\theta + i \sin 2\theta).$$

Similarly, we see that

$$[r(\cos \theta + i \sin \theta)]^3$$
$$= r \cdot r \cdot r \cdot [\cos (\theta + \theta + \theta) + i \sin (\theta + \theta + \theta)]$$
$$= r^3(\cos 3\theta + i \sin 3\theta).$$

DeMoivre's theorem is the generalization of these results.

DeMoivre's Theorem

For any complex number $r(\cos \theta + i \sin \theta)$ and any natural number n,

$$[r(\cos \theta + i \sin \theta)]^n = r^n(\cos n\theta + i \sin n\theta).$$

Example 9   Find each of the following.

a) $(1 + i)^9$

b) $(\sqrt{3} - i)^{10}$

SOLUTION

a) We first find trigonometric notation:

$$1 + i = \sqrt{2}(\cos 45° + i \sin 45°).$$

Then

$$(1 + i)^9 = [\sqrt{2}(\cos 45° + i \sin 45°)]^9$$
$$= (\sqrt{2})^9[\cos (9 \cdot 45°) + i \sin (9 \cdot 45°)] \qquad \text{DeMoivre's theorem}$$
$$= 2^{9/2}(\cos 405° + i \sin 405°)$$
$$= 16\sqrt{2}(\cos 45° + i \sin 45°) \qquad \text{405° has the same terminal side as 45°.}$$

$$= 16\sqrt{2}\left(\frac{\sqrt{2}}{2} + i\frac{\sqrt{2}}{2}\right)$$
$$= 16 + 16i.$$

b) We first convert to trigonometric notation:

$$\sqrt{3} - i = 2(\cos 330° + i \sin 330°).$$

Then

$$(\sqrt{3} - i)^{10} = [2(\cos 330° + i \sin 330°)]^{10}$$
$$= 2^{10}(\cos 3300° + i \sin 3300°)$$
$$= 1024(\cos 60° + i \sin 60°)$$
$$= 1024\left(\frac{1}{2} + i\frac{\sqrt{3}}{2}\right)$$
$$= 512 + 512\sqrt{3}i.$$

Roots of Complex Numbers

As we will see, every nonzero complex number has two square roots. A nonzero complex number has three cube roots, four fourth roots, and so on. In general, a nonzero complex number has n different nth roots. They can be found using the formula that we now state and prove.

Roots of Complex Numbers

The nth roots of a complex number $r(\cos \theta + i \sin \theta)$, $r \neq 0$, are given by

$$r^{1/n}\left[\cos\left(\frac{\theta}{n} + k \cdot \frac{360°}{n}\right) + i \sin\left(\frac{\theta}{n} + k \cdot \frac{360°}{n}\right)\right],$$

where $k = 0, 1, 2, \ldots, n - 1$.

Using DeMoivre's theorem, we show that this formula gives us n different roots. First we take the expression for the nth roots and raise it to the nth power, to show that we get $r(\cos \theta + i \sin \theta)$:

$$\left[r^{1/n}\left[\cos\left(\frac{\theta}{n} + k \cdot \frac{360°}{n}\right) + i \sin\left(\frac{\theta}{n} + k \cdot \frac{360°}{n}\right)\right]\right]^n$$

$$= (r^{1/n})^n\left[\cos\left(n\left(\frac{\theta}{n} + k \cdot \frac{360°}{n}\right)\right) + i \sin\left(n\left(\frac{\theta}{n} + k \cdot \frac{360°}{n}\right)\right)\right]$$

$$= r[\cos(\theta + k \cdot 360°) + i \sin(\theta + k \cdot 360°)]$$

$$= r(\cos \theta + i \sin \theta).$$

Thus we know that the formula gives us nth roots for any nonnegative integer k, $k < n$.

Next, we show that there are at least n different roots. To see this, consider substituting 0, 1, 2, and so on, for k. When $k = n$, the cycle begins to repeat, but from 0 to $n - 1$, the angles obtained and their sines and cosines are all different. There cannot be more than n different nth roots. That fact follows from the fundamental theorem of algebra.

Example 10   Find the square roots of $2 + 2\sqrt{3}i$.

SOLUTION   We first find trigonometric notation:

$$2 + 2\sqrt{3}i = 4(\cos 60° + i \sin 60°).$$

Then

$$[4(\cos 60° + i \sin 60°)]^{1/2}$$

$$= 4^{1/2}\left[\cos\left(\frac{60°}{2} + k \cdot \frac{360°}{2}\right) + i \sin\left(\frac{60°}{2} + k \cdot \frac{360°}{2}\right)\right], \quad k = 0, 1$$

$$= 2[\cos(30° + k \cdot 180°) + i \sin(30° + k \cdot 180°), \quad k = 0, 1.$$

Program

DEMOIVRE: This program graphs *n*th roots of complex numbers.

Thus the roots are

$$2(\cos 30° + i \sin 30°) \text{ for } k = 0$$

and

$$2(\cos 210° + i \sin 210°) \text{ for } k = 1,$$

or

$$\sqrt{3} + i \quad \text{and} \quad -\sqrt{3} - i.$$

In Example 10, we see that the two square roots of the number are opposites of each other. We can illustrate this graphically. We also note that the roots are equally spaced about a circle of radius *r*—in this case, $r = 2$. The roots are 360%2, or 180° apart.

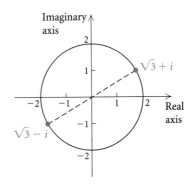

Example 11   Find the cube roots of 1. Then locate them on a graph.

SOLUTION   We begin by finding trigonometric notation:

$$1 = 1(\cos 0° + i \sin 0°).$$

Then

$$[1(\cos 0° + i \sin 0°)]^{1/3}$$

$$= 1^{1/3}\left[\cos\left(\frac{0°}{3} + k \cdot \frac{360°}{3}\right) + i \sin\left(\frac{0°}{3} + k \cdot \frac{360°}{3}\right)\right], \quad k = 0, 1, 2.$$

The roots are

$$1(\cos 0° + i \sin 0°), \quad 1(\cos 120° + i \sin 120°),$$

and

$$1(\cos 240° + i \sin 240°),$$

or

$$1, \quad -\frac{1}{2} + \frac{\sqrt{3}}{2}i, \quad \text{and} \quad -\frac{1}{2} - \frac{\sqrt{3}}{2}i.$$

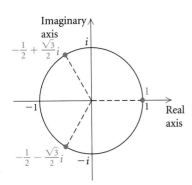

The graphs of the cube roots lie equally spaced about a circle of radius $r = 1$. The roots are 360%3, or 120° apart.

The *n*th roots of 1 are often referred to as the **nth roots of unity**. In Example 11, we found the cube roots of unity.

With a grapher in PARAMETRIC mode, we can approximate the *n*th roots of a number *p*. We use the following window and let

$$X_{1T} = (p^\wedge(1/n)) \cos T \quad \text{and} \quad Y_{1T} = (p^\wedge(1/n)) \sin T.$$

WINDOW

Tmin = 0

Tmax = 360, if in degree mode, or
= 2π, if in radian mode

Tstep = 360/*n*, or $2\pi/n$

Xmin = -3

Xmax = 3

Xscl = 1

Ymin = -2

Ymax = 2

Yscl = 1

To find the fifth roots of 8, enter $X_{1T} = (8^\wedge(1/5)) \cos T$ and $Y_{1T} = (8^\wedge(1/5)) \sin T$. In this case, use DEGREE mode. After the graph has been generated, use the TRACE feature to locate the fifth roots. The T, X, and Y values appear on the screen. What do they represent?

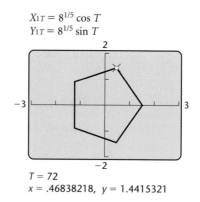

$X_{1T} = 8^{1/5} \cos T$
$Y_{1T} = 8^{1/5} \sin T$

$T = 72$
$x = .46838218, \ y = 1.4415321$

Three of the fifth roots of 8 are approximately

$$1.5157, \quad 0.46838 + 1.44153i, \quad \text{and} \quad -1.22624 + 0.89092i.$$

Find the other two. Then use a grapher to approximate the cube roots of unity that were found in Example 11. Then approximate the fourth roots of 5 and the tenth roots of unity.

4.4 | Exercise Set

Graph each complex number and find its absolute value.

1. $4 + 3i$

2. $-2 - 3i$

3. i

4. $-5 - 2i$

5. $4 - i$

6. $6 + 3i$

7. 3

8. $-2i$

Express the indicated number in both standard notation and trigonometric notation.

9.

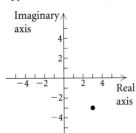

10.

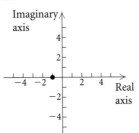

11.

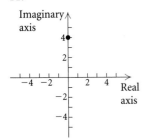

12.

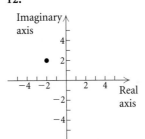

Find trigonometric notation.

13. $1 - i$

14. $-10\sqrt{3} + 10i$

15. $-3i$

16. $-5 + 5i$

17. $\sqrt{3} + i$

18. 4

19. $\dfrac{2}{5}$

20. $7.5i$

Find standard notation, $a + bi$.

21. $3(\cos 30° + i \sin 30°)$

22. $6(\cos 120° + i \sin 120°)$

23. $10(\cos 270° + i \sin 270°)$

24. $3(\cos 0° + i \sin 0°)$

25. $\sqrt{8}\left(\cos \dfrac{\pi}{4} + i \sin \dfrac{\pi}{4}\right)$

26. $5\left(\cos \dfrac{\pi}{3} + i \sin \dfrac{\pi}{3}\right)$

27. $2\left(\cos \dfrac{\pi}{2} + i \sin \dfrac{\pi}{2}\right)$

28. $3\left[\cos\left(-\dfrac{3\pi}{4}\right) + i \sin\left(-\dfrac{3\pi}{4}\right)\right]$

Multiply or divide and leave the answer in trigonometric notation.

29. $\dfrac{12(\cos 48° + i \sin 48°)}{3(\cos 6° + i \sin 6°)}$

30. $5\left(\cos \dfrac{\pi}{3} + i \sin \dfrac{\pi}{3}\right) \cdot 2\left(\cos \dfrac{\pi}{4} + i \sin \dfrac{\pi}{4}\right)$

31. $2.5(\cos 35° + i \sin 35°) \cdot 4.5(\cos 21° + i \sin 21°)$

32. $\dfrac{\dfrac{1}{2}\left(\cos \dfrac{2\pi}{3} + i \sin \dfrac{2\pi}{3}\right)}{\dfrac{3}{8}\left(\cos \dfrac{\pi}{6} + i \sin \dfrac{\pi}{6}\right)}$

Convert to trigonometric notation and then multiply or divide.

33. $(1 - i)(2 + 2i)$

34. $(1 + i\sqrt{3})(1 + i)$

35. $\dfrac{1 - i}{1 + i}$

36. $\dfrac{1 - i}{\sqrt{3} - i}$

37. $(3\sqrt{3} - 3i)(2i)$

38. $(2\sqrt{3} + 2i)(2i)$

39. $\dfrac{2\sqrt{3} - 2i}{1 + \sqrt{3}i}$

40. $\dfrac{3 - 3\sqrt{3}i}{\sqrt{3} - i}$

Raise the number to the given power and write trigonometric notation for the answer.

41. $\left[2\left(\cos \dfrac{\pi}{3} + i \sin \dfrac{\pi}{3}\right)\right]^3$

42. $[2(\cos 120° + i \sin 120°)]^4$

43. $(1 + i)^6$

44. $(-\sqrt{3} + i)^5$

Raise the number to the given power and write standard notation for the answer.

45. $[3(\cos 20° + i \sin 20°)]^3$

46. $[2(\cos 10° + i \sin 10°)]^9$

47. $(1 - i)^5$

48. $(2 + 2i)^4$

49. $\left(\dfrac{1}{\sqrt{2}} - \dfrac{1}{\sqrt{2}}i\right)^{12}$

50. $\left(\dfrac{\sqrt{3}}{2} + \dfrac{1}{2}i\right)^{10}$

Find the square roots of each number.

51. $-i$

52. $1 + i$

53. $2\sqrt{2} - 2\sqrt{2}i$

54. $-\sqrt{3} - i$

Find the cube roots of each number.

55. i

56. $-64i$

57. $2\sqrt{3} - 2i$

58. $1 - \sqrt{3}i$

59. Find and graph the fourth roots of 16.

60. Find and graph the fourth roots of i.

61. Find and graph the fifth roots of -1.

62. Find and graph the sixth roots of 1.

63. Find the tenth roots of 8.

64. Find the ninth roots of -4.

65. Find the sixth roots of -1.

66. Find the fourth roots of 12.

Find all the complex solutions of the equation.

67. $x^3 = 1$

68. $x^5 - 1 = 0$

69. $x^4 + i = 0$

70. $x^4 + 81 = 0$

71. $x^6 + 64 = 0$

72. $x^5 + \sqrt{3} + i = 0$

Skill Maintenance

Convert to degree measure.

73. $\dfrac{\pi}{12}$

74. 3π

Convert to radian measure.

75. $330°$

76. $-225°$

77. Find r.

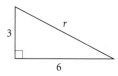

78. Graph these points in the rectangular coordinate system: $(2, -1)$, $(0, 3)$, and $\left(-\frac{1}{2}, -4\right)$.

Synthesis

79. ◆ Find and graph the square roots of $1 - i$. Explain geometrically why they are the opposites of each other.

80. ◆ Explain why trigonometric notation for a complex number is not unique, but rectangular, or standard, notation is unique.

Solve.

81. $x^2 + (1 - i)x + i = 0$

82. $3x^2 + (1 + 2i)x + 1 - i = 0$

83. Find polar notation for $(\cos \theta + i \sin \theta)^{-1}$.

84. Show that for any complex number z,
$$|z| = |-z|.$$
(*Hint*: Let $z = a + bi$.)

85. Show that for any complex number z,
$$|z| = |\bar{z}|.$$
(*Hint*: Let $z = a + bi$.)

86. Show that for any complex number z,
$$|z\bar{z}| = |z^2|.$$

87. Show that for any complex number z,
$$|z^2| = |z|^2.$$

88. Show that for any complex numbers z and w,
$$|z \cdot w| = |z| \cdot |w|.$$
(*Hint*: Let $z = r_1(\cos \theta_1 + i \sin \theta_1)$ and $w = r_2(\cos \theta_2 + i \sin \theta_2)$.)

89. Show that for any complex number z and any nonzero, complex number w,
$$\left|\frac{z}{w}\right| = \frac{|z|}{|w|}.$$ (Use the hint for Exercise 88.)

90. On a complex plane, graph $|z| = 1$.

91. On a complex plane, graph $z + \bar{z} = 3$.

4.5
Vectors and Applications

- *Determine whether two vectors are equivalent.*
- *Find the sum, or resultant, of two vectors.*
- *Resolve a vector into its horizontal and vertical components.*
- *Solve applications involving vectors.*

We measure some quantities using only their magnitudes. For example, we describe time, length, and mass using units like seconds, feet, and kilograms, respectively. However, to measure quantities like **displacement**, **velocity**, or **force**, we need to describe a *magnitude* and a *direction*. Together these describe a **vector**. The following are some examples.

DISPLACEMENT An object moves a certain distance in a certain direction.

A surveyor steps 20 yd to the northeast.

A hiker follows a trail 5 mi to the west.

A batter hits a ball 100 m along the left-field line.

VELOCITY An object travels at a certain speed in a certain direction.

A breeze is blowing 15 mph from the northwest.

An airplane is traveling 450 km/h in a direction of 243°.

FORCE A push or pull is exerted on an object in a certain direction.

A cart is being pulled up a 30° incline, requiring an effort of 200 lb.

A 25-lb force is required to lift a box upward.

A force of 15 newtons is exerted downward on the handle of a jack. (A newton, abbreviated N, is a unit of force used in physics, and 1 N ≈ 0.22 lb.)

Vectors

Vectors can be graphically represented by directed line segments. The length is chosen, according to some scale, to represent the **magnitude of the vector,** and the direction of the directed line segment represents the **direction of the vector.** For example, if we let 1 cm represent 5 km/h, then a 15-km/h wind from the northwest would be represented by a directed line segment 3 cm long, as shown in the figure at left.

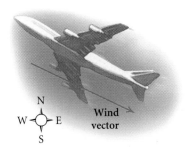

Vector

A *vector* in the plane is a directed line segment. Two vectors are *equivalent* if they have the same magnitude and direction.

Consider a vector drawn from point *A* to point *B*. Point *A* is called the **initial point** of the vector, and point *B* is called the **terminal point**.

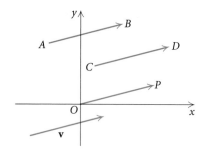

Symbolic notation for this vector is $\overrightarrow{AB}$ (read "vector AB"). Vectors are also denoted by boldface letters such as **u**, **v**, and **w**. The four vectors in the figure at left have the same length and direction. Thus they represent equivalent vectors; that is,

$$\overrightarrow{AB} = \overrightarrow{CD} = \overrightarrow{OP} = \mathbf{v}.$$

These vectors are **equivalent** because they have the *same* length and direction. The length, or **magnitude**, of $\overrightarrow{AB}$ is expressed as $|\overrightarrow{AB}|$. In the context of vectors, we use $=$ to mean equivalent.

Example 1   The vectors **u**, $\overrightarrow{OR}$, and **w** are shown in the figure below. Show that $\mathbf{u} = \overrightarrow{OR} = \mathbf{w}$.

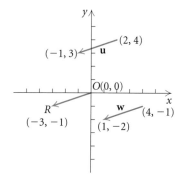

SOLUTION   We first find the length of each vector using the distance formula.

$$|\mathbf{u}| = \sqrt{[2 - (-1)]^2 + (4 - 3)^2} = \sqrt{9 + 1} = \sqrt{10},$$

$$|\overrightarrow{OR}| = \sqrt{[0 - (-3)]^2 + [0 - (-1)]^2} = \sqrt{9 + 1} = \sqrt{10},$$

$$|\mathbf{w}| = \sqrt{(4 - 1)^2 + [-1 - (-2)]^2} = \sqrt{9 + 1} = \sqrt{10}.$$

u ≠ v (not equivalent)
Different magnitudes;
different directions

Thus

$$|\mathbf{u}| = |\overrightarrow{OR}| = |\mathbf{w}|.$$

The vectors **u**, $\overrightarrow{OR}$, and **w** point down and to the left. Thus if the lines that they are on all have the same slope, the vectors have the same direction. We calculate the slopes:

$$\text{Slope} = \underset{\mathbf{u}}{\frac{4 - 3}{2 - (-1)}} = \underset{\overrightarrow{OR}}{\frac{0 - (-1)}{0 - (-3)}} = \underset{\mathbf{w}}{\frac{-1 - (-2)}{4 - 1}} = \frac{1}{3}.$$

u ≠ v
Same magnitude;
different directions

Since **u**, $\overrightarrow{OR}$, and **w** have the *same* magnitude and the *same* direction,

$$\mathbf{u} = \overrightarrow{OR} = \mathbf{w}.$$

u ≠ v
Different magnitudes;
same direction

Keep in mind that the equivalence of vectors requires only the same magnitude and the same direction—not the same location. A few examples are shown at left.

u = v
Same magnitude;
same direction

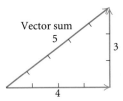

Vector sum
5

3

4

Vector Addition

Suppose a person takes 4 steps east and then 3 steps north. He or she will then be 5 steps from the starting point in the direction shown. The **sum** of the two vectors is the vector 5 steps in magnitude and in the direction shown. The sum is also called the **resultant** of the two vectors.

In general, two nonzero vectors **u** and **v** can be added geometrically by placing the initial point of one vector at the terminal point of the other and then finding the vector that forms the third side of the triangle, as shown in the following figure.

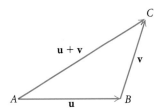

The sum **u** + **v** is the vector represented by the directed line segment from the initial point A of **u** to the terminal point C of **v**. That is, if

$$\mathbf{u} = \overrightarrow{AB} \quad \text{and} \quad \mathbf{v} = \overrightarrow{BC}, \qquad \text{then} \quad \mathbf{u} + \mathbf{v} = \overrightarrow{AB} + \overrightarrow{BC} = \overrightarrow{AC}.$$

We can also describe vector addition by placing the initial points of the vectors together, completing a parallelogram, and finding the diagonal of the parallelogram. (See the figure on the left below.) This description of addition is sometimes called the **parallelogram law** of vector addition because **u** + **v** is given by the diagonal of the parallelogram determined by **u** and **v**. Vector addition is **commutative**. As shown in the figure on the right below, both **u** + **v** and **v** + **u** are represented by the same directed line segment.

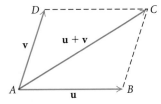

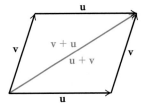

Applications

If two forces F_1 and F_2 act on an object, the *combined* effect is the sum, or resultant, $F_1 + F_2$ of the separate forces.

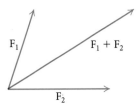

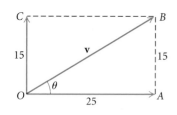

Example 2 Forces of 15 newtons and 25 newtons act on an object at right angles to each other. Find their sum, or resultant, giving the magnitude of the resultant and the angle that it makes with the larger force.

Solution We make a drawing—this time, a rectangle—using **v** or $\overrightarrow{OB}$ to represent the resultant. To find the magnitude, we use the Pythagorean theorem:

$$|\mathbf{v}|^2 = 15^2 + 25^2 \qquad \text{Here } |\mathbf{v}| \text{ denotes the length, or magnitude, of v.}$$
$$|\mathbf{v}| = \sqrt{15^2 + 25^2}$$
$$|\mathbf{v}| \approx 29.2.$$

To find the direction, we note that since OAB is a right triangle,

$$\tan \theta = \tfrac{15}{25} = 0.6.$$

From a grapher, we find θ, the angle that the resultant makes with the larger force:

$$\theta = \tan^{-1}(0.6) \approx 31°.$$

The resultant $\overrightarrow{OB}$ has a magnitude of 29.2 and makes an angle of 31° with the larger force. ▬

Pilots must adjust the direction of their flight when there is a crosswind. Both the wind and the aircraft velocities can be described by vectors.

Example 3 *Airplane Speed and Direction.* An airplane travels on a bearing of 100° at an airspeed of 190 km/h while a wind is blowing 48 km/h from 220°. (For a review of aerial bearings, see p. 151.) Find the ground speed of the airplane and the direction of its track, or course, over the ground.

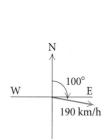

Airplane airspeed

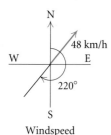

Windspeed

Solution We first make a drawing. The wind is represented by $\overrightarrow{OC}$ and the velocity vector of the airplane by $\overrightarrow{OA}$. The resultant velocity vector is **v**, the sum of the two vectors. The angle θ between **v** and $\overrightarrow{OA}$ is called a **drift angle**.

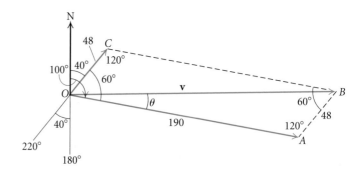

Note that the measure of $\angle COA = 100° - 40° = 60°$. Thus the measure of $\angle CBA$ is also 60° (opposite angles of a parallelogram are equal). Since the sum of all the angles of the parallelogram is 360° and $\angle OCB$ and $\angle OAB$ have the same measure, each must be 120°. By the *law of cosines* in $\triangle OAB$, we have

$$|\mathbf{v}|^2 = 48^2 + 190^2 - 2 \cdot 48 \cdot 190 \cos 120°$$
$$|\mathbf{v}|^2 = 47{,}524$$
$$|\mathbf{v}| = 218.$$

Thus, $|\mathbf{v}|$ is 218 km/h. By the *law of sines* in the same triangle,

$$\frac{48}{\sin \theta} = \frac{218}{\sin 120°},$$

or

$$\sin \theta = \frac{48 \sin 120°}{218}$$
$$\approx 0.1907. \qquad \text{Note: } \sin^{-1} 0.1907 \approx 11°$$

Thus, $\theta = 11°$, to the nearest degree. The ground speed of the airplane is 218 km/h, and its track is in the direction of $100° - 11°$, or 89°. ▬

Components

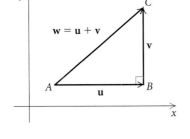

Given a vector $\mathbf{w}$, we may want to find two other vectors $\mathbf{u}$ and $\mathbf{v}$ whose sum is $\mathbf{w}$. The vectors $\mathbf{u}$ and $\mathbf{v}$ are called **components** of $\mathbf{w}$ and the process of finding them is called **resolving**, or **representing**, a vector into its vector components.

When we resolve a vector, we generally look for perpendicular components. Most often, one component will be parallel to the x-axis and the other will be parallel to the y-axis. For this reason, they are often called the **horizontal** and **vertical** components of a vector. In the figure at left, the vector $\mathbf{w} = \overrightarrow{AC}$ is resolved as the sum of $\mathbf{u} = \overrightarrow{AB}$ and $\mathbf{v} = \overrightarrow{BC}$. The horizontal component of $\mathbf{w}$ is $\mathbf{u}$ and the vertical component is $\mathbf{v}$.

Example 4 A certain vector $\mathbf{w}$ has a magnitude of 130 and is inclined 40° with the horizontal. Resolve the vector into horizontal and vertical components.

SOLUTION We first make a drawing showing horizontal and vertical vectors $\mathbf{u}$ and $\mathbf{v}$ whose sum is $\mathbf{w}$.

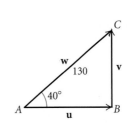

From △ABC, we find |**u**| and |**v**| using the definitions of the cosine and sine functions:

$$\cos 40° = \frac{|\mathbf{u}|}{130}, \quad \text{or} \quad |\mathbf{u}| = 130 \cos 40° \approx 100,$$

$$\sin 40° = \frac{|\mathbf{v}|}{130}, \quad \text{or} \quad |\mathbf{v}| = 130 \sin 40° \approx 84.$$

Thus the horizontal component of **w** is 100 right, and the vertical component of **w** is 84 up. ▬

Example 5 *Shipping Crate.* A wooden shipping crate that weighs 816 lb is placed on a loading ramp that makes an angle of 25° with the horizontal. To keep the crate from sliding, a chain is hooked to the crate and to a pole at the top of the ramp. Find the magnitude of the components of the crate's weight (disregarding friction) perpendicular and parallel to the incline.

SOLUTION   We first make a drawing illustrating the forces with a rectangle. We let

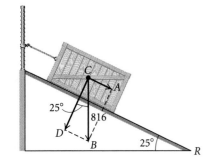

$|\overrightarrow{CB}|$ = the weight of the crate = 816 lb (force of gravity),

$|\overrightarrow{CD}|$ = the magnitude of the component of the crate's weight perpendicular to the incline (force against the ramp), and

$|\overrightarrow{CA}|$ = the magnitude of the component of the crate's weight parallel to the incline (force that pulls the crate down the ramp).

The angle at R is given to be 25° and ∠BCD = ∠R = 25° because the sides of these angles are, respectively, perpendicular. Using the cosine and sine functions, we find

$$\cos 25° = \frac{|\overrightarrow{CD}|}{816}, \quad \text{or} \quad |\overrightarrow{CD}| = 816 \cos 25° \approx 740 \text{ lb}, \quad \text{and}$$

$$\sin 25° = \frac{|\overrightarrow{CA}|}{816}, \quad \text{or} \quad |\overrightarrow{CA}| = 816 \sin 25° \approx 345 \text{ lb}.$$

▬

4.5 | Exercise Set

Sketch the pair of vectors and determine whether they are equivalent. Use the following ordered pairs for the initial and terminal points.

A(−2, 2)   E(−4, 1)   I(−6, −3)
B(3, 4)   F(2, 1)   J(3, 1)
C(−2, 5)   G(−4, 4)   K(−3, −3)
D(−1, −1) H(1, 2)   O(0, 0)

1. $\overrightarrow{GE}$, $\overrightarrow{BJ}$ 2. $\overrightarrow{DJ}$, $\overrightarrow{OF}$

3. $\overrightarrow{DJ}$, $\overrightarrow{AB}$ 4. $\overrightarrow{CG}$, $\overrightarrow{FO}$

5. $\overrightarrow{DK}$, $\overrightarrow{BH}$ 6. $\overrightarrow{BA}$, $\overrightarrow{DI}$

7. $\overrightarrow{EG}$, $\overrightarrow{BJ}$ 8. $\overrightarrow{GC}$, $\overrightarrow{FO}$

9. $\overrightarrow{GA}$, $\overrightarrow{BH}$ 10. $\overrightarrow{JD}$, $\overrightarrow{CG}$

11. $\overrightarrow{AB}$, $\overrightarrow{ID}$ 12. $\overrightarrow{OF}$, $\overrightarrow{HB}$

13.   Two forces of 32 N (newtons) and 45 N act on an object at right angles. Find the magnitude of the resultant and the angle that it makes with the smaller force.

14. Two forces of 50 N and 60 N act on an object at right angles. Find the magnitude of the resultant and the angle that it makes with the larger force.

15. Two forces of 410 N and 600 N act on an object. The angle between the forces is 47°. Find the magnitude of the resultant and the angle that it makes with the larger force.

16. Two forces of 255 N and 325 N act on an object. The angle between the forces is 64°. Find the magnitude of the resultant and the angle that it makes with the smaller force.

In Exercises 17–24, magnitudes of vectors **u** *and* **v** *and the angle* θ *between the vectors are given. Find the sum of* **u** + **v**. *Give the magnitude to the nearest tenth and give the direction by specifying to the nearest degree the angle that the resultant makes with* **u**.

17. $|\mathbf{u}| = 45$, $|\mathbf{v}| = 35$, $\theta = 90°$
18. $|\mathbf{u}| = 54$, $|\mathbf{v}| = 43$, $\theta = 150°$
19. $|\mathbf{u}| = 10$, $|\mathbf{v}| = 12$, $\theta = 67°$
20. $|\mathbf{u}| = 25$, $|\mathbf{v}| = 30$, $\theta = 75°$
21. $|\mathbf{u}| = 20$, $|\mathbf{v}| = 20$, $\theta = 117°$
22. $|\mathbf{u}| = 30$, $|\mathbf{v}| = 30$, $\theta = 123°$
23. $|\mathbf{u}| = 23$, $|\mathbf{v}| = 47$, $\theta = 27°$
24. $|\mathbf{u}| = 32$, $|\mathbf{v}| = 74$, $\theta = 72°$

25. *Hot-air Balloon.* A hot-air balloon is rising vertically 10 ft/sec while the wind is blowing horizontally 5 ft/sec. Find the speed of the balloon and the angle that it makes with the horizontal.

26. *Boat.* A boat heads 35°, propelled by a force of 750 lb. A wind from 320° exerts a force of 150 lb on the boat. How large is the resultant force, and in what direction is the boat moving?

27. *Ship.* A ship sails first N80°E for 120 nautical mi, and then S20°W for 200 nautical mi. How far is the ship, then, from the starting point, and in what direction?

28. *Airplane.* An airplane flies 032° for 210 km, and then 280° for 170 km. How far is the airplane, then, from the starting point, and in what direction?

29. *Airplane.* An airplane has an airspeed of 150 km/h. It is to make a flight in a direction of 070° while there is a 25-km/h wind from 340°. What will the airplane's actual heading be?

30. *Wind.* A wind has an easterly component (*from* the east) of 10 km/h and a southerly component (*from* the south) of 16 km/h. Find the magnitude and the direction of the wind.

31. A vector **w** of magnitude 100 points southeast. Resolve the vector into easterly and southerly components.

32. A vector **u** with a magnitude of 150 lb is inclined to the right and upward 52° from the horizontal. Resolve the vector into components.

33. *Airplane.* An airplane takes off at a speed **S** of 225 mph at an angle of 17° with the horizontal. Resolve the vector **S** into components.

34. *Wheelbarrow.* A wheelbarrow is pushed by applying a 97-lb force **F** that makes a 38° angle with the horizontal. Resolve **F** into its horizontal and vertical components. (The horizontal component is the effective force in the direction of motion and the vertical component adds weight to the wheelbarrow.)

35. *Luggage Wagon.* A luggage wagon is being pulled with vector force **V**, which has a magnitude of 780 lb at an angle of elevation of 60°. Resolve the vector **V** into components.

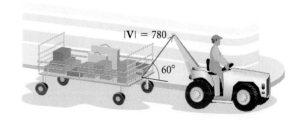

36. *Hot-air Balloon.* A hot-air balloon exerts a 1200-lb pull on a tether line at a 45° angle with the horizontal. Resolve the vector **B** into components.

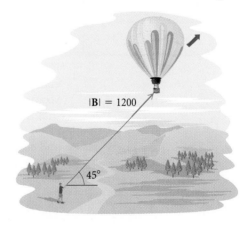

37. *Airplane.*   An airplane is flying at 200 km/h in a direction of 305°. Find the westerly and northerly components of its velocity.

38. *Baseball.*   A baseball player throws a baseball with a speed **S** of 72 mph at an angle of 45° with the horizontal. Resolve the vector **S** into components.

39. A block weighing 100 lb rests on a 25° incline. Find the magnitude of the components of the block's weight perpendicular and parallel to the incline.

40. A shipping crate that weighs 450 kg is placed on a loading ramp that makes an angle of 30° with the horizontal. Find the magnitude of the components of the crate's weight perpendicular and parallel to the incline.

41. An 80-lb block of ice rests on a 37° incline. What force parallel to the incline is necessary in order to keep the ice from sliding down?

42. What force is necessary to pull a 3500-lb truck up a 9° incline?

Skill Maintenance

Find the function value using coordinates of points on the unit circle.

43. $\sin \dfrac{2\pi}{4}$

44. $\cos \dfrac{\pi}{6}$

45. $\cos \dfrac{\pi}{4}$

46. $\sin \dfrac{5\pi}{6}$

Synthesis

47. ◆ Describe the concept of a vector as though you were explaining it to a classmate. What might an arrow shot from a bow have to do with the explanation?

48. ◆ Explain why vectors $\overrightarrow{QR}$ and $\overrightarrow{RQ}$ are not equivalent.

49. *Eagle's Flight.*   An eagle flies from its nest 7 mi in the direction northeast, where it stops to rest on a cliff. It then flies 8 mi in the direction S30°W to land on top of a tree. Place an *xy*-coordinate system so that the origin is the bird's nest, the *x*-axis points east, and the *y*-axis points north.

a) At what point is the cliff located?
b) At what point is the tree?

4.6
Vector Operations

- *Perform calculations with vectors in component form.*
- *Express a vector as a linear combination of unit vectors.*
- *Express a vector in terms of its magnitude and its direction.*
- *Find the angle between two vectors using the dot product.*
- *Solve applications involving forces in equilibrium.*

Position Vectors

Let's consider a vector **v** whose initial point is the *origin* in an *xy*-coordinate system with the terminal point at (a, b). We say that the vector is in **standard position** and refer to it as a position vector. Note that the ordered pair (a, b) defines the vector uniquely. Thus we can use (a, b) to denote the vector. To emphasize that we are thinking of a vector and to avoid the confusion of notation with ordered-pair and interval notation, we generally write

$$\mathbf{v} = \langle a, b \rangle.$$

The coordinate a is the *scalar* **horizontal component** of the vector, and the coordinate b is the *scalar* **vertical component** of the vector. Thus,

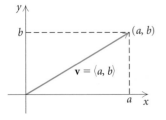

⟨*a*, *b*⟩ is considered to be the *component form* of **v**. Note that *a* and *b* are *not* vectors and should not be confused with the vector component definition given in Section 4.5.

Now consider $\overrightarrow{AC}$ with $A = (x_1, y_1)$ and $C = (x_2, y_2)$. Let's see how to find the position vector equivalent to $\overrightarrow{AC}$. As you can see in the figure below, the initial point A is relocated to the origin $(0, 0)$. The coordinates of P are found by subtracting the coordinates of A from the coordinates of C. Thus, $P = (x_2 - x_1, y_2 - y_1)$ and the position vector is $\overrightarrow{OP}$.

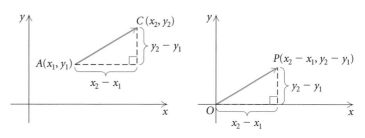

It can be shown that $\overrightarrow{OP}$ and $\overrightarrow{AC}$ have the same magnitude and direction and are therefore equivalent. Thus, $\overrightarrow{AC} = \overrightarrow{OP} = \langle x_2 - x_1, y_2 - y_1 \rangle$.

> ### Component Form of a Vector
>
> The *component form* of $\overrightarrow{AC}$ with $A = (x_1, y_1)$ and $C = (x_2, y_2)$ is
>
> $$\overrightarrow{AC} = \langle x_2 - x_1, y_2 - y_1 \rangle.$$

Example 1 Find the component form of $\overrightarrow{CF}$ if $C = (-4, -3)$ and $F = (1, 5)$.

SOLUTION We have

$$\overrightarrow{CF} = \langle 1 - (-4), 5 - (-3) \rangle = \langle 5, 8 \rangle.$$

Note that vector $\overrightarrow{CF}$ is equivalent to *position vector* $\overrightarrow{OP}$ with $P = (5, 8)$ as shown in the figure at left. ▬

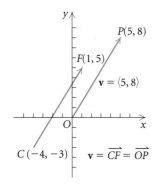

Now that we know how to write vectors in component form, let's restate some definitions that we first considered in Section 4.5.

The length of a vector **v** is easy to determine when the components of the vector are known. For $\mathbf{v} = \langle v_1, v_2 \rangle$, we have

$$|\mathbf{v}|^2 = v_1^2 + v_2^2 \qquad \text{Using the Pythagorean theorem}$$
$$|\mathbf{v}| = \sqrt{v_1^2 + v_2^2}.$$

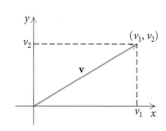

Length of a Vector

The *length*, or *magnitude*, of a vector $\mathbf{v} = \langle v_1, v_2 \rangle$ is given by

$$|\mathbf{v}| = \sqrt{v_1^2 + v_2^2}.$$

Two vectors are **equivalent** if they have the *same* magnitude and the *same* direction.

Equivalent Vectors

Let $\mathbf{u} = \langle u_1, u_2 \rangle$ and $\mathbf{v} = \langle v_1, v_2 \rangle$. Then

$$\langle u_1, u_2 \rangle = \langle v_1, v_2 \rangle \quad \text{if and only if} \quad u_1 = v_1 \quad \text{and} \quad u_2 = v_2.$$

Operations on Vectors

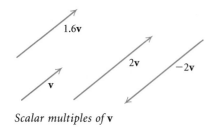

Scalar multiples of **v**

To multiply a vector **v** by a positive real number, we multiply its length by the number. Its direction stays the same. To multiply **v** by 2, we double its length and maintain its direction. To multiply a vector by 1.6, we increase its length by 60% and keep its direction. To multiply a vector by -2, we multiply by 2 and then reverse the direction. Since real numbers work like scaling factors in vector multiplication, we call them **scalars** and the products $k\mathbf{v}$ are called **scalar mutliples** of **v**.

Scalar Multiplication

For a real number k and a vector $\mathbf{v} = \langle v_1, v_2 \rangle$, the *scalar product* of k and **v** is

$$k\mathbf{v} = k\langle v_1, v_2 \rangle = \langle kv_1, kv_2 \rangle.$$

The vector $k\mathbf{v}$ is a *scalar multiple* of the vector **v**.

Example 2   Let $\mathbf{u} = \langle -5, 4 \rangle$ and $\mathbf{w} = \langle 1, -1 \rangle$. Find $-7\mathbf{w}$, $3\mathbf{u}$, and $-1\mathbf{w}$.

SOLUTION

$$-7\mathbf{w} = -7\langle 1, -1 \rangle = \langle -7, 7 \rangle,$$
$$3\mathbf{u} = 3\langle -5, 4 \rangle = \langle -15, 12 \rangle,$$
$$-1\mathbf{w} = -1\langle 1, -1 \rangle = \langle -1, 1 \rangle.$$

In Section 4.5, we used the parallelogram law to add two vectors, but now we can add two vectors using components. To add two vectors given in component form, we add the corresponding components. Let $\mathbf{u} = \langle u_1, u_2 \rangle$ and $\mathbf{v} = \langle v_1, v_2 \rangle$. Then

$$\mathbf{u} + \mathbf{v} = \langle u_1 + v_1, u_2 + v_2 \rangle.$$

For example, if $\mathbf{v} = \langle -3, 2 \rangle$ and $\mathbf{w} = \langle 5, -9 \rangle$, then

$$\mathbf{v} + \mathbf{w} = \langle -3 + 5, 2 + (-9) \rangle = \langle 2, -7 \rangle.$$

Vector Addition

If $\mathbf{u} = \langle u_1, u_2 \rangle$ and $\mathbf{v} = \langle v_1, v_2 \rangle$, then

$$\mathbf{u} + \mathbf{v} = \langle u_1 + v_1, u_2 + v_2 \rangle.$$

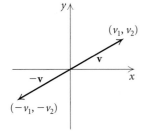

Before we define vector subtraction, we need to define two more vectors: $-\mathbf{v}$ and $\mathbf{u} - \mathbf{v}$. The negative of $\mathbf{v} = \langle v_1, v_2 \rangle$, shown at left, is

$$-\mathbf{v} = (-1)\mathbf{v} = (-1)\langle v_1, v_2 \rangle = \langle -v_1, -v_2 \rangle.$$

Vector subtraction such as $\mathbf{u} - \mathbf{v}$ involves subtracting corresponding components. We show this by rewriting $\mathbf{u} - \mathbf{v}$ as $\mathbf{u} + (-\mathbf{v})$, as shown below. If $\mathbf{u} = \langle u_1, u_2 \rangle$ and $\mathbf{v} = \langle v_1, v_2 \rangle$, then

$$\begin{aligned}
\mathbf{u} - \mathbf{v} = \mathbf{u} + (-\mathbf{v}) &= \langle u_1, u_2 \rangle + \langle -v_1, -v_2 \rangle \\
&= \langle u_1 + (-v_1), u_2 + (-v_2) \rangle \\
&= \langle u_1 - v_1, u_2 - v_2 \rangle.
\end{aligned}$$

We can illustrate vector subtraction with parallelograms, just as we did vector addition.

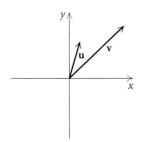

Sketch $\mathbf{u}$ and $\mathbf{v}$.

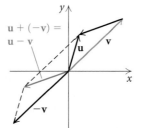

Sketch $-\mathbf{v}$.

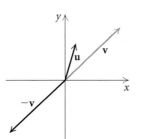

Sketch $\mathbf{u} + (-\mathbf{v})$, or $\mathbf{u} - \mathbf{v}$, using the parallelogram law.

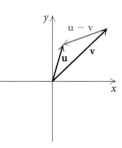

$\mathbf{u} - \mathbf{v}$ is the vector from the terminal point of $\mathbf{v}$ to the terminal point of $\mathbf{u}$.

> **Vector Subtraction**
>
> If $\mathbf{u} = \langle u_1, u_2 \rangle$ and $\mathbf{v} = \langle v_1, v_2 \rangle$, then
>
> $$\mathbf{u} - \mathbf{v} = \langle u_1 - v_1, u_2 - v_2 \rangle.$$

It is interesting to compare the sum of two vectors with the difference of the same two vectors in the same parallelogram. The vectors $\mathbf{u} + \mathbf{v}$ and $\mathbf{u} - \mathbf{v}$ are the diagonals of the parallelogram.

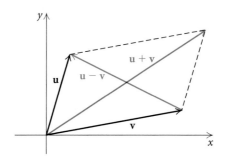

Example 3 Do the following calculations where $\mathbf{u} = \langle 7, 2 \rangle$ and $\mathbf{v} = \langle -3, 5 \rangle$.

a) $\mathbf{u} + \mathbf{v}$ b) $\mathbf{u} - 6\mathbf{v}$

c) $3\mathbf{u} + 4\mathbf{v}$ d) $|5\mathbf{v} - 2\mathbf{u}|$

SOLUTION

a) $\mathbf{u} + \mathbf{v} = \langle 7, 2 \rangle + \langle -3, 5 \rangle = \langle 4, 7 \rangle$

b) $\mathbf{u} - 6\mathbf{v} = \langle 7, 2 \rangle - 6\langle -3, 5 \rangle = \langle 7, 2 \rangle - \langle -18, 30 \rangle = \langle 25, -28 \rangle$

c) $3\mathbf{u} + 4\mathbf{v} = 3\langle 7, 2 \rangle + 4\langle -3, 5 \rangle = \langle 21, 6 \rangle + \langle -12, 20 \rangle = \langle 9, 26 \rangle$

d) $|5\mathbf{v} - 2\mathbf{u}| = |5\langle -3, 5 \rangle - 2\langle 7, 2 \rangle| = |\langle -15, 25 \rangle - \langle 14, 4 \rangle|$

$$= |\langle -29, 21 \rangle|$$
$$= \sqrt{(-29)^2 + 21^2}$$
$$= \sqrt{1282}$$
$$\approx 35.8$$

The magnitude of the vector $5\mathbf{v} - 2\mathbf{u}$ is about 35.8.   ▬

Before we state the properties of vector addition and scalar multiplication, we need to define another special vector—the zero vector. The vector whose initial and terminal points are both $(0, 0)$ is the **zero vector,** denoted by $\mathbf{O} = \langle 0, 0 \rangle$. Its magnitude is 0. In vector addition, the zero vector is the additive identity vector:

$$\mathbf{v} + \mathbf{O} = \mathbf{v}. \qquad \langle v_1, v_2 \rangle + \langle 0, 0 \rangle = \langle v_1, v_2 \rangle$$

Operations on vectors share many of the same properties as operations on real numbers.

Properties of Vector Addition and Scalar Multiplication
For all vectors **u**, **v**, and **w**, and for all scalars b and c:

1. $\mathbf{u} + \mathbf{v} = \mathbf{v} + \mathbf{u}$.
2. $\mathbf{u} + (\mathbf{v} + \mathbf{w}) = (\mathbf{u} + \mathbf{v}) + \mathbf{w}$.
3. $\mathbf{v} + \mathbf{O} = \mathbf{v}$.
4. $1\mathbf{v} = \mathbf{v}; \quad 0\mathbf{v} = \mathbf{O}$.
5. $\mathbf{v} + (-\mathbf{v}) = \mathbf{O}$.
6. $b(c\mathbf{v}) = (bc)\mathbf{v}$.
7. $(b + c)\mathbf{v} = b\mathbf{v} + c\mathbf{v}$.
8. $b(\mathbf{u} + \mathbf{v}) = b\mathbf{u} + b\mathbf{v}$.

Unit Vectors

A vector of magnitude, or length, 1 is called a **unit vector.** The vector
$|\mathbf{v}| = \left\langle -\frac{3}{5}, \frac{4}{5} \right\rangle$ is a unit vector because

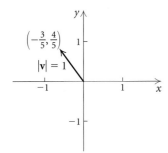

$$|\mathbf{v}| = \left| \left\langle -\tfrac{3}{5}, \tfrac{4}{5} \right\rangle \right| = \sqrt{\left(-\tfrac{3}{5}\right)^2 + \left(\tfrac{4}{5}\right)^2}$$

$$= \sqrt{\tfrac{9}{25} + \tfrac{16}{25}} = \sqrt{\tfrac{25}{25}}$$

$$= \sqrt{1} = 1.$$

Example 4 Find a unit vector that has the same direction as the vector
$\mathbf{w} = \langle -3, 5 \rangle$.

SOLUTION We first find the length of **w**:

$$|\mathbf{w}| = \sqrt{(-3)^2 + 5^2} = \sqrt{34}.$$

Thus we want a vector whose length is $1/\sqrt{34}$ of **w** and whose direction
is the same as vector **w**. That vector is

$$\mathbf{u} = \frac{1}{\sqrt{34}}\mathbf{w} = \frac{1}{\sqrt{34}}\langle -3, 5 \rangle = \left\langle \frac{-3}{\sqrt{34}}, \frac{5}{\sqrt{34}} \right\rangle.$$

The vector **u** is a unit vector because

$$|\mathbf{u}| = \left| \frac{1}{\sqrt{34}}\mathbf{w} \right| = \sqrt{\left(\frac{-3}{\sqrt{34}}\right)^2 + \left(\frac{5}{\sqrt{34}}\right)^2} = \sqrt{\frac{9}{34} + \frac{25}{34}}$$

$$= \sqrt{\frac{34}{34}} = \sqrt{1} = 1.$$

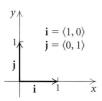

Unit Vector

If **v** is a vector and $\mathbf{v} \neq \mathbf{O}$, then

$$\frac{1}{|\mathbf{v}|} \cdot \mathbf{v}, \quad \text{or} \quad \frac{\mathbf{v}}{|\mathbf{v}|},$$

is a *unit vector* in the direction of **v**.

Although unit vectors can have any direction, the unit vectors parallel to the *x*- and *y*-axes are the most useful. They are defined as

$$\mathbf{i} = \langle 1, 0 \rangle \quad \text{and} \quad \mathbf{j} = \langle 0, 1 \rangle.$$

Any vector can be expressed as a **linear combination** of unit vectors **i** and **j**. For example, let $\mathbf{v} = \langle v_1, v_2 \rangle$. Then

$$\begin{aligned} \mathbf{v} = \langle v_1, v_2 \rangle &= \langle v_1, 0 \rangle + \langle 0, v_2 \rangle \\ &= v_1 \langle 1, 0 \rangle + v_2 \langle 0, 1 \rangle \\ &= v_1 \mathbf{i} + v_2 \mathbf{j}. \end{aligned}$$

Example 5 Express the vector $\mathbf{r} = \langle 2, -6 \rangle$ as a linear combination of **i** and **j**.

SOLUTION

$$\mathbf{r} = \langle 2, -6 \rangle = 2\mathbf{i} + (-6)\mathbf{j} = 2\mathbf{i} - 6\mathbf{j} \qquad \blacksquare$$

Example 6 Write the vector $\mathbf{q} = -\mathbf{i} + 7\mathbf{j}$ in component form.

SOLUTION

$$\mathbf{q} = -\mathbf{i} + 7\mathbf{j} = -1\mathbf{i} + 7\mathbf{j} = \langle -1, 7 \rangle \qquad \blacksquare$$

Vector operations can also be performed when vectors are written as linear combinations of **i** and **j**.

Example 7 If $\mathbf{a} = 5\mathbf{i} - 2\mathbf{j}$ and $\mathbf{b} = -\mathbf{i} + 8\mathbf{j}$, find $3\mathbf{a} - \mathbf{b}$.

SOLUTION

$$\begin{aligned} 3\mathbf{a} - \mathbf{b} &= 3(5\mathbf{i} - 2\mathbf{j}) - (-\mathbf{i} + 8\mathbf{j}) \\ &= 15\mathbf{i} - 6\mathbf{j} + \mathbf{i} - 8\mathbf{j} \\ &= 16\mathbf{i} - 14\mathbf{j} \end{aligned} \qquad \blacksquare$$

Direction Angles

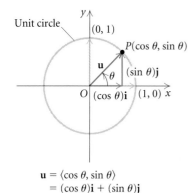

The terminal point *P* of a unit vector in standard position is a point on the unit circle denoted by $(\cos \theta, \sin \theta)$. (It might helpful to review Section 2.5.) Thus the unit vector can be expressed in component form,

$$\mathbf{u} = \langle \cos \theta, \sin \theta \rangle,$$

or as a linear combination of the unit vectors **i** and **j**,

$$\mathbf{u} = (\cos \theta)\mathbf{i} + (\sin \theta)\mathbf{j},$$

where the components of **u** are functions of the **direction angle** θ measured counterclockwise from the x-axis to the vector. As θ varies from 0 to 2π, the point P traces the circle $x^2 + y^2 = 1$. This takes in all possible directions for unit vectors so the equation $\mathbf{u} = (\cos \theta)\mathbf{i} + (\sin \theta)\mathbf{j}$ describes every possible unit vector in the plane.

Example 8 Calculate and sketch the unit vector $\mathbf{u} = (\cos \theta)\mathbf{i} + (\sin \theta)\mathbf{j}$ for $\theta = 2\pi/3$. Include the unit circle in your sketch.

SOLUTION

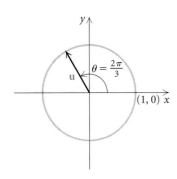

$$\mathbf{u} = \left(\cos \frac{2\pi}{3}\right)\mathbf{i} + \left(\sin \frac{2\pi}{3}\right)\mathbf{j}$$

$$= \left(-\frac{1}{2}\right)\mathbf{i} + \left(\frac{\sqrt{3}}{2}\right)\mathbf{j}$$

Let $\mathbf{v} = \langle v_1, v_2 \rangle$ with direction angle θ. Using the definition of the tangent function, we can determine the direction angle from the components of **v**:

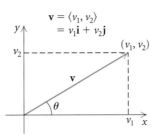

$$\tan \theta = \frac{v_2}{v_1}$$

$$\theta = \tan^{-1} \frac{v_2}{v_1}.$$

Example 9 Determine the direction angle θ of the vector $\mathbf{w} = -4\mathbf{i} - 3\mathbf{j}$.

SOLUTION   We know that

$$\mathbf{w} = -4\mathbf{i} - 3\mathbf{j} = \langle -4, -3 \rangle.$$

Thus we have

$$\tan \theta = \frac{-3}{-4} = \frac{3}{4} \quad \text{and} \quad \theta = \tan^{-1} \frac{3}{4}.$$

Since **w** is in the third quadrant, we know that θ is a third-quadrant angle. Thus the reference angle is

$$\tan^{-1} \frac{3}{4} \approx 37°,$$

and $\theta \approx 180° + 37°$, or $217°$.

It is convenient for work with applications and in subsequent courses, such as calculus, to have a way to express a vector so that both its magnitude and its direction can be determined, or read, easily. Let **v** be a vector with the same direction angle as a unit vector **u**. Then $\mathbf{v}/|\mathbf{v}|$, a unit vector in the same direction as **v**, can be substituted for **u** in the equation $\mathbf{u} = (\cos\theta)\mathbf{i} + (\sin\theta)\mathbf{j}$. Thus we have

$$\frac{\mathbf{v}}{|\mathbf{v}|} = (\cos\theta)\mathbf{i} + (\sin\theta)\mathbf{j}$$

$$\mathbf{v} = |\mathbf{v}|[(\cos\theta)\mathbf{i} + (\sin\theta)\mathbf{j}] \qquad \text{Multiplying by } |\mathbf{v}|$$

$$\mathbf{v} = |\mathbf{v}|(\cos\theta)\mathbf{i} + |\mathbf{v}|(\sin\theta)\mathbf{j}.$$

Let's revisit the application in Example 3 of Section 4.5.

Example 10 *Airplane Speed and Direction.* **An airplane travels on a bearing of 100° at an airspeed of 190 km/h while a wind is blowing 48 km/h from 220°. Find the ground speed of the airplane and the direction of its track, or course, over the ground.**

SOLUTION   We first make a drawing. The wind is represented by $\overrightarrow{OC}$ and the velocity vector of the airplane by $\overrightarrow{OA}$. The resultant velocity vector is **v**, the sum of the two vectors:

$$\mathbf{v} = \overrightarrow{OC} + \overrightarrow{OA}.$$

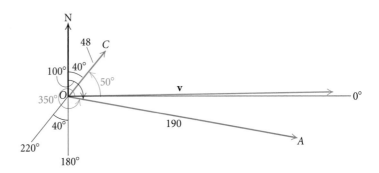

The bearing (measured from north) of the airspeed vector $\overrightarrow{OA}$ is 100°. Its *direction angle* (measured counterclockwise from the positive *x*-axis) is 350°. The bearing (measured from north) of the wind vector $\overrightarrow{OC}$ is 220°. Its direction angle (measured counterclockwise from the positive *x*-axis) is 50°. The magnitudes of $\overrightarrow{OA}$ and $\overrightarrow{OC}$ are 190 and 48, respectively. We have

$$\overrightarrow{OA} = 190(\cos 350°)\mathbf{i} + 190(\sin 350°)\mathbf{j}, \quad \text{and}$$
$$\overrightarrow{OC} = 48(\cos 50°)\mathbf{i} + 48(\sin 50°)\mathbf{j}.$$

Thus,

$$\mathbf{v} = \overrightarrow{OA} + \overrightarrow{OC}$$
$$= [190(\cos 350°)\mathbf{i} + 190(\sin 350°)\mathbf{j}] + [48(\cos 50°)\mathbf{i} + 48(\sin 50°)\mathbf{j}]$$
$$= [190(\cos 350°) + 48(\cos 50°)]\mathbf{i} + [190(\sin 350°) + 48(\sin 50°)]\mathbf{j}$$
$$\approx 217.97\mathbf{i} + 3.78\mathbf{j}.$$

From this form, we can determine the ground speed and the course:

$$\text{Ground speed} \approx \sqrt{(217.97)^2 + (3.78)^2} \approx 218 \text{ km/h.}$$

We let α be the direction angle of **v**. Then

$$\tan \alpha = \frac{3.78}{217.97}$$

$$\alpha = \tan^{-1} \frac{3.78}{217.97} \approx 1°.$$

Thus the course of the airplane (the direction from north) is $90° - 1°$, or $89°$.

Angle Between Vectors

When a vector is multiplied by a scalar, the result is a vector. When two vectors are added, the result is also a vector. Thus we might expect the product of two vectors to be a vector as well, but it is not. The *dot product* of two vectors is a real number, or scalar. This product is useful in finding the angle between two vectors and in determining whether two vectors are perpendicular.

Dot Product

The *dot product* of two vectors $\mathbf{u} = \langle u_1, u_2 \rangle$ and $\mathbf{v} = \langle v_1, v_2 \rangle$ is

$$\mathbf{u} \cdot \mathbf{v} = u_1 v_1 + u_2 v_2.$$

(Note that $u_1 v_1 + u_2 v_2$ is a *scalar*, not a vector.)

Example 11 Find the indicated dot product if

$$\mathbf{u} = \langle 2, -5 \rangle, \quad \mathbf{v} = \langle 0, 4 \rangle, \quad \text{and} \quad \mathbf{w} = \langle -3, 1 \rangle.$$

a) $\mathbf{u} \cdot \mathbf{w}$ **b)** $\mathbf{w} \cdot \mathbf{v}$

SOLUTION

a) $\mathbf{u} \cdot \mathbf{w} = 2(-3) + (-5)1 = -6 - 5 = -11$
b) $\mathbf{w} \cdot \mathbf{v} = -3(0) + 1(4) = 0 + 4 = 4$

The dot product can be used to find the angle between two vectors. The angle *between* two vectors is the smallest positive angle formed by the two directed line segments. Thus the angle θ between **u** and **v** is the same angle as between **v** and **u**, and $0 \leq \theta \leq \pi$.

Angle Between Two Vectors

If θ is the angle between two *nonzero* vectors **u** and **v**, then

$$\cos \theta = \frac{\mathbf{u} \cdot \mathbf{v}}{|\mathbf{u}| \, |\mathbf{v}|}.$$

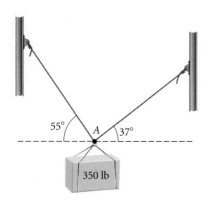

Example 12 Find the angle between $\mathbf{u} = \langle 3, 7 \rangle$ and $\mathbf{v} = \langle -4, 2 \rangle$.

SOLUTION We begin by finding $\mathbf{u} \cdot \mathbf{v}$, $|\mathbf{u}|$, and $|\mathbf{v}|$:

$$\mathbf{u} \cdot \mathbf{v} = 3(-4) + 7(2) = 2,$$
$$|\mathbf{u}| = \sqrt{3^2 + 7^2} = \sqrt{58}, \quad \text{and}$$
$$|\mathbf{v}| = \sqrt{(-4)^2 + 2^2} = \sqrt{20}.$$

Then

$$\cos \alpha = \frac{\mathbf{u} \cdot \mathbf{v}}{|\mathbf{u}||\mathbf{v}|} = \frac{2}{\sqrt{58}\,\sqrt{20}}$$

$$\alpha = \cos^{-1} \frac{2}{\sqrt{58}\,\sqrt{20}}$$

$$\alpha \approx 86.6°.$$

Forces in Equilibrium

When several forces act through the same point on an object, their vector sum must be $\mathbf{O}$ in order for a balance to occur. When a balance occurs, then the object is either stationary or moving in a straight line without acceleration. The fact that the vector sum must be $\mathbf{O}$ for a balance, and vice versa, allows us to solve many applied problems involving forces.

Example 13 *Suspended Block.* A 350-lb block is suspended by two cables, as shown at left. At point A, there are three forces acting: $\mathbf{W}$, the block pulling down, and $\mathbf{R}$ and $\mathbf{S}$, the two cables pulling upward and outward, respectively. Find the tension in each cable.

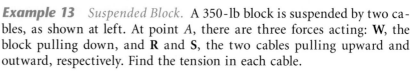

SOLUTION We draw a force diagram with the initial points of each vector at the origin. For there to be a balance, the vector sum must be the vector $\mathbf{O}$:

$$\mathbf{R} + \mathbf{S} + \mathbf{W} = \mathbf{O}.$$

We can express each vector in terms of its magnitude and its direction angle:

$$\mathbf{R} = |\mathbf{R}|[(\cos 125°)\mathbf{i} + (\sin 125°)\mathbf{j}],$$
$$\mathbf{S} = |\mathbf{S}|[(\cos 37°)\mathbf{i} + (\sin 37°)\mathbf{j}], \quad \text{and}$$
$$\mathbf{W} = |\mathbf{W}|[(\cos 270°)\mathbf{i} + (\sin 270°)\mathbf{j}]$$
$$= 350(\cos 270°)\mathbf{i} + 350(\sin 270°)\mathbf{j}$$
$$= -350\mathbf{j}. \qquad \cos 270° = 0; \sin 270° = -1$$

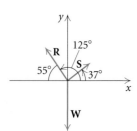

Substituting for $\mathbf{R}$, $\mathbf{S}$, and $\mathbf{W}$ in $\mathbf{R} + \mathbf{S} + \mathbf{W} = \mathbf{O}$, we have

$$[|\mathbf{R}|(\cos 125°) + |\mathbf{S}|(\cos 37°)]\mathbf{i} + [|\mathbf{R}|(\sin 125°) + |\mathbf{S}|(\sin 37°) - 350]\mathbf{j}$$
$$= 0\mathbf{i} + 0\mathbf{j}.$$

This gives us a system of equations:

$$|\mathbf{R}|(\cos 125°) + |\mathbf{S}|(\cos 37°) = 0,$$
$$|\mathbf{R}|(\sin 125°) + |\mathbf{S}|(\sin 37°) = 350.$$

Solving this system, we get

$$|\mathbf{R}| \approx 280 \quad \text{and} \quad |\mathbf{S}| \approx 201.$$

The tensions in the cables are 280 lb and 201 lb.

4.6 | Exercise Set

Find the component form of the vector given the initial and terminal points. Then find the length of the vector.

1. $\overrightarrow{MN}$; $M(6, -7)$, $N(-3, -2)$
2. $\overrightarrow{CD}$; $C(1, 5)$, $D(5, 7)$
3. $\overrightarrow{FE}$; $E(8, 4)$, $F(11, -2)$
4. $\overrightarrow{BA}$; $A(9, 0)$, $B(9, 7)$
5. $\overrightarrow{KL}$; $K(4, -3)$, $L(8, -3)$
6. $\overrightarrow{GH}$; $G(-6, 10)$, $H(-3, 2)$

7. Find the magnitude of vector $\mathbf{u}$ if $\mathbf{u} = \langle -1, 6 \rangle$.
8. Find the magnitude of vector $\overrightarrow{ST}$ if $\overrightarrow{ST} = \langle -12, 5 \rangle$.

Do the indicated calculations in Exercises 9–26 for the vectors

$$\mathbf{u} = \langle 5, -2 \rangle, \quad \mathbf{v} = \langle -4, 7 \rangle, \quad \text{and} \quad \mathbf{w} = \langle -1, -3 \rangle.$$

9. $\mathbf{u} + \mathbf{w}$
10. $\mathbf{w} + \mathbf{u}$
11. $|3\mathbf{w} - \mathbf{v}|$
12. $6\mathbf{v} + 5\mathbf{u}$
13. $\mathbf{v} - \mathbf{u}$
14. $|2\mathbf{w}|$
15. $5\mathbf{u} - 4\mathbf{v}$
16. $-5\mathbf{v}$
17. $|3\mathbf{u}| - |\mathbf{v}|$
18. $|\mathbf{v}| + |\mathbf{u}|$
19. $\mathbf{v} + \mathbf{u} + 2\mathbf{w}$
20. $\mathbf{w} - (\mathbf{u} + 4\mathbf{v})$
21. $2\mathbf{v} + \mathbf{O}$
22. $10|7\mathbf{w} - 3\mathbf{u}|$
23. $\mathbf{u} \cdot \mathbf{w}$
24. $\mathbf{w} \cdot \mathbf{u}$
25. $\mathbf{u} \cdot \mathbf{v}$
26. $\mathbf{v} \cdot \mathbf{w}$

The vectors $\mathbf{u}$, $\mathbf{v}$, and $\mathbf{w}$ are drawn below. Copy them on a sheet of paper. Then sketch each of the vectors in Exercises 27–30.

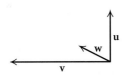

27. $\mathbf{u} + \mathbf{v}$
28. $\mathbf{u} - 2\mathbf{v}$
29. $\mathbf{u} + \mathbf{v} + \mathbf{w}$
30. $\frac{1}{2}\mathbf{u} - \mathbf{w}$

31. Vectors $\mathbf{u}$, $\mathbf{v}$, and $\mathbf{w}$ are determined by the sides of $\triangle ABC$ below.

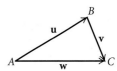

a) Find an expression for $\mathbf{w}$ in terms of $\mathbf{u}$ and $\mathbf{v}$.
b) Find an expression for $\mathbf{v}$ in terms of $\mathbf{u}$ and $\mathbf{w}$.

32. In $\triangle ABC$, vectors $\mathbf{u}$ and $\mathbf{w}$ are determined by the sides shown, where P is the midpoint of side BC. Find an expression for $\mathbf{v}$ in terms of $\mathbf{u}$ and $\mathbf{w}$.

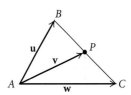

Find a unit vector that has the same direction as the given vector.

33. $\mathbf{v} = \langle -5, 12 \rangle$
34. $\mathbf{u} = \langle 3, 4 \rangle$
35. $\mathbf{w} = \langle 1, -10 \rangle$
36. $\mathbf{a} = \langle 6, -7 \rangle$
37. $\mathbf{r} = \langle -2, -8 \rangle$
38. $\mathbf{t} = \langle -3, -3 \rangle$

Express the vector as a linear combination of the unit vectors $\mathbf{i}$ and $\mathbf{j}$.

39. $\mathbf{w} = \langle -4, 6 \rangle$
40. $\mathbf{r} = \langle -15, 9 \rangle$
41. $\mathbf{s} = \langle 2, 5 \rangle$
42. $\mathbf{u} = \langle 2, -1 \rangle$

Express the vector as a linear combination of $\mathbf{i}$ and $\mathbf{j}$.

43.

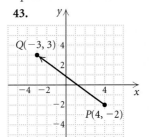

44.

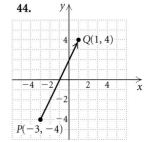

For Exercises 45–48, use the vectors

$$\mathbf{u} = 2\mathbf{i} + \mathbf{j}, \quad \mathbf{v} = -3\mathbf{i} - 10\mathbf{j}, \quad \text{and} \quad \mathbf{w} = \mathbf{i} - 5\mathbf{j}.$$

Perform the indicated vector operations and state the answer in two forms: (a) as a linear combination of $\mathbf{i}$ *and* $\mathbf{j}$ *and (b) in component form.*

45. $4\mathbf{u} - 5\mathbf{w}$ **46.** $\mathbf{v} + 3\mathbf{w}$

47. $\mathbf{u} - (\mathbf{v} + \mathbf{w})$ **48.** $(\mathbf{u} - \mathbf{v}) + \mathbf{w}$

Sketch (include the unit circle) and calculate the unit vector $\mathbf{u} = (\cos \theta)\mathbf{i} + (\sin \theta)\mathbf{j}$ *for the given direction angle.*

49. $\theta = \dfrac{\pi}{2}$ **50.** $\theta = \dfrac{\pi}{3}$

51. $\theta = \dfrac{4\pi}{3}$ **52.** $\theta = \dfrac{3\pi}{2}$

Determine the direction angle θ *of the vector, to the nearest degree.*

53. $\mathbf{u} = \langle -2, -5 \rangle$ **54.** $\mathbf{w} = \langle 4, -3 \rangle$

55. $\mathbf{q} = \mathbf{i} + 2\mathbf{j}$ **56.** $\mathbf{w} = 5\mathbf{i} - \mathbf{j}$

57. $\mathbf{t} = \langle 5, 6 \rangle$ **58.** $\mathbf{b} = \langle -8, -4 \rangle$

Find the magnitude and the direction angle θ *of the vector.*

59. $\mathbf{u} = 3[(\cos 45°)\mathbf{i} + (\sin 45°)\mathbf{j}]$

60. $\mathbf{w} = 6[(\cos 150°)\mathbf{i} + (\sin 150°)\mathbf{j}]$

61. $\mathbf{v} = \left\langle -\dfrac{1}{2}, \dfrac{\sqrt{3}}{2} \right\rangle$

62. $\mathbf{u} = -\mathbf{i} - \mathbf{j}$

Find the angle between the given vectors, to the nearest tenth of a degree.

63. $\mathbf{u} = \langle 2, -5 \rangle, \ \mathbf{v} = \langle 1, 4 \rangle$

64. $\mathbf{a} = \langle -3, -3 \rangle, \ \mathbf{b} = \langle -5, 2 \rangle$

65. $\mathbf{w} = \langle 3, 5 \rangle, \ \mathbf{r} = \langle 5, 5 \rangle$

66. $\mathbf{v} = \langle -4 \ 2 \rangle, \ \mathbf{t} = \langle 1, -4 \rangle$

67. $\mathbf{a} = \mathbf{i} + \mathbf{j}, \ \mathbf{b} = 2\mathbf{i} - 3\mathbf{j}$

68. $\mathbf{u} = 3\mathbf{i} + 2\mathbf{j}, \ \mathbf{v} = -\mathbf{i} + 4\mathbf{j}$

Express each vector in Exercises 69–72 in the form $a\mathbf{i} + b\mathbf{j}$ *and sketch each as an arrow in the coordinate plane.*

69. The unit vectors $\mathbf{u} = (\cos \theta)\mathbf{i} + (\sin \theta)\mathbf{j}$ for $\theta = \pi/6$ and $\theta = 3\pi/4$. Include the unit circle $x^2 + y^2 = 1$ in your sketch.

70. The unit vectors $\mathbf{u} = (\cos \theta)\mathbf{i} + (\sin \theta)\mathbf{j}$ for $\theta = -\pi/4$ and $\theta = -3\pi/4$. Include the unit circle $x^2 + y^2 = 1$ in your sketch.

71. The unit vector obtained by rotating $\mathbf{j}$ counterclockwise $3\pi/4$ radians about the origin

72. The unit vector obtained by rotating $\mathbf{j}$ clockwise $2\pi/3$ radians about the origin

For the vectors in Exercises 73 and 74, find the unit vectors $\mathbf{u} = (\cos \theta)\mathbf{i} + (\sin \theta)\mathbf{j}$ *in the same direction.*

73. $-\mathbf{i} + 3\mathbf{j}$ **74.** $6\mathbf{i} - 8\mathbf{j}$

For the vectors in Exercises 75 and 76, express each vector in terms of its magnitude and its direction.

75. $2\mathbf{i} - 3\mathbf{j}$ **76.** $5\mathbf{i} + 12\mathbf{j}$

77. Use a sketch to show that

$$\mathbf{v} = 3\mathbf{i} - 6\mathbf{j} \quad \text{and} \quad \mathbf{u} = -\mathbf{i} + 2\mathbf{j}$$

have opposite directions.

78. Use a sketch to show that

$$\mathbf{v} = 3\mathbf{i} - 6\mathbf{j} \quad \text{and} \quad \mathbf{u} = \tfrac{1}{2}\mathbf{i} - \mathbf{j}$$

have the same direction.

Exercises 79–82 appeared first in Exercise Set 4.5, where we used the law of cosines and the law of sines to solve the applications. For this exercise set, solve the problem using the vector form

$$\mathbf{v} = |\mathbf{v}|[(\cos \theta)\mathbf{i} + (\sin \theta)\mathbf{j}].$$

79. *Ship.*   A ship sails first N80°E for 120 nautical mi, and then S20°W for 200 nautical mi. How far is the ship, then, from the starting point, and in what direction?

80. *Boat.*   A boat heads 35°, propelled by a force of 750 lb. A wind from 320° exerts a force of 150 lb on the boat. How large is the resultant force, and in what direction is the boat moving?

81. *Airplane.*   An airplane has an airspeed of 150 km/h. It is to make a flight in a direction of 070° while there is a 25-km/h wind from 340°. What will the airplane's actual heading be?

82. *Airplane.*   An airplane flies 032° for 210 mi, and then 280° for 170 mi. How far is the airplane, then, from the starting point, and in what direction?

83. Two cables support a 1000-lb weight, as shown. Find the tension in each cable.

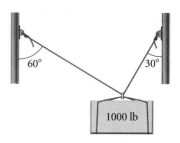

84. A 2500-kg block is suspended by two ropes, as shown. Find the tension in each rope.

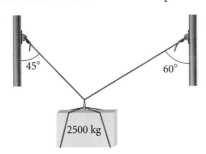

85. A 150-lb sign is hanging from the end of a hinged boom, supported by a cable inclined 42° with the horizontal. Find the tension in the cable and the compression in the boom.

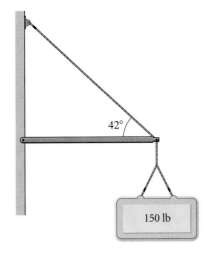

86. A weight of 200 lb is supported by a frame made of two rods and hinged at points *A*, *B*, and *C*. Find the forces exerted by the two rods.

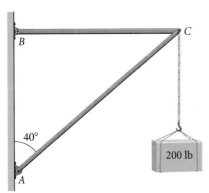

Let $\mathbf{u} = \langle u_1, u_2 \rangle$ *and* $\mathbf{v} = \langle v_1, v_2 \rangle$. *Prove each of the following properties.*

87. $\mathbf{u} + \mathbf{v} = \mathbf{v} + \mathbf{u}$

88. $\mathbf{u} \cdot \mathbf{v} = \mathbf{v} \cdot \mathbf{u}$

Skill Maintenance

Find the point(s) of intersection of each pair of equations.

89. $y_1 = 3x + 2,\ y_2 = -\frac{1}{2}x + 1$

90. $y_1 = 1.1 + x,\ y_2 = 3.4 - 0.9x$

91. $y_1 = 5 + 2x - x^2,\ y_2 = x^2 + 4x - 7$

92. $y_1 = 5 + 2x - x^2,\ y_2 = x^4 - 7x^3 + 5x + 4$

Synthesis

93. ◈ Explain how unit vectors are related to the unit circle.

94. ◈ Write a vector sum problem for a classmate for which the answer is $\mathbf{v} = 5\mathbf{i} - 8\mathbf{j}$.

95. If the dot product of two nonzero vectors **u** and **v** is 0, then the vectors are perpendicular (**orthogonal**). Let $\mathbf{u} = \langle u_1, u_2 \rangle$ and $\mathbf{v} = \langle v_1, v_2 \rangle$.

    a) Prove that if $\mathbf{u} \cdot \mathbf{v} = 0$, then **u** and **v** are perpendicular.

    b) Given an example of two perpendicular vectors and show that their dot product is 0.

96. If $\overrightarrow{PQ}$ is any vector, what is $\overrightarrow{PQ} + \overrightarrow{QP}$?

97. Find all the unit vectors that are parallel to the vector $\langle 3, -4 \rangle$.

98. Find a vector of length 2 whose direction is the opposite of the direction of the vector $\mathbf{v} = -\mathbf{i} + 2\mathbf{j}$. How many such vectors are there?

99. Given the vector $\overrightarrow{AB} = 3\mathbf{i} - \mathbf{j}$ and *A* is the point $(2, 9)$, find the point *B*.

100. Find vector **v** from point *A* to the origin, where $\overrightarrow{AB} = 4\mathbf{i} - 2\mathbf{j}$ and *B* is the point $(-2, 5)$.

4   Summary and Review

Important Properties and Formulas

The Law of Sines

$$\frac{a}{\sin A} = \frac{b}{\sin B} = \frac{c}{\sin C}$$

The Law of Cosines

$$a^2 = b^2 + c^2 - 2bc \cos A,$$
$$b^2 = a^2 + c^2 - 2ac \cos B,$$
$$c^2 = a^2 + b^2 - 2ab \cos C$$

The Area of a Triangle

$$K = \frac{1}{2} bc \sin A = \frac{1}{2} ab \sin C = \frac{1}{2} ac \sin B$$

Complex Numbers

Complex Number:	$a + bi$, a, b real, $i^2 = -1$
Imaginary Number:	$a + bi$, $b \neq 0$
Complex Conjugates:	$a + bi$, $a - bi$
Absolute Value:	$|a + bi| = \sqrt{a^2 + b^2}$
Trigonometric Notation:	$a + bi = r(\cos \theta + i \sin \theta)$
Multiplication:	$r_1(\cos \theta_1 + i \sin \theta_1) \cdot r_2(\cos \theta_2 + i \sin \theta_2)$
	$= r_1 r_2 [\cos (\theta_1 + \theta_2) + i \sin (\theta_1 + \theta_2)]$
Division:	$\dfrac{r_1(\cos \theta_1 + i \sin \theta_1)}{r_2(\cos \theta_2 + i \sin \theta_2)} = \dfrac{r_1}{r_2} [\cos (\theta_1 - \theta_2) + i \sin (\theta_1 - \theta_2)], \quad r_2 \neq 0$

DeMoivre's Theorem

$$[r(\cos \theta + i \sin \theta)]^n = r^n(\cos n\theta + i \sin n\theta)$$

Roots of Complex Numbers

The nth roots of $r(\cos \theta + i \sin \theta)$ are

$$r^{1/n}\left[\cos \left(\frac{\theta}{n} + k \cdot \frac{360°}{n} \right) + i \sin \left(\frac{\theta}{n} + k \cdot \frac{360°}{n} \right) \right], \quad r \neq 0, k = 0, 1, 2, \ldots, n - 1.$$

Vectors

If $\mathbf{u} = \langle u_1, u_2 \rangle$ and $\mathbf{v} = \langle v_1, v_2 \rangle$ and k is a scalar, then:

Length:	$|\mathbf{v}| = \sqrt{v_1^2 + v_2^2}$
Addition:	$\mathbf{u} + \mathbf{v} = \langle u_1 + v_1, u_2 + v_2 \rangle$

Subtraction: $\mathbf{u} - \mathbf{v} = \langle u_1 - v_1, u_2 - v_2 \rangle$

Scalar Multiplication: $k\mathbf{v} = \langle kv_1, kv_2 \rangle$

Dot Product: $\mathbf{u} \cdot \mathbf{v} = u_1v_1 + u_2v_2$

Angle Between Two Vectors: $\cos \theta = \dfrac{\mathbf{u} \cdot \mathbf{v}}{|\mathbf{u}|\,|\mathbf{v}|}$

REVIEW EXERCISES

Solve $\triangle ABC$, if possible.

1. $a = 23.4$ ft, $b = 15.7$ ft, $c = 8.3$ ft

2. $B = 27°$, $C = 35°$, $b = 19$ in.

3. $A = 133°28'$, $C = 31°42'$, $b = 890$ m

4. $B = 37°$, $b = 4$ yd, $c = 8$ yd

5. Find the area of $\triangle ABC$ if $b = 9.8$ m, $c = 7.3$ m, and $A = 67.3°$.

6. A parallelogram has sides of lengths 3.21 ft and 7.85 ft. One of its angles measures 147°. Find the area of the parallelogram.

7. *Sandbox.* A child-care center has a triangular-shaped sandbox. Two of the three sides measure 15 ft and 12.5 ft and form an included angle of 42°. To determine the amount of sand that is needed to fill the box, the director must determine the area of the triangular area. Find the area of the floor of the box to the nearest square foot.

15 ft

42°

12.5 ft

8. *Flower Garden.* A triangular flower garden has sides of lengths 11 m, 9 m, and 6 m. Find the angles of the garden to the nearest tenth of a degree.

9. In an isosceles triangle, the base angles each measure 52.3° and the base is 513 ft long. Find the lengths of the other two sides to the nearest foot.

10. *Airplanes.* Two airplanes leave an airport at the same time. The first flies 175 km/h in a direction of 305.6°. The second flies 220 km/h in a direction of 195.5°. After 2 hr, how far apart are the planes?

Express in terms of i.

11. $-\sqrt{-48}$

12. $\sqrt{-3}\,\sqrt{-7}$

Simplify. Write answers in the form $a + bi$, where a and b are real numbers.

13. $(-2 + 8i) + (3 - 7i)$

14. $(-4 - 2i) - (1 - 5i)$

15. $9i(2 - 3i)$

16. $(7 - 4i)(7 + 4i)$

17. $(5 - 2i)^2$

18. $\dfrac{2 - i}{3 + 2i}$

Simplify.

19. i^{65}

20. i^{42}

21. Solve: $x^2 + 4x + 20 = 0$.

Graph each complex number and find its absolute value.

22. $2 - 5i$

23. 4

24. $2i$

25. $-3 + i$

Find trigonometric notation.

26. $1 + i$

27. $-4i$

28. $-5\sqrt{3} + 5i$

29. $\dfrac{3}{4}$

Find standard notation, $a + bi$.

30. $4(\cos 60° + i \sin 60°)$

31. $7(\cos 0° + i \sin 0°)$

32. $5\left(\cos \dfrac{2\pi}{3} + i \sin \dfrac{2\pi}{3}\right)$

33. $2\left[\cos\left(-\dfrac{\pi}{3}\right) + i \sin\left(-\dfrac{\pi}{3}\right)\right]$

Convert to trigonometric notation and then multiply or divide, expressing the answer in standard notation.

34. $(1 + i\sqrt{3})(1 - i)$

35. $\dfrac{2 - 2i}{2 + 2i}$

36. $\dfrac{2 + 2\sqrt{3}i}{\sqrt{3} - i}$

37. $i(3 - 3\sqrt{3}i)$

Raise the number to the given power and write trigonometric notation for the answer.

38. $[2(\cos 60° + i \sin 60°)]^3$

39. $(1 - i)^4$

Raise the number to the given power and write standard notation for the answer.

40. $(1 + i)^6$

41. $\left(\dfrac{1}{2} + \dfrac{\sqrt{3}}{2}i\right)^{10}$

42. Find the square roots of $-1 + i$.

43. Find the cube roots of $3\sqrt{3} - 3i$.

44. Find and graph the fourth roots of 81.

45. Find and graph the fifth roots of 1.

Find all the complex solutions of the equation.

46. $x^4 - i = 0$

47. $x^3 + 1 = 0$

*Magnitudes of vectors **u** and **v** and the angle θ between the vectors are given. Find the magnitude of the sum, **u** + **v**, to the nearest tenth and give the direction by specifying to the nearest degree the angle that it makes with the vector **u**.*

48. $|\mathbf{u}| = 12$, $|\mathbf{v}| = 15$, $\theta = 120°$

49. $|\mathbf{u}| = 41$, $|\mathbf{v}| = 60$, $\theta = 25°$

*The vectors **u**, **v**, and **w** are drawn below. Copy them on a sheet of paper. Then sketch each of the vectors in Exercises 50 and 51.*

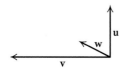

50. $\mathbf{u} - \mathbf{v}$

51. $\mathbf{u} + \frac{1}{2}\mathbf{w}$

52. Forces of 230 N and 500 N act on an object. The angle between the forces is 52°. Find the resultant, giving the angle that it makes with the smaller force.

53. *Wind.* A wind has an easterly component of 15 km/h and a southerly component of 25 km/h. Find the magnitude and the direction of the wind.

54. *Ship.* A ship sails N75°E for 90 nautical mi, and then S10°W for 100 nautical mi. How far is the ship, then, from the starting point, and in what direction?

Find the component form of the vector given the initial and terminal points.

55. $\overrightarrow{AB}$; $A(2, -8)$, $B(-2, -5)$

56. $\overrightarrow{TR}$; $R(0, 7)$, $T(-2, 13)$

57. Find the magnitude of vector **u** if $\mathbf{u} = \langle 5, -6 \rangle$.

Do the calculations in Exercises 58–61 for the vectors
$$\mathbf{u} = \langle 3, -4 \rangle, \quad \mathbf{v} = \langle -3, 9 \rangle, \quad \text{and} \quad \mathbf{w} = \langle -2, -5 \rangle.$$

58. $4\mathbf{u} + \mathbf{w}$

59. $2\mathbf{w} - 6\mathbf{v}$

60. $|\mathbf{u}| + |2\mathbf{w}|$

61. $\mathbf{u} \cdot \mathbf{w}$

62. Find a unit vector that has the same direction as $\mathbf{v} = \langle -6, -2 \rangle$.

63. Express the vector $\mathbf{t} = \langle -9, 4 \rangle$ as a linear combination of the unit vectors **i** and **j**.

64. Determine the direction angle θ of the vector $\mathbf{w} = \langle -4, -1 \rangle$ to the nearest degree.

65. Find the magnitude and the direction angle θ of $\mathbf{u} = -5\mathbf{i} - 3\mathbf{j}$.

66. Find the angle between $\mathbf{u} = \langle 3, -7 \rangle$ and $\mathbf{v} = \langle 2, 2 \rangle$ to the nearest tenth of a degree.

67. *Airplane.* An airplane has an airspeed of 160 mph. It is to make a flight in a direction of 080° while there is a 20-mph wind from 310°. What will the airplane's actual heading be?

Do the calculations in Exercises 68–71 for the vectors
$$\mathbf{u} = 2\mathbf{i} + 5\mathbf{j}, \quad \mathbf{v} = -3\mathbf{i} + 10\mathbf{j}, \quad \text{and} \quad \mathbf{w} = 4\mathbf{i} + 7\mathbf{j}.$$

68. $5\mathbf{u} - 8\mathbf{v}$

69. $\mathbf{u} - (\mathbf{v} + \mathbf{w})$

70. $|\mathbf{u} - \mathbf{v}|$

71. $3|\mathbf{w}| + |\mathbf{v}|$

72. Express the vector $\overrightarrow{PQ}$ in the form $a\mathbf{i} + b\mathbf{j}$, if P is the point $(1, -3)$ and Q is the point $(-4, 2)$.

Express each vector in Exercises 73 and 74 in the form $a\mathbf{i} + b\mathbf{j}$ and sketch each as an arrow in the coordinate plane.

73. The unit vectors $\mathbf{u} = (\cos \theta)\mathbf{i} + (\sin \theta)\mathbf{j}$ for $\theta = \pi/4$ and $\theta = 5\pi/4$. Include the unit circle $x^2 + y^2 = 1$ in your sketch.

74. The unit vector obtained by rotating **j** counterclockwise $2\pi/3$ radians about the origin.

75. Express the vector $3\mathbf{i} - \mathbf{j}$ as a product of its magnitude and its direction.

Synthesis

76. ◆ Explain why the law of sines cannot be used to find the first angle when solving a triangle given three sides.

77. ◆ Summarize how you can tell algebraically when solving triangles whether there is no solution, one solution, or two solutions.

78. Let $\mathbf{u} = 12\mathbf{i} + 5\mathbf{j}$. Find a vector that has the same direction as **u** but has length 3.

79. A parallelogram has sides of lengths 3.42 and 6.97. Its area is 18.4. Find the sizes of its angles.

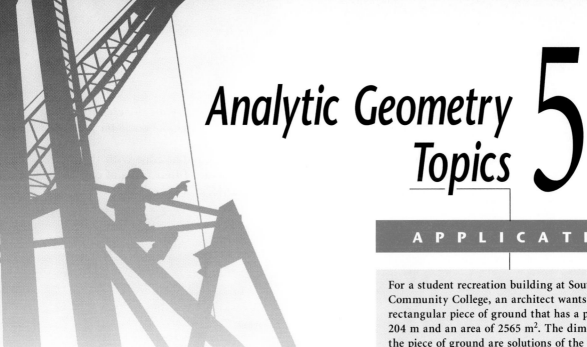

Analytic Geometry Topics 5

A P P L I C A T I O N

For a student recreation building at Southport Community College, an architect wants to lay out a rectangular piece of ground that has a perimeter of 204 m and an area of 2565 m². The dimensions of the piece of ground are solutions of the nonlinear system of equations

$$2x + 2y = 204,$$
$$xy = 2565.$$

I n this chapter, we study *conic sections*. These curves are formed by the intersection of a cone and a plane. Conic sections and their properties were first studied by the Greeks. Today they have many applications, as we will see. We also introduce the polar coordinate system, and we study conics in this coordinate system.

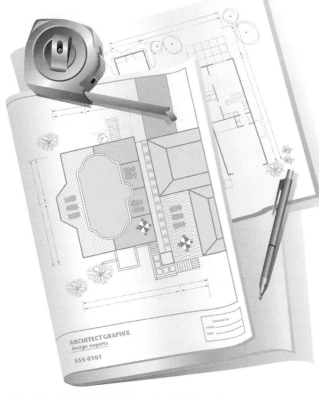

X	Y₁	Y₂
42	60	61.071
43	59	59.651
44	58	58.295
45	57	57
46	56	55.761
47	55	54.574
48	54	53.438

X = 45

$y_1 = (204 - 2x)/2, \; y_2 = 2565/x$

(57, 45)
(45, 57)

5.1 The Parabola
5.2 The Circle and the Ellipse
5.3 The Hyperbola
5.4 Nonlinear Systems of Equations
5.5 Rotation of Axes
5.6 Polar Coordinates and Graphs
5.7 Polar Equations of Conics
5.8 Parametric Equations and Graphs
SUMMARY AND REVIEW

5.1
The Parabola

- *Given an equation of a parabola, complete the square, if necessary, and then find the vertex, the focus, and the directrix and graph the parabola.*

A **conic section** is formed when a right circular cone with two parts, called *nappes*, is intersected by a plane. One of four types of curves can be formed: a parabola, a circle, an ellipse, or a hyperbola.

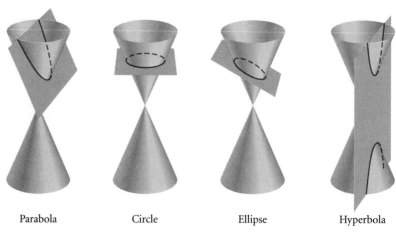

| Parabola | Circle | Ellipse | Hyperbola |

Conic Sections

Parabolas

Conic sections can be defined algebraically using second-degree equations of the form $Ax^2 + Bxy + Cy^2 + Dx + Ey + F = 0$. In addition, they can be defined geometrically as a set of points that satisfy certain conditions.

In Section 1.6, we saw that the graph of the quadratic function $f(x) = ax^2 + bx + c$, $a \neq 0$, is a parabola. A parabola can be defined geometrically.

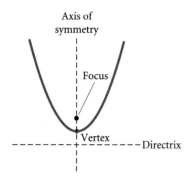

> #### Parabola
>
> A *parabola* is the set of all points in a plane equidistant from a fixed line (the *directrix*) and a fixed point not on the line (the *focus*).

The line that is perpendicular to the directrix and contains the focus is the **axis of symmetry**. The **vertex** is the midpoint of the segment between the focus and the directrix. (See the figure at left.)

Let's derive the standard equation of a parabola with vertex $(0, 0)$ and directrix $y = -p$, $p > 0$. We place the coordinate axes as shown in the figure at the top of the following page. The y-axis contains the focus F. The distance from the focus to the vertex is the same as the distance from the vertex to the directrix. Thus the coordinates of F are $(0, p)$.

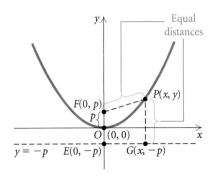

Let $P(x, y)$ be any point of the parabola and consider $\overline{PG}$ perpendicular to the line $y = -p$. The coordinates of G are $(x, -p)$. By the definition of a parabola,

$$PF = PG.$$ The distance from P to the focus is the same as the distance from P to the directrix.

Then using the distance formula (See Appendix A.2.), we have

$$\sqrt{(x - 0)^2 + (y - p)^2} = \sqrt{(x - x)^2 + (y + p)^2}$$
$$x^2 + y^2 - 2py + p^2 = y^2 + 2py + p^2 \quad \text{Squaring both sides}$$
$$x^2 = 4py.$$

We have shown that if $P(x, y)$ is on the parabola shown above, then its coordinates satisfy this equation. The converse is also true, but we will not prove it here.

Note that if $p > 0$, as above, the graph opens up. If $p < 0$, the graph opens down.

The equation of a parabola with vertex $(0, 0)$ and directrix $x = -p$ is derived similarly. Such a parabola opens either right $(p > 0)$ or left $(p < 0)$.

Standard Equation of a Parabola with Vertex at the Origin

The standard equation of a parabola with vertex $(0, 0)$ and directrix $y = -p$ is

$$x^2 = 4py.$$

The focus is $(0, p)$ and the y-axis is the axis of symmetry.

The standard equation of a parabola with vertex $(0, 0)$ and directrix $x = -p$ is

$$y^2 = 4px.$$

The focus is $(p, 0)$ and the x-axis is the axis of symmetry.

Example 1 Find the focus and the directrix of the parabola $y = -\frac{1}{12}x^2$. Then graph the parabola.

SOLUTION We write $y = -\frac{1}{12}x^2$ in the form $x^2 = 4py$:

$$-\frac{1}{12}x^2 = y \qquad \text{Given equation}$$
$$x^2 = -12y \qquad \text{Multiplying both sides by } -12$$
$$x^2 = 4(-3)y. \qquad \text{Standard form}$$

Thus, $p = -3$, so the focus is $(0, p)$, or $(0, -3)$. The directrix is $y = -p = -(-3) = 3$.

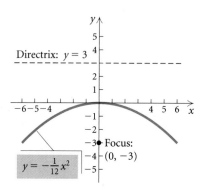

Example 2 Find an equation of the parabola with vertex $(0, 0)$ and focus $(5, 0)$. Then graph the parabola.

SOLUTION The focus is on the x-axis so the line of symmetry is the x-axis. Thus the equation is of the type

$$y^2 = 4px.$$

Since the focus is 5 units to the right of the vertex, $p = 5$ and the equation is

$$y^2 = 4(5)x, \quad \text{or}$$
$$y^2 = 20x.$$

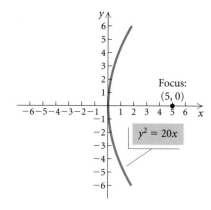

We can check the graph on a grapher using a squared viewing window. But first it might be necessary to solve for y:

$$y^2 = 20x$$
$$y = \pm\sqrt{20x}. \qquad \text{Using the principle of square roots}$$

We now graph $y_1 = \sqrt{20x}$ and $y_2 = -\sqrt{20x}$. On some graphers, it is possible to graph $y_1 = \sqrt{20x}$ and $y_2 = -y_1$ by using the Y-VARS menu.

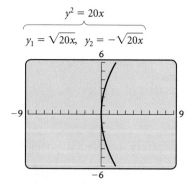

Finding Standard Form by Completing the Square

If a parabola with vertex at the origin is translated $|h|$ units horizontally and $|k|$ units vertically, it has an equation as follows.

Standard Equation of a Parabola with Vertex (h, k) and Vertical Axis of Symmetry

The standard equation of a parabola with vertex (h, k) and vertical axis of symmetry is

$$(x - h)^2 = 4p(y - k),$$

where the vertex is (h, k), the focus is $(h, k + p)$, and the directrix is $y = k - p$.

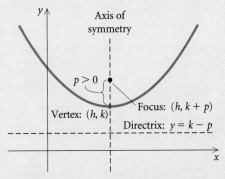

(When $p < 0$, the parabola opens down.)

Standard Equation of a Parabola with Vertex (h, k) and Horizontal Axis of Symmetry

The standard equation of a parabola with vertex (h, k) and horizontal axis of symmetry is

$$(y - k)^2 = 4p(x - h),$$

where the vertex is (h, k), the focus is $(h + p, k)$, and the directrix is $x = h - p$.

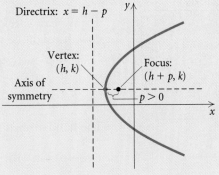

(When $p < 0$, the parabola opens left.)

We can complete the square on equations of the form

$$y = ax^2 + bx + c \quad \text{or} \quad x = ay^2 + by + c$$

in order to write them in standard form.

Example 3   For the parabola

$$x^2 + 6x + 4y + 5 = 0,$$

find the vertex, the focus, and the directrix. Then draw the graph.

SOLUTION   We first complete the square:

$$x^2 + 6x + 4y + 5 = 0$$

$$x^2 + 6x \qquad = -4y - 5 \qquad \text{Subtracting } 4y \text{ and } 5 \text{ on both sides}$$

$$x^2 + 6x + 9 = -4y - 5 + 9 \qquad \text{Adding 9 on both sides to complete the square on the left side}$$

$$x^2 + 6x + 9 = -4y + 4$$

$$(x + 3)^2 = -4(y - 1) \qquad \text{Factoring}$$

$$[(x - (-3)]^2 = 4(-1)(y - 1). \qquad \text{Writing standard form:} \\ (x - h)^2 = 4p(y - k)$$

We now have the following:

Vertex (h, k): $(-3, 1)$;
Focus $(h, k + p)$: $(-3, 1 + (-1))$, or $(-3, 0)$;
Directrix $y = k - p$: $y = 1 - (-1)$, or $y = 2$.

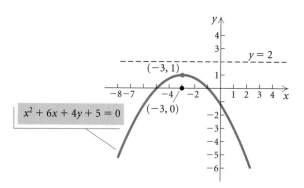

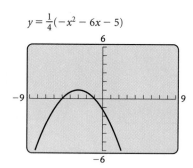

$y = \frac{1}{4}(-x^2 - 6x - 5)$

We can check the graph on a grapher using a squared viewing window. It might be necessary to solve for y first:

$$x^2 + 6x + 4y + 5 = 0$$

$$4y = -x^2 - 6x - 5$$

$$y = \tfrac{1}{4}(-x^2 - 6x - 5).$$

The hand-drawn graph appears to be correct.

Example 4   For the parabola

$$y^2 - 2y - 8x - 31 = 0,$$

find the vertex, the focus, and the directrix. Then draw the graph.

SOLUTION   We first complete the square:

$$y^2 - 2y - 8x - 31 = 0$$
$$y^2 - 2y = 8x + 31 \qquad \text{Adding } 8x \text{ and } 31 \text{ on both sides}$$
$$y^2 - 2y + 1 = 8x + 31 + 1 \qquad \text{Adding 1 on both sides to complete the square on the left side}$$

$$y^2 - 2y + 1 = 8x + 32$$
$$(y - 1)^2 = 8(x + 4)$$
$$(y - 1)^2 = 4(2)[x - (-4)]. \qquad \text{Writing standard form: } (y - k)^2 = 4p(x - h)$$

We now have the following:

Vertex (h, k): $(-4, 1)$;
Focus $(h + p, k)$: $(-4 + 2, 1)$, or $(-2, 1)$;
Directrix $x = h - p$: $x = -4 - 2$, or $x = -6$.

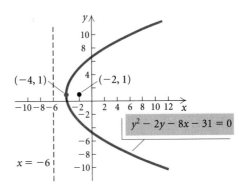

We can check the graph on a grapher using a squared window. We solve the original equation for y using the quadratic formula:

$$y^2 - 2y - 8x - 31 = 0$$
$$y^2 - 2y + (-8x - 31) = 0$$
$$a = 1, \quad b = -2, \quad c = -8x - 31$$
$$y = \frac{-(-2) \pm \sqrt{(-2)^2 - 4 \cdot 1(-8x - 31)}}{2 \cdot 1}$$
$$y = \frac{2 \pm \sqrt{32x + 128}}{2}.$$

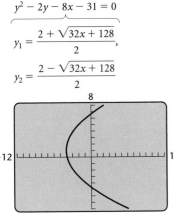

We now graph

$$y_1 = \frac{2 + \sqrt{32x + 128}}{2} \quad \text{and} \quad y_2 = \frac{2 - \sqrt{32x + 128}}{2}.$$

The hand-drawn graph appears to be correct.

Applications

Parabolas have many applications. For example, cross sections of car headlights, flashlights, and searchlights are parabolas. The bulb is located at the focus and light from that point is reflected outward parallel to the axis of symmetry. Satellite dishes and field microphones used at sporting events often have parabolic cross sections. Incoming radio waves or sound waves, parallel to the axis, are reflected into the focus. Cables hung between structures in suspension bridges, such as the Golden Gate Bridge, form parabolas. When a cable supports only its own weight, however, it forms a curve called a *catenary* rather than a parabola.

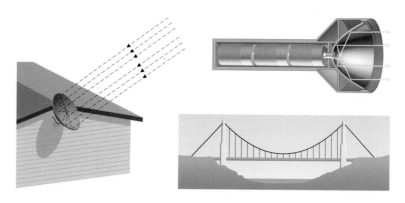

5.1 | Exercise Set

In Exercises 1–6, match each equation with one of the graphs (a)–(f), which follow.

a)
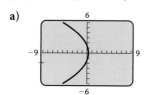

b)
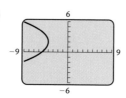

c)
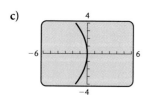

d)
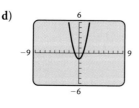

e)
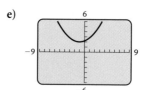

f)
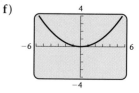

1. $x^2 = 8y$

2. $y^2 = -10x$

3. $(y - 2)^2 = -3(x + 4)$

4. $(x + 1)^2 = 5(y - 2)$

5. $13x^2 - 8y - 9 = 0$

6. $41x + 6y^2 = 12$

Find the vertex, the focus, and the directrix. Then draw the graph.

7. $x^2 = 20y$ **8.** $x^2 = 16y$

9. $y^2 = -6x$ **10.** $y^2 = -2x$

11. $x^2 - 4y = 0$ **12.** $y^2 + 4x = 0$

13. $x = 2y^2$ **14.** $y = \frac{1}{2}x^2$

Find an equation of a parabola satisfying the given conditions.

15. Focus $(4, 0)$, directrix $x = -4$

16. Focus $\left(0, \frac{1}{4}\right)$, directrix $y = -\frac{1}{4}$

17. Focus $(0, -\pi)$, directrix $y = \pi$

18. Focus $(-\sqrt{2}, 0)$, directrix $x = \sqrt{2}$

19. Focus $(3, 2)$, directrix $x = -4$

20. Focus $(-2, 3)$, directrix $y = -3$

Find the vertex, the focus, and the directrix. Then draw the graph.

21. $(x + 2)^2 = -6(y - 1)$

22. $(y - 3)^2 = -20(x + 2)$

23. $x^2 + 2x + 2y + 7 = 0$

24. $y^2 + 6y - x + 16 = 0$

25. $x^2 - y - 2 = 0$

26. $x^2 - 4x - 2y = 0$

27. $y = x^2 + 4x + 3$

28. $y = x^2 + 6x + 10$

29. $y^2 - y - x + 6 = 0$

30. $y^2 + y - x - 4 = 0$

31. *Satellite Dish.*   An engineer designs a satellite dish with a parabolic cross section. The dish is 15 ft wide at the opening and the focus is placed 4 ft from the vertex.

   a) Position a coordinate system with the origin at the vertex and the x-axis on the parabola's axis of symmetry and find an equation of the parabola.
   b) Find the depth of the satellite dish at the vertex.

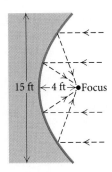

15 ft ⟵ 4 ft ⟶ Focus

32. *Headlight Mirror.*   A car headlight mirror has a parabolic cross section with diameter 6 in. and depth 1 in.

   a) Position a coordinate system with the origin at the vertex and the x-axis on the parabola's axis of symmetry and find an equation of the parabola.

   b) How far from the vertex should the bulb be positioned if it is to be placed at the focus?

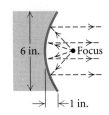

6 in.   •Focus

⟶ ⟵1 in.

33. *Spotlight.*   A spotlight has a parabolic cross section that is 4 ft wide at the opening and 1.5 ft deep at the vertex. How far from the vertex is the focus?

34. *Field Microphone.*   A field microphone used at a football game has a parabolic cross section and is 18 in. deep. The focus is 4 in. from the vertex. Find the width of the microphone at the opening.

Skill Maintenance

Complete the square and write the result as the square of a binomial.

35. $x^2 + 10x$

36. $y^2 - 9y$

Find the center and the radius of each circle.

37. $(x - 1)^2 + (y + 2)^2 = 9$

38. $(x + 3)^2 + (y - 5)^2 = 36$

Synthesis

39. ◈ Is a parabola always the graph of a function? Why or why not?

40. ◈ Explain how the distance formula is used to find the standard equation of a parabola.

41. Find an equation of the parabola with a vertical axis of symmetry and vertex $(-1, 2)$ and containing the point $(-3, 1)$.

42. Find an equation of a parabola with a horizontal axis of symmetry and vertex $(-2, 1)$ and containing the point $(-3, 5)$.

Use a grapher to find the vertex, the focus, and the directrix of each of the following.

43. $4.5x^2 - 7.8x + 9.7y = 0$

44. $134.1y^2 + 43.4x - 316.6y - 122.4 = 0$

45. *Suspension Bridge.* The cables of a suspension bridge are 50 ft above the roadbed at the ends of the bridge and 10 ft above it in the center of the bridge. The roadbed is 200 ft long. Vertical cables are to be spaced every 20 ft along the bridge. Calculate the lengths of these vertical cables.

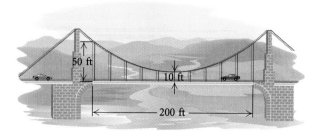

5.2
The Circle and the Ellipse

- *Given an equation of a circle, complete the square, if necessary, and then find the center and the radius and graph the circle.*
- *Given an equation of an ellipse, complete the square, if necessary, and then find the center, the vertices, and the foci and graph the ellipse.*

Circles

We can define a circle geometrically.

> **Circle**
>
> A *circle* is the set of all points in a plane that are at a fixed distance from a fixed point (the *center*) in the plane.

Recall the standard equation of a circle with center (h, k) and radius r. (See Appendix A.2 for a review of circles.)

> **Standard Equation of a Circle**
>
> The standard equation of a circle with center (h, k) and radius r is
>
> $$(x - h)^2 + (y - k)^2 = r^2.$$

Example 1 For the circle

$$x^2 + y^2 - 16x + 14y + 32 = 0,$$

find the center and the radius. Then graph the circle.

SOLUTION   First we complete the square twice:

$$x^2 + y^2 - 16x + 14y + 32 = 0$$
$$x^2 - 16x \qquad + y^2 + 14y \qquad = -32$$
$$x^2 - 16x + 64 + y^2 + 14y + 49 = -32 + 64 + 49$$

Adding 64 and 49 on both sides to complete
the square twice on the left side

$$(x - 8)^2 + (y + 7)^2 = 81$$
$$(x - 8)^2 + [y - (-7)]^2 = 9^2. \qquad \text{Writing standard form}$$

The center is $(8, -7)$ and the radius is 9.

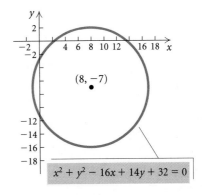

To use a grapher to graph the circle, it might be necessary to solve for y first. The original equation can be solved using the quadratic formula, or the standard form of the equation can be solved using the principle of square roots. The second alternative is illustrated here:

$$(x - 8)^2 + (y + 7)^2 = 81$$
$$(y + 7)^2 = 81 - (x - 8)^2$$
$$y + 7 = \pm\sqrt{81 - (x - 8)^2} \qquad \text{Using the principle of square roots}$$
$$y = -7 \pm \sqrt{81 - (x - 8)^2}.$$

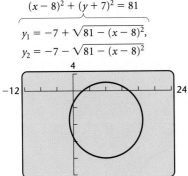

Then we graph

$$y_1 = -7 + \sqrt{81 - (x - 8)^2}$$

and

$$y_2 = -7 - \sqrt{81 - (x - 8)^2}$$

in a squared viewing window.

The hand-drawn graph appears to be correct.   ▬

Some graphers have a DRAW feature that provides a quick way to graph a circle when the center and the radius are known.

Ellipses

We have studied two conic sections, the parabola and the circle. Now we turn our attention to a third, the *ellipse*.

Ellipse

An *ellipse* is the set of all points in a plane, the sum of whose distances from two fixed points (the *foci*) is constant. The *center* of an ellipse is the midpoint of the segment between the foci.

We can draw an ellipse by first placing two thumbtacks in a piece of cardboard. These are the foci (singular, *focus*). We then attach a piece of string to the tacks. Its length is the constant sum of the distances $d_1 + d_2$ from the foci to any point on the ellipse. Next, we trace a curve with a pencil held tight against the string. The figure traced is an ellipse.

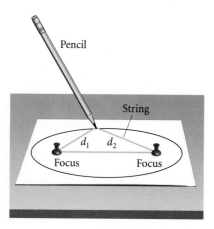

Let's first consider the ellipse shown below with center at the origin. The points F_1 and F_2 are the foci. The segment $\overline{A'A}$ is the **major axis**, and the points A' and A are the **vertices**. The segment $\overline{B'B}$ is the **minor axis**, and the points B and B' are the **y-intercepts**. Note that the major axis of an ellipse is always longer than the minor axis.

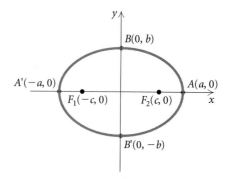

Standard Equation of an Ellipse with Center at the Origin

Major Axis Horizontal

$$\frac{x^2}{a^2} + \frac{y^2}{b^2} = 1, \ a > b > 0$$

Vertices: $(-a, 0), (a, 0)$

y-intercepts: $(0, -b), (0, b)$

Foci: $(-c, 0), (c, 0)$, where $c^2 = a^2 - b^2$

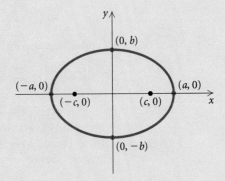

Major Axis Vertical

$$\frac{x^2}{b^2} + \frac{y^2}{a^2} = 1, \ a > b > 0$$

Vertices: $(0, -a), (0, a)$

x-intercepts: $(-b, 0), (b, 0)$

Foci: $(0, -c), (0, c)$, where $c^2 = a^2 - b^2$

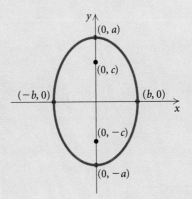

Example 2   Find the standard equation of the ellipse with vertices $(-5, 0)$ and $(5, 0)$ and foci $(-3, 0)$ and $(3, 0)$. Then graph the ellipse.

SOLUTION   Since the foci are on the x-axis and the origin is the midpoint of the segment between them, the major axis is horizontal and $(0, 0)$ is the center of the ellipse. Thus the equation is of the form

$$\frac{x^2}{a^2} + \frac{y^2}{b^2} = 1.$$

Since the vertices are $(-5, 0)$ and $(5, 0)$ and the foci are $(-3, 0)$ and $(3, 0)$, we know that $a = 5$ and $c = 3$. These values can be used to find b^2:

$$c^2 = a^2 - b^2$$
$$3^2 = 5^2 - b^2$$
$$9 = 25 - b^2$$
$$b^2 = 16.$$

Thus the equation of the ellipse is

$$\frac{x^2}{25} + \frac{y^2}{16} = 1.$$

To graph the ellipse, we plot the vertices $(-5, 0)$ and $(5, 0)$. Since $b^2 = 16$, we know that the y-intercepts are $(0, -4)$ and $(0, 4)$. We plot these points as well and connect the four points we have plotted with a smooth curve.

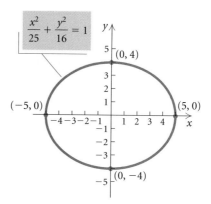

To draw the graph using a grapher, it might be necessary to solve for y first:

$$y = \pm \sqrt{\frac{400 - 16x^2}{25}}.$$

Then we graph

$$y_1 = -\sqrt{\frac{400 - 16x^2}{25}} \quad \text{and} \quad y_2 = \sqrt{\frac{400 - 16x^2}{25}}$$

or

$$y_1 = -\sqrt{\frac{400 - 16x^2}{25}} \quad \text{and} \quad y_2 = -y_1$$

in a squared viewing window.

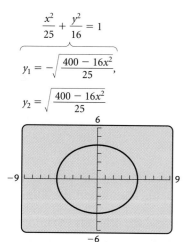

Example 3   For the ellipse

$$9x^2 + 4y^2 = 36,$$

find the vertices and the foci. Then draw the graph.

SOLUTION   We first find standard form:

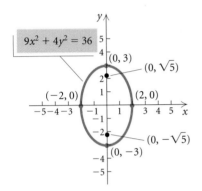

$$9x^2 + 4y^2 = 36$$

$$\frac{9x^2}{36} + \frac{4y^2}{36} = \frac{36}{36} \qquad \text{Dividing by 36 on both sides to get 1 on the right side}$$

$$\frac{x^2}{4} + \frac{y^2}{9} = 1$$

$$\frac{x^2}{2^2} + \frac{y^2}{3^2} = 1. \qquad \text{Writing standard form}$$

Thus, $a = 3$ and $b = 2$. The major axis is vertical, so the vertices are $(0, -3)$ and $(0, 3)$. Since we know that $c^2 = a^2 - b^2$, we have $c^2 = 9 - 4$, so $c = \sqrt{5}$ and the foci are $(0, -\sqrt{5})$ and $(0, \sqrt{5})$.

To graph the ellipse, we plot the vertices. Note also that since $b = 2$, the x-intercepts are $(-2, 0)$ and $(2, 0)$. We plot these points as well and connect the four points we have plotted with a smooth curve.   ▬

If the center of an ellipse is not at the origin but at some point (h, k), then we can think of an ellipse with center at the origin being translated $|h|$ units left or right and $|k|$ units up or down.

Standard Equation of an Ellipse with Center at (h, k)
Major Axis Horizontal

$$\frac{(x - h)^2}{a^2} + \frac{(y - k)^2}{b^2} = 1, \ a > b > 0$$

Vertices: $(h - a, k), (h + a, k)$
Length of minor axis: $2b$
Foci: $(h - c, k), (h + c, k)$, where $c^2 = a^2 - b^2$

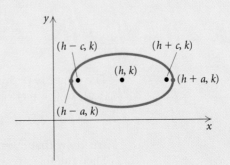

(continued)

Major Axis Vertical

$$\frac{(x-h)^2}{b^2} + \frac{(y-k)^2}{a^2} = 1, \ a > b > 0$$

Vertices: $(h, k-a), (h, k+a)$

Length of minor axis: $2b$

Foci: $(h, k-c), (h, k+c)$, where $c^2 = a^2 - b^2$

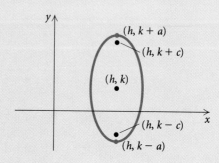

Example 4   For the ellipse

$$4x^2 + y^2 + 24x - 2y + 21 = 0,$$

find the center, the vertices, and the foci. Then draw the graph.

SOLUTION   First we complete the square to get standard form:

$$4x^2 + y^2 + 24x - 2y + 21 = 0$$
$$4(x^2 + 6x \qquad) + (y^2 - 2y \qquad) = -21$$
$$4(x^2 + 6x + 9) + (y^2 - 2y + 1) = -21 + 4 \cdot 9 + 1$$

Completing the square twice by adding $4 \cdot 9$ and 1 on both sides

$$4(x+3)^2 + (y-1)^2 = 16$$
$$\frac{1}{16}[4(x+3)^2 + (y-1)^2] = \frac{1}{16} \cdot 16$$
$$\frac{(x+3)^2}{4} + \frac{(y-1)^2}{16} = 1$$
$$\frac{[x-(-3)]^2}{2^2} + \frac{(y-1)^2}{4^2} = 1.$$

Writing standard form:
$$\frac{(x-h)^2}{b^2} + \frac{(y-k)^2}{a^2} = 1$$

The center is $(-3, 1)$. Note that $a = 4$ and $b = 2$. The major axis is vertical, so the vertices are 4 units above and below the center:

$$(-3, 1+4) \text{ and } (-3, 1-4), \quad \text{or} \quad (-3, 5) \text{ and } (-3, -3).$$

We know that $c^2 = a^2 - b^2$, so $c^2 = 16 - 4 = 12$ and $c = \sqrt{12} = 2\sqrt{3}$. Then the foci are $2\sqrt{3}$ units above and below the center:

$$(-3, 1 + 2\sqrt{3}) \quad \text{and} \quad (-3, 1 - 2\sqrt{3}).$$

To graph the ellipse, we plot the vertices. Note also that since $b = 2$, two other points on the graph are the endpoints of the minor axis, 2 units right and left of the center:

$$(-3 + 2, 1) \text{ and } (-3 - 2, 1), \quad \text{or} \quad (-1, 1) \text{ and } (-5, 1).$$

We plot these points as well and connect the four points with a smooth curve.

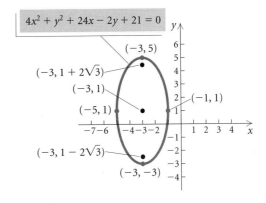

Applications

One of the most exciting recent applications of an ellipse is a medical device called a *lithotripter*. This machine uses underwater shock waves to crush kidney stones. The waves originate at one focus of an ellipse and are reflected to the kidney stone, which is positioned at the other focus. Recovery time following the use of this technique is much shorter than with conventional surgery and the mortality rate is far lower.

A room with an ellipsoidal ceiling is known as a *whispering gallery*. In such a room, a word whispered at one focus can be clearly heard at the other. Whispering galleries are found in the rotunda of the Capitol Building in Washington, D.C., and in the Mormon Tabernacle in Salt Lake City.

Ellipses have many other applications. Planets travel around the sun in elliptical orbits with the sun at one focus, for example, and satellites travel around the earth in elliptical orbits as well.

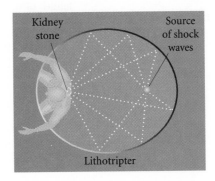

Lithotripter

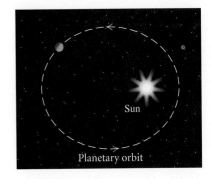

Planetary orbit

5.2 | Exercise Set

In Exercises 1–6, match each equation with one of the graphs (a)–(f), which follow.

a)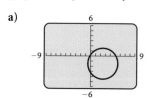

b)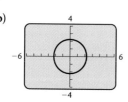

c)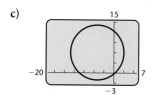

d)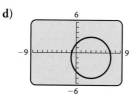

e)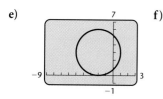

f)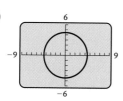

1. $x^2 + y^2 = 5$

2. $y^2 = 20 - x^2$

3. $x^2 + y^2 - 6x + 2y = 6$

4. $x^2 + y^2 + 10x - 12y = 3$

5. $x^2 + y^2 - 5x + 3y = 0$

6. $x^2 + 4x - 2 = 6y - y^2 - 6$

Find the center and the radius of the circle with the given equation. Then draw the graph.

7. $x^2 + y^2 - 14x + 4y = 11$

8. $x^2 + y^2 + 2x - 6y = -6$

9. $x^2 + y^2 + 4x - 6y - 12 = 0$

10. $x^2 + y^2 - 8x - 2y - 19 = 0$

11. $x^2 + y^2 + 6x - 10y = 0$

12. $x^2 + y^2 - 7x - 2y = 0$

13. $x^2 + y^2 - 9x = 7 - 4y$

14. $y^2 - 6y - 1 = 8x - x^2 + 3$

In Exercises 15–18, match each equation with one of the graphs (a)–(d), which follow.

a)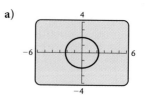

b)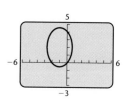

c)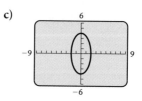

d)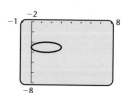

15. $16x^2 + 4y^2 = 64$

16. $4x^2 + 5y^2 = 20$

17. $x^2 + 9y^2 - 6x + 90y = -225$

18. $9x^2 + 4y^2 + 18x - 16y = 11$

Find the vertices and the foci of the ellipse with the given equation. Then draw the graph.

19. $\dfrac{x^2}{4} + \dfrac{y^2}{1} = 1$

20. $\dfrac{x^2}{25} + \dfrac{y^2}{36} = 1$

21. $16x^2 + 9y^2 = 144$

22. $9x^2 + 4y^2 = 36$

23. $2x^2 + 3y^2 = 6$

24. $5x^2 + 7y^2 = 35$

25. $4x^2 + 9y^2 = 1$

26. $25x^2 + 16y^2 = 1$

Find an equation of an ellipse satisfying the given conditions.

27. Vertices: $(-7, 0)$ and $(7, 0)$;
foci: $(-3, 0)$ and $(3, 0)$

28. Vertices: $(0, -6)$ and $(0, 6)$;
foci: $(0, -4)$ and $(0, 4)$

29. Vertices: $(0, -8)$ and $(0, 8)$;
length of minor axis: 10

30. Vertices: $(-5, 0)$ and $(5, 0)$;
length of minor axis: 6

31. Foci: $(-2, 0)$ and $(2, 0)$;
length of major axis: 6

32. Foci: $(0, -3)$ and $(0, 3)$;
length of major axis: 10

Find the center, the vertices, and the foci of each ellipse. Then draw the graph.

33. $\dfrac{(x - 1)^2}{9} + \dfrac{(y - 2)^2}{4} = 1$

34. $\dfrac{(x - 1)^2}{1} + \dfrac{(y - 2)^2}{4} = 1$

35. $\dfrac{(x + 3)^2}{25} + \dfrac{(y - 5)^2}{36} = 1$

36. $\dfrac{(x-2)^2}{16} + \dfrac{(y+3)^2}{25} = 1$

37. $3(x+2)^2 + 4(y-1)^2 = 192$

38. $4(x-5)^2 + 3(y-4)^2 = 48$

39. $4x^2 + 9y^2 - 16x + 18y - 11 = 0$

40. $x^2 + 2y^2 - 10x + 8y + 29 = 0$

41. $4x^2 + y^2 - 8x - 2y + 1 = 0$

42. $9x^2 + 4y^2 + 54x - 8y + 49 = 0$

*The **eccentricity** of an ellipse is defined as $e = c/a$. For an ellipse, $0 < c < a$, so $0 < e < 1$. When e is close to 0, an ellipse appears to be nearly circular. When e is close to 1, an ellipse is very flat.*

43. Observe the shapes of the ellipses in Examples 2 and 4. Which ellipse has the smaller eccentricity? Confirm your answer by computing the eccentricity of each ellipse.

44. Which ellipse below has the smaller eccentricity? (Assume that the coordinate systems have the same scale.)

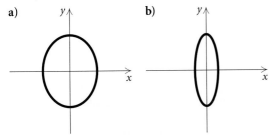

a)   **b)**

45. Find an equation of an ellipse with vertices $(0, -4)$ and $(0, 4)$ and $e = \frac{1}{4}$.

46. Find an equation of an ellipse with vertices $(-3, 0)$ and $(3, 0)$ and $e = \frac{7}{10}$.

47. *Bridge Supports.* The bridge support shown in the figure below is the top half of an ellipse. Assuming that a coordinate system is superimposed on the drawing in such a way that the center of the ellipse is at point Q, find an equation of the ellipse.

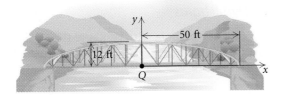

48. *The Ellipse.* In Washington, D.C., there is a large grassy area south of the White House known as the Ellipse. It is actually an ellipse with major axis of length 1048 ft and minor axis of length 898 ft. Assuming that a coordinate system is superimposed on the area in such a way that the center is at the origin and the major and minor axes are on the x- and y-axes of the coordinate system, respectively, find an equation of the ellipse.

49. *The Earth's Orbit.* The maximum distance of the earth from the sun is 9.3×10^7 miles. The minimum distance is 9.1×10^7 miles. The sun is at one focus of the elliptical orbit. Find the distance from the sun to the other focus.

50. *Carpentry.* A carpenter is cutting a 3-ft by 4-ft elliptical sign from a 3-ft by 4-ft piece of plywood. The ellipse will be drawn using a string attached to the board at the foci of the ellipse.

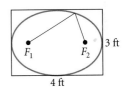

a) How far from the ends of the board should the string be attached?

b) How long should the string be?

Skill Maintenance

Graph.

51. $y = \frac{5}{2}x$

52. $y = -\frac{3}{4}x$

53. $y - 1 = \frac{1}{3}(x + 3)$

54. $y + 2 = -\frac{5}{4}(x - 5)$

Synthesis

55. ◆ Explain why function notation is not used in this section.

56. ◆ Would you prefer to graph the ellipse

$$\frac{(x-3)^2}{4} + \frac{(y+5)^2}{9} = 1$$

by hand or using a grapher? Explain why you answered as you did.

Find an equation of an ellipse satisfying the given conditions.

57. Vertices: $(3, -4)$, $(3, 6)$; endpoints of minor axis: $(1, 1)$, $(5, 1)$

58. Vertices: $(-1, -1)$, $(-1, 5)$;
endpoints of minor axis: $(-3, 2)$, $(1, 2)$

59. Vertices: $(-3, 0)$ and $(3, 0)$; passing through $\left(2, \frac{22}{3}\right)$

60. Center: $(-2, 3)$; major axis vertical;
length of major axis: 4;
length of minor axis: 1

Use a grapher to find the center and the vertices of each of the following.

61. $4x^2 + 9y^2 - 16.025x + 18.0927y - 11.346 = 0$

62. $9x^2 + 4y^2 + 54.063x - 8.016y + 49.872 = 0$

63. *Bridge Arch.* A bridge with a semielliptical arch spans a river as shown below. What is the clearance 6 ft from the riverbank?

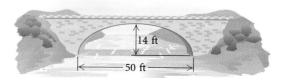

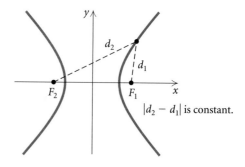

14 ft

50 ft

5.3

The Hyperbola

- *Given an equation of a hyperbola, complete the square, if necessary, and then find the center, the vertices, and the foci and graph the hyperbola.*

The last type of conic section that we will study is the *hyperbola*.

Hyperbola

A *hyperbola* is the set of all points in a plane for which the absolute value of the difference of the distances from two fixed points (the *foci*) is constant. The midpoint of the segment between the foci is the *center* of the hyperbola.

d_2

d_1

F_2

F_1

$|d_2 - d_1|$ is constant.

Standard Equations of Hyperbolas

We first consider the equation of a hyperbola with center at the origin. In the figure at the top of the following page, F_1 and F_2 are the foci. The segment $\overline{V_2 V_1}$ is the **transverse axis** and the points V_2 and V_1 are the **vertices**.

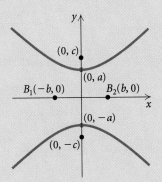

Standard Equation of a Hyperbola with Center at the Origin

Transverse Axis Horizontal

$$\frac{x^2}{a^2} - \frac{y^2}{b^2} = 1$$

Vertices: $(-a, 0)$, $(a, 0)$

Foci: $(-c, 0)$, $(c, 0)$, where $c^2 = a^2 + b^2$

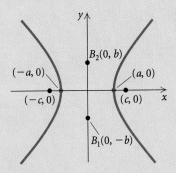

Transverse Axis Vertical

$$\frac{y^2}{a^2} - \frac{x^2}{b^2} = 1$$

Vertices: $(0, -a)$, $(0, a)$

Foci: $(0, -c)$, $(0, c)$, where $c^2 = a^2 + b^2$

The segment $\overline{B_1B_2}$ is the **conjugate axis** of the hyperbola.

To graph a hyperbola with a horizontal transverse axis, it is helpful to begin by graphing the lines $y = -(b/a)x$ and $y = (b/a)x$. These are the **asymptotes** of the hyperbola. For a hyperbola with a vertical transverse axis, the asymptotes are $y = -(a/b)x$ and $y = (a/b)x$. As $|x|$ gets larger and larger, the graph of the hyperbola gets closer and closer to the asymptotes.

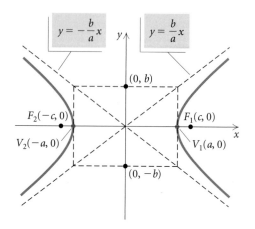

Example 1 Find an equation of the hyperbola with vertices $(0, -4)$ and $(0, 4)$ and foci $(0, -6)$ and $(0, 6)$.

SOLUTION   We know that $a = 4$ and $c = 6$. We find b^2.

$$c^2 = a^2 + b^2$$
$$6^2 = 4^2 + b^2$$
$$36 = 16 + b^2$$
$$20 = b^2$$

Since the vertices and the foci are on the y-axis, we know that the transverse axis is vertical. We can now write the equation of the hyperbola:

$$\frac{y^2}{a^2} - \frac{x^2}{b^2} = 1$$
$$\frac{y^2}{16} - \frac{x^2}{20} = 1.$$

Example 2 For the hyperbola given by

$$9x^2 - 16y^2 = 144,$$

find the vertices, the foci, and the asymptotes. Then graph the hyperbola.

SOLUTION   First, we find standard form:

$$9x^2 - 16y^2 = 144$$

$$\frac{1}{144}(9x^2 - 16y^2) = \frac{1}{144} \cdot 144 \qquad \text{Multiplying by } \tfrac{1}{144} \text{ to get 1 on the right side}$$

$$\frac{x^2}{16} - \frac{y^2}{9} = 1$$

$$\frac{x^2}{4^2} - \frac{y^2}{3^2} = 1. \qquad \text{Writing standard form}$$

The hyperbola has a horizontal transverse axis, so the vertices are $(-a, 0)$ and $(a, 0)$, or $(-4, 0)$ and $(4, 0)$. From the standard form of the equation, we know that $a^2 = 4^2$, or 16, and $b^2 = 3^2$, or 9. We find the foci:

$$c^2 = a^2 + b^2$$
$$c^2 = 16 + 9$$
$$c^2 = 25$$
$$c = 5.$$

Thus the foci are $(-5, 0)$ and $(5, 0)$.

Next we find the asymptotes:

$$y = \frac{b}{a}x = \frac{3}{4}x$$

and

$$y = -\frac{b}{a}x = -\frac{3}{4}x.$$

To draw the graph, we sketch the asymptotes first. This is easily done by drawing the rectangle with horizontal sides passing through $(0, 3)$ and $(0, -3)$ and vertical sides through $(4, 0)$ and $(-4, 0)$. Then we draw and extend the diagonals of this rectangle. The two extended diagonals are the asymptotes of the hyperbola. Next, we plot the vertices and draw the branches of the hyperbola outward from the vertices toward the asymptotes.

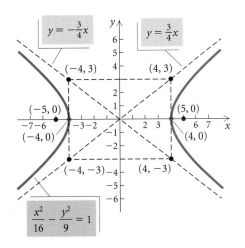

We can use a grapher as a check. It might be necessary to solve for y first and then graph the top and bottom halves of the hyperbola in the same squared viewing window.

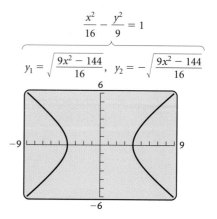

$$\frac{x^2}{16} - \frac{y^2}{9} = 1$$

$$y_1 = \sqrt{\frac{9x^2 - 144}{16}}, \quad y_2 = -\sqrt{\frac{9x^2 - 144}{16}}$$

If a hyperbola with center at the origin is translated $|h|$ units left or right and $|k|$ units up or down, the center is at the point (h, k).

Standard Equation of a Hyperbola with Center (h, k)
Transverse Axis Horizontal

$$\frac{(x - h)^2}{a^2} - \frac{(y - k)^2}{b^2} = 1$$

Vertices: $(h - a, k)$, $(h + a, k)$

Asymptotes: $y - k = \dfrac{b}{a}(x - h)$, $y - k = -\dfrac{b}{a}(x - h)$

Foci: $(h - c, k)$, $(h + c, k)$, where $c^2 = a^2 + b^2$

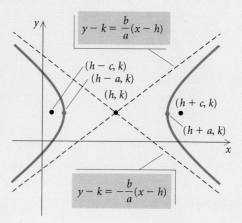

(continued)

Transverse Axis Vertical

$$\frac{(y-k)^2}{a^2} - \frac{(x-h)^2}{b^2} = 1$$

Vertices: $(h, k-a)$, $(h, k+a)$

Asymptotes: $y - k = \dfrac{a}{b}(x-h)$, $y - k = -\dfrac{a}{b}(x-h)$

Foci: $(h, k-c)$, $(h, k+c)$, where $c^2 = a^2 + b^2$

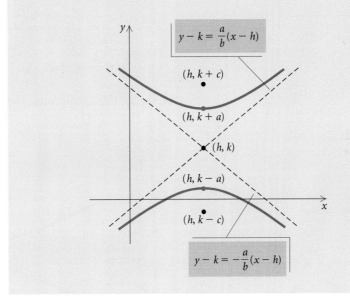

Example 3   For the hyperbola given by

$$4y^2 - x^2 + 24y + 4x + 28 = 0,$$

find the center, the vertices, and the foci. Then draw the graph.

SOLUTION   First, we complete the square to get standard form:

$$4y^2 - x^2 + 24y + 4x + 28 = 0$$
$$4(y^2 + 6y \qquad) - (x^2 - 4x \qquad) = -28$$
$$4(y^2 + 6y + 9 - 9) - (x^2 - 4x + 4 - 4) = -28$$
$$4(y^2 + 6y + 9) - (x^2 - 4x + 4) = -28 + 36 - 4$$
$$4(y+3)^2 - (x-2)^2 = 4$$
$$\frac{(y+3)^2}{1} - \frac{(x-2)^2}{4} = 1 \qquad \text{Dividing by 4}$$
$$\frac{[y-(-3)]^2}{1^2} - \frac{(x-2)^2}{2^2} = 1. \qquad \text{Standard form}$$

The center is $(2, -3)$. Note that $a = 1$ and $b = 2$. The transverse axis is vertical, so the vertices are 1 unit below and above the center:

$$(2, -3 - 1) \text{ and } (2, -3 + 1), \quad \text{or} \quad (2, -4) \text{ and } (2, -2).$$

We know that $c^2 = a^2 + b^2$, so $c^2 = 1 + 4 = 5$ and $c = \sqrt{5}$. Thus the foci are $\sqrt{5}$ units below and above the center:

$$(2, -3 - \sqrt{5}) \quad \text{and} \quad (2, -3 + \sqrt{5}).$$

The asymptotes are

$$y - (-3) = \frac{1}{2}(x - 2) \quad \text{and} \quad y - (-3) = -\frac{1}{2}(x - 2),$$

or

$$y + 3 = \frac{1}{2}(x - 2) \quad \text{and} \quad y + 3 = -\frac{1}{2}(x - 2).$$

We sketch the asymptotes, plot the vertices, and draw the graph.

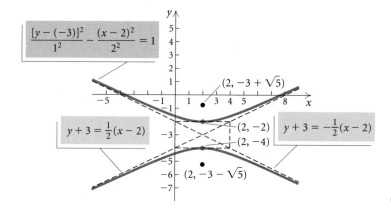

Applications

Some comets travel in hyperbolic paths with the sun at one focus. Such comets pass by the sun only one time unlike those with elliptical orbits, which reappear at intervals. A cross section of an amphitheater might be one branch of a hyperbola. A cross section of a nuclear cooling tower might also be a hyperbola.

One other important application of hyperbolas is in the long range navigation system LORAN. This system uses transmitting stations in three locations to send out simultaneous signals to a ship or aircraft. The difference in the arrival times of the signals from one pair of transmitters is recorded on the ship or aircraft. This difference is also recorded for signals from another pair of transmitters. For each pair, a computation is performed to determine the difference in the distances from each member of the pair to the ship or aircraft. If each pair of differences is kept constant, two hyperbolas can be drawn. Each has one of the pairs of transmitters as foci and the ship or aircraft lies on the intersection of two of their branches.

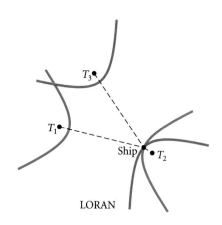

LORAN

5.3 | Exercise Set

In Exercises 1–6, match each equation with one of the graphs (a)–(f), which follow.

a)

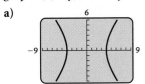

b)

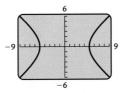

c)

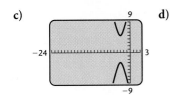

d)

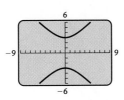

e)

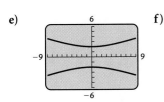

f)

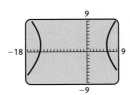

1. $\dfrac{x^2}{25} - \dfrac{y^2}{9} = 1$

2. $\dfrac{y^2}{4} - \dfrac{x^2}{36} = 1$

3. $\dfrac{(y - 1)^2}{16} - \dfrac{(x + 3)^2}{1} = 1$

4. $\dfrac{(x + 4)^2}{100} - \dfrac{(y - 2)^2}{81} = 1$

5. $25x^2 - 16y^2 = 400$

6. $y^2 - x^2 = 9$

Find an equation of a hyperbola satisfying the given conditions.

7. Vertices at $(0, 3)$ and $(0, -3)$; foci at $(0, 5)$ and $(0, -5)$

8. Vertices at $(1, 0)$ and $(-1, 0)$; foci at $(2, 0)$ and $(-2, 0)$

9. Asymptotes $y = \frac{3}{2}x$, $y = -\frac{3}{2}x$; one vertex $(2, 0)$

10. Asymptotes $y = \frac{5}{4}x$, $y = -\frac{5}{4}x$; one vertex $(0, 3)$

Find the center, the vertices, the foci, and the asymptotes. Then draw the graph.

11. $\dfrac{x^2}{4} - \dfrac{y^2}{4} = 1$

12. $\dfrac{x^2}{1} - \dfrac{y^2}{9} = 1$

13. $\dfrac{(x - 2)^2}{9} - \dfrac{(y + 5)^2}{1} = 1$

14. $\dfrac{(x - 5)^2}{16} - \dfrac{(y + 2)^2}{9} = 1$

15. $\dfrac{(y + 3)^2}{4} - \dfrac{(x + 1)^2}{16} = 1$

16. $\dfrac{(y + 4)^2}{25} - \dfrac{(x + 2)^2}{16} = 1$

17. $x^2 - 4y^2 = 4$

18. $4x^2 - y^2 = 16$

19. $9y^2 - x^2 = 81$

20. $y^2 - 4x^2 = 4$

21. $x^2 - y^2 = 2$

22. $x^2 - y^2 = 3$

23. $y^2 - x^2 = \frac{1}{4}$

24. $y^2 - x^2 = \frac{1}{9}$

Find the center, the vertices, the foci, and the asymptotes of each hyperbola. Then draw the graph.

25. $x^2 - y^2 - 2x - 4y - 4 = 0$

26. $4x^2 - y^2 + 8x - 4y - 4 = 0$

27. $36x^2 - y^2 - 24x + 6y - 41 = 0$

28. $9x^2 - 4y^2 + 54x + 8y + 41 = 0$

29. $9y^2 - 4x^2 - 18y + 24x - 63 = 0$

30. $x^2 - 25y^2 + 6x - 50y = 41$

31. $x^2 - y^2 - 2x - 4y = 4$

32. $9y^2 - 4x^2 - 54y - 8x + 41 = 0$

33. $y^2 - x^2 - 6x - 8y - 29 = 0$

34. $x^2 - y^2 = 8x - 2y - 13$

The **eccentricity** of a hyperbola is defined as $e = c/a$. For a hyperbola, $c > a > 0$, so $e > 1$. When e is close to 1, a hyperbola appears to be very narrow. As the eccentricity increases, the hyperbola becomes "wider."

35. Observe the shapes of the hyperbolas in Examples 2 and 3. Which hyperbola has the larger eccentricity? Confirm your answer by computing the eccentricity of each hyperbola.

36. Which hyperbola below has the larger eccentricity? (Assume that the coordinate systems have the same scale.)

a)

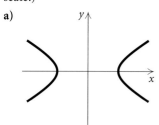

b)

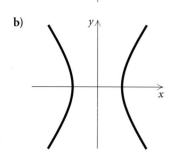

37. Find an equation of a hyperbola with vertices $(3, 7)$ and $(-3, 7)$ and $e = \frac{5}{3}$.

38. Find an equation of a hyperbola with vertices $(-1, 3)$ and $(-1, 7)$ and $e = 4$.

39. *Hyperbolic Mirror.* Certain telescopes contain both a parabolic and a hyperbolic mirror. In the telescope shown in the figure below, the parabola and the hyperbola share focus F_1, which is 14 m above the vertex of the parabola. The hyperbola's second focus F_2 is 2 m above the parabola's vertex. The vertex of the hyperbolic mirror is 1 m below F_1. Position a coordinate system with the origin at the center of the hyperbola and with the foci on the y-axis. Then find the equation of the hyperbola.

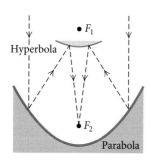

40. *Nuclear Cooling Tower.* A cross section of a nuclear cooling tower is a hyperbola with equation

$$\frac{x^2}{90^2} - \frac{y^2}{130^2} = 1.$$

The tower is 450 ft tall and the distance from the top of the tower to the center of the hyperbola is half the distance from the base of the tower to the center of the hyperbola. Find the diameter of the top and the base of the tower.

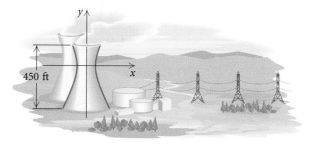

Skill Maintenance

Solve.

41. $x + y = 5,$
$\quad x - y = 7$

42. $3x - 2y = 5,$
$\quad 5x + 2y = 3$

43. $2x - 3y = 7,$
$\quad 3x + 5y = 1$

44. $3x + 2y = -1,$
$\quad 2x + 3y = 6$

Synthesis

45. ◆ How does the graph of a parabola differ from the graph of one branch of a hyperbola?

46. ◆ Would you prefer to graph the hyperbola

$$\frac{(y + 5)^2}{16} + \frac{(x - 4)^2}{36} = 1$$

by hand or using a grapher? Explain why you answered as you did.

Find an equation of a hyperbola satisfying the given conditions.

47. Vertices at $(3, -8)$ and $(3, -2)$;
asymptotes $y = 3x - 14$, $y = -3x + 4$

48. Vertices at $(-9, 4)$ and $(-5, 4)$;
asymptotes $y = 3x + 25$, $y = -3x - 17$

Use a grapher to find the center, the vertices, and the asymptotes.

49. $5x^2 - 3.5y^2 + 14.6x - 6.7y + 3.4 = 0$

50. $x^2 - y^2 - 2.046x - 4.088y - 4.228 = 0$

51. *Navigation.*   Two radio transmitters positioned 300 mi apart send simultaneous signals to a ship that is 200 mi offshore, sailing parallel to the shoreline. The signal from transmitter S reaches the ship 200 microseconds later than the signal from transmitter T. The signals travel at a speed of 186,000 miles per second, or 0.186 mile per microsecond. Find the equation of the hyperbola with foci S and T on which the ship is located. (*Hint:* For any point on the hyperbola, the absolute value of the difference of its distances from the foci is $2a$.)

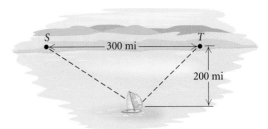

5.4
Nonlinear Systems of Equations

- *Solve a nonlinear system of equations.*
- *Use nonlinear systems of equations to solve applied problems.*

A system of equations is composed of two or more equations considered simultaneously. We now consider systems of two equations in two variables in which at least one equation is not linear.

Interactive Discovery

Graph the circle $x^2 + y^2 = 5$ and the line $y = x - 4$ in the window $[-6, 6, -4, 4]$. How many points of intersection are there? What conclusion can you draw about the number of real-number solutions of the system of equations composed of $x^2 + y^2 = 5$ and $y = x - 4$?

Now graph $x^2 + y^2 = 5$ and $y = -0.5x + 2.5$ in the same window and determine the number of points of intersection. What does each point of intersection represent?

Finally, graph $x^2 + y^2 = 5$ and $y = x$ in the same window. How many points of intersection are there? What does each represent?

Nonlinear Systems of Equations

The graph of a nonlinear system of equations can have no point of intersection or one or more points of intersection. The coordinates of each point of intersection are a solution of the system of equations. When no point of intersection exists, the system of equations has no real-number solution.

Real-number solutions of nonlinear systems of equations can be found algebraically, using the substitution or elimination method, or graphically. However, since a graph shows *only* real-number solutions, solutions involving imaginary numbers must always be found algebraically.

Example 1 Solve the following system of equations:

$$x^2 + y^2 = 25, \quad (1) \qquad \text{(The graph is a circle.)}$$
$$3x - 4y = 0. \quad (2) \qquad \text{(The graph is a line.)}$$

ALGEBRAIC SOLUTION

We use the substitution method. First, we solve equation (2) for x:

$$x = \tfrac{4}{3}y. \quad (3) \qquad \text{We could have solved for } y \text{ instead.}$$

Next, we substitute $\tfrac{4}{3}y$ for x in equation (1) and solve for y:

$$\left(\tfrac{4}{3}y\right)^2 + y^2 = 25$$
$$\tfrac{16}{9}y^2 + y^2 = 25$$
$$\tfrac{25}{9}y^2 = 25$$
$$y^2 = 9 \qquad \text{Multiplying by } \tfrac{9}{25}$$
$$y = \pm 3.$$

Now we substitute these numbers for y in equation (3) and solve for x:

$$x = \tfrac{4}{3}(3) = 4, \qquad \text{The pair } (4, 3) \text{ appears to be a solution.}$$

$$x = \tfrac{4}{3}(-3) = -4. \qquad \text{The pair } (-4, -3) \text{ appears to be a solution.}$$

CHECK: For (4, 3):

$$\begin{array}{c|c}
x^2 + y^2 = 25 & 3x - 4y = 0 \\
\hline
4^2 + 3^2 \ ? \ 25 & 3(4) - 4(3) \ ? \ 0 \\
16 + 9 & 12 - 12 \\
25 \ | \ 25 \quad \text{TRUE} & 0 \ | \ 0 \quad \text{TRUE}
\end{array}$$

For (−4, −3):

$$\begin{array}{c|c}
x^2 + y^2 = 25 & 3x - 4y = 0 \\
\hline
(-4)^2 + (-3)^2 \ ? \ 25 & 3(-4) - 4(-3) \ ? \ 0 \\
16 + 9 & -12 + 12 \\
25 \ | \ 25 \quad \text{TRUE} & 0 \ | \ 0 \quad \text{TRUE}
\end{array}$$

The pairs (4, 3) and (−4, −3) check, so they are the solutions.

GRAPHICAL SOLUTION

We graph both equations in the same viewing window and note that there are two points of intersection. We can find their coordinates using the INTERSECT feature or the TRACE and ZOOM features.

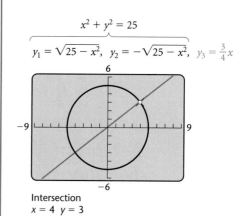

$$y_1 = \sqrt{25 - x^2}, \quad y_2 = -\sqrt{25 - x^2}, \quad y_3 = \tfrac{3}{4}x$$

Intersection
$x = 4 \quad y = 3$

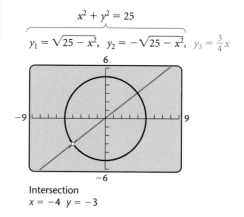

$$y_1 = \sqrt{25 - x^2}, \quad y_2 = -\sqrt{25 - x^2}, \quad y_3 = \tfrac{3}{4}x$$

Intersection
$x = -4 \quad y = -3$

The graph can also be used to check the solutions that we found algebraically.

In the algebraic solution above, suppose that to find x, we had substituted 3 and −3 in equation (1) rather than equation (3). If $y = 3$, $y^2 = 9$, and if $y = -3$, $y^2 = 9$, so both substitutions can be performed at the same time:

$$x^2 + y^2 = 25 \quad (1)$$
$$x^2 + (\pm 3)^2 = 25$$
$$x^2 + 9 = 25$$
$$x^2 = 16$$
$$x = \pm 4.$$

Thus, if $y = 3$, $x = 4$ or $x = -4$, and if $y = -3$, $x = 4$ or $x = -4$. The possible solutions are $(4, 3)$, $(-4, 3)$, $(4, -3)$, and $(-4, -3)$. A check reveals that $(4, -3)$ and $(-4, 3)$ are not solutions of equation (2). Had we graphed the system of equations before solving algebraically, it would have been clear that there are only two real-number solutions.

Example 2   Solve the following system of equations:

$$x + y = 5, \quad (1) \qquad \text{(The graph is a line.)}$$
$$y = 3 - x^2. \quad (2) \qquad \text{(The graph is a parabola.)}$$

ALGEBRAIC SOLUTION

We use the substitution method, substituting $3 - x^2$ for y in the first equation:

$$x + 3 - x^2 = 5$$
$$-x^2 + x - 2 = 0$$
$$x^2 - x + 2 = 0.$$

Next, we use the quadratic formula:

$$x = \frac{-b \pm \sqrt{b^2 - 4ac}}{2a} = \frac{-(-1) \pm \sqrt{(-1)^2 - 4(1)(2)}}{2(1)}$$
$$= \frac{1 \pm \sqrt{1 - 8}}{2} = \frac{1 \pm \sqrt{-7}}{2} = \frac{1 \pm i\sqrt{7}}{2} = \frac{1}{2} \pm \frac{\sqrt{7}}{2}i.$$

Now, we substitute these values for x in equation (1) and solve for y:

$$\frac{1}{2} + \frac{\sqrt{7}}{2}i + y = 5$$
$$y = 5 - \frac{1}{2} - \frac{\sqrt{7}}{2}i = \frac{9}{2} - \frac{\sqrt{7}}{2}i$$

and $\dfrac{1}{2} - \dfrac{\sqrt{7}}{2}i + y = 5$

$$y = 5 - \frac{1}{2} + \frac{\sqrt{7}}{2}i = \frac{9}{2} + \frac{\sqrt{7}}{2}i.$$

The solutions are

$$\left(\frac{1}{2} + \frac{\sqrt{7}}{2}i, \frac{9}{2} - \frac{\sqrt{7}}{2}i \right) \quad \text{and} \quad \left(\frac{1}{2} - \frac{\sqrt{7}}{2}i, \frac{9}{2} + \frac{\sqrt{7}}{2}i \right).$$

There are no real-number solutions.

GRAPHICAL SOLUTION

We graph both equations in the same viewing window.

$y_1 = 5 - x, \quad y_2 = 3 - x^2$

Note that there are no points of intersection. This indicates that there are no real-number solutions. Algebra must be used, as at left, to find the complex-number solutions.

Example 3   Solve the following system of equations:

$$2x^2 + 5y^2 = 39, \quad (1) \qquad \text{(The graph is an ellipse.)}$$
$$3x^2 - y^2 = -1. \quad (2) \qquad \text{(The graph is a hyperbola.)}$$

ALGEBRAIC SOLUTION

We use the elimination method. First we multiply equation (2) by 5 and add to eliminate the y^2-term:

$$
\begin{array}{ll}
2x^2 + 5y^2 = 39 & (1) \\
\underline{15x^2 - 5y^2 = -5} & \text{Multiplying (2) by 5} \\
17x^2 \qquad\ = 34 & \text{Adding} \\
\qquad x^2 = 2 \\
\qquad x = \pm\sqrt{2}.
\end{array}
$$

If $x = \sqrt{2}$, $x^2 = 2$, and if $x = -\sqrt{2}$, $x^2 = 2$. Thus substituting $\sqrt{2}$ or $-\sqrt{2}$ for x in equation (2) gives us

$$
\begin{aligned}
3(\pm\sqrt{2})^2 - y^2 &= -1 \\
3 \cdot 2 - y^2 &= -1 \\
6 - y^2 &= -1 \\
-y^2 &= -7 \\
y^2 &= 7 \\
y &= \pm\sqrt{7}.
\end{aligned}
$$

Thus, for $x = \sqrt{2}$, we have $y = \sqrt{7}$ or $y = -\sqrt{7}$, and for $x = -\sqrt{2}$, we have $y = \sqrt{7}$ or $y = -\sqrt{7}$. The possible solutions are $(\sqrt{2}, \sqrt{7})$, $(\sqrt{2}, -\sqrt{7})$, $(-\sqrt{2}, \sqrt{7})$, and $(-\sqrt{2}, -\sqrt{7})$. All four pairs check, so they are the solutions.

GRAPHICAL SOLUTION

We graph both equations in the same viewing window and note that there are four points of intersection. We can use the INTERSECT feature or the TRACE and ZOOM features to find their coordinates.

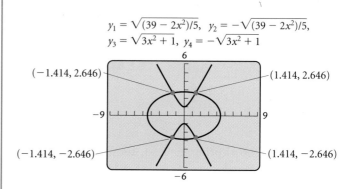

$$y_1 = \sqrt{(39 - 2x^2)/5}, \quad y_2 = -\sqrt{(39 - 2x^2)/5},$$
$$y_3 = \sqrt{3x^2 + 1}, \quad y_4 = -\sqrt{3x^2 + 1}$$

Note that the algebraic method yields exact solutions whereas the graphical method yields decimal approximations of the solutions.

Example 4 Solve the following system of equations:

$$x^2 - 3y^2 = 6, \quad (1)$$
$$xy = 3. \quad (2)$$

r--*ALGEBRAIC SOLUTION*

We use the substitution method. First, we solve equation (2) for y:

$$xy = 3$$
$$y = \frac{3}{x}. \quad (3)$$

Then we substitute $3/x$ for y in equation (1) and solve for x:

$$x^2 - 3\left(\frac{3}{x}\right)^2 = 6$$

$$x^2 - 3 \cdot \frac{9}{x^2} = 6$$

$$x^2 - \frac{27}{x^2} = 6$$

$$x^4 - 27 = 6x^2 \qquad \text{Multiplying by } x^2$$

$$x^4 - 6x^2 - 27 = 0$$

$$u^2 - 6u - 27 = 0 \qquad \text{Letting } u = x^2$$

$$(u - 9)(u + 3) = 0 \qquad \text{Factoring}$$

$$u = 9 \quad \text{or} \quad u = -3 \qquad \text{Principle of zero products}$$

$$x^2 = 9 \quad \text{or} \quad x^2 = -3$$

$$x = \pm 3 \quad \text{or} \quad x = \pm i\sqrt{3}.$$

Since $y = 3/x$,

when $x = 3$, $\qquad y = \dfrac{3}{3} = 1;$

when $x = -3$, $\qquad y = \dfrac{3}{-3} = -1;$

when $x = i\sqrt{3}$, $\qquad y = \dfrac{3}{i\sqrt{3}} = \dfrac{3}{i\sqrt{3}} \cdot \dfrac{-i\sqrt{3}}{-i\sqrt{3}} = -i\sqrt{3};$

when $x = -i\sqrt{3}$, $\quad y = \dfrac{3}{-i\sqrt{3}} = \dfrac{3}{-i\sqrt{3}} \cdot \dfrac{i\sqrt{3}}{i\sqrt{3}} = i\sqrt{3}.$

The pairs $(3, 1)$, $(-3, -1)$, $(i\sqrt{3}, -i\sqrt{3})$, and $(-i\sqrt{3}, i\sqrt{3})$ check, so they are the solutions.

r--*GRAPHICAL SOLUTION*

We graph both equations in the same viewing window and find the coordinates of their points of intersection.

$$x^2 - 3y^2 = 6$$
$$y_1 = \sqrt{(x^2 - 6)/3}, \quad y_2 = -\sqrt{(x^2 - 6)/3}, \quad y_3 = 3/x$$

Again, note that the graphical method yields only the real-number solutions of the system of equations. The algebraic method must be used to find *all* the solutions.

Modeling and Problem Solving

Example 5 *Dimensions of a Piece of Land.* For a student recreation building at Southport Community College, an architect wants to lay out a rectangular piece of land that has a perimeter of 204 m and an area of 2565 m^2. Find the dimensions of the piece of land.

SOLUTION

1. **Familiarize.** We make a drawing and label it, letting l = the length, in meters, and w = the width, in meters.

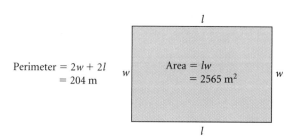

$$l$$

$$\text{Perimeter} = 2w + 2l$$
$$= 204 \text{ m}$$

$$w \quad \begin{array}{c}\text{Area} = lw \\ = 2565 \text{ m}^2\end{array} \quad w$$

$$l$$

2. **Translate.** We now have the following:

Perimeter: $2w + 2l = 204$,   (1)

Area: $lw = 2565.$ (2)

3. **Carry out.** We solve the system of equations both algebraically and graphically.

ALGEBRAIC SOLUTION

We solve the system of equations

$$2w + 2l = 204,$$
$$lw = 2565.$$

Solving the second equation for l gives us $l = 2565/w$. We then substitute $2565/w$ for l in the first equation and solve for w:

$$2w + 2\left(\frac{2565}{w}\right) = 204$$

$$2w^2 + 2(2565) = 204w \qquad \text{Multiplying by } w$$

$$2w^2 - 204w + 2(2565) = 0$$

$$w^2 - 102w + 2565 = 0 \qquad \text{Multiplying by } \tfrac{1}{2}$$

$$(w - 57)(w - 45) = 0$$

$$w = 57 \quad or \quad w = 45.$$

                     Principle of zero products

If $w = 57$, then $l = 2565/w = 2565/57 = 45$. If $w = 45$, then $l = 2565/w = 2565/45 = 57$. Since length is generally considered to be longer than width, we have the solution $l = 57$ and $w = 45$, or $(57, 45)$.

GRAPHICAL SOLUTION

We replace l with x and w with y, graph $y_1 = (204 - 2x)/2$ and $y_2 = 2565/x$, and find the point(s) of intersection of the graphs.

$$y_1 = (204 - 2x)/2, \ y_2 = 2565/x$$

(57, 45)

(45, 57)

As in the algebraic solution, we have two possible solutions: $(45, 57)$ and $(57, 45)$. Since length (x) is generally considered to be longer than width (y), we have the solution $(57, 45)$.

4. **Check.** If $l = 57$ and $w = 45$, the perimeter is $2 \cdot 57 + 2 \cdot 45$, or 204. The area is $57 \cdot 45$, or 2565. The numbers check.

5. **State.** The length of the piece of land is 57 m and the width is 45 m.

5.4 Exercise Set

In Exercises 1–6, match each system of equations with one of the graphs (a)–(f), which follow.

a)

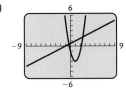

b)

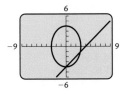

c)

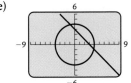

d)

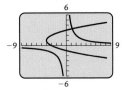

e)

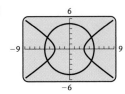

f)

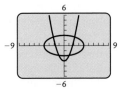

1. $x^2 + y^2 = 16,$
$x + y = 3$

2. $16x^2 + 9y^2 = 144,$
$x - y = 4$

3. $y = x^2 - 4x - 2,$
$2y - x = 1$

4. $4x^2 - 9y^2 = 36,$
$x^2 + y^2 = 25$

5. $y = x^2 - 3,$
$x^2 + 4y^2 = 16$

6. $y^2 - 2y = x + 3,$
$xy = 4$

Solve.

7. $x^2 + y^2 = 25,$
$y - x = 1$

8. $x^2 + y^2 = 100,$
$y - x = 2$

9. $4x^2 + 9y^2 = 36,$
$3y + 2x = 6$

10. $9x^2 + 4y^2 = 36,$
$3x + 2y = 6$

11. $x^2 + y^2 = 25,$
$y^2 = x + 5$

12. $y = x^2,$
$x = y^2$

13. $x^2 + y^2 = 9,$
$x^2 - y^2 = 9$

14. $y^2 - 4x^2 = 4,$
$4x^2 + y^2 = 4$

15. $y^2 - x^2 = 9,$
$2x - 3 = y$

16. $x + y = -6,$
$xy = -7$

17. $y^2 = x + 3,$
$2y = x + 4$

18. $y = x^2,$
$3x = y + 2$

19. $x^2 + y^2 = 25,$
$xy = 12$

20. $x^2 - y^2 = 16,$
$x + y^2 = 4$

21. $x^2 + y^2 = 4,$
$16x^2 + 9y^2 = 144$

22. $x^2 + y^2 = 25,$
$25x^2 + 16y^2 = 400$

23. $x^2 + 4y^2 = 25,$
$x + 2y = 7$

24. $y^2 - x^2 = 16,$
$2x - y = 1$

25. $x^2 - xy + 3y^2 = 27,$
$x - y = 2$

26. $2y^2 + xy + x^2 = 7,$
$x - 2y = 5$

27. $x^2 + y^2 = 16,$
$y^2 - 2x^2 = 10$

28. $x^2 + y^2 = 14,$
$x^2 - y^2 = 4$

29. $x^2 + y^2 = 5,$
$xy = 2$

30. $x^2 + y^2 = 20,$
$xy = 8$

31. $3x + y = 7,$
$4x^2 + 5y = 56$

32. $2y^2 + xy = 5,$
$4y + x = 7$

33. $a + b = 7,$
$ab = 4$

34. $p + q = -4,$
$pq = -5$

35. $x^2 + y^2 = 13,$
$xy = 6$

36. $x^2 + 4y^2 = 20,$
$xy = 4$

37. $x^2 + y^2 + 6y + 5 = 0,$
$x^2 + y^2 - 2x - 8 = 0$

38. $2xy + 3y^2 = 7,$
$3xy - 2y^2 = 4$

39. $2a + b = 1,$
$b = 4 - a^2$

40. $4x^2 + 9y^2 = 36,$
$x + 3y = 3$

41. $a^2 + b^2 = 89,$
$a - b = 3$

42. $xy = 4,$
$x + y = 5$

43. $xy - y^2 = 2,$
$2xy - 3y^2 = 0$

44. $4a^2 - 25b^2 = 0,$
$2a^2 - 10b^2 = 3b + 4$

45. $m^2 - 3mn + n^2 + 1 = 0,$
$3m^2 - mn + 3n^2 = 13$

46. $ab - b^2 = -4,$
$ab - 2b^2 = -6$

47. $x^2 + y^2 = 5,$
$x - y = 8$

48. $4x^2 + 9y^2 = 36,$
$y - x = 8$

49. $a^2 + b^2 = 14,$
$ab = 3\sqrt{5}$

50. $x^2 + xy = 5,$
$2x^2 + xy = 2$

51. $x^2 + y^2 = 25,$
$9x^2 + 4y^2 = 36$

52. $x^2 + y^2 = 1,$
$9x^2 - 16y^2 = 144$

53. $5y^2 - x^2 = 1,$
$xy = 2$

54. $x^2 - 7y^2 = 6,$
$xy = 1$

55. *Picture Frame Dimensions.* Frank's Frame Shop is building a frame for a rectangular oil painting with a perimeter of 28 cm and a diagonal of 10 cm. Find the dimensions of the painting.

56. *Landscaping.*   Green Leaf Landscaping is planting a rectangular wildflower garden with a perimeter of 6 m and a diagonal of $\sqrt{5}$ m. Find the dimensions of the garden.

57. *Pamphlet Design.*   A graphic artist is designing a rectangular advertising pamphlet that is to have an area of 20 in² and a perimeter of 18 in. Find the dimensions of the pamphlet.

58. *Sign Dimensions.*   Peden's Advertising is building a rectangular sign with an area of 2 yd² and a perimeter of 6 yd. Find the dimensions of the sign.

59. *Banner Design.*   A rectangular banner with an area of $\sqrt{3}$ m² is being designed to advertise an exhibit at the Madison Art League. The length of a diagonal is 2 m. Find the dimensions of the banner.

Area = $\sqrt{3}$ m²

60. *Carpentry.*   A carpenter wants to make a rectangular tabletop with an area of $\sqrt{2}$ m² and a diagonal of $\sqrt{3}$ m. Find the dimensions of the tabletop.

61. *Fencing.*   It will take 210 yd of fencing to enclose a rectangular dog run. The area of the run is 2250 yd². What are the dimensions of the run?

62. *Office Dimensions.*   The diagonal of the floor of a rectangular office cubicle is 1 ft longer than the length of the cubicle and 3 ft longer than twice the width. Find the dimensions of the cubicle.

63. *Seed Test Plots.*   The Burton Seed Company has two square test plots. The sum of their areas is 832 ft² and the difference of their areas is 320 ft². Find the length of a side of each plot.

64. *Investment.*   Jenna made an investment for 1 yr that earned $7.50 simple interest. If the principal had been $25 more and the interest rate 1% less, the interest would have been the same. Find the principal and the rate.

Skill Maintenance

Solve, finding all solutions in [0°, 360°).

65. $\cot \theta = -\dfrac{1}{\sqrt{3}}$

66. $4 \cot \theta = 4$

67. $\dfrac{1}{2} \cot 2\theta = \dfrac{\sqrt{3}}{2}$

68. $\cot 2\theta = 0$

Synthesis

69. ◆ What would you say to a classmate who tells you that any system of nonlinear equations can be solved graphically?

70. ◆ Write a problem that can be translated to a system of nonlinear equations, and ask a classmate to solve it. Devise the problem so that the solution is "The dimensions of the rectangle are 6 ft by 8 ft."

71. Find an equation of the circle that passes through the points (2, 4) and (3, 3) and whose center is on the line $3x - y = 3$.

72. Find an equation of the circle that passes through the points (−2, 3) and (−4, 1) and whose center is on the line $5x + 8y = -2$.

73. Find an equation of an ellipse centered at the origin that passes through the points (1, $\sqrt{3}/2$) and ($\sqrt{3}$, 1/2).

74. Find an equation of a hyperbola of the type
$$\frac{x^2}{b^2} - \frac{y^2}{a^2} = 1$$
that passes through the points (−3, −3$\sqrt{5}$/2) and (−3/2, 0).

75. Find an equation of the circle that passes through the points (4, 6), (−6, 2), and (1, −3).

76. Find an equation of the circle that passes through the points (2, 3), (4, 5), and (0, −3).

77. Show that a hyperbola does not intersect its asymptotes. That is, solve the system of equations
$$\frac{x^2}{a^2} - \frac{y^2}{b^2} = 1,$$
$$y = \frac{b}{a}x \quad \left(\text{or } y = -\frac{b}{a}x\right).$$

78. *Numerical Relationship.*   Find two numbers whose product is 2 and the sum of whose reciprocals is $\frac{33}{8}$.

79. *Numerical Relationship.*   The square of a number exceeds twice the square of another number by $\frac{1}{8}$. The sum of their squares is $\frac{5}{16}$. Find the numbers.

80. *Box Dimensions.*   Four squares with sides 5 in. long are cut from the corners of a rectangular metal sheet that has an area of 340 in². The edges are bent up to form an open box with a volume of 350 in³. Find the dimensions of the box.

81. *Numerical Relationship.*   The sum of two numbers is 1, and their product is 1. Find the sum of their cubes. There is a method to solve this problem that is easier than solving a nonlinear system of equations. Can you discover it?

82. Solve for x and y:

$$x^2 - y^2 = a^2 - b^2,$$
$$x - y = a - b.$$

Solve.

83. $x^3 + y^3 = 72,$
$x + y = 6$

84. $a + b = \dfrac{5}{6},$

$\dfrac{a}{b} + \dfrac{b}{a} = \dfrac{13}{6}$

85. $p^2 + q^2 = 13,$
$\dfrac{1}{pq} = -\dfrac{1}{6}$

86. $x^2 + y^2 = 4,$
$(x - 1)^2 + y^2 = 4$

Solve using a grapher.

87. $x^2 + y^2 = 19{,}380{,}510.36,$
$27{,}942.25x - 6.125y = 0$

88. $2x + 2y = 1660,$
$xy = 35{,}325$

89. $14.5x^2 - 13.5y^2 - 64.5 = 0,$
$5.5x - 6.3y - 12.3 = 0$

90. $13.5xy + 15.6 = 0,$
$5.6x - 6.7y - 42.3 = 0$

91. $0.319x^2 + 2688.7y^2 = 56{,}548,$
$0.306x^2 - 2688.7y^2 = 43{,}452$

92. $18.465x^2 + 788.723y^2 = 6408,$
$106.535x^2 - 788.723y^2 = 2692$

5.5
Rotation of Axes

- *Use rotation of axes to graph conic sections.*
- *Use the discriminant to determine the type of conic represented by a given equation.*

In Section 5.1, we saw that conic sections can be defined algebraically using a second-degree equation of the form $Ax^2 + Bxy + Cy^2 + Dx + Ey + F = 0$. Up to this point, we have considered only equations of this form for which $B = 0$. Now we turn our attention to equations of conics that contain an xy-term.

Rotation of Axes

When B is non-zero, the graph of $Ax^2 + Bxy + Cy^2 + Dx + Ey + F = 0$ is a conic section with an axis that is not parallel to the x- or y-axis. We use a technique called **rotation of axes** when we graph such an equation. The goal is to rotate the x- and y-axes through a positive angle θ to yield an $x'y'$-coordinate system, as shown below. For the appropriate choice of θ, the graph of any conic section with an xy-term will have its axis parallel to the x'-axis or the y'-axis.

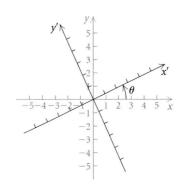

Algebraically we want to rewrite an equation

$$Ax^2 + Bxy + Cy^2 + Dx + Ey + F = 0$$

in the xy-coordinate system in the form

$$A'(x')^2 + C'(y')^2 + D'x' + E'y' + F' = 0$$

in the $x'y'$-coordinate system. Equations of this second type were graphed in Sections 5.1–5.3.

To achieve our goal, we find formulas relating the xy-coordinates of a point and the $x'y'$-coordinates of the same point. We begin by letting P be a point with coordinates (x, y) in the xy-coordinate system and (x', y') in the $x'y'$-coordinate system.

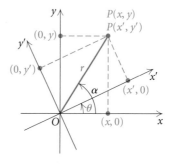

Let r represent the distance OP, and let α represent the angle from the x-axis to OP. Then

$$\cos \alpha = \frac{x}{r} \quad \text{and} \quad \sin \alpha = \frac{y}{r},$$

so

$$x = r \cos \alpha \quad \text{and} \quad y = r \sin \alpha.$$

We also see from the figure above that

$$\cos (\alpha - \theta) = \frac{x'}{r} \quad \text{and} \quad \sin (\alpha - \theta) = \frac{y'}{r},$$

so

$$x' = r \cos (\alpha - \theta) \quad \text{and} \quad y' = r \sin (\alpha - \theta).$$

Then

$$x' = r \cos \alpha \cos \theta + r \sin \alpha \sin \theta$$

and

$$y' = r \sin \alpha \cos \theta - r \cos \alpha \sin \theta.$$

Substituting x for $r \cos \alpha$ and y for $r \sin \alpha$ gives us

$$x' = x \cos \theta + y \sin \theta \tag{1}$$

and

$$y' = y \cos \theta - x \sin \theta. \tag{2}$$

We can use these formulas to find the $x'y'$-coordinates of any point given that point's xy-coordinates and an angle of rotation θ. To express xy-coordinates in terms of $x'y'$-coordinates and an angle of rotation θ, we solve the system composed of equations (1) and (2) above for x and y. (See Exercise 47.) We get

$$x = x' \cos \theta - y' \sin \theta$$

and

$$y = x' \sin \theta + y' \cos \theta.$$

Rotation of Axes Formulas

If the x- and y-axes are rotated about the origin through a positive acute angle θ, then the coordinates (x, y) and (x', y') of a point P in the xy- and $x'y'$-coordinate systems are related by the following formulas:

$$x' = x \cos \theta + y \sin \theta, \quad y' = -x \sin \theta + y \cos \theta;$$
$$x = x' \cos \theta - y' \sin \theta, \quad y = x' \sin \theta + y' \cos \theta.$$

Example 1 Suppose that the xy-axes are rotated through an angle of 45°. Write the equation $xy = 1$ in the $x'y'$-coordinate system.

SOLUTION   We substitute 45° for θ in the rotation of axes formulas for x and y:

$$x = x' \cos 45° - y' \sin 45°,$$
$$y = x' \sin 45° + y' \cos 45°.$$

Then we have

$$x = x'\left(\frac{\sqrt{2}}{2}\right) - y'\left(\frac{\sqrt{2}}{2}\right) = \frac{\sqrt{2}}{2}(x' - y')$$

and

$$y = x'\left(\frac{\sqrt{2}}{2}\right) + y'\left(\frac{\sqrt{2}}{2}\right) = \frac{\sqrt{2}}{2}(x' + y').$$

Next, we substitute these expressions for x and y in the equation $xy = 1$:

$$\frac{\sqrt{2}}{2}(x' - y') \cdot \frac{\sqrt{2}}{2}(x' + y') = 1$$

$$\frac{1}{2}[(x')^2 - (y')^2] = 1$$

$$\frac{(x')^2}{2} - \frac{(y')^2}{2} = 1, \quad \text{or} \quad \frac{(x')^2}{(\sqrt{2})^2} - \frac{(y')^2}{(\sqrt{2})^2} = 1.$$

We have the equation of a hyperbola in the $x'y'$-coordinate system with its axis on the x'-axis and with vertices $(-\sqrt{2}, 0)$ and $(\sqrt{2}, 0)$. Its asymptotes are $y' = -x'$ and $y' = x'$. These correspond to the axes of the xy-coordinate system.

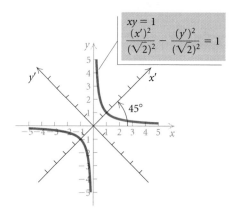

Now let's substitute the rotation of axes formulas for x and y in the equation

$$Ax^2 + Bxy + Cy^2 + Dx + Ey + F = 0.$$

We have

$$A(x' \cos \theta - y' \sin \theta)^2 + B(x' \cos \theta - y' \sin \theta)(x' \sin \theta + y' \cos \theta)$$
$$+ C(x' \sin \theta + y' \cos \theta)^2$$
$$+ D(x' \cos \theta - y' \sin \theta) + E(x' \sin \theta + y' \cos \theta) + F = 0.$$

Performing the operations indicated and collecting like terms yields the equation

$$A'(x')^2 + B'x'y' + C'(y')^2 + D'x' + E'y' + F' = 0, \qquad (3)$$

where

$$A' = A \cos^2 \theta + B \sin \theta \cos \theta + C \sin^2 \theta,$$
$$B' = 2(C - A) \sin \theta \cos \theta + B(\cos^2 \theta - \sin^2 \theta),$$
$$C' = A \sin^2 \theta - B \sin \theta \cos \theta + C \cos^2 \theta,$$
$$D' = D \cos \theta + E \sin \theta,$$
$$E' = -D \sin \theta + E \cos \theta, \quad \text{and}$$
$$F' = F.$$

Recall that our goal is to produce an equation without an $x'y'$-term, or with $B' = 0$. Then we must have

$$2(C - A) \sin \theta \cos \theta + B(\cos^2 \theta - \sin^2 \theta) = 0$$
$$(C - A) \sin 2\theta + B \cos 2\theta = 0 \qquad \text{Using double-angle formulas}$$
$$B \cos 2\theta = (A - C) \sin 2\theta$$
$$\frac{\cos 2\theta}{\sin 2\theta} = \frac{A - C}{B}$$
$$\cot 2\theta = \frac{A - C}{B}.$$

Thus, when θ is chosen so that

$$\cot 2\theta = \frac{A - C}{B},$$

equation (3) will have no $x'y'$-term. Although we will not do so here, it can be shown that we can always find θ such that $0° < 2\theta < 180°$, or $0° < \theta < 90°$.

Eliminating the xy-Term

To eliminate the xy-term from the equation

$$Ax^2 + Bxy + Cy^2 + Dx + Ey + F = 0, \quad B \neq 0,$$

select an angle θ such that

$$\cot 2\theta = \frac{A - C}{B}, \quad 0° < 2\theta < 180°,$$

and use the rotation of axes formulas.

Example 2 Graph the equation

$$3x^2 - 2\sqrt{3}xy + y^2 + 2x + 2\sqrt{3}y = 0.$$

SOLUTION We have

$$A = 3, \quad B = -2\sqrt{3}, \quad C = 1, \quad D = 2, \quad E = 2\sqrt{3}, \quad \text{and}$$
$$F = 0.$$

To select the angle of rotation θ, we must have

$$\cot 2\theta = \frac{A - C}{B} = \frac{3 - 1}{-2\sqrt{3}} = -\frac{1}{\sqrt{3}}.$$

Thus, $2\theta = 120°$, and $\theta = 60°$. We substitute this value for θ in the rotation of axes formulas for x and y:

$$x = x' \cos 60° - y' \sin 60°,$$
$$y = x' \sin 60° + y' \cos 60°.$$

This gives us

$$x = x' \cdot \frac{1}{2} - y' \cdot \frac{\sqrt{3}}{2} = \frac{x'}{2} - \frac{y'\sqrt{3}}{2}$$

and

$$y = x' \cdot \frac{\sqrt{3}}{2} + y' \cdot \frac{1}{2} = \frac{x'\sqrt{3}}{2} + \frac{y'}{2}.$$

Now we substitute these expressions for x and y in the given equation:

$$3\left(\frac{x'}{2} - \frac{y'\sqrt{3}}{2}\right)^2 - 2\sqrt{3}\left(\frac{x'}{2} - \frac{y'\sqrt{3}}{2}\right)\left(\frac{x'\sqrt{3}}{2} + \frac{y'}{2}\right) +$$
$$\left(\frac{x'\sqrt{3}}{2} + \frac{y'}{2}\right)^2 + 2\left(\frac{x'}{2} - \frac{y'\sqrt{3}}{2}\right) + 2\sqrt{3}\left(\frac{x'\sqrt{3}}{2} + \frac{y'}{2}\right) = 0.$$

After simplifying, we get

$$4(y')^2 + 4x' = 0, \quad \text{or}$$
$$x' = -(y')^2.$$

This is the equation of a parabola with its vertex at $(0, 0)$ of the $x'y'$-coordinate system and axis of symmetry $x' = 0$. We sketch the graph.

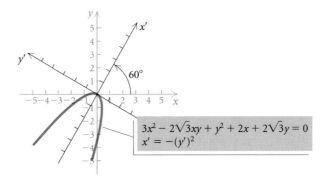

$$3x^2 - 2\sqrt{3}xy + y^2 + 2x + 2\sqrt{3}y = 0$$
$$x' = -(y')^2$$

The Discriminant

It is possible to determine the type of conic represented by the equation $Ax^2 + Bxy + Cy^2 + Dx + Ey + F = 0$ before rotating the axes. Using the expressions for A', B', and C' in terms of A, B, C, and θ developed earlier, it can be shown that

$$(B')^2 - 4A'C' = B^2 - 4AC.$$

Now when θ is chosen so that

$$\cot 2\theta = \frac{A - C}{B},$$

rotation of axes gives us an equation

$$A'(x')^2 + C'(y')^2 + D'x' + E'y' + F' = 0.$$

If A' and C' have the same sign, or $A'C' > 0$, then the graph of this equation is an ellipse or a circle. If A' and C' have different signs, or $A'C' < 0$, then the graph is a hyperbola. And, if either $A' = 0$ or $C' = 0$, or $A'C' = 0$, the graph is a parabola.

Since $B' = 0$ and $(B')^2 - 4A'C' = B^2 - 4AC$, it follows that $B^2 - 4AC = -4A'C'$. Then the graph is an ellipse or a circle if $B^2 - 4AC < 0$, a hyperbola if $B^2 - 4AC > 0$, or a parabola if $B^2 - 4AC = 0$. (There are certain special cases, called *degenerate conics*, where these statements do not hold, but we will not concern ourselves with these here.) The expression $B^2 - 4AC$ is the **discriminant** of the equation $Ax^2 + Bxy + Cy^2 + Dx + Ey + F = 0$.

The graph of the equation $Ax^2 + Bxy + Cy^2 + Dx + Ey + F = 0$ is, except in degenerate cases,

1. an ellipse or a circle if $B^2 - 4AC < 0$,
2. a hyperbola if $B^2 - 4AC > 0$, and
3. a parabola if $B^2 - 4AC = 0$.

Example 3   Graph the equation $3x^2 + 2xy + 3y^2 = 16$.

SOLUTION   We have

$$A = 3, \quad B = 2, \quad \text{and} \quad C = 3, \quad \text{so}$$
$$B^2 - 4AC = 2^2 - 4 \cdot 3 \cdot 3 = 4 - 36 = -32.$$

Since the discriminant is negative, the graph is an ellipse or a circle. Now, to rotate the axes, we begin by determining θ:

$$\cot 2\theta = \frac{A - C}{B} = \frac{3 - 3}{2} = \frac{0}{2} = 0.$$

Then $2\theta = 90°$ and $\theta = 45°$, so

$$\sin \theta = \frac{\sqrt{2}}{2} \quad \text{and} \quad \cos \theta = \frac{\sqrt{2}}{2}.$$

Substituting in the rotation of axes formulas gives

$$x = x' \cos \theta - y' \sin \theta = x'\left(\frac{\sqrt{2}}{2}\right) - y'\left(\frac{\sqrt{2}}{2}\right) = \frac{\sqrt{2}}{2}(x' - y')$$

and

$$y = x' \sin \theta + y' \cos \theta = x'\left(\frac{\sqrt{2}}{2}\right) + y'\left(\frac{\sqrt{2}}{2}\right) = \frac{\sqrt{2}}{2}(x' + y').$$

Now we substitute for x and y in the given equation:

$$3\left[\frac{\sqrt{2}}{2}(x' - y')\right]^2 + 2\left[\frac{\sqrt{2}}{2}(x' - y')\right]\left[\frac{\sqrt{2}}{2}(x' + y')\right] + 3\left[\frac{\sqrt{2}}{2}(x' + y')\right]^2 = 16.$$

After simplifying, we have

$$4(x')^2 + 2(y')^2 = 16, \quad \text{or}$$
$$\frac{(x')^2}{4} + \frac{(y')^2}{8} = 1.$$

This is the equation of an ellipse with vertices $(0, -\sqrt{8})$ and $(0, \sqrt{8})$, or $(0, -2\sqrt{2})$ and $(0, 2\sqrt{2})$, on the y'-axis. The x'-intercepts are $(-2, 0)$ and $(2, 0)$. We sketch the graph.

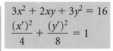

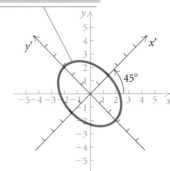

Example 4 Graph the equation $4x^2 - 24xy - 3y^2 - 156 = 0$.

SOLUTION We have

$$A = 4, \qquad B = -24, \quad \text{and} \quad C = -3, \qquad \text{so}$$
$$B^2 - 4AC = (-24)^2 - 4 \cdot 4(-3) = 576 + 48 = 624.$$

Since the discriminant is positive, the graph is a hyperbola. To rotate the axes, we begin by determining θ:

$$\cot 2\theta = \frac{A - C}{B} = \frac{4 - (-3)}{-24} = -\frac{7}{24}.$$

Since $\cot 2\theta < 0$, we have $90° < 2\theta < 180°$. From the triangle at left, we see that $\cos 2\theta = -\frac{7}{25}$.

Using half-angle formulas, we have

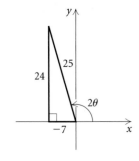

$$\sin \theta = \sqrt{\frac{1 - \cos 2\theta}{2}} = \sqrt{\frac{1 - \left(-\frac{7}{25}\right)}{2}} = \frac{4}{5}$$

and

$$\cos \theta = \sqrt{\frac{1 + \cos 2\theta}{2}} = \sqrt{\frac{1 + \left(-\frac{7}{25}\right)}{2}} = \frac{3}{5}.$$

Substituting in the rotation of axes formulas gives us

$$x = x' \cos \theta - y' \sin \theta = \tfrac{3}{5}x' - \tfrac{4}{5}y'$$

and

$$y = x' \sin \theta + y' \cos \theta = \tfrac{4}{5}x' + \tfrac{3}{5}y'.$$

Now we substitute for x and y in the given equation:

$$4\left(\tfrac{3}{5}x' - \tfrac{4}{5}y'\right)^2 - 24\left(\tfrac{3}{5}x' - \tfrac{4}{5}y'\right)\left(\tfrac{4}{5}x' + \tfrac{3}{5}y'\right) - 3\left(\tfrac{4}{5}x' + \tfrac{3}{5}y'\right)^2 - 156 = 0.$$

After simplifying, we have

$$13(y')^2 - 12(x')^2 - 156 = 0$$
$$13(y')^2 - 12(x')^2 = 156$$
$$\frac{(y')^2}{12} - \frac{(x')^2}{13} = 1.$$

The graph of this equation is a hyperbola with vertices $(0, -\sqrt{12})$ and $(0, \sqrt{12})$, or $(0, -2\sqrt{3})$ and $(0, 2\sqrt{3})$, on the y'-axis. Since we know that $\sin\theta = \frac{4}{5}$ and $0° < \theta < 90°$, we can use a grapher to find that $\theta \approx 53.1°$. Thus the xy-axes are rotated through an angle of about $53.1°$ in order to obtain the $x'y'$-axes. We sketch the graph.

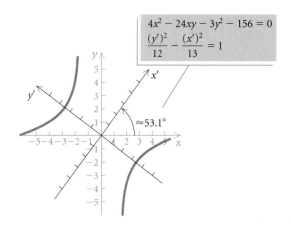

$$4x^2 - 24xy - 3y^2 - 156 = 0$$
$$\frac{(y')^2}{12} - \frac{(x')^2}{13} = 1$$

5.5 Exercise Set

For the given angle of rotation and coordinates of a point in the xy-coordinate system, find the coordinates of the point in the x'y'-coordinate system.

1. $\theta = 45°$, $(\sqrt{2}, -\sqrt{2})$ **2.** $\theta = 45°$, $(-1, 3)$

3. $\theta = 30°$, $(0, 2)$ **4.** $\theta = 60°$, $(0, \sqrt{3})$

For the given angle of rotation and coordinates of a point in the x'y'-coordinate system, find the coordinates of the point in the xy-coordinate system.

5. $\theta = 45°$, $(1, -1)$ **6.** $\theta = 45°$, $(-3\sqrt{2}, \sqrt{2})$

7. $\theta = 30°$, $(2, 0)$ **8.** $\theta = 60°$, $(-1, -\sqrt{3})$

Use the discriminant to determine whether the graph of the equation is an ellipse (or a circle), a hyperbola, or a parabola.

9. $3x^2 - 5xy + 3y^2 - 2x + 7y = 0$

10. $5x^2 + 6xy - 4y^2 + x - 3y + 4 = 0$

11. $x^2 - 3xy - 2y^2 + 12 = 0$

12. $4x^2 + 7xy + 2y^2 - 3x + y = 0$

13. $4x^2 - 12xy + 9y^2 - 3x + y = 0$

14. $6x^2 + 5xy + 6y^2 + 15 = 0$

15. $2x^2 - 8xy + 7y^2 + x - 2y + 1 = 0$

16. $x^2 + 6xy + 9y^2 - 3x + 4y = 0$

17. $8x^2 - 7xy + 5y^2 - 17 = 0$

18. $x^2 + xy - y^2 - 4x + 3y - 2 = 0$

Graph the equation.

19. $3x^2 + 2xy + 3y^2 = 16$

20. $3x^2 + 10xy + 3y^2 + 8 = 0$

21. $x^2 - 10xy + y^2 + 36 = 0$

22. $x^2 + 2xy + y^2 + 4\sqrt{2}x - 4\sqrt{2}y = 0$

23. $x^2 - 2\sqrt{3}xy + 3y^2 - 12\sqrt{3}x - 12y = 0$

24. $13x^2 + 6\sqrt{3}xy + 7y^2 - 16 = 0$

25. $7x^2 + 6\sqrt{3}xy + 13y^2 - 32 = 0$

26. $x^2 + 4xy + y^2 - 9 = 0$

27. $11x^2 + 10\sqrt{3}xy + y^2 = 32$

28. $5x^2 - 8xy + 5y^2 = 81$

29. $\sqrt{2}x^2 + 2\sqrt{2}xy + \sqrt{2}y^2 - 8x + 8y = 0$

30. $x^2 + 2\sqrt{3}xy + 3y^2 - 8x + 8\sqrt{3}y = 0$

31. $x^2 + 6\sqrt{3}xy - 5y^2 + 8x - 8\sqrt{3}y - 48 = 0$

32. $3x^2 - 2xy + 3y^2 - 6\sqrt{2}x + 2\sqrt{2}y - 26 = 0$

33. $x^2 + xy + y^2 = 24$

34. $4x^2 + 3\sqrt{3}xy + y^2 = 55$

35. $4x^2 - 4xy + y^2 - 8\sqrt{5}x - 16\sqrt{5}y = 0$

36. $9x^2 - 24xy + 16y^2 - 400x - 300y = 0$

37. $11x^2 + 7xy - 13y^2 = 621$

38. $3x^2 + 4xy + 6y^2 = 28$

Skill Maintenance

Convert to radian measure.

39. 120°

40. −315°

Convert to degree measure.

41. $\dfrac{\pi}{3}$

42. $\dfrac{3\pi}{4}$

43. Find *r*.

44. Graph the points $(-1, 3)$, $(0, 2)$, $(-4, -3)$, and $(1, 0)$.

Synthesis

45. ◆ Explain how the procedure you would follow for graphing an equation of the form $Ax^2 + Bxy + Cy^2 + Dx + Ey + F = 0$ when $B \neq 0$ differs from the procedure you would follow when $B = 0$.

46. ◆ Discuss some circumstances under which you might use rotation of axes.

47. Solve this system of equations for *x* and *y*:
$$x' = x \cos \theta + y \sin \theta,$$
$$y' = y \cos \theta - x \sin \theta.$$
Show your work.

48. Show that substituting $x' \cos \theta - y' \sin \theta$ for *x* and $x' \sin \theta + y' \cos \theta$ for *y* in the equation
$$Ax^2 + Bxy + Cy^2 + Dx + Ey + F = 0$$
yields the equation
$$A'(x')^2 + B'x'y' + C'(y')^2 + D'x' + E'y' + F' = 0,$$
where
$A' = A \cos^2 \theta + B \sin \theta \cos \theta + C \sin^2 \theta,$
$B' = 2(C - A) \sin \theta \cos \theta + B(\cos^2 \theta - \sin^2 \theta),$
$C' = A \sin^2 \theta - B \sin \theta \cos \theta + C \cos^2 \theta,$
$D' = D \cos \theta + E \sin \theta,$
$E' = -D \sin \theta + E \cos \theta,$ and
$F' = F.$

49. Show that $A + C = A' + C'$.

50. Show that for any angle θ, the equation $x^2 + y^2 = r^2$ becomes $(x')^2 + (y')^2 = r^2$ when the rotation of axes formulas are applied.

5.6

Polar Coordinates and Graphs

- *Graph points given their polar coordinates.*
- *Convert from rectangular to polar coordinates and from polar to rectangular coordinates.*
- *Convert from rectangular to polar equations and from polar to rectangular equations.*
- *Graph polar equations.*

Polar Coordinates

All graphing throughout this text has been done with rectangular coordinates, (x, y), in the Cartesian coordinate system. We now introduce the polar coordinate system. As shown in the diagram below, any point *P* has rectangular coordinates (x, y) and polar coordinates (r, θ). On a polar graph, the origin is called the **pole** and the positive half of the *x*-axis is called the **polar axis**. The point *P* can be plotted given the directed angle θ from the polar axis to the ray *OP* and the directed distance *r* from the pole to the point. The angle θ can be expressed in degrees or radians.

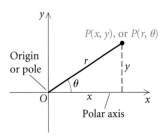

To plot points on a polar graph:

1. Locate the directed angle θ.

2. Move a directed distance r from the pole. If $r > 0$, move along ray OP. If $r < 0$, move in the opposite direction of ray OP.

Polar graph paper, shown below, facilitates plotting. Points B and G illustrate that θ may be in radians. Points E and F illustrate that the polar coordinates of a point are not unique.

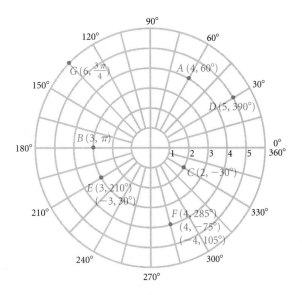

Example 1 Graph each of the following points.

a) $A(3, 60°)$ 　　　　　　　　　　**b)** $B(0, 10°)$

c) $C(-5, 120°)$ 　　　　　　　　**d)** $D(1, -60°)$

e) $E\left(2, \dfrac{3\pi}{2}\right)$ 　　　　　　　　**f)** $F\left(-4, \dfrac{\pi}{3}\right)$

SOLUTION

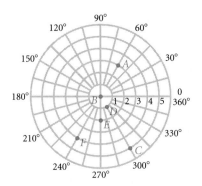

To convert from rectangular to polar coordinates and from polar to rectangular coordinates, we need to recall the following relationships.

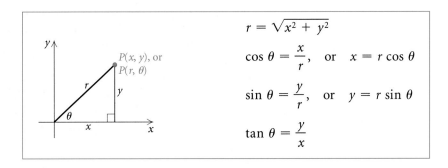

$$r = \sqrt{x^2 + y^2}$$

$$\cos\theta = \frac{x}{r}, \quad \text{or} \quad x = r\cos\theta$$

$$\sin\theta = \frac{y}{r}, \quad \text{or} \quad y = r\sin\theta$$

$$\tan\theta = \frac{y}{x}$$

Example 2   Convert each of the following to polar coordinates.

a) (3, 3)

b) $(2\sqrt{3}, -2)$

SOLUTION

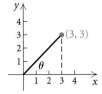

a) We first find r:

$$r = \sqrt{3^2 + 3^2}$$
$$= \sqrt{18} = 3\sqrt{2}.$$

Then we determine θ:

$$\tan\theta = \frac{3}{3} = 1; \quad \text{therefore,} \quad \theta = 45°, \text{ or } \frac{\pi}{4}.$$

We know that $\theta = \pi/4$ and not $5\pi/4$ since (3, 3) is in quadrant I. Thus, $(r, \theta) = (3\sqrt{2}, 45°)$, or $(3\sqrt{2}, \pi/4)$. Other possibilities for polar coordinates include $(3\sqrt{2}, -315°)$ and $(-3\sqrt{2}, 5\pi/4)$.

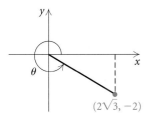

(2√3, −2)

b) We first find r:

$$r = \sqrt{(2\sqrt{3})^2 + (-2)^2} = \sqrt{12 + 4} = \sqrt{16} = 4.$$

Then we determine θ:

$$\tan \theta = \frac{-2}{2\sqrt{3}} = -\frac{1}{\sqrt{3}}; \quad \text{therefore,} \quad \theta = 330°, \text{ or } \frac{11\pi}{6}.$$

Thus, $(r, \theta) = (4, 330°)$, or $(4, 11\pi/6)$. Other possibilities for polar coordinates for this point include $(4, -\pi/6)$ and $(-4, 150°)$. ▬

It is easier to convert from polar to rectangular coordinates than from rectangular to polar coordinates.

Example 3   Convert each of the following to rectangular coordinates.

a) $\left(10, \dfrac{\pi}{3} \right)$ **b)** $(-5, 135°)$

SOLUTION

a) The ordered pair $(10, \pi/3)$ gives us $r = 10$ and $\theta = \pi/3$. We now find x and y:

$$x = r \cos \theta = 10 \cos \frac{\pi}{3} = 10 \cdot \frac{1}{2} = 5$$

and

$$y = r \sin \theta = 10 \sin \frac{\pi}{3} = 10 \cdot \frac{\sqrt{3}}{2} = 5\sqrt{3}.$$

Thus, $(x, y) = (5, 5\sqrt{3})$.

b) From the ordered pair $(-5, 135°)$, we know that $r = -5$ and $\theta = 135°$. We now find x and y:

$$x = -5 \cos 135° = -5 \cdot \left(-\frac{\sqrt{2}}{2} \right) = \frac{5\sqrt{2}}{2}$$

and

$$y = -5 \sin 135° = -5 \cdot \left(\frac{\sqrt{2}}{2} \right) = -\frac{5\sqrt{2}}{2}.$$

Thus, $(x, y) = (5\sqrt{2}/2, -5\sqrt{2}/2)$. ▬

The above conversions can be easily made with some graphers.

Polar and Rectangular Equations

Some curves have simpler equations in polar coordinates than in rectangular coordinates. For others, the reverse is true.

Example 4   Convert each of the following to a polar equation.

a) $x^2 + y^2 = 25$ **b)** $2x - y = 5$

SOLUTION

a) We have

$$x^2 + y^2 = 25$$
$$(r \cos \theta)^2 + (r \sin \theta)^2 = 25 \qquad \text{Substituting for } x \text{ and } y$$
$$r^2 \cos^2 \theta + r^2 \sin^2 \theta = 25$$
$$r^2(\cos^2 \theta + \sin^2 \theta) = 25$$
$$r^2 = 25 \qquad \cos^2 \theta + \sin^2 \theta = 1$$
$$r = 5.$$

This example illustrates that the polar equation of a circle centered at the origin is much simpler than the rectangular equation.

b) We have

$$2x - y = 5$$
$$2(r \cos \theta) - (r \sin \theta) = 5$$
$$r(2 \cos \theta - \sin \theta) = 5.$$

In this example, we see that the rectangular equation is simpler than the polar equation.

Example 5   Convert each of the following to a rectangular equation.

a) $r = 4$

b) $r \cos \theta = 6$

c) $r = 2 \cos \theta + 3 \sin \theta$

SOLUTION

a) We have

$$r = 4$$
$$\sqrt{x^2 + y^2} = 4 \qquad \text{Substituting for } r$$
$$x^2 + y^2 = 16. \qquad \text{Squaring}$$

In squaring, we must be careful not to introduce solutions of the equation that are not already present. In this case, we did not, because the graph of either equation is a circle of radius 4 centered at the origin.

b) We have

$$r \cos \theta = 6$$
$$x = 6. \qquad x = r \cos \theta$$

The graph of $r \cos \theta = 6$, or $x = 6$, is a vertical line.

c) We have

$$r = 2 \cos \theta + 3 \sin \theta$$
$$r^2 = 2r \cos \theta + 3r \sin \theta \qquad \text{Multiplying both sides by } r$$
$$x^2 + y^2 = 2x + 3y. \qquad \text{Substituting } x^2 + y^2 \text{ for } r^2, x$$
$$\text{for } r \cos \theta, \text{ and } y \text{ for } r \sin \theta.$$

Graphing Polar Equations

To graph a polar equation, we can make a table of values, choosing values of θ and calculating corresponding values of r. We plot the points and complete the graph, as in the rectangular case. A difference occurs in the case of a polar equation, because as θ increases sufficiently, points may, in some cases, begin to repeat and the curve will be traced again and again. If such a point is reached, the curve is complete.

Example 6   Make a hand-drawn graph of $r = 1 - \sin \theta$.

SOLUTION   We first make a table of values. The TABLE feature on a grapher is the most efficient way to create this list. Note that the points begin to repeat at $\theta = 360°$. We plot these points and draw the curve, as shown below.

θ	r
0°	1
15°	0.74118
30°	0.5
45°	0.29289
60°	0.13397
75°	0.03407
90°	0
105°	0.03407
120°	0.13397
135°	0.29289
150°	0.5
165°	0.74118
180°	1

θ	r
195°	1.2588
210°	1.5
225°	1.7071
240°	1.866
255°	1.9659
270°	2
285°	1.9659
300°	1.866
315°	1.7071
330°	1.5
345°	1.2588
360°	1
375°	0.74118
390°	0.5

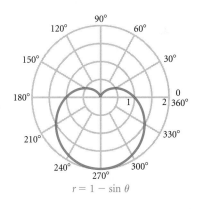

$r = 1 - \sin \theta$

Because of its heart shape, this curve is called a *cardioid*.

We plotted points in Example 6 because we feel that it is important to understand how these curves are developed. In the age of graphers, most of our work in graphing polar equations is done with technology. Nearly all graphers allow the graphing of polar equations. In general, the equation must be written first in the form $r = f(\theta)$. When graphing polar equations, it is necessary to decide on not only the best window dimensions but also the range of values for θ. Typically, we begin with a range of 0 to 2π for θ in radians and 0° to 360° for θ in degrees. Because most polar graphs are curved, it is important to square the window to minimize distortion.

Interactive Discovery

Graph $r = 4 \sin 3\theta$. Begin by setting the grapher in POLAR mode and use either of the following windows:

Window
(Radians)
  θmin $= 0$
  θmax $= 2\pi$
  θstep $= \pi/24$
  Xmin $= -9$
  Xmax $= 9$
  Xscl $= 1$
  Ymin $= -6$
  Ymax $= 6$
  Yscl $= 1$

Window
(Degrees)
  θmin $= 0$
  θmax $= 360$
  θstep $= 1$
  Xmin $= -9$
  Xmax $= 9$
  Xscl $= 1$
  Ymin $= -6$
  Ymax $= 6$
  Yscl $= 1$

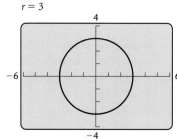

We observe the same graph in both windows. The grapher allows us to view the curve as it is formed.

Now graph each of the following equations and observe the effect of changing the coefficient of $\sin 3\theta$ and the coefficient of θ:

$$r = 2 \sin 3\theta, \qquad r = 6 \sin 3\theta, \qquad r = 4 \sin \theta,$$
$$r = 4 \sin 5\theta, \qquad r = 4 \sin 2\theta, \qquad r = 4 \sin 4\theta.$$

Polar equations of the form $r = a \cos n\theta$ and $r = a \sin n\theta$ have rose-shaped curves. **The number a determines the length of the petals, and the number n determines the number of petals.** If n is odd, there are n petals. If n is even, there are $2n$ petals.

Example 7   Graph each of the following polar equations. Try to visualize the shape of the curve before graphing it.

a) $r = 3$ **b)** $r = 5 \sin \theta$ **c)** $r = \cos \theta$ **d)** $r = 2 \csc \theta$

SOLUTION

a) $r = 3$

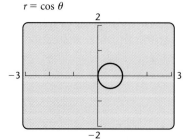

b) $r = 5 \sin \theta$

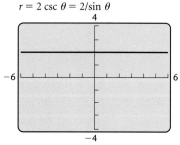

c) $r = \cos \theta$

d) $r = 2 \csc \theta = 2/\sin \theta$

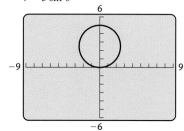

We can verify our graphs in parts (a)–(d) by converting the polar equation to the equivalent rectangular equation and visualizing the graph. For $r = 3$, we substitute $\sqrt{x^2 + y^2}$ for r and square. The resulting equation, $x^2 + y^2 = 3^2$, is the equation of a circle with radius 3 centered at the origin.

In part (d), we have

$$r = 2 \csc \theta$$

$$r = \frac{2}{\sin \theta}$$

$$r \sin \theta = 2$$

$$y = 2. \qquad \text{Substituting } y \text{ for } r \sin \theta$$

The graph of $y = 2$ is a horizontal line passing through $(0, 2)$.

Parts (b) and (c) can also be checked by rewriting the equations in rectangular form. ▬

Example 8 Graph: $r + 1 = 2 \cos 2\theta$.

SOLUTION We first solve for r:

$$r = 2 \cos 2\theta - 1.$$

We then obtain the following graph.

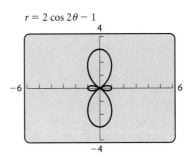

$r = 2 \cos 2\theta - 1$

▬

5.6 | *Exercise Set*

Graph each point on a polar grid.

1. $(2, 45°)$ **2.** $(4, \pi)$ **3.** $(3.5, 210°)$

4. $(-3, 135°)$ **5.** $\left(1, \dfrac{\pi}{6}\right)$ **6.** $(2.75, 150°)$

7. $\left(-5, \dfrac{\pi}{2}\right)$ **8.** $(0, 15°)$ **9.** $(3, -315°)$

10. $\left(1.2, -\dfrac{2\pi}{3}\right)$ **11.** $(4.3, -60°)$ **12.** $(3, 405°)$

Find polar coordinates of each of these points. Give three answers for each point.

13.   **14.**

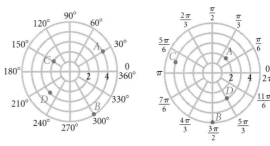

Find the polar coordinates of the point. Express the

angle in degrees and then in radians, using the smallest possible positive angle.

15. $(0, -3)$ **16.** $(-4, 4)$

17. $(3, -3\sqrt{3})$ **18.** $(-\sqrt{3}, 1)$

19. $(4\sqrt{3}, -4)$ **20.** $(2\sqrt{3}, 2)$

21. $(-\sqrt{2}, -\sqrt{2})$ **22.** $(-3, 3\sqrt{3})$

Use a grapher to convert from rectangular to polar coordinates. Express the answer in both degrees and radians, using the smallest possible positive angle.

23. $(3, 7)$ **24.** $(-2, -\sqrt{5})$

25. $(-\sqrt{10}, 3.4)$ **26.** $(0.9, -6)$

Find the rectangular coordinates of the point.

27. $(5, 60°)$ **28.** $(0, -23°)$

29. $(-3, 45°)$ **30.** $(6, 30°)$

31. $(3, -120°)$ **32.** $\left(7, \dfrac{\pi}{6}\right)$

33. $\left(-2, \dfrac{5\pi}{3}\right)$ **34.** $(1.4, 225°)$

Use a grapher to convert from polar to rectangular coordinates. Round the coordinates to the nearest hundredth.

35. $(3, -43°)$ **36.** $\left(-5, \dfrac{\pi}{7}\right)$

37. $\left(-4.2, \dfrac{3\pi}{5}\right)$ **38.** $(2.8, 166°)$

Convert to a polar equation.

39. $3x + 4y = 5$ **40.** $5x + 3y = 4$

41. $x = 5$ **42.** $y = 4$

43. $x^2 + y^2 = 36$ **44.** $x^2 - 4y^2 = 4$

45. $x^2 = 25y$ **46.** $2x - 9y + 3 = 0$

47. $y^2 - 5x - 25 = 0$ **48.** $x^2 + y^2 = 8y$

Convert to a rectangular equation.

49. $r = 5$ **50.** $\theta = \dfrac{3\pi}{4}$

51. $r \sin \theta = 2$ **52.** $r = -3 \sin \theta$

53. $r + r \cos \theta = 3$ **54.** $r = \dfrac{2}{1 - \sin \theta}$

55. $r - 9 \cos \theta = 7 \sin \theta$ **56.** $r + 5 \sin \theta = 7 \cos \theta$

57. $r = 5 \sec \theta$ **58.** $r = 3 \cos \theta$

Graph the equation by plotting points. Then check your work with a grapher.

59. $r = \sin \theta$ **60.** $r = 1 - \cos \theta$

61. $r = 4 \cos 2\theta$ **62.** $r = 1 - 2 \sin \theta$

In Exercises 63–74, use a grapher to match the equation with figures (a)–(l), which follow. Try matching the graphs mentally before using a grapher.

a)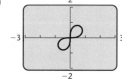

b)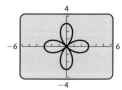

c)

d)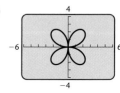

e)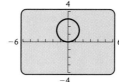

f)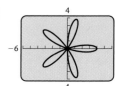

g)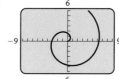

h)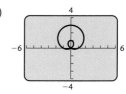

i)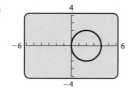

j)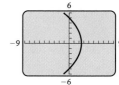

k)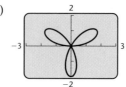

l)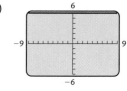

63. $r = 3 \sin 2\theta$ **64.** $r = 4 \cos \theta$

65. $r = \theta$ **66.** $r^2 = \sin 2\theta$

67. $r = \dfrac{5}{1 + \cos \theta}$ **68.** $r = 1 + 2 \sin \theta$

69. $r = 3 \cos 2\theta$ **70.** $r = 3 \sec \theta$

71. $r = 3 \sin \theta$ **72.** $r = 4 \cos 5\theta$

73. $r = 2 \sin 3\theta$ **74.** $r \sin \theta = 6$

Skill Maintenance

Graph.

75. $x = -3$

76. $y = 2$

77. $x^2 - y + 1 = 0$

78. $16x^2 + 4y^2 = 64$

79. $4x^2 + 9y^2 = 36$

80. $9x^2 - y^2 = 9$

Synthesis

81. ◈ Explain why the rectangular coordinates of a point are unique and the polar coordinates of a point are not unique.

82. ◈ Give an example of an equation that is easier to graph on a grapher in polar notation than in rectangular notation and explain why.

83. Convert to a rectangular equation:

$$r = \sec^2 \frac{\theta}{2}.$$

84. The center of a regular hexagon is at the origin, and one vertex is the point $(4, 0°)$. Find the coordinates of the other vertices.

Graph.

85. $r = \sin \theta \tan \theta$ (Cissoid)

86. $r = 3\theta$ (Spiral of Archimedes)

87. $r = e^{\theta/10}$ (Logarithmic spiral)

88. $r = 10^{2\theta}$ (Logarithmic spiral)

89. $r = \cos 2\theta \sec \theta$ (Strophoid)

90. $r = \cos 2\theta - 2$ (Peanut)

91. $r = \frac{1}{4} \tan^2 \theta \sec \theta$ (Semicubical parabola)

92. $r = \sin 2\theta + \cos \theta$ (Twisted sister)

5.7

Polar Equations of Conics

- *Graph polar equations of conics.*
- *Convert from polar to rectangular equations of conics.*
- *Find polar equations of conics.*

In Sections 5.1–5.3, we saw that the parabola, the ellipse, and the hyperbola have different definitions in rectangular coordinates. When polar coordinates are used, we can give a single definition that applies to all three conics.

An Alternate Definition of Conics

Let L be a fixed line (the *directrix*); let F be a fixed point (the *focus*), not on L; and let e be a positive constant (the *eccentricity*). A *conic* is the set of all points P in the plane such that

$$\frac{PF}{PL} = e,$$

where PF represents the distance from P to F and PL represents the distance from P to L. The conic is a parabola if $e = 1$, an ellipse if $e < 1$, and a hyperbola if $e > 1$.

Note that if $e = 1$, then $PF = PL$ and the alternate definition of a parabola is identical to the definition presented in Section 5.1.

Polar Equations of Conics

To derive equations for the conics in polar coordinates, we position the focus F at the pole and position the directrix L either perpendicular to the polar axis or parallel to it. In the figure below, we place L perpendicular to the polar axis and p units to the right of the focus, or pole.

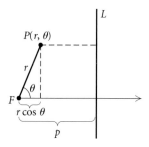

Note that $PL = p - r \cos \theta$. Then if P is any point on the conic, we have

$$\frac{PF}{PL} = e$$

$$\frac{r}{p - r \cos \theta} = e$$

$$r = ep - er \cos \theta$$

$$r + er \cos \theta = ep$$

$$r(1 + e \cos \theta) = ep$$

$$r = \frac{ep}{1 + e \cos \theta}.$$

Thus we see that the polar equation of a conic with focus at the pole and directrix perpendicular to the polar axis and p units to the right of the pole is

$$r = \frac{ep}{1 + e \cos \theta},$$

where e is the eccentricity of the conic.

For an ellipse and a hyperbola, we can make the following statement regarding eccentricity.

For an ellipse and a hyperbola, the *eccentricity* e is given by

$$e = \frac{c}{a},$$

where c is the distance from the center to a focus and a is the distance from the center to a vertex.

Interactive Discovery

Graph each of the following equations in POLAR mode in the square window $[-12, 12, -8, 8]$. Graph the third equation in DOT mode.

$$r = \frac{3}{1 + \cos \theta}, \qquad r = \frac{5}{1 + 0.4 \cos \theta}, \qquad r = \frac{8}{1 + 4 \cos \theta}$$

Describe each curve and use its equation to find its eccentricity. Do your graphs confirm the relationship between the graph of a conic and its eccentricity?

Example 1   Describe and graph the conic $r = \dfrac{18}{6 + 3 \cos \theta}$.

SOLUTION   We begin by dividing the numerator and the denominator by 6 to obtain a constant term of 1 in the denominator:

$$r = \frac{3}{1 + 0.5 \cos \theta}.$$

This equation is in the form

$$r = \frac{ep}{1 + e \cos \theta}$$

with $e = 0.5$. Since $e < 1$, the graph is an ellipse. Also, since $e = 0.5$ and $ep = 0.5p = 3$, we have $p = 6$. Thus the ellipse has a vertical directrix that lies 6 units to the right of the pole. We graph the equation in the square window $[-10, 5, -5, 5]$.

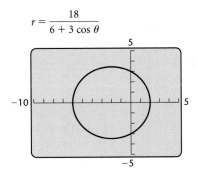

$$r = \frac{18}{6 + 3 \cos \theta}$$

Note that the major axis is horizontal and lies on the polar axis. The vertices are found by letting $\theta = 0$ and $\theta = \pi$. They are $(2, 0)$ and $(6, \pi)$. The center of the ellipse is at the midpoint of the segment connecting the vertices, or at $(2, \pi)$.

The length of the major axis is 8, so we have $2a = 8$, or $a = 4$. From the equation of the conic, we know that $e = 0.5$. Using the equation $e = c/a$, we can find that $c = 2$. Finally, using $a = 4$ and $c = 2$ in $b^2 = a^2 - c^2$ gives us

$$b^2 = 4^2 - 2^2 = 16 - 4 = 12$$
$$b = \sqrt{12}, \text{ or } 2\sqrt{3},$$

so the length of the minor axis is $\sqrt{12}$, or $2\sqrt{3}$. This is useful to know when sketching a hand-drawn graph of the conic.

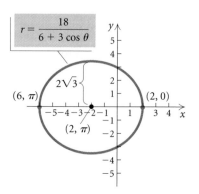

Other derivations similar to the one on page 378 lead to the following result.

Polar Equations of Conics

A polar equation of any of the four forms

$$r = \frac{ep}{1 \pm e \cos \theta}, \qquad r = \frac{ep}{1 \pm e \sin \theta}$$

is a conic section. The conic is a parabola if $e = 1$, an ellipse if $0 < e < 1$, and a hyperbola if $e > 1$.

The table below describes the polar equations of conics with a focus at the pole and the directrix either perpendicular to or parallel to the polar axis.

EQUATION	DESCRIPTION
$r = \dfrac{ep}{1 + e \cos \theta}$	Vertical directrix p units to the right of the pole (or focus)
$r = \dfrac{ep}{1 - e \cos \theta}$	Vertical directrix p units to the left of the pole (or focus)
$r = \dfrac{ep}{1 + e \sin \theta}$	Horizontal directrix p units above the pole (or focus)
$r = \dfrac{ep}{1 - e \sin \theta}$	Horizontal directrix p units below the pole (or focus)

Example 2   Describe and graph the conic $r = \dfrac{10}{5 - 5\sin\theta}$.

SOLUTION   We first divide the numerator and the denominator by 5:

$$r = \frac{2}{1 - \sin\theta}.$$

This equation is in the form

$$r = \frac{ep}{1 - e\sin\theta}$$

with $e = 1$, so the graph is a parabola. Since $e = 1$ and $ep = 1 \cdot p = 2$, we have $p = 2$. Thus the parabola has a horizontal directrix 2 units below the pole. We graph the equation in the square window $[-9, 9, -4, 8]$.

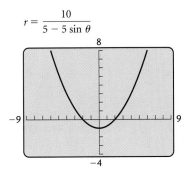

The parabola has a vertical axis of symmetry. We find the vertex of the parabola by letting $\theta = 3\pi/2$. It is $(1, 3\pi/2)$.   ▬

Example 3   Describe and graph the conic $r = \dfrac{4}{2 + 6\sin\theta}$.

SOLUTION   We first divide the numerator and the denominator by 2:

$$r = \frac{2}{1 + 3\sin\theta}.$$

This equation is in the form

$$r = \frac{ep}{1 + e\sin\theta}$$

with $e = 3$. Since $e > 1$, the graph is a hyperbola. We have $e = 3$ and $ep = 3p = 2$, so $p = \frac{2}{3}$. Thus the hyperbola has a horizontal directrix that lies $\frac{2}{3}$ unit above the pole. We graph the equation in the square window $[-6, 6, -3, 5]$ using DOT mode.

Note that the transverse axis is vertical. To find the vertices, we let $\theta = \pi/2$ and $\theta = 3\pi/2$. The vertices are $(1/2, \pi/2)$ and $(-1, 3\pi/2)$.

$$r = \frac{4}{2 + 6 \sin \theta}$$

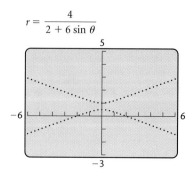

The center of the hyperbola is the midpoint of the segment connecting the vertices, or $(3/4, \pi/2)$. Thus the distance c from the center to a focus is $3/4$. Using $c = 3/4$, $e = 3$, and $e = c/a$, we have $a = 1/4$. Then since $c^2 = a^2 + b^2$, we have

$$b^2 = \left(\frac{3}{4}\right)^2 - \left(\frac{1}{4}\right)^2 = \frac{9}{16} - \frac{1}{16} = \frac{1}{2}$$

$$b = \frac{1}{\sqrt{2}}, \text{ or } \frac{\sqrt{2}}{2}.$$

Knowing the values of a and b allows us to sketch the asymptotes if we are graphing the hyperbola by hand. We can also easily plot the points $(2, 0)$ and $(2, \pi)$ on the polar axis.

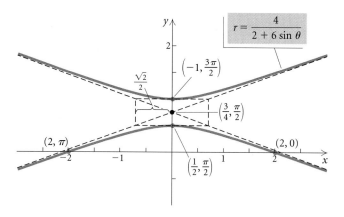

Converting from Polar to Rectangular Equations

We can use the relationships between polar and rectangular coordinates that were developed in Section 5.6 to convert polar equations of conics to rectangular equations.

Example 4   Convert to a rectangular equation: $r = \dfrac{2}{1 - \sin \theta}$.

SOLUTION   We have

$$r = \frac{2}{1 - \sin \theta}$$

$$r - r \sin \theta = 2$$

$$r = r \sin \theta + 2$$

$$\sqrt{x^2 + y^2} = y + 2 \qquad \text{Substituting } \sqrt{x^2 + y^2} \text{ for } r \text{ and } y \text{ for } r \sin \theta$$

$$x^2 + y^2 = y^2 + 4y + 4 \qquad \text{Squaring both sides}$$

$$x^2 = 4y + 4, \quad \text{or}$$

$$x^2 - 4y - 4 = 0.$$

This is the equation of a parabola, as we should have anticipated, since $e = 1$.　　　　■

Finding Polar Equations of Conics

We can find the polar equation of a conic with a focus at the pole if we know its eccentricity and the equation of the directrix.

Example 5   Find a polar equation of the conic with a focus at the pole, eccentricity $\frac{1}{3}$, and directrix $r = 2 \csc \theta$.

SOLUTION   The equation of the directrix can be written

$$r = \frac{2}{\sin \theta}, \quad \text{or} \quad r \sin \theta = 2.$$

This corresponds to the equation $y = 2$ in rectangular coordinates, so the directrix is a horizontal line 2 units above the polar axis. Using the table on page 380, we see that the equation is of the form

$$r = \frac{ep}{1 + e \sin \theta}.$$

Substituting $\frac{1}{3}$ for e and 2 for p gives us

$$r = \frac{\frac{1}{3} \cdot 2}{1 + \frac{1}{3} \sin \theta} = \frac{\frac{2}{3}}{1 + \frac{1}{3} \sin \theta} = \frac{2}{3 + \sin \theta}.$$　　　　■

5.7 | *Exercise Set*

In Exercises 1–6, match each equation with graphs (a)–(f), which follow.

a)

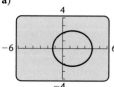

b)

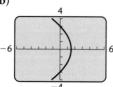

c)

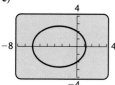

d)

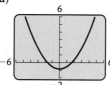

e)

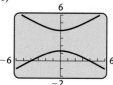

f)

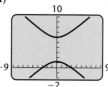

1. $r = \dfrac{3}{1 + \cos \theta}$

2. $r = \dfrac{4}{1 + 2 \sin \theta}$

3. $r = \dfrac{8}{4 - 2 \cos \theta}$

4. $r = \dfrac{12}{4 + 6 \sin \theta}$

5. $r = \dfrac{5}{3 - 3 \sin \theta}$

6. $r = \dfrac{6}{3 + 2 \cos \theta}$

For each equation:

a) *Tell whether the equation describes a parabola, an ellipse, or a hyperbola.*
b) *State whether the directrix is vertical or horizontal and give its location in relation to the pole.*
c) *Find the vertex or vertices.*
d) *Graph the equation.*

7. $r = \dfrac{1}{1 + \cos \theta}$

8. $r = \dfrac{4}{2 + \cos \theta}$

9. $r = \dfrac{15}{5 - 10 \sin \theta}$

10. $r = \dfrac{12}{4 + 8 \sin \theta}$

11. $r = \dfrac{8}{6 - 3 \cos \theta}$

12. $r = \dfrac{6}{2 + 2 \sin \theta}$

13. $r = \dfrac{20}{10 + 15 \sin \theta}$

14. $r = \dfrac{10}{8 - 2 \cos \theta}$

15. $r = \dfrac{9}{6 + 3 \cos \theta}$

16. $r = \dfrac{4}{3 - 9 \sin \theta}$

17. $r = \dfrac{3}{2 - 2 \sin \theta}$

18. $r = \dfrac{12}{3 + 9 \cos \theta}$

19. $r = \dfrac{4}{2 - \cos \theta}$

20. $r = \dfrac{5}{1 - \sin \theta}$

21. $r = \dfrac{7}{2 + 10 \sin \theta}$

22. $r = \dfrac{3}{8 - 4 \cos \theta}$

23–38. Convert the equations in Exercises 7–22 to rectangular equations.

Find a polar equation of the conic with a focus at the pole and the given eccentricity and directrix.

39. $e = 2,\ r = 3 \csc \theta$

40. $e = \frac{2}{3},\ r = -\sec \theta$

41. $e = 1,\ r = 4 \sec \theta$

42. $e = 3,\ r = 2 \csc \theta$

43. $e = \frac{1}{2},\ r = -2 \sec \theta$

44. $e = 1,\ r = 4 \csc \theta$

45. $e = \frac{3}{4},\ r = 5 \csc \theta$

46. $e = \frac{4}{5},\ r = 2 \sec \theta$

47. $e = 4,\ r = -2 \csc \theta$

48. $e = 3,\ r = 3 \csc \theta$

Skill Maintenance

For $f(x) = (x - 3)^2 + 4$, find each of the following.

49. $f(t)$

50. $f(2t)$

51. $f(t - 1)$

52. $f(t + 2)$

Synthesis

53. ◈ Consider the graphs of

$$r = \frac{e}{1 - e \sin \theta}$$

for $e = 0.2, 0.4, 0.6,$ and 0.8. Explain the effect of the value of e on the graph.

54. ◈ When using a grapher, would you prefer to graph a conic in rectangular form or in polar form? Why?

Parabolic Orbit. *Suppose that a comet travels in a parabolic orbit with the sun as its focus. Position a polar coordinate system with the pole at the sun and the axis of the orbit perpendicular to the polar axis. When the comet is the given distance from the sun, the segment from the comet to the sun makes the given angle with the polar axis. Find a polar equation of the orbit, assuming the directrix lies above the pole.*

55. 100 million miles, $\dfrac{\pi}{6}$

56. 120 million miles, $\dfrac{\pi}{4}$

5.8
Parametric Equations and Graphs

- *Graph parametric equations and determine an equivalent rectangular equation.*
- *Determine parametric equations for a rectangular equation.*
- *Determine the location of a moving object at a specific time.*

Graphing Parametric Equations

Much of our graphing of curves in this text has been with rectangular equations involving only two variables, x and y. We graphed sets of ordered pairs, (x, y), where y is a function of x. In this section, we introduce a third variable, t, such that x and y are each a function of t.

Consider a point P with coordinates (x, y) in the rectangular coordinate plane. As P moves in the plane, its location at time t is given by two functions

$$x = f(t) \quad \text{and} \quad y = g(t).$$

For example, let the equations be

$$x = \tfrac{1}{2}t \quad \text{and} \quad y = t^2 - 3, \quad t \geq 0.$$

We restrict t to nonnegative values since t represents time. When $t = 2$, we have

$$x = \tfrac{1}{2} \cdot 2 = 1 \quad \text{and} \quad y = 2^2 - 3 = 1.$$

Thus after 2 sec, the coordinates of P are $(1, 1)$. The table below lists other ordered pairs. We plot them and draw the curve for $t \geq 0$.

t	x	y	(x, y)
0	0	-3	$(0, -3)$
1	$\tfrac{1}{2}$	-2	$\left(\tfrac{1}{2}, -2\right)$
2	1	1	$(1, 1)$
3	$\tfrac{3}{2}$	6	$\left(\tfrac{3}{2}, 6\right)$

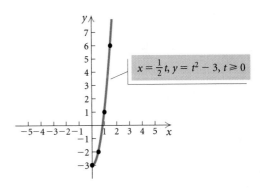

$x = \tfrac{1}{2}t, y = t^2 - 3, t \geq 0$

This curve appears to be part of a parabola. Let's verify this by finding the equivalent rectangular equation. Solving $x = \tfrac{1}{2}t$ for t, we get $t = 2x$. Substituting $2x$ for t in $y = t^2 - 3$, we have

$$y = (2x)^2 - 3 = 4x^2 - 3.$$

We know from Section 1.6 that this is a quadratic equation and that the graph is a parabola. Thus the curve is part of the parabola $y = 4x^2 - 3$. The equations $x = \tfrac{1}{2}t$ and $y = t^2 - 3$ are called the **parametric equations** for the curve. The variable t is called the **parameter**. The equation $y = 4x^2 - 3$, with the restriction $x \geq 0$, is the equivalent rectangular equation for the curve. Since $t \geq 0$ and $x = \tfrac{1}{2}t$, the restriction

$x \geq 0$ must be included. In this example, t represents time, but t, in general, can represent any real number in a specified interval. One advantage of parametric equations is that the restriction on the parameter t allows us to investigate a portion of the curve.

Parametric Equations

If f and g are continuous functions of t on an interval I, then the set of ordered pairs (x, y) such that $x = f(t)$ and $y = g(t)$ is a *plane curve*. The equations $x = f(t)$ and $y = f(t)$ are *parametric equations* for the curve. The variable t is the *parameter*.

Plane curves described with parametric equations can also be graphed on graphers. Consult your grapher's manual or the Graphing Calculator Manual that accompanies this text for specific instructions.

Example 1   Using a grapher, graph each of the following plane curves given their respective parametric equations and the restriction for the parameter. Then find the equivalent rectangular equation.

a) $x = t^2$, $y = t - 1$; $-1 \leq t \leq 4$
b) $x = \sqrt{t}$, $y = 2t + 3$; $0 \leq t \leq 3$

SOLUTION

a) When using a grapher set in PARAMETRIC mode, we must set minimum and maximum values for x, y, and t.

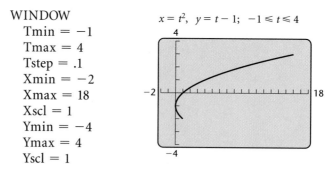

WINDOW
 Tmin = -1
 Tmax = 4
 Tstep = .1
 Xmin = -2
 Xmax = 18
 Xscl = 1
 Ymin = -4
 Ymax = 4
 Yscl = 1

Using the TRACE feature, we can see the T, X, and Y values along the curve. To find an equivalent rectangular equation, we can solve either equation for t. Let's select the simpler equation $y = t - 1$ and solve for t:

$$y = t - 1$$
$$y + 1 = t.$$

We then substitute $y + 1$ for t in $x = t^2$:

$$x = t^2$$
$$x = (y + 1)^2. \quad \text{Parabola}$$

This is an equation of a parabola that opens to the right. Given that $-1 \leq t \leq 4$, we have the corresponding restrictions on x and y: $0 \leq x \leq 16$ and $-2 \leq y \leq 3$. Thus the equivalent rectangular equation is

$$x = (y + 1)^2; \quad 0 \leq x \leq 16, \quad -2 \leq y \leq 3.$$

b) Using a grapher set in PARAMETRIC mode, we have

WINDOW
  Tmin = 0
  Tmax = 3
  Tstep = .1
  Xmin = −3
  Xmax = 3
  Xscl = 1
  Ymin = −2
  Ymax = 10
  Yscl = 1

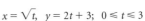

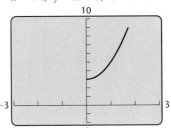

To find an equivalent rectangular equation, we first solve $x = \sqrt{t}$ for t:

$$x = \sqrt{t}$$
$$x^2 = t.$$

Then we substitute x^2 for t in $y = 2t + 3$:

$$y = 2t + 3$$
$$y = 2x^2 + 3. \qquad \text{Parabola}$$

When $0 \leq t \leq 3$, $0 \leq x \leq \sqrt{3}$ and $3 \leq y \leq 9$. The equivalent rectangular equation is

$$y = 2x^2 + 3; \quad 0 \leq x \leq \sqrt{3}, \quad 3 \leq y \leq 9. \qquad \rule[0.5ex]{1.2em}{0.12ex}$$

We first graphed in PARAMETRIC mode in Section 2.5 (p. 172). There we used an angle measure, usually in radians, as the parameter.

Example 2 Graph the plane curve described by $x = \cos t$ and $y = \sin t$, with t in $[0, 2\pi]$. Then determine an equivalent rectangular equation.

SOLUTION Using a squared window and a Tstep of $\pi/48$, we obtain the graph at left. It appears to be a circle of radius 1.

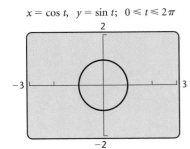

The equivalent rectangular equation can be obtained by squaring each parametric equation:

$$x^2 = \cos^2 t \quad \text{and} \quad y^2 = \sin^2 t.$$

This allows us to use the trigonometric identity $\sin^2 \theta + \cos^2 \theta = 1$. Substituting, we get

$$x^2 + y^2 = 1.$$

As expected, this is an equation of a circle with radius 1. $\qquad \rule[0.5ex]{1.2em}{0.12ex}$

One advantage of graphing the unit circle parametrically is that it provides a method of finding trigonometric function values.

Example 3   Using the CALC menu and the parametric graph of the unit circle, find each of the following function values.

a) $\cos \dfrac{7\pi}{6}$
b) $\sin 4.13$

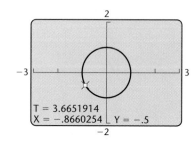

T = 3.6651914
X = −.8660254   Y = −.5

SOLUTION

a) Using the CALC menu, we enter $7\pi/6$ for t. The value of x, which is the $\cos t$, is shown on the screen. Thus,

$$\cos \frac{7\pi}{6} \approx -0.8660.$$

b) We enter 4.13 for t. The value of y, which is the $\sin t$, is shown on the screen. Thus $\sin 4.13 \approx -0.8352$.   ▬

Example 4   Graph the plane curve represented by

$$x = 5 \cos t \quad \text{and} \quad y = 3 \sin t, \quad 0 \le t \le 2\pi.$$

Then eliminate the parameter to find the rectangular equation.

SOLUTION

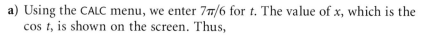

This appears to be the graph of an ellipse. To find the rectangular equation, we first solve for $\cos t$ and $\sin t$ in the parametric equations:

$$x = 5 \cos t \qquad y = 3 \sin t$$

$$\frac{x}{5} = \cos t, \qquad \frac{y}{3} = \sin t.$$

Using the identity $\sin^2 \theta + \cos^2 \theta = 1$, we can substitute to eliminate the parameter:

$$\sin^2 t + \cos^2 t = 1$$

$$\left(\frac{y}{3}\right)^2 + \left(\frac{x}{5}\right)^2 = 1 \qquad \text{Substituting}$$

$$\frac{x^2}{25} + \frac{y^2}{9} = 1. \qquad \text{Ellipse}$$

The rectangular form of the equation confirms that the graph is an ellipse centered at the origin with vertices at $(5, 0)$ and $(-5, 0)$.   ▬

Determining Parametric Equations for a Given Rectangular Equation

Many sets of parametric equations can represent the same plane curve. In fact, there are infinitely many such equations.

Example 5 Find four sets of parametric equations for the parabola

$$y = 4 - (x + 3)^2.$$

SOLUTION

If $x = t$, then $y = 4 - (t + 3)^2$.

If $x = t - 3$, then $y = 4 - (t - 3 + 3)^2$, or $4 - t^2$.

If $x = \dfrac{t}{3}$, then $y = 4 - \left(\dfrac{t}{3} + 3\right)^2$, or $-\dfrac{t^2}{9} - 2t - 5$.

In each of the above three sets, t is any real number. The following equations restrict t to any number greater than or equal to 0.

If $x = \sqrt{t} - 3$, then $y = 4 - (\sqrt{t} - 3 + 3)^2$, or $4 - t$; $t \geq 0$.

Applications

The motion of an object can be described with parametric equations.

Example 6 *Motion of a Baseball.* A baseball player can throw a ball from a height of 5.9 ft with an initial velocity of 100 ft/sec at an angle of 41° with the horizontal. Neglecting air resistance, we know that the parametric equations for the path of the ball are

$$x = (100 \cos 41°)t \quad \text{and} \quad y = 5.9 + (100 \sin 41°)t - 16t^2,$$

where x and y are in feet and t is in seconds. Since t represents the time that the ball is in the air, we must have t greater than or equal to 0.

a) Graph the plane curve.

b) Determine the location of the ball after 0.5 sec, 1 sec, and 2 sec.

c) Find a rectangular equation that describes the curve.

SOLUTION

a) Set the grapher to DEGREE mode and enter the parametric equations.

$x = (100 \cos 41°)t, \quad y = 5.9 + (100 \sin 41°)t - 16t^2$

b) Using the TRACE feature, we find that

when $T = 0.5$, $(x, y) = (37.735479, 34.702951)$,

when $T = 1$, $(x, y) = (75.470958, 55.505903)$, and

when $T = 2$, $(x, y) = (150.94192, 73.111806)$.

In each ordered pair, the x-coordinate represents the horizontal distance and the y-coordinate represents the corresponding vertical height.

c) To find the equivalent rectangular equation, we solve $x = (100 \cos 41°)t$ for t to obtain

$$t = \frac{x}{100 \cos 41°}.$$

Then we substitute $x/(100 \cos 41°)$ for t in the equation for y:

$$y = 5.9 + (100 \sin 41°)\frac{x}{100 \cos 41°} - 16\left(\frac{x}{100 \cos 41°}\right)^2$$

$$y = 5.9 + x \tan 41° - \frac{16}{100^2 \cos^2 41°}x^2. \qquad \text{Rectangular equation}$$

Note that this is a quadratic equation, as expected, since the graph is a parabola.   ▬

   The path of a fixed point on the circumference of a circle as it rolls along a line is called a **cycloid**. For example, a point on the rim of a bicycle wheel traces out a cycloid curve.

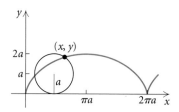

Example 7   Graph the cycloid described by the parametric equations

$$x = 3(t - \sin t) \quad \text{and} \quad y = 3(1 - \cos t); \quad 0 \le t \le 6\pi.$$

SOLUTION

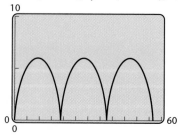

5.8 | Exercise Set

Using a grapher, graph the plane curve given by the set of parametric equations and the restriction for the parameter. Then find the equivalent rectangular equation.

1. $x = \frac{1}{2}t$, $y = 6t - 7$; $-1 \le t \le 6$
2. $x = t$, $y = 5 - t$; $-2 \le t \le 3$
3. $x = t^3$, $y = t - 4$; $-1 \le t \le 10$
4. $x = \sqrt{t}$, $y = 2t + 3$; $0 \le t \le 8$
5. $x = t^2$, $y = \sqrt{t}$; $0 \le t \le 4$
6. $x = t^3 + 1$, $y = t$; $-3 \le t \le 3$
7. $x = t + 3$, $y = \dfrac{1}{t + 3}$; $-2 \le t \le 2$
8. $x = 2t^3 + 1$, $y = 2t^3 - 1$; $-4 \le t \le 4$
9. $x = 2t - 1$, $y = t^2$; $-3 \le t \le 3$
10. $x = \frac{1}{3}t$, $y = t$; $-5 \le t \le 5$
11. $x = e^{-t}$, $y = e^t$; $-\infty < t < \infty$
12. $x = 2 \ln t$, $y = t^2$; $0 < t < \infty$
13. $x = 3 \cos t$, $y = 3 \sin t$; $0 \le t \le 2\pi$
14. $x = 2 \cos t$, $y = 4 \sin t$; $0 \le t \le 2\pi$
15. $x = \cos t$, $y = 2 \sin t$; $0 \le t \le 2\pi$
16. $x = 2 \cos t$, $y = 2 \sin t$; $0 \le t \le 2\pi$
17. $x = \sec t$, $y = \cos t$; $-\dfrac{\pi}{2} < t < \dfrac{\pi}{2}$
18. $x = \sin t$, $y = \csc t$; $0 < t < \pi$
19. $x = 1 + 2 \cos t$, $y = 2 + 2 \sin t$; $0 \le t \le 2\pi$
20. $x = 2 + \sec t$, $y = 1 + 3 \tan t$; $0 < t < \frac{\pi}{2}$

Using a parametric graph of the unit circle, find the function value.

21. $\sin \dfrac{\pi}{4}$
22. $\cos \dfrac{2\pi}{3}$
23. $\cos \dfrac{17\pi}{12}$
24. $\sin \dfrac{4\pi}{5}$
25. $\tan \dfrac{\pi}{5}$
26. $\tan \dfrac{2\pi}{7}$
27. $\cos 5.29$
28. $\sin 1.83$

Find two sets of parametric equations for the rectangular equation.

29. $y = 4x - 3$
30. $y = x^2 - 1$
31. $y = (x - 2)^2 - 6x$
32. $y = x^3 + 3$

33. *Motion of a Ball.* A ball is thrown from a height of 6.25 ft with an initial velocity of 80 ft/sec at an angle of 35° with the horizontal. The parametric equations for the path of the ball are

$x = (80 \cos 35°)t$, and
$y = 6.25 + (80 \sin 35°)t - 16t^2$,

where x and y are in feet and t is in seconds.
a) Find the height of the ball after 1 sec.
b) Find the total horizontal distance traveled by the ball.
c) How long does it take the ball to hit the ground?

34. For Exercise 33, determine the rectangular equation describing the path of the ball.

35. In Example 6, use a grapher to determine the maximum height that the ball reaches.

36. The path of a projectile ejected with an initial velocity of 200 ft/sec at an angle of 55° with the horizontal is described with the equations

$x = (210 \cos 55°)t$ and $y = (210 \sin 55°)t - 16t^2$,

where $t \ge 0$.
a) Find the height after 4 sec.
b) Determine the equivalent rectangular equation for this curve.

Graph each cycloid for t in the indicated interval.

37. $x = 2(t - \sin t)$, $y = 2(1 - \cos t)$; $0 \le t \le 4\pi$
38. $x = 4t - 4 \sin t$, $y = 4 - 4 \cos t$; $0 \le t \le 6\pi$
39. $x = t - \sin t$, $y = 1 - \cos t$; $-2\pi \le t \le 2\pi$
40. $x = 5(t - \sin t)$, $y = 5(1 - \cos t)$; $-4\pi \le t \le 4\pi$

Skill Maintenance

Graph.

41. $y = x^3$
42. $x = y^3$
43. $f(x) = \sqrt{x - 2}$
44. $f(x) = \dfrac{3}{x^2 - 1}$

Synthesis

45. ◆ Show that $x = a \cos t + h$ and $y = b \sin t + k$, $0 \le t \le 2\pi$, are parametric equations of an ellipse with center (h, k).

46. ◆ Consider the graph in Example 3. Explain how the values of T, X, and Y displayed relate to both the pararametric and rectangular equations of the unit circle.

47. Graph the curve described by

$x = 3 \cos t$ and $y = 3 \sin t$, $0 \le t \le 2\pi$.

As t increases, the path of the curve is generated counterclockwise. How can this set of equations be changed so that the curve is generated clockwise?

48. Graph the plane curve described by

$x = \cos^3 t$ and $y = \sin^3 t$, $0 \le t \le 2\pi$.

Then find the equivalent rectangular equation.

5 Summary and Review

Important Properties and Formulas

Standard Equation of a Parabola with Vertex at the Origin

The standard equation of a parabola with vertex $(0, 0)$ and directrix $y = -p$ is

$$x^2 = 4py.$$

The focus is $(0, p)$ and the y-axis is the axis of symmetry.

The standard equation of a parabola with vertex $(0, 0)$ and directrix $x = -p$ is

$$y^2 = 4px.$$

The focus is $(p, 0)$ and the x-axis is the axis of symmetry.

Standard Equation of a Parabola with Vertex (h, k) and Vertical Axis of Symmetry

The standard equation of a parabola with vertex (h, k) and vertical axis of symmetry is

$$(x - h)^2 = 4p(y - k),$$

where the vertex is (h, k), the focus is $(h, k + p)$, and the directrix is $y = k - p$.

Standard Equation of a Parabola with Vertex (h, k) and Horizontal Axis of Symmetry

The standard equation of a parabola with vertex (h, k) and horizontal axis of symmetry is

$$(y - k)^2 = 4p(x - h),$$

where the vertex is (h, k), the focus is $(h + p, k)$, and the directrix is $x = h - p$.

Standard Equation of a Circle

The standard equation of a circle with center (h, k) and radius r is

$$(x - h)^2 + (y - k)^2 = r^2.$$

Standard Equation of an Ellipse with Center at the Origin

Major axis horizontal

$$\frac{x^2}{a^2} + \frac{y^2}{b^2} = 1, \, a > b > 0$$

Vertices: $(-a, 0)$, $(a, 0)$
y-intercepts: $(0, -b)$, $(0, b)$
Foci: $(-c, 0)$, $(c, 0)$, where $c^2 = a^2 - b^2$

Major axis vertical

$$\frac{x^2}{b^2} + \frac{y^2}{a^2} = 1, \, a > b > 0$$

Vertices: $(0, -a)$, $(0, a)$
x-intercepts: $(-b, 0)$, $(b, 0)$
Foci: $(0, -c)$, $(0, c)$, where $c^2 = a^2 - b^2$

Standard Equation of an Ellipse with Center at (h, k)

Major axis horizontal

$$\frac{(x - h)^2}{a^2} + \frac{(y - k)^2}{b^2} = 1, \, a > b > 0$$

Vertices: $(h + a, k)$, $(h - a, k)$
Length of minor axis: $2b$
Foci: $(h - c, k)$, $(h + c, k)$, where $c^2 = a^2 - b^2$

Major axis vertical

$$\frac{(x - h)^2}{b^2} + \frac{(y - k)^2}{a^2} = 1, \, a > b > 0$$

Vertices: $(h, k - a)$, $(h, k + a)$
Length of minor axis: $2b$
Foci: $(h, k - c)$, $(h, k + c)$, where $c^2 = a^2 - b^2$

Standard Equation of a Hyperbola with Center at the Origin

Transverse axis horizontal

$$\frac{x^2}{a^2} - \frac{y^2}{b^2} = 1$$

Vertices: $(-a, 0), (a, 0)$

Foci: $(-c, 0), (c, 0)$, where $c^2 = a^2 + b^2$

Transverse axis vertical

$$\frac{y^2}{a^2} - \frac{x^2}{b^2} = 1$$

Vertices: $(0, -a), (0, a)$

Foci: $(0, -c), (0, c)$, where $c^2 = a^2 + b^2$

Standard Equation of a Hyperbola with Center at (h, k)

Transverse axis horizontal

$$\frac{(x - h)^2}{a^2} - \frac{(y - k)^2}{b^2} = 1$$

Vertices: $(h - a, k), (h + a, k)$

Asymptotes: $y - k = \dfrac{b}{a}(x - h),$

$$y - k = -\frac{b}{a}(x - h)$$

Foci: $(h - c, k), (h + c, k)$, where $c^2 = a^2 + b^2$

Transverse axis vertical

$$\frac{(y - k)^2}{a^2} - \frac{(x - h)^2}{b^2} = 1$$

Vertices: $(h, k - a), (h, k + a)$

Asymptotes: $y - k = \dfrac{a}{b}(x - h),$

$$y - k = -\frac{a}{b}(x - h)$$

Foci: $(h, k - c), (h, k + c)$, where $c^2 = a^2 + b^2$

Rotation of Axes Formulas

$$x' = x \cos \theta + y \sin \theta,$$
$$y' = -x \sin \theta + y \cos \theta$$
$$x = x' \cos \theta - y' \sin \theta,$$
$$y = x' \sin \theta + y' \cos \theta$$

Eliminating the xy-Term

To eliminate the xy-term from the equation

$$Ax^2 + Bxy + Cy^2 + Dx + Ey + F = 0, \qquad B \neq 0,$$

select an angle θ such that

$$\cot 2\theta = \frac{A - C}{B}, \quad 0 < 2\theta < 180°,$$

and use the rotation of axes formulas.

The Discriminant

The graph of the equation $Ax^2 + Bxy + Cy^2 + Dx + Ey + F = 0$ is, except in degenerate cases,

1. an ellipse or a circle if $B^2 - 4AC < 0$,
2. a hyperbola if $B^2 - 4AC > 0$,   and
3. a parabola if $B^2 - 4AC = 0$.

Polar Equations of Conics

A polar equation of any of the four forms

$$r = \frac{ep}{1 \pm e \cos \theta}, \qquad r = \frac{ep}{1 \pm e \sin \theta}$$

is a conic section. The conic is a parabola if $e = 1$, an ellipse if $0 < e < 1$, and a hyperbola if $e > 1$.

REVIEW EXERCISES

In Exercises 1–8, match each equation with one of the graphs (a)–(h), which follow.

a)
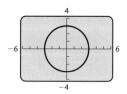

b)
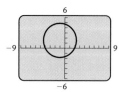

c)
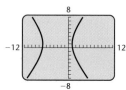

d)
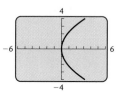

e)
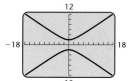

f)
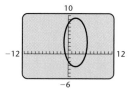

g)
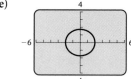

h)
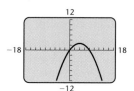

1. $y^2 = 5x$

2. $y^2 = 9 - x^2$

3. $3x^2 + 4y^2 = 12$

4. $9y^2 - 4x^2 = 36$

5. $x^2 + y^2 + 2x - 3y = 8$

6. $4x^2 + y^2 - 16x - 6y = 15$

7. $x^2 - 8x + 6y = 0$

8. $\dfrac{(x+3)^2}{16} - \dfrac{(y-1)^2}{25} = 1$

9. Find an equation of the parabola with directrix $y = \frac{3}{2}$ and focus $\left(0, -\frac{3}{2}\right)$.

10. Find the focus, the vertex, and the directrix of the parabola given by
$$y^2 = -12x.$$

11. Find the vertex, the focus, and the directrix of the parabola given by
$$x^2 + 10x + 2y + 9 = 0.$$

12. Find the center, the vertices, and the foci of the ellipse given by
$$16x^2 + 25y^2 - 64x + 50y - 311 = 0.$$
Then draw the graph.

13. Find an equation of the ellipse having vertices $(0, -4)$ and $(0, 4)$ with minor axis of length 6.

14. Find the center, the vertices, the foci, and the asymptotes of the hyperbola given by
$$x^2 - 2y^2 + 4x + y - \tfrac{1}{8} = 0.$$

15. *Spotlight.*   A spotlight has a parabolic cross section that is 2 ft wide at the opening and 1.5 ft deep at the vertex. How far from the vertex is the focus?

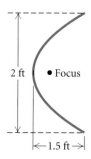

Solve.

16. $x^2 - 16y = 0,$
$x^2 - y^2 = 64$

17. $4x^2 + 4y^2 = 65,$
$6x^2 - 4y^2 = 25$

18. $x^2 - y^2 = 33,$
$x + y = 11$

19. $x^2 - 2x + 2y^2 = 8,$
$2x + y = 6$

20. $x^2 - y = 3,$
$2x - y = 3$

21. $x^2 + y^2 = 25,$
$x^2 - y^2 = 7$

22. $x^2 - y^2 = 3,$
$y = x^2 - 3$

23. $x^2 + y^2 = 18,$
$2x + y = 3$

24. $x^2 + y^2 = 100,$
$2x^2 - 3y^2 = -120$

25. $x^2 + 2y^2 = 12,$
$xy = 4$

26. *Numerical Relationship.*   The sum of two numbers is 11 and the sum of their squares is 65. Find the numbers.

27. *Dimensions of a Rectangle.*   A rectangle has a perimeter of 38 m and an area of 84 m². What are the dimensions of the rectangle?

28. *Numerical Relationship.*   Find two positive integers whose sum is 12 and the sum of whose reciprocals is $\frac{3}{8}$.

29. *Perimeter.*   The perimeter of a square is 12 cm more than the perimeter of another square. The area of the first square exceeds the area of the other by 39 cm². Find the perimeter of each square.

30. *Radius of a Circle.* The sum of the areas of two circles is 130π ft². The difference of the areas is 112π ft². Find the radius of each circle.

Graph each equation.

31. $5x^2 - 2xy + 5y^2 - 24 = 0$

32. $x^2 - 10xy + y^2 + 12 = 0$

33. $5x^2 + 6\sqrt{3}xy - y^2 = 16$

34. $x^2 + 2xy + y^2 - \sqrt{2}x + \sqrt{2}y = 0$

35. Find the polar coordinates of each of these points. Give three answers for each point.

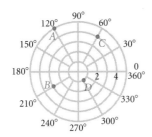

Find the polar coordinates of the point. Express the answer in degrees and then in radians.

36. $(-4\sqrt{2}, 4\sqrt{2})$ **37.** $(0, -5)$

Use a grapher to convert from rectangular to polar coordinates. Express the answer in degrees and then in radians.

38. $(-2, 5)$ **39.** $(-4.2, \sqrt{7})$

Find the rectangular coordinates of the point.

40. $\left(3, \dfrac{\pi}{4}\right)$ **41.** $(-6, -120°)$

Use a grapher to convert from polar to rectangular coordinates. Round the coordinates to the nearest hundredth.

42. $(2, -15°)$ **43.** $\left(-2.3, \dfrac{\pi}{5}\right)$

Convert to a polar equation.

44. $5x - 2y = 6$ **45.** $y = 3$

46. $x^2 + y^2 = 9$ **47.** $y^2 - 4x - 16 = 0$

Convert to a rectangular equation.

48. $r = 6$ **49.** $r + r\sin\theta = 1$

50. $r = \dfrac{3}{1 - \cos\theta}$ **51.** $r - 2\cos\theta = 3\sin\theta$

In Exercises 52–55, use a grapher to match the equation with figures (a)–(d), which follow. Try matching the graphs mentally before using a grapher.

a)
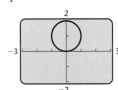

b)
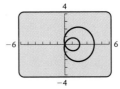

c)
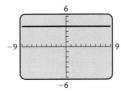

d)
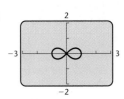

52. $r = 2\sin\theta$ **53.** $r^2 = \cos 2\theta$

54. $r = 1 + 3\cos\theta$ **55.** $r\sin\theta = 4$

Graph each equation. State whether the directrix is vertical or horizontal, describe its location in relation to the pole, and find the vertex or vertices.

56. $r = \dfrac{6}{3 - 3\sin\theta}$ **57.** $r = \dfrac{8}{2 + 4\cos\theta}$

58. $r = \dfrac{4}{2 - \cos\theta}$ **59.** $r = \dfrac{18}{9 + 6\sin\theta}$

60–63. Convert the equations in Exercises 56–59 to rectangular equations.

Find a polar equation of the conic with a focus at the pole and the given eccentricity and directrix.

64. $e = \frac{1}{2}$, $r = 2\sec\theta$ **65.** $e = 3$, $r = -6\csc\theta$

66. $e = 1$, $r = -4\sec\theta$ **67.** $e = 2$, $r = 3\csc\theta$

Using a grapher, graph the plane curve given by the set of parametric equations and the restrictions for the parameter. Then find the equivalent rectangular equation.

68. $x = t$, $y = 2 + t$, $-3 \le t \le 3$

69. $x = \sqrt{t}$, $y = t - 1$, $0 \le t \le 9$

70. $x = 2\cos t$, $y = 2\sin t$, $0 \le t \le 2\pi$

71. $x = 3\sin t$, $y = \cos t$, $0 \le t \le 2\pi$

Find two sets of parametric equations for the given rectangular equation.

72. $y = 2x - 3$ **73.** $y = x^2 + 4$

Synthesis

74. ◈ What would you say to a classmate who tells you that an algebraic solution of a nonlinear system of equations is always preferable to a graphical solution?

75. ◆ Is a circle a special type of ellipse? Why or why not?

76. Find an equation of the ellipse containing the point $(-1/2, 3\sqrt{3}/2)$ and with vertices $(0, -3)$ and $(0, 3)$.

77. Find two numbers whose product is 4 and the sum of whose reciprocals is $\frac{65}{56}$.

78. Find an equation of the circle that passes through the points $(10, 7)$, $(-6, 7)$, and $(-8, 1)$.

79. *Navigation.*   Two radio transmitters positioned 400 mi apart send simultaneous signals to a ship that is 250 mi offshore, sailing parallel to the shoreline. The signal from transmitter A reaches the ship 300 microseconds before the signal from transmitter B. The signals travel at a speed of 186,000 miles per second, or 0.186 mile per microsecond. Find the equation of the hyperbola with foci A and B on which the ship is located. (*Hint:* For any point on the hyperbola, the absolute value of the difference of its distances from the foci is $2a$.)

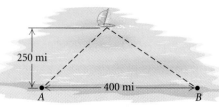

80. Convert to a rectangular equation:

$$r = \csc^2 \frac{\theta}{2}.$$

Exponential and Logarithmic Functions 6

APPLICATION

The number of cellular phones in this country is modeled by an exponential function, where $y =$ the number of telephones, in millions, in the year x. Here $x = 0$ corresponds to 1985. (*Source:* Cellular Telecommunications Industry Association)

X	Y_1
0	.2
1	.4
2	1.0
3	1.8
5	4.1
7	8.6
9	19.3

X = 5

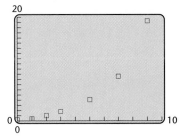

I n this chapter, we will consider two kinds of closely related functions. The first, called *exponential functions,* are those that have a variable in the exponent. Such functions have many applications to the growth of populations, commodities, and investments. The inverses of exponential functions, called *logarithmic functions,* or *logarithm functions,* are also important in many applications such as earthquake magnitude, sound level, and chemical pH.

397

6.1
Exponential Functions and Graphs

- *Graph exponential equations and functions.*
- *Solve problems involving applications of exponential functions and their graphs.*

We now turn our attention to the study of a set of functions very rich in application. Consider the following graphs.

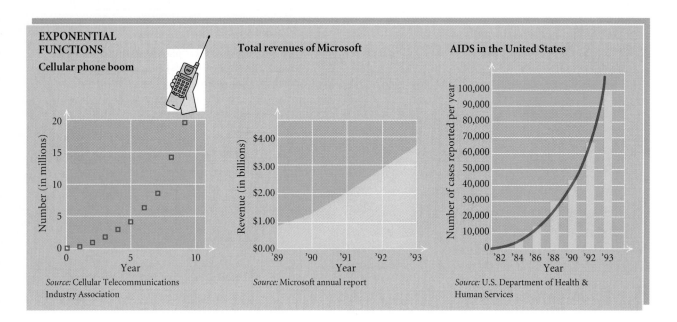

EXPONENTIAL FUNCTIONS

Cellular phone boom

Source: Cellular Telecommunications Industry Association

Total revenues of Microsoft

Source: Microsoft annual report

AIDS in the United States

Source: U.S. Department of Health & Human Services

Each of these graphs illustrates the idea of an *exponential function*. The graph on the right shows the numbers of cases of AIDS reported in various years. The curve drawn along the graph approximates an exponential function. In this section, we consider such graphs, their inverses, and some important applications of exponential functions.

Graphing Exponential Functions

We now define exponential functions. We assume that a^x has meaning for any real number x and any positive real number a and that the laws of exponents still hold, though we will not prove them here.

Exponential Function

The function $f(x) = a^x$, where x is a real number, $a > 0$ and $a \neq 1$, is called the *exponential function*, base a.

We require the **base** to be positive in order to avoid the complex numbers that would occur by taking even roots of negative numbers—

an example is the square root of -1, $(-1)^{1/2}$, which is not a real number. The restriction $a \neq 1$ is made to exclude the constant function $f(x) = 1^x = 1$, which does not have an inverse because it is not one-to-one.

The following are examples of exponential functions:

$$f(x) = 2^x, \qquad f(x) = \left(\frac{1}{2}\right)^x, \qquad f(x) = (3.57)^x.$$

Note that, in contrast to functions like $f(x) = x^5$ and $f(x) = x^{1/2}$, the variable in an exponential function is *in the exponent.* Let's now consider graphs of exponential functions.

Example 1 Graph the exponential function: $y = f(x) = 2^x$.

SOLUTION We compute some function values and list the results in a table (at left).

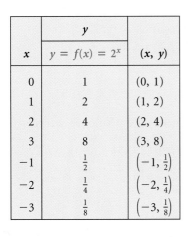

x	$y = f(x) = 2^x$	(x, y)
0	1	$(0, 1)$
1	2	$(1, 2)$
2	4	$(2, 4)$
3	8	$(3, 8)$
-1	$\frac{1}{2}$	$\left(-1, \frac{1}{2}\right)$
-2	$\frac{1}{4}$	$\left(-2, \frac{1}{4}\right)$
-3	$\frac{1}{8}$	$\left(-3, \frac{1}{8}\right)$

$$f(0) = 2^0 = 1; \qquad f(-1) = 2^{-1} = \frac{1}{2^1} = \frac{1}{2};$$

$$f(1) = 2^1 = 2; \qquad f(-2) = 2^{-2} = \frac{1}{2^2} = \frac{1}{4};$$

$$f(2) = 2^2 = 4; \qquad f(-3) = 2^{-3} = \frac{1}{2^3} = \frac{1}{8}.$$

$$f(3) = 2^3 = 8;$$

Next, we plot these points and connect them with a smooth curve. Be sure to plot enough points to determine how steeply the curve rises.

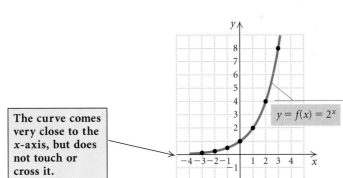

The curve comes very close to the x-axis, but does not touch or cross it.

$$y = f(x) = 2^x$$

Note that as x increases, the function values increase without bound. As x decreases, the function values decrease, getting close to 0. That is, as $x \to -\infty$, $y \to 0$. The x-axis, or the line $y = 0$, is a horizontal asymptote. As the x-inputs decrease, the curve gets closer and closer to this line, but does not cross it.

Check the graph of $y = f(x) = 2^x$ on a grapher. In general, this function is entered as $y = 2 \wedge x$ or $f(x) = 2 \wedge x$. Then graph $y = f(x) = 3^x$. Use the TRACE and ZOOM features to confirm that the graph never touches the x-axis.

Example 2   Graph the exponential function: $y = f(x) = \left(\dfrac{1}{2}\right)^x$.

SOLUTION   We compute some function values and list the results in a table (at left). Before we plot these points and draw the curve, note that

$$y = f(x) = \left(\frac{1}{2}\right)^x = (2^{-1})^x = 2^{-x}.$$

This tells us, before we begin graphing, that this graph is a reflection of the graph of $y = 2^x$ across the y-axis.

	y	
x	$y = f(x) = 2^{-x}$	(x, y)
0	1	$(0, 1)$
1	$\frac{1}{2}$	$\left(1, \frac{1}{2}\right)$
2	$\frac{1}{4}$	$\left(2, \frac{1}{4}\right)$
3	$\frac{1}{8}$	$\left(3, \frac{1}{8}\right)$
−1	2	$(-1, 2)$
−2	4	$(-2, 4)$
−3	8	$(-3, 8)$

$$f(0) = 2^{-0} = 1; \qquad\qquad f(-1) = 2^{-(-1)} = 2^1 = 2;$$

$$f(1) = 2^{-1} = \frac{1}{2^1} = \frac{1}{2}; \qquad f(-2) = 2^{-(-2)} = 2^2 = 4;$$

$$f(2) = 2^{-2} = \frac{1}{2^2} = \frac{1}{4}; \qquad f(-3) = 2^{-(-3)} = 2^3 = 8.$$

$$f(3) = 2^{-3} = \frac{1}{2^3} = \frac{1}{8};$$

Next, we plot these points and connect them with a smooth curve.

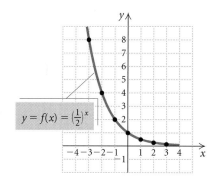

$$y = f(x) = \left(\tfrac{1}{2}\right)^x$$

Check the graph in Example 2 on a grapher. Then graph $y = f(x) = \left(\frac{1}{3}\right)^x$. Also, try using the TABLE feature to make a table simultaneously for each of the functions $f(x) = 3^x$ and $g(x) = 3^{-x}$.

Interactive Discovery

Graph each of the following functions using the same set of axes and a viewing window of $[-10, 10, -10, 10]$. Look for patterns in the graphs.

$$f(x) = 1.1^x, \qquad f(x) = 2.7^x, \qquad f(x) = 4^x$$

Next, clear the screen and graph each of the following functions using the same set of axes and viewing window. Again, look for patterns in the graphs.

$$f(x) = 0.1^x, \qquad f(x) = 0.6^x, \qquad f(x) = 0.23^x$$

What relationship do you see between the base a and the shape of the resulting graph of a^x? What do all the graphs have in common? How do they differ?

The preceding examples and interactive discovery illustrate exponential functions with various bases. Let's list some characteristics, keeping in mind that the definition of an exponential function, $f(x) = a^x$, requires that a be positive and different from 1.

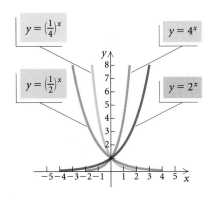

> **Properties of Exponential Functions**
>
> $f(x) = a^x$, $a > 1$:
>
> > Continuous
> > One-to-one
> > Domain: All real numbers, $\mathbb{R}$
> > Range: All positive real numbers, $(0, \infty)$
> > Increasing
> > Horizontal asymptote is x-axis: ($a^x \to 0$ as $x \to -\infty$)
> > y-intercept: $(0, 1)$
>
> $f(x) = a^x$ for $0 < a < 1$ or, equivalently, $f(x) = a^{-x}$, $a > 1$:
>
> > Continuous
> > One-to-one
> > Domain: All real numbers, $\mathbb{R}$
> > Range: All positive real numbers, $(0, \infty)$
> > Decreasing
> > Horizontal asymptote is x-axis: ($a^x \to 0$ as $x \to \infty$)
> > y-intercept: $(0, 1)$

To graph other types of exponential functions, keep in mind the ideas of translation, stretching, reflection, and combinations of these ideas. All these concepts allow us to visualize the graph before drawing it.

Example 3 Graph each of the following. Before doing so, describe how each graph can be obtained from $f(x) = 2^x$.

a) $f(x) = 2^{x-2}$ **b)** $f(x) = 2^x - 4$ **c)** $f(x) = 5 - 2^{-x}$

SOLUTION

a) The graph of this function is the graph of $y = 2^x$ shifted *right* 2 units.

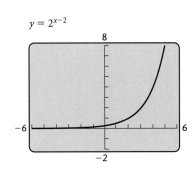

$y = 2^{x-2}$

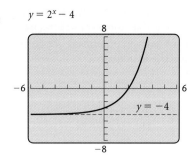

$y = 2^x - 4$

$y = -4$

b) The graph is the graph of $y = 2^x$ shifted *down* 4 units (see the graph at left).

c) The graph is a reflection of the graph of $y = 2^x$ across the y-axis, followed by a reflection across the x-axis and then a shift *up* 5 units.

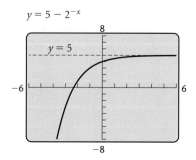

$y = 5 - 2^{-x}$

$y = 5$

Graphs of Inverses of Exponential Functions

We have noted that every exponential function (with $a > 0$ and $a \neq 1$) is one-to-one. Thus each such function has an inverse that is a function. In the next section, we will name these inverse functions and use them in applications. For now, we draw their graphs by interchanging x and y.

Example 4   Graph: $x = 2^y$.

SOLUTION   Note that x is alone on one side of the equation. We can find ordered pairs that are solutions by choosing values for y and then computing the corresponding x-values.

x $x = 2^y$	y	x, y
1	0	$(1, 0)$
2	1	$(2, 1)$
4	2	$(4, 2)$
8	3	$(8, 3)$
$\frac{1}{2}$	-1	$\left(\frac{1}{2}, -1\right)$
$\frac{1}{4}$	-2	$\left(\frac{1}{4}, -2\right)$
$\frac{1}{8}$	-3	$\left(\frac{1}{8}, -3\right)$

For $y = 0$, $x = 2^0 = 1$.

For $y = 1$, $x = 2^1 = 2$.

For $y = 2$, $x = 2^2 = 4$.

For $y = 3$, $x = 2^3 = 8$.

For $y = -1$, $x = 2^{-1} = \dfrac{1}{2^1} = \dfrac{1}{2}$.

For $y = -2$, $x = 2^{-2} = \dfrac{1}{2^2} = \dfrac{1}{4}$.

For $y = -3$, $x = 2^{-3} = \dfrac{1}{2^3} = \dfrac{1}{8}$.

(1) Choose values for y.

(2) Compute values for x.

We plot the points and connect them with a smooth curve. Note that the curve does not touch or cross the y-axis. The y-axis is a vertical asymptote.

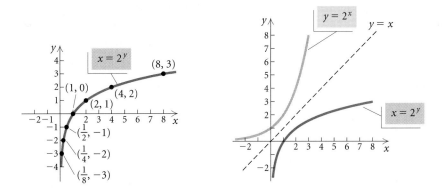

Note too that this curve looks just like the graph of $y = 2^x$, except that it is reflected across the line $y = x$, as we would expect for an inverse. The inverse of $y = 2^x$ is $x = 2^y$. We will explore the inverses of exponential functions further in the next section.

Applications

Graphers are especially helpful when working with exponential functions. They not only facilitate computations but they also allow us to visualize the function. It is worthwhile to get in the habit of creating such a graph, and perhaps looking at a table, even if an exercise or application may not specifically request a graph or table.

Example 5 *Compound Interest.* The amount of money A that a principal P will be worth after t years at interest rate i, compounded n times per year, is given by the formula

$$A = P\left(1 + \frac{i}{n}\right)^{nt}.$$

Suppose that \$100,000 is invested at 8% interest, compounded semiannually.

a) Find a function for the amount of money after t years.
b) Find the amount of money in the account at $t = 0, 4, 8,$ and 10 yr.
c) Graph the function.
d) When will the amount of money in the account reach \$400,000?

SOLUTION

a) Since $P = \$100{,}000$, $i = 8\% = 0.08$, and $n = 2$, we can substitute these values and form the following function:

$$A(t) = 100{,}000\left(1 + \frac{0.08}{2}\right)^{2 \cdot t}$$

$$= 100{,}000(1.04)^{2t}.$$

b) We can enter the function into a grapher, using a TABLE feature set in ASK mode, to compute function values. We can also calculate the values directly on a grapher by substituting in the expression for $A(t)$.

$$A(0) = 100{,}000(1.04)^{2 \cdot 0} = 100{,}000;$$
$$A(4) = 100{,}000(1.04)^{2 \cdot 4} \approx 136{,}856.91;$$
$$A(8) = 100{,}000(1.04)^{2 \cdot 8} \approx 187{,}298.12;$$
$$A(10) = 100{,}000(1.04)^{2 \cdot 10} \approx 219{,}112.31.$$

c) For the graph, we use the viewing window [0, 20, 0, 500,000] because of the large numbers and the fact that negative time values and amounts of money have no meaning in this application.

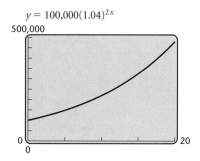

$y = 100{,}000(1.04)^{2x}$

d) To find the amount of time it takes for the account to grow to $400,000, we set

$$100{,}000(1.04)^{2t} = 400{,}000$$

and solve. One way we can do this is by graphing the equations

$$y_1 = 100{,}000(1.04)^{2x} \quad \text{and} \quad y_2 = 400{,}000.$$

Then we can use the TRACE and ZOOM features or the INTERSECT feature as discussed in the Introduction to Graphs and Graphers to estimate the point of intersection.

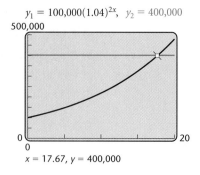

$y_1 = 100{,}000(1.04)^{2x}, \quad y_2 = 400{,}000$

$x = 17.67, y = 400{,}000$

We can also use the SOLVE feature on a grapher. It might be necessary to write the equation as $100{,}000(1.04)^{2x} - 400{,}000 = 0$. In Section 6.4, we will see how to solve equations such as this algebraically.

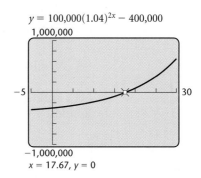

$y = 100{,}000(1.04)^{2x} - 400{,}000$

$x = 17.67,\ y = 0$

Regardless of the method we use to find the solution, we see that the account grows to \$400,000 after about 17.67 yr.

Interactive Discovery

Take a sheet of $8\frac{1}{2}$-in. by 11-in. paper and cut it into two equal pieces. Then cut each of these in half. Next, cut each of the four pieces in half again. Continue this process.

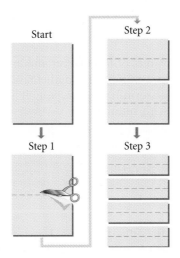

Start

Step 1

Step 2

Step 3

	t	$f(t) = 0.004 \cdot 2^t$
Start	0	$0.004 \cdot 2^0$, or 0.004
Step 1	1	$0.004 \cdot 2^1$, or 0.008
Step 2	2	$0.004 \cdot 2^2$, or 0.016
Step 3	3	
Step 4	4	
Step 5	5	

a) Place all the pieces in a stack and measure the thickness with a micrometer or other measuring device such as a dial caliper.

b) A piece of paper is typically 0.004 in. thick. Check the calculation in part (a) by completing the table shown here.

c) Graph the function $f(t) = 0.004(2)^t$.

d) Compute the thickness of the paper (in miles) after 25 steps.

The Number $e = 2.7182818284\ldots$

We now consider a very special number in mathematics. Though you may not have encountered it before, you will see here and in future mathematics that it has many important applications. To derive this number, we use the compound interest formula $A = P(1 + i/n)^{nt}$ in Example 5. Suppose that \$1 is an initial investment at 100% interest for

1 yr (no bank would pay this). The formula above becomes a function A defined in terms of the number of compounding periods n:

$$A(n) = \left(1 + \frac{1}{n}\right)^n.$$

Let's find some function values using the scientific keys on a grapher. Rounding to six decimal places gives us the following table.

n	$A(n) = \left(1 + \dfrac{1}{n}\right)^n$
1 (compounded annually)	$2.00
2 (compounded semiannually)	$2.25
3	$2.370370
4 (compounded quarterly)	$2.441406
5	$2.488320
100	$2.704814
365 (compounded daily)	$2.714567
8760 (compounded hourly)	$2.718127

As the values of n get larger and larger, the function values get closer and closer to a number in mathematics, called e. Its decimal representation does not terminate or repeat; it is irrational. In 1741, Leonhard Euler named this number e.

$$e = 2.7182818284\ldots$$

Interactive Discovery

Graph the function $y = (1 + 1/x)^x$. Consider the graph for larger and larger values of x. Does this function have a horizontal asymptote? Explore with a grapher to determine the asymptote, if it exists. You might also try to use the TABLE feature.

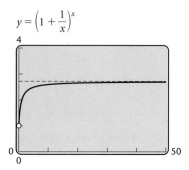

$$y = \left(1 + \frac{1}{x}\right)^x$$

We can find values of the exponential function $f(x) = e^x$ using the scientific $\boxed{e^x}$ key on a grapher.

Example 6   Find each value of e^x, to four decimal places, on a grapher.

a) e^3 **b)** $e^{-0.23}$ **c)** e^0 **d)** $100e^{5.8}$ **e)** e^1

SOLUTION

FUNCTION VALUE	READOUT	ROUNDED
a) e^3	20.08553692	20.0855
b) $e^{-0.23}$	0.7945336025	0.7945
c) e^0	1	1
d) $100e^{5.8}$	33029.95599	33029.9560
e) e^1	2.718281828	2.7183

Graphs of Exponential Functions, Base e

We demonstrate ways in which to graph exponential functions.

Example 7   Graph $f(x) = e^x$ and $g(x) = e^{-x}$.

SOLUTION

Method 1. We can compute points for each equation using the $\boxed{e^x}$ key on a grapher. Then we plot these points and draw the graphs of the functions.

x	$f(x) = e^x$	$g(x) = e^{-x}$
-4	0.0183	54.5982
-3	0.0498	20.0855
-2	0.1353	7.3891
-1	0.3679	2.7183
0	1	1
1	2.7183	0.3679
2	7.3891	0.1353
3	20.0855	0.0498
4	54.5982	0.0183

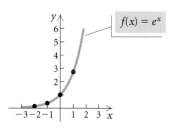
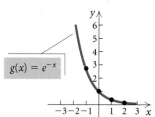

Note that the graph of g is a reflection of the graph of f across the y-axis.

Method 2. On a grapher, we simply enter the equations $y_1 = e^x$ and $y_2 = e^{-x}$. We may need to enter these equations as $y_1 = \exp(x)$ and $y_2 = \exp(-x)$.

Example 8   Graph each of the following using a grapher. Briefly describe how each graph can be obtained from $y = e^x$.

a) $f(x) = e^{-0.5x}$
b) $f(x) = 1 - e^{-2x}$
c) $f(x) = e^{x+3}$

SOLUTION

a) We note that the graph of this function is a horizontal stretching of the graph of $y = e^x$ followed by a reflection across the y-axis.

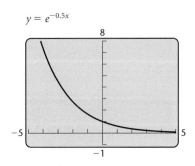

$y = e^{-0.5x}$

b) The graph is a horizontal shrinking of the graph of $y = e^x$, followed by a reflection across the y-axis, then across the x-axis, followed by a translation up 1 unit. (See the figure on the left below.)

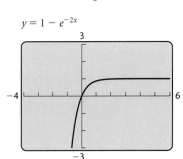

$y = 1 - e^{-2x}$

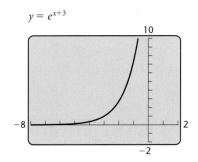

$y = e^{x+3}$

c) The graph is a translation of the graph of $y = e^x$ to the left 3 units. (See the figure on the right above.)

6.1 Exercise Set

Find each of the following, to four decimal places, on a grapher.

1. e^4 **2.** e^{10}

3. $e^{-2.458}$ **4.** $e^{1.0345}$

Make a hand-drawn graph of each function. Then check your work using a grapher, if possible.

5. $f(x) = 3^x$ **6.** $f(x) = 5^x$

7. $f(x) = 6^x$ **8.** $f(x) = 3^{-x}$

9. $f(x) = \left(\frac{1}{4}\right)^x$ **10.** $f(x) = \left(\frac{2}{3}\right)^x$

11. $x = 3^y$ **12.** $x = 4^y$

13. $x = \left(\frac{1}{2}\right)^y$ **14.** $x = \left(\frac{4}{3}\right)^y$

15. $y = \frac{1}{4}e^x$ **16.** $y = 2e^{-x}$

17. $f(x) = 1 - e^{-x}$ **18.** $f(x) = e^x - 2$

Graph each function using a grapher. Describe how each graph can be obtained from a basic exponential function.

19. $f(x) = 2^{x+1}$ **20.** $f(x) = 2^{x-1}$

21. $f(x) = 2^x - 3$ **22.** $f(x) = 2^x + 1$

23. $f(x) = 4 - 3^{-x}$ **24.** $f(x) = 2^{x-1} - 3$

25. $f(x) = \left(\frac{3}{2}\right)^{x-1}$ **26.** $f(x) = 3^{4-x}$

27. $f(x) = 2^{x+3} - 5$ **28.** $f(x) = -3^{x-2}$

29. $f(x) = e^{2x}$ **30.** $f(x) = e^{-0.2x}$

31. $y = e^{-x+1}$ **32.** $y = e^{2x} + 1$

33. $f(x) = 2(1 - e^{-x})$

34. $f(x) = 1 - e^{-0.01x}$

35. *Growth of AIDS.* The total number of Americans who have contracted AIDS is approximated by the exponential function

$$N(t) = 100,000(1.4)^t,$$

where $t = 0$ corresponds to 1989. (*Source:* U.S. Department of Health & Human Services)

a) According to this function, how many Americans had been infected as of 1997?
b) Predict the number of Americans who will have been infected by 2001.
c) Graph the function.
d) Include the graph of the equation $y = 1,000,000$ with the graph in part (c). Find the length of time it took for the number of Americans who contracted AIDS to reach 1 million.

36. *Growth of Bacteria* Escherichia coli. The bacteria *Escherichia coli* are commonly found in the human bladder. Suppose that 3000 of the bacteria are present at time $t = 0$. Then under certain conditions, t minutes later, the number of bacteria present is

$$N(t) = 3000(2)^{t/20}.$$

a) How many bacteria will be present after 10 min? 20 min? 30 min? 40 min? 60 min?
b) Graph the function.
c) This bacteria can cause bladder infections in humans when the number of bacteria reaches 100,000,000. Use a grapher to include the graph of the equation $y = 100,000,000$ with the graph in part (b). Find the length of time it takes for a bladder infection to be possible.

37. *Recycling Aluminum Cans.* It is estimated that two thirds of all aluminum cans distributed will be recycled each year (*Source:* Alcoa Corporation). A beverage company distributes 350,000 cans. The number still in use after time t, in years, is given by the exponential function

$$N(t) = 350,000\left(\frac{2}{3}\right)^t.$$

a) How many cans are still in use after 0 yr? 1 yr? 4 yr? 10 yr?
b) Graph the function.
c) After how long will 2000 of the cans still be in use?

38. *Interest in a College Trust Fund.* Following the birth of a child, Juan deposits $10,000 in a college trust fund where interest is 6.4%, compounded semiannually.

a) Find a function for the amount in the account after t years.
b) Find the amount of money in the account at $t = 0, 4, 8, 10,$ and 18 yr.

c) Graph the function.
d) After how long will the account contain $100,000?

39. *Salvage Value.* A top-quality fax–copying machine is purchased for $5800. Its value each year is about 80% of the value of the preceding year. After t years, its value, in dollars, is given by the exponential function

$$V(t) = 5800(0.8)^t.$$

a) Find the value of the machine after 0 yr, 1 yr, 2 yr, 5 yr, and 10 yr.
b) Graph the function.
c) The company decides to replace the machine when its value has declined to $500. After how long will the machine be replaced?

40. *Revenues of Packard Bell.* Sales of Packard Bell computers have grown exponentially in recent years. The total revenue R, in billions, is given by

$$R(t) = 0.518(1.42)^t,$$

where $t =$ the number of years since 1990.

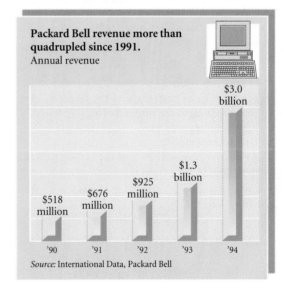

Packard Bell revenue more than quadrupled since 1991.
Annual revenue

$3.0 billion
$1.3 billion
$925 million
$676 million
$518 million

'90 '91 '92 '93 '94

Source: International Data, Packard Bell

a) Find the total revenue in 1990, 1994, 1998, and 2004.
b) Graph the function.
c) When will revenues be $8 billion?

41. *Timber Demand.* World demand for timber is increasing exponentially. The demand N, in billions of cubic feet, purchased is given by

$$N(t) = 46.6(1.018)^t,$$

where t = the number of years since 1981. (*Source:* U.N. Food and Agricultural Organization, American Forest and Paper Association)

a) Find the demand for timber in 1985, 1997, and 2010.
b) Graph the function.
c) After how many years will the demand for timber be 93.4 billion cubic feet?

42. *Typing Speed.* Sarah is taking typing in college. After she practices for t hours, her speed, in words per minute, is given by the function

$$S(t) = 200[1 - (0.86)^t].$$

a) What is Sarah's speed after practicing for 10 hr? 20 hr? 40 hr? 100 hr?
b) Graph the function.
c) How much time passes before Sarah's speed is 100 words per minute?
d) Does this graph have an asymptote? If so, what is it, and what is its significance to Sarah's learning?

43. *Advertising.* A company begins a radio advertising campaign in New York City to market a new CD-ROM video game. The percentage of the target market that buys a game is generally a function of the length of the advertising campaign. The estimated percentage is given by

$$f(t) = 100(1 - e^{-0.04t}),$$

where t = the number of days of the campaign.

a) Find $f(25)$, the percentage of the target market that has bought the product after a 25-day advertising campaign.
b) Graph the function.
c) After how long will 90% of the target market have bought the product?

44. *Growth of a Stock.* The value of a stock is given by the function

$$V(t) = 58(1 - e^{-1.1t}) + 20,$$

where V = the value of the stock after time t, in months.

a) Find $V(1)$, $V(2)$, $V(4)$, $V(6)$, and $V(12)$.
b) Graph the function.
c) After how long will the value of the stock be $75?

In Exercises 45–58, use a grapher to match the equation with one of figures (a)–(n), which follow.

a)
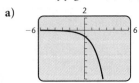

b)
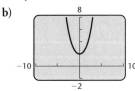

c)
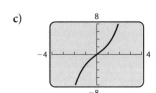

d)
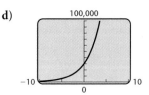

e)
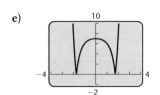

f)
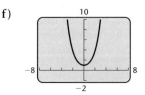

g)
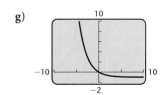

h)
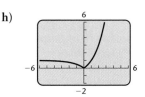

i)
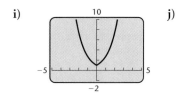

j)
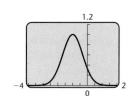

k)
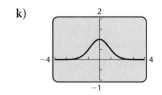

l)
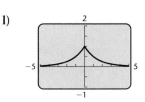

m)
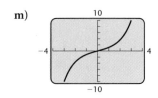

n)
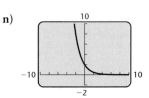

45. $y = 3^x - 3^{-x}$

46. $y = 3^{-(x+1)^2}$

47. $f(x) = -2.3^x$

48. $f(x) = 30,000(1.4)^x$

49. $y = 2^{-|x|}$

50. $y = 2^{-(x-1)}$

51. $f(x) = (0.58)^x - 1$

52. $y = 2^x + 2^{-x}$

53. $g(x) = e^{|x|}$

54. $f(x) = |2^x - 1|$

55. $y = 2^{-x^2}$

56. $y = |2^{x^2} - 8|$

57. $g(x) = \dfrac{e^x - e^{-x}}{2}$

58. $f(x) = \dfrac{e^x + e^{-x}}{2}$

Use a grapher to find the point(s) of intersection of the graphs of each of the following pairs of equations.

59. $y = |1 - 3^x|,$
$y = 4 + 3^{-x^2}$

60. $y = 4^x + 4^{-x},$
$y = 8 - 2x - x^2$

61. $y = 2e^x - 3, \ y = \dfrac{e^x}{x}$

62. $y = \dfrac{1}{e^x + 1}, \ y = 0.3x + \dfrac{7}{9}$

Use a grapher. Solve graphically.

63. $5.3^x - 4.2^x = 1073$

64. $e^x = x^3$

65. $2^x > 1$

66. $3^x \le 1$

67. $2^x + 3^x = x^2 + x^3$

68. $31{,}245e^{-3x} = 523{,}467$

Skill Maintenance

Simplify.

69. $a^0, \ a \ne 0$ **70.** a^1 **71.** 10^3 **72.** 10^{-2}

73. To what power must you raise 5 in order to get 25?

74. To what power must you raise 10 in order to get 0.001?

Synthesis

75. ◆ Describe the differences between the graphs of $f(x) = x^3$ and $g(x) = 3^x$.

76. ◆ Suppose that $10,000 is invested for 8 yr at 6.4% interest, compounded annually. In what year will the most interest be earned? Why?

77. ◆ Graph each pair of equations using the same set of axes. Then compare the results in parts (a)

and (b).

a) $y = 3^x, \ x = 3^y$ **b)** $y = 1^x, \ x = 1^y$

78. Which is larger, 7^π or π^7?

79. Graph $f(x) = x^{1/(x-1)}$. Use a grapher and the TABLE feature to identify the horizontal asymptote.

In Exercises 80 and 81:

a) *Graph using a grapher.*
b) *Approximate the zeros.*
c) *Approximate the relative maximum and minimum values. If your grapher has a MAX–MIN feature, use it.*

80. $f(x) = x^2e^{-x}$ **81.** $f(x) = e^{-x^2}$

82. Consider each of the following functions:

$$y_1 = e^x, \qquad y_2 = 1 + x + \frac{x^2}{2} + \frac{x^3}{6} + \frac{x^4}{24}.$$

a) Use a grapher to graph both functions using the viewing window $[0, 1, -1, 4]$, with Xscl = 1 and Yscl = 1.

b) On the basis of your graphs, would you consider

$$e^x = 1 + x + \frac{x^2}{2} + \frac{x^3}{6} + \frac{x^4}{24}$$

an identity?

c) See if you can prove the equation in part (b) to be an identity. Substitute $x = 1$ in each expression. What is the result?

d) Now go back to the original equations. Change the viewing window, use TRACE and ZOOM and the TABLE feature to examine the graphs in more detail. What do you discover about the graphs?

e) What caution must you be aware of using a grapher to determine whether an equation is an identity?

6.2
Logarithmic Functions and Graphs

• *Graph logarithmic functions.*
• *Convert between exponential and logarithmic equations.*
• *Find common and natural logarithms on a grapher.*

We now consider *logarithmic*, or *logarithm*, *functions*. These functions are inverses of exponential functions and have many applications.

Logarithmic Functions

Consider the function $f(x) = 2^x$ graphed below. We see from the graph that this function passes the horizontal-line test and is one-to-one. Thus f has an inverse that is a function.

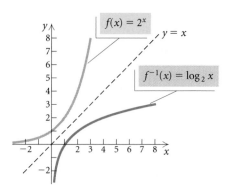

To find a formula for f^{-1} when $f(x) = 2^x$, we try to use the method of Section 1.8:

1. Replace $f(x)$ with y: $y = 2^x$
2. Interchange x and y: $x = 2^y$
3. Solve for y: $y =$ the power to which we raise 2 to get x
4. Replace y with $f^{-1}(x)$: $f^{-1}(x) =$ the power to which we raise 2 to get x.

Mathematicians have defined a new symbol to replace the words "the power to which we raise 2 to get x." That symbol is "$\log_2 x$," read "the logarithm, base 2, of x."

Logarithmic Function, Base 2

"$\log_2 x$," read "the logarithm, base 2, of x," means "the power to which we raise 2 to get x."

Thus if $f(x) = 2^x$, then $f^{-1}(x) = \log_2 x$. For example,

$$f^{-1}(8) = \log_2 8$$
$$= 3,$$

because

3 is the power to which we raise 2 to get 8.

Similarly, $\log_2 13$ is the power to which we raise 2 to get 13. As yet, we have no simpler way to say this other than "$\log_2 13$ is the power to which we raise 2 to get 13." Later, however, we will learn how to approximate this expression using a grapher.

For any exponential function $f(x) = a^x$, its inverse is called a **logarithmic function, base a**. The graph of the inverse can be obtained by reflecting the graph of $y = a^x$ across the line $y = x$, to obtain $x = a^y$. Then $x = a^y$ is equivalent to $y = \log_a x$. We read $\log_a x$ as "the logarithm, base a, of x."

The inverse of $f(x) = a^x$ is given by $f^{-1}(x) = \log_a x$.

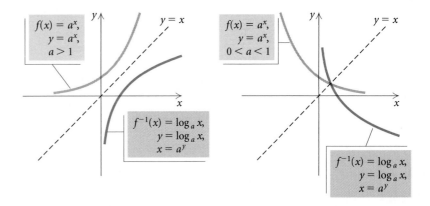

Logarithmic Function, Base a

We define $y = \log_a x$ as that number y such that $x = a^y$, where $x > 0$ and a is a positive constant other than 1.

In the following table, we compare exponential and logarithmic functions. In general, we use a number a that is greater than 1 for the logarithm base.

Exponential Function	Logarithmic Function
$y = a^x$	$x = a^y$
$f(x) = a^x$	$f^{-1}(x) = \log_a x$
$a > 1$	$a > 1$
Continuous	Continuous
One-to-one	One-to-one
Domain: All real numbers, $\mathbb{R}$	Domain: All positive real numbers, $(0, \infty)$
Range: All positive real numbers, $(0, \infty)$	Range: All real numbers, $\mathbb{R}$
Increasing	Increasing
Horizontal asymptote is x-axis: $(a^x \to 0$ as $x \to -\infty)$	Vertical asymptote is y-axis: $(\log_a x \to -\infty$ as $x \to 0^+)$
y-intercept: $(0, 1)$	x-intercept: $(1, 0)$
There is no x-intercept.	There is no y-intercept.

Converting Between Exponential and Logarithmic Equations

It is helpful in dealing with logarithmic functions to remember that a logarithm of a number is an *exponent*. It is the exponent y in $x = a^y$. You might think to yourself, "the logarithm, base a, of a number x is the power to which a must be raised to get x."

We are led to the following. (The symbol $\longleftrightarrow$ means that the two statements are equivalent; that is, when one is true, the other is true. The words "if and only if" can be used in place of $\longleftrightarrow$.)

$$\log_a x = y \longleftrightarrow x = a^y \qquad \text{A logarithm is an exponent!}$$

Be sure to memorize this relationship! It is probably the most important definition in the chapter. Many times this definition will be justification for a proof or a procedure.

Example 1 Convert each of the following to a logarithmic equation.

a) $16 = 2^x$　　　　　b) $10^{-3} = 0.001$　　　　　c) $e^t = 70$

SOLUTION

The exponent is the logarithm.

a) $16 = 2^x \qquad \log_2 16 = x$

The base remains the same.

b) $10^{-3} = 0.001 \longrightarrow \log_{10} 0.001 = -3$

c) $e^t = 70 \longrightarrow \log_e 70 = t$

Example 2 Convert each of the following to an exponential equation.

a) $\log_2 32 = 5$　　　　　b) $\log_a Q = 8$　　　　　c) $x = \log_t M$

SOLUTION

The logarithm is the exponent.

a) $\log_2 32 = 5 \qquad 2^5 = 32$

The base remains the same.

b) $\log_a Q = 8 \longrightarrow a^8 = Q$

c) $x = \log_t M \longrightarrow t^x = M$

Finding Certain Logarithms

Let's use the definition of logarithms to find some logarithmic values.

Example 3 Find each of the following logarithms.

a) $\log_{10} 10{,}000$　　　　　b) $\log_{10} 0.01$
c) $\log_2 8$　　　　　d) $\log_9 3$
e) $\log_6 1$　　　　　f) $\log_8 8$

SOLUTION

RESULT	REASON
a) $\log_{10} 10{,}000 = 4$	Think of the meaning of $\log_{10} 10{,}000$. It is the exponent to which we raise 10 to get 10,000. That exponent is 4.

b) $\log_{10} 0.01 = -2$

We have $0.01 = \dfrac{1}{100} = \dfrac{1}{10^2} = 10^{-2}$. The exponent to which we raise 10 to get 0.01 is -2.

c) $\log_2 8 = 3$

$8 = 2^3$. The exponent to which we raise 2 to get 8 is 3.

d) $\log_9 3 = \frac{1}{2}$

$3 = \sqrt{9} = 9^{1/2}$. The exponent to which we raise 9 to get 3 is $\frac{1}{2}$.

e) $\log_6 1 = 0$

$1 = 6^0$. The exponent to which we raise 6 to get 1 is 0.

f) $\log_8 8 = 1$

$8 = 8^1$. The exponent to which we raise 8 to get 8 is 1. —

Examples 3(e) and 3(f) illustrate two important properties of logarithms.

$\log_a 1 = 0$ and $\log_a a = 1$, for any logarithmic base a.

$\log_a 1 = 0$ follows from the fact that $a^0 = 1$. Thus, $\log_7 1 = 0$, $\log_{10} 1 = 0$, and so on. $\log_a a = 1$ follows from the fact that $a^1 = a$. Thus, $\log_5 5 = 1$, $\log_{10} 10 = 1$, and so on.

Finding Logarithms on a Grapher

Before calculators became so widely available, base-10, or **common logarithms**, were used extensively to simplify complicated calculations. In fact, that is why logarithms were invented. The abbreviation **log**, with no base written, is used to represent the common logarithms, or base 10 logarithms. Thus,

$$\log 29 \quad \text{means} \quad \log_{10} 29.$$

We can approximate $\log 29$ as follows:

$$\left.\begin{array}{l} \log 100 = \log_{10} 100 = 2 \\ \qquad\qquad \log 29 = ? \\ \log 10 = \log_{10} 10 = 1 \end{array}\right\} \quad \begin{array}{l}\text{It seems reasonable that } \log 29 \text{ is between} \\ 1 \text{ and } 2.\end{array}$$

On a grapher, the scientific key for common logarithms is generally marked $\boxed{\text{LOG}}$. Using that key, we find that

$$\log 29 \approx 1.462397998 \approx 1.4624$$

rounded to four decimal places. This also tells us that $10^{1.4624} \approx 29$.

Interactive Discovery

Find log 1000 and log 10,000 without using a grapher. Between what two whole numbers is log 9874? Now find log 9874, rounded to four decimal places, on a grapher.

Example 4 Find each of the following common logarithms on a grapher. Round to four decimal places.

a) log 645,778   **b)** log 0.0000239   **c)** log (−3)

SOLUTION

FUNCTION VALUE	READOUT	ROUNDED
a) log 645,778	5.810083246	5.8101
b) log 0.0000239	−4.621602099	−4.6216
c) log (−3)	ERROR*	Does not exist

In Example 4(c), log (−3) does not exist as a real number because there is no real-number power to which we can raise 10 to get −3. The number 10 raised to any real-number power is positive. The common logarithm of a negative number does not exist as a real number. ▬

Natural Logarithms

Logarithms, base *e*, are called **natural logarithms**. The abbreviation "ln" is generally used for natural logarithms. Thus

     ln 53   means $\log_e$ 53.

On a grapher, the scientific key for natural logarithms is generally marked $\boxed{\text{LN}}$. Using that key, we find that

     ln 53 ≈ 3.970291914 ≈ 3.9703

rounded to four decimal places.

Example 5 Find each of the following natural logarithms on a grapher. Round to four decimal places.

a) ln 645,778   **b)** ln 0.0000239   **c)** ln (−5)
d) ln *e*   **e)** ln 1

SOLUTION

FUNCTION VALUE	READOUT	ROUNDED
a) ln 645,778	13.37821107	13.3782
b) ln 0.0000239	−10.64163210	−10.6416
c) ln (−5)	ERROR	Does not exist
d) ln *e*	1	1
e) ln 1	0	0

Note that ln *e* = $\log_e$ *e* = 1 and ln 1 = $\log_e$ 1 = 0. ▬

---

*On some graphers, it would be expressed as a complex number—about 0.477 + 1.364*i*.

Interactive Discovery

In some textbooks and computer applications, log x is used to represent $\log_e x$ rather than $\log_{10} x$. Explore some methods you might use with a grapher to discover what the base actually is.

Changing Logarithmic Bases

Most graphers give the values of both common logarithms and natural logarithms. To find a logarithm with a base other than 10 or e, we can use the following conversion formula.

The Change-of-Base Formula

For any logarithmic bases a and b, and any positive number M,

$$\log_b M = \frac{\log_a M}{\log_a b}.$$

We will prove this result in the next section.

Example 6   Find $\log_5 8$ using common logarithms.

SOLUTION   First we let $a = 10$, $b = 5$, and $M = 8$. Then we substitute into the change-of-base formula:

$$\log_5 8 = \frac{\log_{10} 8}{\log_{10} 5} \qquad \text{Substituting}$$

$$\approx \frac{0.903090}{0.698970}$$

$$\approx 1.2920.$$

To check, we use the power key to verify that $5^{1.2920} \approx 8$.

We can also use base e for a conversion.

Example 7   Find $\log_4 31$ using natural logarithms.

SOLUTION   Substituting e for a, 4 for b, and 31 for M, we have

$$\log_4 31 = \frac{\log_e 31}{\log_e 4} = \frac{\ln 31}{\ln 4} \approx \frac{3.433987}{1.386294} \approx 2.4771.$$

Graphs of Logarithmic Functions

We demonstrate several ways to graph logarithmic functions.

Example 8   Graph: $g(x) = \ln x$.

SOLUTION

Method 1. To graph $y = g(x) = \ln x$, we first write its equivalent exponential equation, $x = e^y$. Then we select values for y and use a grapher to

---

Program

LOGBASE: This program provides a formula for changing the base of logarithms.

find the corresponding values of e^y. We then plot points, remembering that x still is the first coordinate.

x, or e^y	y
1	0
2.7	1
7.4	2
20	3
0.1	−2
0.4	−1

(1) Select y.
(2) Compute x.

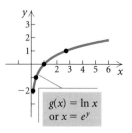

$g(x) = \ln x$
or $x = e^y$

Connect the points with a smooth curve.

Method 2. We can also use the $\boxed{\text{LN}}$ key on a grapher to find function values. Then we plot points and draw the graph.

Method 3. We use a grapher to graph $y = \ln x$.

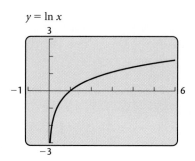

$y = \ln x$

Example 9   Graph $y = f(x) = \log_5 x$.

SOLUTION

Method 1. The equation $y = \log_5 x$ is equivalent to $x = 5^y$. We can find ordered pairs that are solutions by choosing values for y and computing the x-values.

For $y = 0, x = 5^0 = 1.$

For $y = 1, x = 5^1 = 5.$

For $y = 2, x = 5^2 = 25.$

For $y = 3, x = 5^3 = 125.$

For $y = -1, x = 5^{-1} = \dfrac{1}{5}.$

For $y = -2, x = 5^{-2} = \dfrac{1}{25}.$

x, or 5^y	y
1	0
5	1
25	2
125	3
$\dfrac{1}{5}$	−1
$\dfrac{1}{25}$	−2

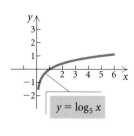

$y = \log_5 x$

$y = \log_5 x = \dfrac{\ln x}{\ln 5}$

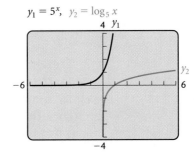

Method 2. To use a grapher, we must first change the base. Here we change from base 5 to base e:

$$y = \log_5 x = \frac{\ln x}{\ln 5}. \qquad \text{Using } \log_b M = \frac{\log_a M}{\log_a b}$$

The graph is as shown at left.

Method 3. Some graphers have a feature that graphs inverses automatically. If we begin with $y_1 = 5^x$, the graphs of both y_1 and its inverse $y_2 = \log_5 x$ will be drawn.

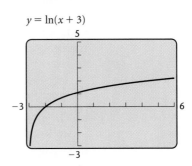

The graph of $f(x) = \log_a x$, for any base a, has the x-intercept $(1, 0)$. The domain is the set of positive real numbers. The range is the set of all real numbers.

Example 10 Graph each of the following using a grapher. Describe how each graph can be obtained from $y = \ln x$. Give the domain of each function and discuss the vertical asymptotes.

a) $f(x) = \ln (x + 3)$
b) $f(x) = 3 - \frac{1}{2} \ln x$
c) $f(x) = |\ln (x - 1)|$

SOLUTION

a) The graph is a shift of the graph of $y = \ln x$ left 3 units.

$y = \ln(x + 3)$

The domain is the set of all real numbers greater than -3, $(-3, \infty)$. The line $x = -3$ is a vertical asymptote.

b) The graph is a shrinking of the graph of $y = \ln x$ in the y-direction, followed by a reflection across the x-axis, and then a translation 3 units up.

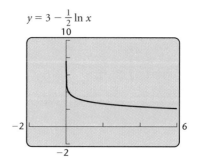

$$y = 3 - \frac{1}{2}\ln x$$

The domain is the set of all positive real numbers, $(0, \infty)$. The y-axis is a vertical asymptote.

c) The graph is a translation of the graph of $y = \ln x$, 1 unit to the right. Then the absolute value has the effect of reflecting negative outputs across the x-axis.

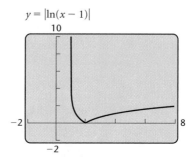

$$y = |\ln(x - 1)|$$

The domain is the set of all real numbers greater than 1, $(1, \infty)$. The line $x = 1$ is a vertical asymptote.

Applications

Example 11 *Walking Speed.* In a study by psychologists Bornstein and Bornstein, it was found that the average walking speed w, in feet per second, of a person living in a city of population P, in thousands, is given by the function

$$w(P) = 0.37 \ln P + 0.05.$$

(*Source: International Journal of Psychology*)

a) The population of Orlando, Florida, is 1,073,000. Find the average walking speed of people living in Orlando.

b) Graph the equation.

c) A sociologist computes the average walking speed in a city to be about 2.0 ft/sec. Use this information to estimate the population of the city.

SOLUTION

a) We substitute 1073 for P, since P is in thousands:

$$w(1073) = 0.37 \ln 1073 + 0.05 \quad \text{Substituting}$$
$$\approx 2.6 \text{ ft/sec.} \quad \text{Finding the natural logarithm and simplifying}$$

The average walking speed of people living in Orlando is about 2.6 ft/sec.

b) We graph with a viewing window of $[0, 600, 0, 4]$ because inputs are very large and outputs are very small by comparison.

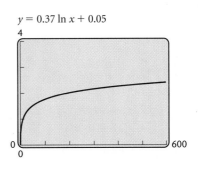

$y = 0.37 \ln x + 0.05$

c) To find the population for which the average walking speed is 2.0 ft/sec, we set

$$2.0 = 0.37 \ln P + 0.05$$

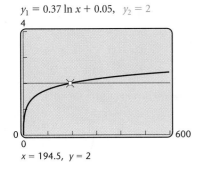

$y_1 = 0.37 \ln x + 0.05, \quad y_2 = 2$

$x = 194.5, \ y = 2$

and solve. We begin by graphing the equations $y_1 = 0.37 \ln x + 0.05$ and $y_2 = 2$. Then we can use the TRACE and ZOOM features, or the INTERSECT feature, to approximate the point of intersection. See one such TRACE at left. We see that a city with a population of about 194,500 would have an average walking speed of 2.0 ft/sec. Some graphers can solve the equation directly.

Example 12 *Earthquake Magnitude.* The magnitude R, measured on the Richter scale, of an earthquake of intensity I is defined as

$$R = \log \frac{I}{I_0},$$

where I_0 is a minimum intensity used for comparison. We can think of I_0 as a threshold intensity that is the weakest earthquake that can be recorded on a seismograph. If one earthquake is 10 times as intense as another, its magnitude on the Richter scale is 1 greater than that of the other. If one earthquake is 100 times as intense as another, its magnitude on the Richter scale is 2 higher, and so on. Thus an earthquake whose magnitude is 7 on the Richter scale is 10 times as intense as an earthquake whose magnitude is 6. Earthquakes can be interpreted as multiples of the minimum intensity I_0.

The Kobe, Japan, earthquake of 1995 had an intensity of $10^{6.8} \cdot I_0$. What was its magnitude on the Richter scale?

SOLUTION   We substitute into the formula:

$$R = \log \frac{I}{I_0} = \log \frac{10^{6.8}I_0}{I_0} = \log 10^{6.8} = 6.8.$$

The magnitude of the earthquake was 6.8 on the Richter scale.   ▬

$\dfrac{6.2}{}$ **Exercise Set**

Make a hand-drawn graph of each of the following. Then check your work using a grapher.

1. $y = \log_3 x$ **2.** $y = \log_4 x$

3. $f(x) = \log x$ **4.** $f(x) = \ln x$

Find each of the following. Do not use a grapher.

5. $\log_2 16$ **6.** $\log_3 9$

7. $\log_5 125$ **8.** $\log_2 64$

9. $\log 0.001$ **10.** $\log 100$

11. $\log_2 \frac{1}{4}$ **12.** $\log_8 2$

13. $\ln 1$ **14.** $\ln e$

15. $\log 10$ **16.** $\log 1$

Convert to a logarithmic equation.

17. $10^3 = 1000$ **18.** $5^{-3} = \frac{1}{125}$

19. $8^{1/3} = 2$ **20.** $10^{0.3010} = 2$

21. $e^3 = t$ **22.** $Q^t = x$

23. $e^2 = 7.3891$ **24.** $e^{-1} = 0.3679$

25. $p^k = 3$ **26.** $e^{-t} = 4000$

Convert to an exponential equation.

27. $\log_5 5 = 1$ **28.** $t = \log_4 7$

29. $\log 0.01 = -2$ **30.** $\log 7 = 0.845$

31. $\ln 30 = 3.4012$ **32.** $\ln 0.38 = -0.9676$

33. $\log_a M = -x$ **34.** $\log_t Q = k$

35. $\log_a T^3 = x$ **36.** $\ln W^5 = t$

Find each of the following on a grapher. Round to four decimal places.

37. $\log 3$ **38.** $\log 8$

39. $\log 532$ **40.** $\log 93{,}100$

41. $\log 0.57$ **42.** $\log 0.082$

43. $\log (-2)$ **44.** $\ln 50$

45. $\ln 2$ **46.** $\ln (-4)$

47. $\ln 809.3$ **48.** $\ln 0.00037$

49. $\ln (-1.32)$ **50.** $\ln 0$

Find the logarithm using the change-of-base formula.

51. $\log_4 100$ **52.** $\log_3 20$

53. $\log_{100} 0.3$ **54.** $\log_\pi 100$

55. $\log_{200} 50$ **56.** $\log_{5.3} 1700$

For each of the following functions, briefly describe how each graph can be obtained from a basic logarithmic function. Then graph the function using a grapher. Give the domain of each function and discuss the vertical asymptotes.

57. $f(x) = \log_2 (x + 3)$ **58.** $f(x) = \log_3 (x - 2)$

59. $y = \log_3 x - 1$ **60.** $y = 3 + \log_2 x$

61. $f(x) = 4 \ln x$ **62.** $f(x) = \frac{1}{2} \ln x$

63. $y = 2 - \ln x$ **64.** $y = \ln (x + 1)$

Graph each function and its inverse using the same set of axes. Use any method.

65. $f(x) = 3^x, \; f^{-1}(x) = \log_3 x$

66. $f(x) = \log_4 x, \; f^{-1}(x) = 4^x$

67. $f(x) = \log x, \; f^{-1}(x) = 10^x$

68. $f(x) = e^x, \; f^{-1}(x) = \ln x$

69. *Walking Speed.*   Refer to Example 11. Various cities and their populations are given below. Find the average walking speed in each city.

a) Los Angeles, California: 3,486,000
b) Chicago, Illinois: 3,005,000
c) Tucson, Arizona: 406,000
d) Rome, New York: 50,400

70. *Earthquake Magnitude.*   Refer to Example 12. Various locations of earthquakes and their intensities are given below. What was the magnitude on the Richter scale?

a) Mexico City, 1978: $10^{7.85} \cdot I_0$
b) San Francisco, 1906: $10^{8.25} \cdot I_0$
c) Chile, 1960: $10^{9.6} \cdot I_0$
d) Italy, 1980: $10^{7.85} \cdot I_0$
e) San Francisco, 1989: $10^{6.9} \cdot I_0$

71. *Forgetting.* Students in an accounting class took a final exam. They took equivalent forms of the exam in monthly intervals thereafter. The average score $S(t)$, in percent, after t months was found to be given by the function

$$S(t) = 78 - 15 \log (t + 1), \quad t \geq 0.$$

a) What was the average score when they initially took the test, $t = 0$?
b) What was the average score after 4 months? 24 months?
c) Graph the function.
d) After what time t was the average score 50?

72. *pH of Substances in Chemistry.* In chemistry, the pH of a substance is defined as

$$pH = -\log [H^+],$$

where H^+ is the hydrogen ion concentration, in moles per liter. Find the pH of each substance.

Litmus paper is used to test pH.

SUBSTANCE	HYDROGEN ION CONCENTRATION
a) Pineapple juice	1.6×10^{-4}
b) Hair rinse	0.0013
c) Mouthwash	6.3×10^{-7}
d) Eggs	1.6×10^{-8}
e) Tomatoes	6.3×10^{-5}

73. Find the hydrogen ion concentration of each substance, given the pH. Express the answer in scientific notation.

SUBSTANCE	pH
a) Tap water	7
b) Rainwater	5.4
c) Orange juice	3.2
d) Wine	4.8

74. *Advertising.* A model for advertising response is given by the function

$$N(a) = 1000 + 200 \ln a, \quad a \geq 1,$$

where $N(a)$ = the number of units sold and a = the amount spent on advertising, in thousands of dollars.

a) How many units were sold after spending $1000 ($a = 1$) on advertising?

b) How many units were sold after spending $5000?
c) Graph the function.
d) How much would have to be spent in order to sell 2000 units?

75. *Loudness of Sound.* The **loudness** L, in bels (after Alexander Graham Bell), of a sound of intensity I is defined to be

$$L = \log \frac{I}{I_0},$$

where I_0 = the minimum intensity detectable by the human ear (such as the tick of a watch at 20 ft under quiet conditions). If a sound is 10 times as intense as another, its loudness is 1 bel greater than that of the other. If a sound is 100 times as intense as another, its loudness is 2 bels greater, and so on. The bel is a large unit, so a subunit, the **decibel**, is generally used. For L, in decibels, the formula is

$$L = 10 \log \frac{I}{I_0}.$$

Find the loudness, in decibels, of each sound with the given intensity.

SOUND	INTENSITY
a) Library	$2510 \cdot I_0$
b) Dishwater	$2{,}500{,}000 \cdot I_0$
c) Conversational speech	$10^6 \cdot I_0$
d) Heavy truck	$10^9 \cdot I_0$

Skill Maintenance

76. Multiply and simplify: $x^{-5t} \cdot x^{7t}$.

77. Divide and simplify: $\dfrac{y^{-52}}{y^{-76}}$.

Simplify.
78. $(e^{-12})^{-5}$
79. $(10^4)^t$

Synthesis

80. ◆ Explain how the graph of $f(x) = \ln x$ can be used to obtain the graph of $g(x) = e^{x-2}$.

81. ◆ Explain how the graph of $f(x) = e^x$ can be used to obtain the graph of $g(x) = 3 + \ln x$.

Simplify.
82. $\dfrac{\log_3 64}{\log_3 16}$
83. $\dfrac{\log_5 8}{\log_5 2}$

Find the domain of each function.
84. $f(x) = \log_4 x^2$
85. $f(x) = \log_5 x^3$
86. $f(x) = \log (3x - 4)$
87. $f(x) = \ln |x|$

Solve.

88. $\log_2 (x - 3) \geq 4$

89. $\log_2 (2x + 5) < 0$

90. Use a grapher. Find the point(s) of intersection of the graphs.

$$y = 4 \ln x, \qquad y = \frac{4}{e^x + 1}$$

In Exercises 91–94, match the equation with one of figures (a)–(d), which follow. If needed, use a grapher.

a)

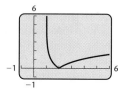

b)

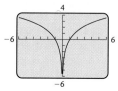

c)

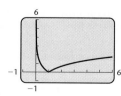

d)

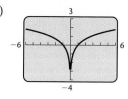

91. $f(x) = \ln |x|$

92. $f(x) = |\ln x|$

93. $f(x) = \ln x^2$

94. $g(x) = |\ln (x - 1)|$

For Exercises 95–98:

a) *Graph the function.*

b) *Estimate the zeros.*

c) *Estimate the relative maximum and the minimum values.*

95. $f(x) = x \ln x$

96. $f(x) = x^2 \ln x$

97. $f(x) = \dfrac{\ln x}{x^2}$

98. $f(x) = e^{-x} \ln x$

6.3
Properties of Logarithmic Functions

• *Convert from logarithms of products, powers, and quotients to expressions in terms of individual logarithms, and conversely.*
• *Simplify expressions of the type $\log_a a^x$ and $a^{\log_a x}$.*

We now establish some properties of logarithmic functions that are fundamental to their use here and in subsequent mathematics. These properties are based on corresponding rules for exponents.

Logarithms of Products

Interactive Discovery

Graph each of the following functions. Use the graphs to discover an identity.

$$y_1 = \log (4x^2 \cdot x^3), \quad y_2 = \log (4x^2) + \log (x^3),$$
$$y_3 = \log (4x^2 + x^3), \quad y_4 = [\log (4x^2)][\log (x^3)]$$

Test your result by using a TABLE feature to complete this table.

X	Y₁	Y₂	Y₃	Y₄
1				
10				
20				
300				

X = 1

Graph some other functions to test your discovery.

The first property of logarithms corresponds to the rule of exponents: $a^m \cdot a^n = a^{m+n}$.

The Product Rule

For any positive numbers M and N and any logarithmic base a,

$$\log_a MN = \log_a M + \log_a N.$$

(The logarithm of a product is the sum of the logarithms of the factors.)

Example 1   Express as a sum of logarithms: $\log_3 (9 \cdot 27)$.

SOLUTION   We have

$$\log_3 (9 \cdot 27) = \log_3 9 + \log_3 27. \qquad \text{Using the product rule}$$

As a check, note that

$$\log_3 (9 \cdot 27) = \log_3 243 = 5$$

and that

$$\log_3 9 + \log_3 27 = 2 + 3 = 5.$$

Example 2   Express as a single logarithm: $\log_2 p^3 + \log_2 q$.

SOLUTION   We have

$$\log_2 p^3 + \log_2 q = \log_2 (p^3 q).$$

A Proof of the Product Rule: Let $\log_a M = x$ and $\log_a N = y$. Converting to exponential equations, we have $a^x = M$ and $a^y = N$. Then

$$MN = a^x \cdot a^y$$
$$= a^{x+y}.$$

Converting back to a logarithmic equation, we get

$$\log_a MN = x + y.$$

Remembering what x and y represent, it follows that

$$\log_a MN = \log_a M + \log_a N.$$

Logarithms of Powers

Interactive Discovery

Graph each of the following functions. Use the graphs to discover an identity.

$$y_1 = \log x^5, \qquad y_2 = \log x + \log 5,$$
$$y_3 = 5 \log x, \qquad y_4 = \log (x + 5).$$

Test your result by using a TABLE feature to complete this table.

(continued)

X	Y1	Y2	Y3	Y4
1				
10				
20				
300				

X = 1

Graph some other functions to test your discovery.

The second property of logarithms corresponds to the rule of exponents: $(a^m)^n = a^{mn}$.

The Power Rule

For any positive number M, any logarithmic base a, and any real number p,

$$\log_a M^p = p \log_a M.$$

(The logarithm of a power of M is the exponent times the logarithm of M.)

Example 3 Express each of the following as a product.

a) $\log_a 11^{-3}$ **b)** $\log_a \sqrt[4]{7}$

SOLUTION

a) $\log_a 11^{-3} = -3 \log_a 11$ Using the power rule

b) $\log_a \sqrt[4]{7} = \log_a 7^{1/4}$ Writing exponential notation

$\qquad\qquad = \frac{1}{4} \log_a 7$ Using the power rule

A Proof of the Power Rule: Let $x = \log_a M$. The equivalent exponential equation is $a^x = M$. Raising both sides to the power p, we obtain

$$(a^x)^p = M^p,$$

or

$$a^{xp} = M^p.$$

Converting back to a logarithmic equation, we get

$$\log_a M^p = xp.$$

But $x = \log_a M$, so substituting gives us

$$\log_a M^p = (\log_a M)p$$
$$= p \log_a M.$$

Logarithms of Quotients

Interactive Discovery

Graph each of the following functions. Use the graphs to discover an identity.

$$y_1 = \log\left(\frac{4x^5}{x^2}\right), \qquad y_2 = \log(4x^5) + \log(x^2),$$

$$y_3 = \log(4x^5) - \log(x^2), \qquad y_4 = \log(4x + 5x^2)$$

Test your result by using a TABLE feature to complete this table.

X	Y1	Y2	Y3	Y4
1				
10				
20				
300				

X = 1

Graph some other functions to test your discovery.

The third property of logarithms corresponds to the rule of exponents:

$$\frac{a^m}{a^n} = a^{m-n}.$$

The Quotient Rule

For any positive numbers M and N, and any logarithmic base a,

$$\log_a \frac{M}{N} = \log_a M - \log_a N.$$

(The logarithm of a quotient is the logarithm of the numerator minus the logarithm of the denominator.)

Example 4 Express as a difference of logarithms: $\log_t \frac{8}{w}$.

SOLUTION

$$\log_t \frac{8}{w} = \log_t 8 - \log_t w \qquad \text{Using the quotient rule}$$

Example 5 Express as a single logarithm: $\log_b 64 - \log_b 16$.

SOLUTION

$$\log_b 64 - \log_b 16 = \log_b \frac{64}{16} = \log_b 4$$

A Proof of the Quotient Rule: The proof follows from both the product and the power rules:

$$\log_a \frac{M}{N} = \log_a MN^{-1}$$

$$= \log_a M + \log_a N^{-1} \qquad \text{Using the product rule}$$

$$= \log_a M + (-1) \log_a N \qquad \text{Using the power rule}$$

$$= \log_a M - \log_a N.$$

Note the following.

$\log_a MN \neq (\log_a M)(\log_a N)$	The logarithm of a product is *not* the product of the logarithms.
$\log_a (M + N) \neq \log_a M + \log_a N$	The logarithm of a sum is *not* the sum of the logarithms.
$\log_a \dfrac{M}{N} \neq \dfrac{\log_a M}{\log_a N}$	The logarithm of a quotient is *not* the quotient of the logarithms.

Using the Properties Together

Example 6 Express each of the following in terms of sums and differences of logarithms.

a) $\log_a \dfrac{x^2 y^5}{z^4}$

b) $\log_a \sqrt[3]{\dfrac{a^2 b}{c^5}}$

c) $\log_b \dfrac{a y^5}{m^3 n^4}$

SOLUTION

a) $\log_a \dfrac{x^2 y^5}{z^4} = \log_a (x^2 y^5) - \log_a z^4$ Using the quotient rule

$$= \log_a x^2 + \log_a y^5 - \log_a z^4 \qquad \text{Using the product rule}$$

$$= 2 \log_a x + 5 \log_a y - 4 \log_a z \qquad \text{Using the power rule}$$

b) $\log_a \sqrt[3]{\dfrac{a^2 b}{c^5}} = \log_a \left(\dfrac{a^2 b}{c^5}\right)^{1/3}$ Writing exponential notation

$$= \frac{1}{3} \log_a \frac{a^2 b}{c^5} \qquad \text{Using the power rule}$$

$$= \frac{1}{3} (\log_a a^2 b - \log_a c^5) \qquad \text{Using the quotient rule}$$

$$= \frac{1}{3} (2 \log_a a + \log_a b - 5 \log_a c) \qquad \text{Using the product and power rules. The parentheses are important.}$$

$$= \frac{1}{3} (2 + \log_a b - 5 \log_a c) \qquad \log_a a = 1$$

$$= \frac{2}{3} + \frac{1}{3} \log_a b - \frac{5}{3} \log_a c \qquad \text{Multiplying to remove parentheses}$$

c) $\log_b \dfrac{ay^5}{m^3 n^4} = \log_b ay^5 - \log_b m^3 n^4$ Using the quotient rule

$$= (\log_b a + \log_b y^5) - (\log_b m^3 + \log_b n^4)$$ Using the product rule

$$= \log_b a + \log_b y^5 - \log_b m^3 - \log_b n^4$$ Removing parentheses

$$= \log_b a + 5 \log_b y - 3 \log_b m - 4 \log_b n$$ Using the power rule

Example 7 Express as a single logarithm:

$$5 \log_b x - \log_b y + \frac{1}{4} \log_b z.$$

SOLUTION

$$5 \log_b x - \log_b y + \frac{1}{4} \log_b z = \log_b x^5 - \log_b y + \log_b z^{1/4}$$

                 Using the power rule

$$= \log_b \frac{x^5}{y} + \log_b z^{1/4}$$

                 Using the quotient rule

$$= \log_b \frac{x^5 z^{1/4}}{y}, \text{ or } \log_b \frac{x^5 \sqrt[4]{z}}{y}$$

                 Using the product rule

Example 8 Given that $\log_a 2 = 0.301$ and $\log_a 3 = 0.477$, find each of the following.

a) $\log_a 6$ **b)** $\log_a \dfrac{2}{3}$ **c)** $\log_a 81$

d) $\log_a \sqrt{a}$ **e)** $\log_a 5$ **f)** $\dfrac{\log_a 3}{\log_a 2}$

SOLUTION

a) $\log_a 6 = \log_a (2 \cdot 3) = \log_a 2 + \log_a 3$ Using the product rule

$$= 0.301 + 0.477$$

$$= 0.778$$

b) $\log_a \frac{2}{3} = \log_a 2 - \log_a 3$ Using the quotient rule

$$= 0.301 - 0.477 = -0.176$$

c) $\log_a 81 = \log_a 3^4 = 4 \log_a 3$ Using the power rule

$$= 4(0.477) = 1.908$$

d) $\log_a \sqrt{a} = \log_a a^{1/2} = \frac{1}{2} \log_a a$ Using the power rule

$$= \frac{1}{2} \cdot 1 = \frac{1}{2}$$

e) $\log_a 5$ *cannot be found using these properties and the given information.*

$$(\log_a 5 \neq \log_a 2 + \log_a 3)$$

f) $\dfrac{\log_a 3}{\log_a 2} = \dfrac{0.477}{0.301} \approx 1.585$ We simply divided, not using any of the properties.

In Example 8, a is actually 10 so we have common logarithms. Check as many results as possible using a grapher.

Simplifying Expressions of the Type $\log_a a^x$ and $a^{\log_a x}$

We have two final properties to consider. The first follows from the product rule: Since $\log_a a^x = x \log_a a = x \cdot 1 = x$, we have $\log_a a^x = x$. This property also follows from the definition of a logarithm: x is the power to which we raise a in order to get a^x.

The Logarithm of a Base to a Power

For any base a and any real number x,

$$\log_a a^x = x.$$

(The logarithm, base a, of a to a power is the power.)

Example 9   Simplify each of the following.

a) $\log_a a^8$ **b)** $\ln e^{-t}$ **c)** $\log 10^{3k}$

SOLUTION

a) $\log_a a^8 = 8$ 8 is the power to which we raise a in order to get a^8.
b) $\ln e^{-t} = \log_e e^{-t} = -t$
c) $\log 10^{3k} = \log_{10} 10^{3k} = 3k$

Let $M = \log_a x$. Then $a^M = x$. Substituting $\log_a x$ for M, we obtain $a^{\log_a x} = x$. This also follows from the definition of a logarithm: $\log_a x$ is the power to which a is raised in order to get x.

A Base to a Logarithmic Power

For any base a and any positive real number x,

$$a^{\log_a x} = x.$$

(The number a raised to the power $\log_a x$ is x.)

Example 10   Simplify each of the following.

a) $4^{\log_4 k}$ **b)** $e^{\ln 5}$ **c)** $10^{\log 7t}$

SOLUTION

a) $4^{\log_4 k} = k$

b) $e^{\ln 5} = e^{\log_e 5} = 5$

c) $10^{\log 7t} = 10^{\log_{10} 7t} = 7t$

   ━

A Proof of the Change-of-Base Formula: We close this section by proving the change-of-base formula and summarizing the properties of logarithms considered thus far in this chapter. In Section 6.2, we used the change-of-base formula,

$$\log_b M = \frac{\log_a M}{\log_a b},$$

to make base conversions in order to graph logarithmic functions on a grapher. Let $x = \log_b M$. Then by the definition of logarithms, $b^x = M$. We have established that logarithmic functions are one-to-one. This says that $M = N \longleftrightarrow \log_a M = \log_a N$ for any positive numbers M and N. Thus we can take the logarithm base a on both sides of an expression such as $b^x = M$. Then we get $\log_a b^x = \log_a M$. Using the power rule, we obtain $x \log_a b = \log_a M$. Solving for x gives us

$$x = \frac{\log_a M}{\log_a b}, \quad \text{so} \quad x = \log_b M = \frac{\log_a M}{\log_a b}.$$

Following is a summary of the properties of logarithms.

Summary of the Properties of Logarithms

The Product Rule:	$\log_a MN = \log_a M + \log_a N$
The Power Rule:	$\log_a M^p = p \log_a M$
The Quotient Rule:	$\log_a \dfrac{M}{N} = \log_a M - \log_a N$
The Change-of-Base Formula:	$\log_b M = \dfrac{\log_a M}{\log_a b}$
Other Formulas:	$\log_a a = 1, \qquad \log_a 1 = 0,$
	$\log_a a^x = x, \qquad a^{\log_a x} = x$

6.3 Exercise Set

Express as a sum of logarithms.

1. $\log_3 (81 \cdot 27)$ **2.** $\log_2 (8 \cdot 64)$

3. $\log_5 (5 \cdot 125)$ **4.** $\log_4 (64 \cdot 32)$

5. $\log_t 8Y$ **6.** $\log_e Qx$

Express as a product.

7. $\log_b t^3$ **8.** $\log_a x^4$

9. $\log y^8$ **10.** $\ln y^5$

11. $\log_c K^{-6}$ **12.** $\log_b Q^{-8}$

Express as a difference of logarithms.

13. $\log_t \dfrac{M}{8}$ **14.** $\log_a \dfrac{76}{13}$

15. $\log_a \dfrac{x}{y}$

16. $\log_b \dfrac{3}{w}$

Express in terms of sums and differences of logarithms.

17. $\log_a 6xy^5z^4$

18. $\log_a x^3y^2z$

19. $\log_b \dfrac{p^2q^5}{m^4b^9}$

20. $\log_b \dfrac{x^2y}{b^3}$

21. $\log_a \sqrt{\dfrac{x^6}{p^5q^8}}$

22. $\log_c \sqrt[3]{\dfrac{y^3z^2}{x^4}}$

23. $\log_a \sqrt[4]{\dfrac{m^8n^{12}}{a^3b^5}}$

24. $\log_a \sqrt{\dfrac{a^6b^8}{a^2b^5}}$

Express as a single logarithm and simplify, if possible.

25. $\log_a 75 + \log_a 2$

26. $\log 0.01 + \log 1000$

27. $\log 10,000 - \log 100$

28. $\ln 54 - \ln 6$

29. $\frac{1}{2}\log_a x + 4\log_a y - 3\log_a x$

30. $\frac{2}{5}\log_a x - \frac{1}{3}\log_a y$

31. $\ln x^2 - 2\ln \sqrt{x}$

32. $\ln 2x + 3(\ln x - \ln y)$

33. $\ln (x^2 - 4) - \ln (x + 2)$

34. $\log_a \dfrac{a}{\sqrt{x}} - \log_a \sqrt{ax}$

35. $\ln x - 3[\ln (x - 5) + \ln (x + 5)]$

36. $\frac{2}{3}[\ln (x^2 - 9) - \ln (x + 3)] + \ln (x + y)$

37. $\frac{3}{2}\ln 4x^6 - \frac{4}{5}\ln 2y^{10}$

38. $120(\ln \sqrt[5]{x^3} + \ln \sqrt[3]{y^2} - \ln \sqrt[4]{16z^5})$

Given that $\log_b 3 = 1.0986$ *and* $\log_b 5 = 1.6094$, *find each of the following.*

39. $\log_b \frac{3}{5}$

40. $\log_b 15$

41. $\log_b \frac{1}{5}$

42. $\log_b \frac{5}{3}$

43. $\log_b \sqrt{b}$

44. $\log_b \sqrt{b^3}$

45. $\log_b 5b$

46. $\log_b 9$

47. $\log_b 75$

48. $\log_b \dfrac{1}{b}$

Simplify.

49. $\log_p p^3$

50. $\log_t t^{2713}$

51. $\log_e e^{|x-4|}$

52. $\log_q q^{\sqrt{3}}$

53. $3^{\log_3 4x}$

54. $5^{\log_5 (4x-3)}$

55. $10^{\log w}$

56. $e^{\ln x^3}$

57. $\ln e^{8t}$

58. $\log 10^{-k}$

Skill Maintenance

Solve.

59. $3x - 7 = 5$

60. $x^4 - 6x^2 + 7 = 0$

61. $\dfrac{2x - 1}{x - 4} = 9$

62. $x(x - 3) = 10$

Synthesis

63. ◆ Given that $f(x) = a^x$ and $g(x) = \log_a x$, find $(f \circ g)(x)$ and $(g \circ f)(x)$. These results are alternative proofs of what properties of logarithms already proven in this section? Explain.

64. ◆ Explain the errors, if any, in the following:

$$\log_a ab^3 = (\log_a a)(\log_a b^3)$$
$$= 3\log_a b.$$

Solve for x. Do an algebraic solution.

65. $5^{\log_5 8} = 2x$

66. $\ln e^{3x-5} = -8$

Express as a single logarithm and simplify, if possible.

67. $\log_a (x^2 + xy + y^2) + \log_a (x - y)$

68. $\log_a (a^{10} - b^{10}) - \log_a (a + b)$

Express as a sum or a difference of logarithms.

69. $\log_a \dfrac{x - y}{\sqrt{x^2 - y^2}}$

70. $\log_a \sqrt{9 - x^2}$

71. Given that $\log_a x = 2$, $\log_a y = 3$, and $\log_a z = 4$, find

$$\log_a \dfrac{\sqrt[4]{y^2z^5}}{\sqrt[4]{x^3z^{-2}}}.$$

Determine whether each of the following is true. Assume that a, x, M, and N are positive.

72. $\log_a M + \log_a N = \log_a (M + N)$

73. $\log_a M - \log_a N = \log_a \dfrac{M}{N}$

74. $\dfrac{\log_a M}{\log_a N} = \log_a M - \log_a N$

75. $\dfrac{\log_a M}{x} = \log_a M^{1/x}$

76. $\log_a x^3 = 3\log_a x$

77. $\log_a 8x = \log_a x + \log_a 8$

78. $\log_N (MN)^x = x\log_N M + x$

Suppose that $\log_a x = 2$. *Find each of the following.*

79. $\log_a \left(\dfrac{1}{x} \right)$ **80.** $\log_{1/a} x$

81. Simplify:

$\log_{10} 11 \cdot \log_{11} 12 \cdot \log_{12} 13 \cdots \log_{998} 999 \cdot \log_{999} 1000.$

Prove each of the following for any base a and any positive number x.

82. $\log_a \left(\dfrac{1}{x} \right) = -\log_a x = \log_{1/a} x$

83. $\log_a \left(\dfrac{x + \sqrt{x^2 - 5}}{5} \right) = -\log_a \left(x - \sqrt{x^2 - 5} \right)$

6.4
Solving Exponential and Logarithmic Equations

- Solve exponential and logarithmic equations.

Solving Exponential Equations

Equations with variables in the exponents, such as $3^x = 20$ and $2^{5x} = 64$, are called **exponential equations**. We now consider solving exponential equations.

Sometimes, as is the case with the equation $2^{5x} = 64$, we can write each side as a power of the same number:

$$2^{5x} = 2^6.$$

We can then set the exponents equal and solve:

$$5x = 6$$
$$x = \tfrac{6}{5}.$$

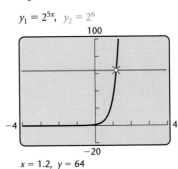

$y_1 = 2^{5x}, \ y_2 = 2^6$

$x = 1.2, \ y = 64$

We use the following property.

Base–Exponent Property

For any $a > 0$, $a \neq 1$,

$$a^x = a^y \longleftrightarrow x = y.$$

This property follows from the fact that for any $a > 0$, $a \neq 1$, $f(x) = a^x$ is a one-to-one function. If $a^x = a^y$, then $f(x) = f(y)$. Then since f is one-to-one (see the definition in Section 3.1), it follows that $x = y$. Conversely, if $x = y$, it follows that $a^x = a^y$, since we are raising a to the same power.

Example 1   Solve: $2^{3x-7} = 32$.

┌── *ALGEBRAIC SOLUTION*

Note that $32 = 2^5$. Thus we can write each side as a power of the same number:

$$2^{3x-7} = 2^5.$$

Since the bases are the same number, 2, we can use the base–exponent property and set the exponents equal:

$$3x - 7 = 5$$
$$3x = 12$$
$$x = 4$$

CHECK: $2^{3x-7} = 32$
─────────────
         $2^{3(4)-7}$? 32
         2^{12-7}
         2^5
         32 | 32 TRUE

The solution is 4. The solution set is {4}.

┌── *GRAPHICAL SOLUTION*

To solve on a grapher, we graph the equations

$$y_1 = 2^{3x-7} \quad \text{and} \quad y_2 = 32.$$

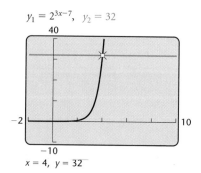

$y_1 = 2^{3x-7}, \quad y_2 = 32$

$x = 4, \ y = 32$

Then we can use the TRACE and ZOOM features or the INTERSECT feature to approximate the point of intersection.

We see that the solution is 4. On some graphers, we can get the solution by solving the equation $2^{3x-7} - 32 = 0$ using a SOLVE feature.

When it does not seem possible to write each side as a power of the same base, we can take the common or natural logarithm on each side and use the power rule for logarithms.

Example 2   Solve: $3^x = 20$.

┌── *ALGEBRAIC SOLUTION*

We have

$$3^x = 20$$

$\log 3^x = \log 20$ **Taking the common logarithm on both sides**

$x \log 3 = \log 20$ **Using the power rule**

$x = \dfrac{\log 20}{\log 3}$.   **Dividing by log 3**

This is an exact answer. We cannot simplify further, but we can approximate using a grapher:

$$x = \frac{\log 20}{\log 3} \approx \frac{1.3010}{0.4771} \approx 2.7268.$$

We can check this by finding $3^{2.7268}$. The solution is about 2.727.

┌── *GRAPHICAL SOLUTION*

We graph the equations

$$y_1 = 3^x \quad \text{and} \quad y_2 = 20$$

and use TRACE and ZOOM or INTERSECT to determine a point of intersection.

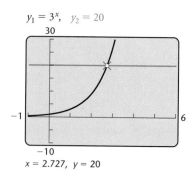

$y_1 = 3^x, \quad y_2 = 20$

$x = 2.727, \ y = 20$

The solution is about 2.727.

It will ease our work if we take the natural logarithm when working with equations that have e as a base.

Example 3   Solve: $e^{0.08t} = 2500$.

ALGEBRAIC SOLUTION

We have

$$e^{0.08t} = 2500$$

$\ln e^{0.08t} = \ln 2500$ **Taking the natural logarithm on both sides**

$0.08t = \ln 2500$ **Finding the logarithm of a base to a power: $\log_a a^x = x$**

$$t = \frac{\ln 2500}{0.08} \approx 97.8.$$

The solution is about 97.8.

GRAPHICAL SOLUTION

We graph the equations

$$y_1 = e^{0.08x} \quad \text{and} \quad y_2 = 2500$$

and determine a point of intersection.

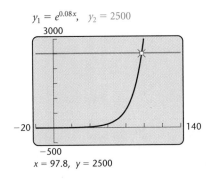

$y_1 = e^{0.08x}, \quad y_2 = 2500$

$x = 97.8, \ y = 2500$

The solution is about 97.8.

Example 4   Solve: $e^x + e^{-x} - 6 = 0$.

ALGEBRAIC SOLUTION

In this case, we have more than one term with x in the exponent. To get a single expression with x in the exponent, we do the following:

$$e^x + e^{-x} - 6 = 0$$

$e^x + \dfrac{1}{e^x} - 6 = 0$ **Rewriting e^{-x} with a positive exponent**

$e^{2x} + 1 - 6e^x = 0.$ **Multiplying on both sides by e^x**

This equation is reducible to quadratic with $u = e^x$:

$$u^2 - 6u + 1 = 0.$$

The coefficients of the reduced quadratic equation are $a = 1$, $b = -6$, and $c = 1$. Using the quadratic formula, we obtain

$$u = e^x = 3 \pm \sqrt{8}.$$

We now take the natural logarithm on both sides:

$$\ln e^x = \ln (3 \pm \sqrt{8})$$
$x = \ln (3 \pm \sqrt{8}).$ **Using $\ln e^x = x$**

Approximating each of the solutions, we obtain 1.76 and −1.76.

GRAPHICAL SOLUTION

We begin by graphing $f(x) = e^x + e^{-x} - 6$. We look for the first coordinates of the points where the graph of $y = e^x + e^{-x} - 6$ crosses the x-axis. The zeros of the function are the solutions of the equation.

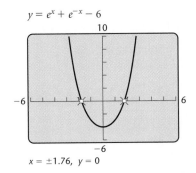

$y = e^x + e^{-x} - 6$

$x = \pm 1.76, \ y = 0$

We obtain the approximate solutions −1.76 and 1.76.

It is possible that when encountering an equation like the one in Example 4, you might not recognize that it could be solved in the algebraic manner shown. This points out the value of the graphical solution.

Solving Logarithmic Equations

Equations containing variables in logarithmic expressions, such as $\log_2 x = 4$ and $\log x + \log (x + 3) = 1$, are called **logarithmic equations**.

> To solve logarithmic equations algebraically, first try to obtain a single logarithmic expression on one side and then write an equivalent exponential equation.

Example 5   Solve: $\log_3 x = -2$.

ALGEBRAIC SOLUTION

We have

$\log_3 x = -2$

$3^{-2} = x$ Converting to an exponential equation

$\dfrac{1}{3^2} = x$

$\dfrac{1}{9} = x.$

CHECK:

$$\log_3 x = -2$$

$$\log_3 \frac{1}{9} \ ? \ -2$$

$$\log_3 3^{-2}$$

$$\qquad\quad -2 \ \Big| \ -2 \quad \text{TRUE}$$

The solution is $\frac{1}{9}$.

GRAPHICAL SOLUTION

Use the change-of-base formula and graph the equations

$$y_1 = \log_3 x = \frac{\ln x}{\ln 3}$$

and

$$y_2 = -2.$$

Then look for the point(s) of intersection.

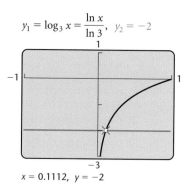

$$y_1 = \log_3 x = \frac{\ln x}{\ln 3}, \ y_2 = -2$$

x = 0.1112, y = −2

The approximate solution is 0.111, which is about $\frac{1}{9}$.

Example 6   Solve: $\log x + \log (x + 3) = 1$.

In this case, we have common logarithms. Including the base 10's will help us understand the problem:

$$\log_{10} x + \log_{10} (x + 3) = 1$$

$\log_{10} [x(x + 3)] = 1$ Using the product rule to obtain a single logarithm

$x(x + 3) = 10^1$ Writing an equivalent exponential equation

$$x^2 + 3x = 10$$

$$x^2 + 3x - 10 = 0$$

$(x - 2)(x + 5) = 0$ Factoring

$$x - 2 = 0 \quad or \quad x + 5 = 0$$

$$x = 2 \quad or \quad x = -5.$$

CHECK: For 2:

$$\log x + \log (x + 3) = 1$$

$$\log 2 + \log (2 + 3) \; ? \; 1$$

$$\log 2 + \log 5$$

$$\log 10$$

$$1 \mid 1 \quad \text{TRUE}$$

For -5:

$$\log x + \log (x + 3) = 1$$

$$\log (-5) + \log (-5 + 3) \; ? \; 1 \quad \text{FALSE}$$

The number -5 is not a solution because negative numbers do not have real-number logarithms. The solution is 2.

We graph the equations

$$y_1 = \log x + \log (x + 3)$$

and

$$y_2 = 1$$

and find the point(s) of intersection.

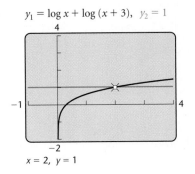

$y_1 = \log x + \log (x + 3), \; y_2 = 1$

$x = 2, \; y = 1$

The solution is 2.

Example 7   Solve: $\log_3 (2x - 1) - \log_3 (x - 4) = 2$.

ALGEBRAIC SOLUTION

We have

$\log_3 (2x - 1) - \log_3 (x - 4) = 2$

$$\log_3 \frac{2x - 1}{x - 4} = 2 \qquad \text{Using the quotient rule}$$

$$\frac{2x - 1}{x - 4} = 3^2 \qquad \text{Writing an equivalent exponential equation}$$

$$\frac{2x - 1}{x - 4} = 9$$

$$2x - 1 = 9(x - 4) \qquad \text{Multiplying by the LCD, } x - 4$$

$$2x - 1 = 9x - 36$$

$$35 = 7x$$

$$5 = x.$$

CHECK: $\dfrac{\log_3 (2x - 1) - \log_3 (x - 4) = 2}{}$

$\log_3 (2 \cdot 5 - 1) - \log_3 (5 - 4) \; ? \; 2$

$\log_3 9 - \log_3 1$

$2 - 0$

$2 \; \bigm| \; 2 \qquad$ TRUE

The solution is 5.

GRAPHICAL SOLUTION

We use the change-of-base formula and graph the equations

$$y_1 = \frac{\ln (2x - 1)}{\ln 3} - \frac{\ln (x - 4)}{\ln 3}$$

and

$$y_2 = 2.$$

Then we find the point(s) of intersection.

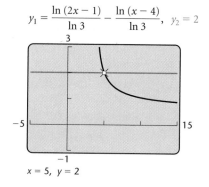

$$y_1 = \frac{\ln (2x - 1)}{\ln 3} - \frac{\ln (x - 4)}{\ln 3}, \; y_2 = 2$$

$x = 5, \; y = 2$

The solution is 5.

Sometimes we encounter equations for which an algebraic solution seems difficult or impossible.

Example 8   Solve: $e^{0.5x} - 7.3 = 2.08x + 6.2$.

ALGEBRAIC SOLUTION

In this case, we have an equation for which an algebraic solution seems difficult or impossible.

GRAPHICAL SOLUTION

We graph the equations

$$y_1 = e^{0.5x} - 7.3 \quad \text{and} \quad y_2 = 2.08x + 6.2$$

and look for points of intersection. (See the figure at left.)
We can also consider the equation

$$y = e^{0.5x} - 7.3 - 2.08x - 6.2, \quad \text{or} \quad y = e^{0.5x} - 2.08x - 13.5,$$

and look for zeros using TRACE and ZOOM or a SOLVE feature. The approximate solutions are -6.471 and 6.610.

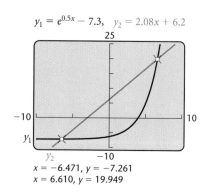

$y_1 = e^{0.5x} - 7.3, \quad y_2 = 2.08x + 6.2$

$x = -6.471, y = -7.261$
$x = 6.610, y = 19.949$

6.4 | *Exercise Set*

Solve each exponential equation algebraically. Then check on a grapher.

1. $3^x = 81$

2. $2^x = 32$

3. $2^{2x} = 8$

4. $3^{7x} = 27$

5. $2^x = 33$

6. $2^x = 40$

7. $5^{4x-7} = 125$

8. $4^{3x-5} = 16$

9. $27 = 3^{5x} \cdot 9^{x^2}$

10. $3^{x^2+4x} = \frac{1}{27}$

11. $84^x = 70$

12. $28^x = 10^{-3x}$

13. $e^t = 1000$

14. $e^{-t} = 0.04$

15. $e^{-0.03t} = 0.08$

16. $1000e^{0.09t} = 5000$

17. $3^x = 2^{x-1}$

18. $5^{x+2} = 4^{1-x}$

19. $(3.9)^x = 48$

20. $250 - (1.87)^x = 0$

21. $e^x + e^{-x} = 5$

22. $e^x - 6e^{-x} = 1$

23. $\dfrac{e^x + e^{-x}}{e^x - e^{-x}} = 3$

24. $\dfrac{5^x - 5^{-x}}{5^x + 5^{-x}} = 8$

Solve each logarithmic equation algebraically. Then check on a grapher.

25. $\log_5 x = 4$

26. $\log_2 x = -3$

27. $\log x = -4$

28. $\log x = 1$

29. $\ln x = 1$

30. $\ln x = -2$

31. $\log_2 (10 + 3x) = 5$

32. $\log_5 (8 - 7x) = 3$

33. $\log x + \log (x - 9) = 1$

34. $\log_2 (x + 1) + \log_2 (x - 1) = 3$

35. $\log_8 (x + 1) - \log_8 x = 2$

36. $\log x - \log (x + 3) = -1$

37. $\log_4 (x + 3) + \log_4 (x - 3) = 2$

38. $\ln (x + 1) - \ln x = \ln 4$

39. $\log (2x + 1) - \log (x - 2) = 1$

40. $\log_5 (x + 4) + \log_5 (x - 4) = 2$

Use only a grapher. Find approximate solutions of each equation or approximate the point(s) of intersection of a pair of equations.

41. $e^{7.2x} = 14.009$

42. $0.082e^{0.05x} = 0.034$

43. $xe^{3x} - 1 = 3$

44. $5e^{5x} + 10 = 3x + 40$

45. $4 \ln (x + 3.4) = 2.5$

46. $\ln x^2 = -x^2$

47. $\log_8 x + \log_8 (x + 2) = 2$

48. $\log_3 x + 7 = 4 - \log_5 x$

49. $\log_5 (x + 7) - \log_5 (2x - 3) = 1$

50. $y = \ln 3x, \ y = 3x - 8$

51. $2.3x + 3.8y = 12.4, \ y = 1.1 \ln (x - 2.05)$

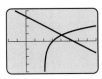

52. $y = 2.3 \ln (x + 10.7), \ y = 10e^{-0.07x^2}$

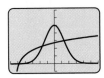

53. $y = 2.3 \ln (x + 10.7), \ y = 10e^{-0.007x^2}$

Skill Maintenance

54. Solve $K = \frac{1}{2}mv^2$ for *v*.

Solve.

55. $x^4 + 5x^2 = 36$

56. $t^{2/3} - 10 = 3t^{1/3}$

57. *Total Sales of Goodyear.* The following table shows factual data regarding total sales of The Goodyear Tire and Rubber Company.

YEAR, *x*	TOTAL SALES, *y* (IN MILLIONS)
1. 1991	$10,906.8
2. 1992	11,784.9
3. 1993	11,643.4
4. 1994	12,288.2

Source: The Goodyear Tire and Rubber Company Annual Report.

a) Use linear regression on a grapher to fit an equation $y = mx + b$, where $x = 1$ corresponds to 1991, to the data points. Predict total sales in 1999. (See Appendix A.3.)

b) Use quadratic regression on a grapher to fit an equation $y = ax^2 + bx + c$ to the data points. Predict total sales in 1999. (See Appendix A.3.)

Synthesis

58. ◈ In Example 3, we took the natural logarithm on both sides. What would have happened had we taken the common logarithm? Explain which seems best to you and why.

59. ◈ Explain how Exercises 29 and 30 could be solved using the graph of $f(x) = \ln x$.

Solve using any method.

60. $\ln (\ln x) = 2$

61. $\ln (\log x) = 0$

62. $\ln \sqrt[4]{x} = \sqrt{\ln x}$

63. $\sqrt{\ln x} = \ln \sqrt{x}$

64. $\log_3 (\log_4 x) = 0$

65. $(\log_3 x)^2 - \log_3 x^2 = 3$

66. $(\log x)^2 - \log x^2 = 3$

67. $\ln x^2 = (\ln x)^2$

68. $e^{2x} - 9 \cdot e^x + 14 = 0$

69. $5^{2x} - 3 \cdot 5^x + 2 = 0$

70. $x \left(\ln \frac{1}{6} \right) = \ln 6$

71. $\log_3 |x| = 2$

72. $x^{\log x} = \dfrac{x^3}{100}$

73. $\ln x^{\ln x} = 4$

74. $\dfrac{(e^{3x+1})^2}{e^4} = e^{10x}$

75. $\dfrac{\sqrt{(e^{2x} \cdot e^{-5x})^{-4}}}{e^x \div e^{-x}} = e^7$

76. $|\log_a x| = \log_a |x|$

77. $\ln (x - 2) > 4$

78. $e^x < \dfrac{4}{5}$

79. $|\log_5 x| + 3 \log_5 |x| = 4$

80. $|2^{x^2} - 8| = 3$

81. Given that $a = \log_8 225$ and $b = \log_2 15$, express a as a function of b.

82. Given that $a = (\log_{125} 5)^{\log_5 125}$, find the value of $\log_3 a$.

83. Given that

$$\log_2 [\log_3 (\log_4 x)] = \log_3 [\log_2 (\log_4 y)]$$
$$= \log_4 [\log_3 (\log_2 z)]$$
$$= 0,$$

find $x + y + z$.

84. Given that $f(x) = e^x - e^{-x}$, find $f^{-1}(x)$ if it exists.

6.5

Applications and Models: Growth and Decay

- Solve applications involving exponential growth and decay.
- Find models involving exponential and logarithmic functions.

Exponential and logarithmic functions with base e are rich in applications to many fields such as business, science, psychology, and sociology. In this section, we consider some basic applications and then use curve fitting to do others.

Population Growth

The function

$$P(t) = P_0 e^{kt}$$

is a model of many kinds of population growth, whether it be a population of people, bacteria, cellular phones, or money. In this function, P_0 is the population at time 0, P is the population after time t, and k is called the **exponential growth rate**. The graph of such an equation is shown at right.

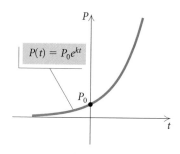

Example 1 *Population Growth of the United States.* In 1997, the population of the United States was about 266 million and the exponential growth rate was 0.9% per year. (*Source:* Statistical Abstract of the United States)

a) Find the exponential growth function.

b) What will the population be in 2002?

c) Graph the exponential growth function.

d) When will the population be double what it was in 1997?

SOLUTION

a) At $t = 0$ (1997), the population was 266 million. We substitute 266 for P_0 and 0.9%, or 0.009, for k to obtain the exponential growth function

$$P(t) = 266e^{0.009t}.$$

b) In 2002, $t = 5$; that is, 5 yr have passed. To find the population in 2002, we substitute 5 for t:

$$P(5) = 266e^{0.009(5)} = 266e^{0.045} \approx 278.$$

In 2002, the population of the United States will be about 278 million.

c) Using a grapher, we obtain the graph (at left) of the exponential growth function.

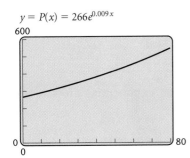

$y = P(x) = 266e^{0.009x}$

d) We are looking for the time T for which $P(T) = 2 \cdot 266$, or 532. The number T is called the **doubling time**. To find T, we solve the equation $532 = 266e^{0.009T}$.

ALGEBRAIC SOLUTION

We have

$532 = 266e^{0.009T}$	Substituting 532 for $P(T)$
$2 = e^{0.009T}$	Dividing by 266
$\ln 2 = \ln e^{0.009T}$	Taking the natural logarithm on both sides
$\ln 2 = 0.009T$	$\ln e^x = x$
$\dfrac{\ln 2}{0.009} = T$	Dividing by 0.009
$77 \approx T.$	

The population of the United States will be double what it was in 1997 in about 77 yr.

GRAPHICAL SOLUTION

To solve on a grapher, we graph the equations

$$y_1 = 266e^{0.009x} \quad \text{and} \quad y_2 = 532$$

and find the first coordinate of their point of intersection.

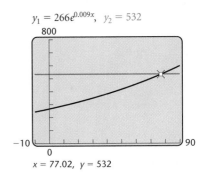

$y_1 = 266e^{0.009x}, \quad y_2 = 532$

$x = 77.02, \quad y = 532$

On some graphers, we can get the solution by having the grapher solve the equation directly. We see that the solution is about 77 yr.

No wonder ecologists are so concerned about population growth.

Interest Compounded Continuously

Here we explore the mathematics behind the concept of **interest compounded continuously**. Suppose that an amount P_0 is invested in a savings account at interest rate k *compounded continuously*. The amount $P(t)$ in the account after t years is given by the exponential function

$$P(t) = P_0 e^{kt}.$$

Example 2 *Interest Compounded Continuously.* Suppose $2000 is invested at interest rate k, compounded continuously, and grows to $2983.65 in 5 yr.

a) What is the interest rate?

b) Find the exponential growth function.

c) What will the balance be after 10 yr?

d) After how long will the $2000 have doubled?

Solution

a) At $t = 0$, $P(0) = P_0 = \$2000$. Thus the exponential growth function is

$$P(t) = 2000e^{kt}.$$

We know that $P(5) = \$2983.65$. We substitute and solve for k, as shown below.

¬ ¬ ¬ *Algebraic Solution*

We have

$$2983.65 = 2000e^{k(5)} \qquad \text{Substituting 2983.65 for } P(t)$$

$$2983.65 = 2000e^{5k}$$

$$\frac{2983.65}{2000} = e^{5k} \qquad \text{Dividing by 2000}$$

$$\ln \frac{2983.65}{2000} = \ln e^{5k} \qquad \text{Taking the natural logarithm}$$

$$\ln \frac{2983.65}{2000} = 5k \qquad \text{Using } \ln e^x = x$$

$$\frac{\ln \dfrac{2983.65}{2000}}{5} = k$$

$$0.08 \approx k.$$

The interest rate is about 0.08, or 8%.

¬ ¬ ¬ *Graphical Solution*

We can solve by graphing the equations

$$y_1 = 2000e^{5x} \quad \text{and} \quad y_2 = 2983.65$$

on a grapher. Then we can use TRACE and ZOOM or other features of a grapher to find an approximation for the first coordinate of the point of intersection.

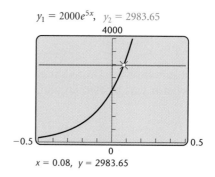

$y_1 = 2000e^{5x}, \; y_2 = 2983.65$

$x = 0.08, \; y = 2983.65$

The solution is about 0.08, or 8%.

b) The exponential growth function is

$$P(t) = 2000e^{0.08t}.$$

c) The balance after 10 yr is

$$P(10) = 2000e^{0.08(10)}$$
$$= 2000e^{0.8}$$
$$\approx \$4451.08.$$

d) We solve using both the algebraic method and a grapher.

r-- *ALGEBRAIC SOLUTION*

To find the doubling time T, we set $P(T) = \$4000$ and solve for T:

$$4000 = 2000e^{0.08T}$$

$2 = e^{0.08T}$	Dividing by 2000
$\ln 2 = \ln e^{0.08T}$	Taking the natural logarithm
$\ln 2 = 0.08T$	$\ln e^x = x$
$\dfrac{\ln 2}{0.08} = T$	Dividing by 0.08
$8.7 \approx T.$	

Thus the original investment of \$2000 will double in about 8.7 yr.

r-- *GRAPHICAL SOLUTION*

To solve on a grapher, we graph the equations

$$y_1 = 2000e^{0.08x} \quad \text{and} \quad y_2 = 4000$$

and find the first coordinate of their point of intersection.

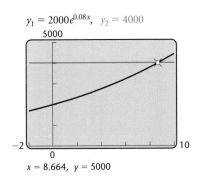

$y_1 = 2000e^{0.08x}, \; y_2 = 4000$

$x = 8.664, \; y = 5000$

The solution is about 8.7.

We can find a general expression relating the growth rate k and the doubling time T by solving the following equation:

$2P_0 = P_0 e^{kT}$	Substituting $2P_0$ for P and T for t
$2 = e^{kT}$	Dividing by P_0
$\ln 2 = \ln e^{kt}$	Taking the natural logarithm
$\ln 2 = kT$	Using $\ln e^x = x$
$\dfrac{\ln 2}{k} = T.$	

Growth Rate and Doubling Time

The *growth rate* k and the *doubling time* T are related by

$$kT = \ln 2, \quad \text{or} \quad k = \frac{\ln 2}{T}, \quad \text{or} \quad T = \frac{\ln 2}{k}.$$

Note that the relationship between k and T does not depend on P_0.

Example 3 *World Population Growth.* The population of the world is now doubling every 24.8 yr. What is the exponential growth rate?

SOLUTION   We have

$$k = \frac{\ln 2}{T} \approx \frac{0.693147}{24.8} \approx 2.8\%.$$

The growth rate of the world population is about 2.8% per year. ▬

Models of Limited Growth

The model $P(t) = P_0 e^{kt}$ has many applications involving what may seem to be unlimited population growth. There can be factors that prevent a population from exceeding some limiting value—perhaps a limitation on food, living space, or other natural resources. One model of such growth is

$$P(t) = \frac{a}{1 + be^{-kt}},$$

which is called a **logistic equation**. This function increases toward a *limiting value a* as $t \to \infty$.

Example 4   *Limited Population Growth.* A ship carrying 1000 passengers has the misfortune to be shipwrecked on a small island from which the passengers are never rescued. The natural resources of the island limit the population to 5780. The population gets closer and closer to this limiting value, but never reaches it. The population of the island after time t, in years, is given by the logistic equation

$$P(t) = \frac{5780}{1 + 4.78e^{-0.4t}}.$$

a) Find the population after 0, 1, 2, 5, 10, and 20 yr.

b) Graph the function.

SOLUTION

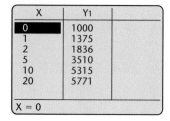

a) We use a grapher to find the function values, listing them in a table as shown at left. (We round to the nearest unit.) A grapher with a TABLE feature set in ASK mode can also be used. Thus the population will be 1000 after 0 yr, 1375 after 1 yr, 1836 after 2 yr, 3510 after 5 yr, 5315 after 10 yr, and 5771 after 20 yr.

b) We use a grapher to graph the function. The graph is the S-shaped curve shown below. Note that this function increases toward a limiting value of 5780.

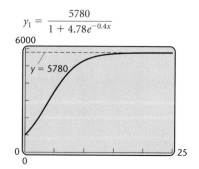

$$y_1 = \frac{5780}{1 + 4.78e^{-0.4x}}$$

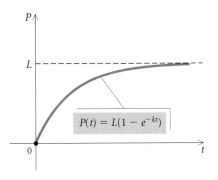

Another model of limited growth is provided by the function

$$P(t) = L(1 - e^{-kt}),$$

which is shown graphed at left. This function also increases toward a limiting value L, as $x \to \infty$.

Exponential Decay

The function

$$P(t) = P_0 e^{-kt}$$

is an effective model of the decline, or decay, of a population. An example is the decay of a radioactive substance. In this case, P_0 is the amount of the substance at time $t = 0$, and $P(t)$ is the amount of the substance left after time t, where k is a positive constant that depends on the situation. The constant k is called the **decay rate**.

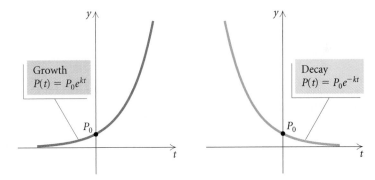

How can scientists determine that an animal bone has lost 30% of its carbon-14? The assumption is that the percentage of carbon-14 in the atmosphere and in living plants and animals is the same. When a plant or an animal dies, the amount of carbon-14 decays exponentially. The scientist burns the animal bone and uses a Geiger counter to determine the percentage of the smoke that is carbon-14. It is the amount that this varies from the percentage in the atmosphere that tells how much carbon-14 has been lost.

The process of carbon-14 dating was developed by the American chemist Willard E. Libby in 1952. It is known that the radioactivity in a living plant is 16 disintegrations per gram per minute. Since the half-life of carbon-14 is 5750 years, an object with an activity of 8 disintegrations per gram per minute is 5750 years old, one with an activity of 4 disintegrations per gram per minute is 11,500 years old, and so on. Carbon-14 dating can be used to measure the age of objects from 30,000 to 40,000 years old. Beyond such an age, it is too difficult to measure the radioactivity and some other method would have to be used.

Carbon-14 was indeed used to find the age of the Dead Sea Scrolls. It was used recently to refute the authenticity of the Shroud of Turin, presumed to have covered the body of Christ.

The **half-life** of bismuth is 5 days. This means that half of an amount of bismuth will cease to be radioactive in 5 days. The effect of half-life T is shown in the graph below for nonnegative inputs. The exponential function gets close to 0, but never reaches 0, as t gets very large. Thus, according to an exponential decay model, a radioactive substance never completely decays.

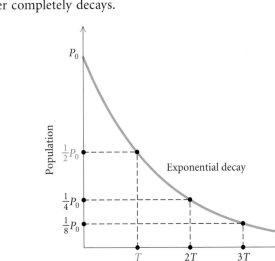

Radioactive decay curve

In 1947, a Bedouin youth looking for a stray goat climbed into a cave at Kirbet Qumran on the shores of the Dead Sea near Jericho and came upon earthenware jars containing an incalculable treasure of ancient manuscripts. Shown here are fragments of those so-called Dead Sea Scrolls, a portion of some 600 or so texts found so far and which concern the Jewish books of the Bible. Officials date them before 70 A.D., making them the oldest Biblical manuscripts by 1000 years.

Example 5   *Carbon Dating.* The radioactive element carbon-14 has a half-life of 5750 yr. The percentage of carbon-14 present in the remains of organic matter can be used to determine the age of that organic matter. Archeologists discovered that the linen wrapping from one of the Dead Sea Scrolls had lost 22.3% of its carbon-14 at the time it was found. How old was the linen wrapping?

SOLUTION   We first find k. When $t = 5750$ (the half-life), $P(t)$ will be half of P_0. We substitute $\frac{1}{2} P_0$ for $P(t)$ and 5750 for t and solve for k. Then

$$\tfrac{1}{2} P_0 = P_0 e^{-k(5750)}$$

or

$$\tfrac{1}{2} = e^{-5750k}.$$

We take the natural logarithm on both sides:

$$\ln \tfrac{1}{2} = \ln e^{-5750k}$$
$$= -5750k.$$

Then

$$k = \frac{\ln 0.5}{-5750}$$
$$\approx 0.00012.$$

We could also solve the equation $\frac{1}{2} = e^{-5750k}$ using a grapher. Now we have the function

$$P(t) = P_0 e^{-0.00012t}.$$

(This equation can be used for any subsequent carbon-dating problem.) If the linen wrapping has lost 22.3% of its carbon-14 from an initial amount P_0, then $77.7\% P_0$ is the amount present. To find the age t of the wrapping, we solve the following equation for t:

$$\begin{aligned}
77.7\% P_0 &= P_0 e^{-0.00012t} \qquad \text{Substituting } 77.7\% P_0 \text{ for } P \\
0.777 &= e^{-0.00012t} \\
\ln 0.777 &= \ln e^{-0.00012t} \\
\ln 0.777 &= -0.00012t \qquad \ln e^x = x \\
\frac{\ln 0.777}{-0.00012} &= t \\
2103 &\approx t.
\end{aligned}$$

Thus the linen wrapping on the Dead Sea Scrolls is about 2103 yr old.   ▬

Exponential and Logarithmic Curve Fitting

We have added several new functions to our candidates for curve fitting. Let's review some of them.

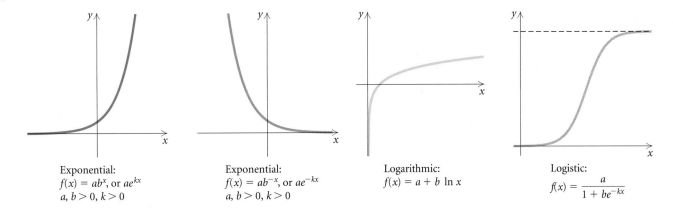

Exponential:
$f(x) = ab^x$, or ae^{kx}
$a, b > 0, k > 0$

Exponential:
$f(x) = ab^{-x}$, or ae^{-kx}
$a, b > 0, k > 0$

Logarithmic:
$f(x) = a + b \ln x$

Logistic:
$f(x) = \dfrac{a}{1 + be^{-kx}}$

Now, when we analyze data, these models can be considered as well as linear, quadratic, polynomial, and rational models. Some graphers can use *regression* to fit an exponential, a logarithmic, or a logistic* equation to a set of data. (See Appendix A.3 for a discussion of regression.)

Example 6   *Cellular Phones.* The number of cellular phones in use has grown dramatically in recent years. Let's examine the following data.

x, YEAR	NUMBER OF CELLULAR PHONES, y (IN MILLIONS)
0. 1985	0.2
1. 1986	0.4
2. 1987	1.0
3. 1988	1.8
4. 1989	2.8
5. 1990	4.1
6. 1991	6.2
7. 1992	8.6
8. 1993	13.0
9. 1994	19.3

Source: Cellular Telecommunications
Industry Association

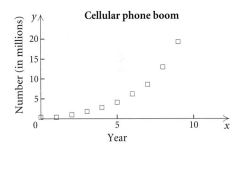

From the scatterplot on the right above, it would appear that we have exponential growth.

a) Use a grapher and regression to fit an exponential function to the data.

b) Graph the function.

c) Predict the number of cellular phones in use in the year 2000.

---

*The TI-83 is an example.

SOLUTION

a) We are fitting the data to an equation of the type $y = a \cdot b^x$. Entering the data into the grapher and carrying out the regression procedure, we find that

$$a = 0.3039703339,$$
$$b = 1.627328317,$$
$$r = 0.9860606791.$$

This tells us that the equation is

$$y = 0.3039703339(1.627328317)^x,$$

or about

$$y = 0.304(1.627)^x.$$

The *correlation coefficient* is very close to 1. This gives us a good indication that the data fit an exponential equation.

b) The graph is shown at left.

c) To predict the number of cellular phones in use in 2000, we substitute 15 for x in the exponential equation:

$$y = 0.304(1.627)^x$$
$$= 0.304(1.627)^{15} \approx 450.$$

Thus according to this model, there will be about 450 million cellular phones in use in 2000. ▬

On some graphers, there may be a REGRESSION feature that yields an exponential function, base e. If not, and you wish to find such a function, a conversion can be done using the following.

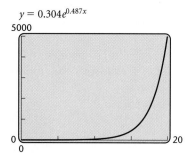

$y = 0.304e^{0.487x}$

> ### Converting from Base b to Base e
> $$b^x = e^{x(\ln b)}$$

Then, for the equation in Example 6, we have

$$y = 0.304(1.627)^x$$
$$= 0.304e^{x(\ln 1.627)} = 0.304e^{0.487x}.$$

We can prove this conversion formula using properties of logarithms, as follows:

$$e^{x(\ln b)} = e^{\ln b^x} = b^x.$$

Power Models

There are many situations in which so-called **power models** $y = ax^b$ can be fit to data using regression. Note that the constants to be determined are a and b. The base is the variable x.

Example 7 *Cholesterol Level and the Risk of Heart Attack.* The data in the following table show the relationship of cholesterol level in men to the risk of a heart attack.

CHOLESTEROL LEVEL, x	MEN, PER 10,000, WHO SUFFER A HEART ATTACK, y
100	30
200	65
250	100
275	130
300	180

Source: Nutrition Action Healthletter.

a) Use the REGRESSION feature on a grapher to find a power function to fit the data.

b) Graph the function.

c) Use the answer to part (b) to predict the heart attack rate for men with cholesterol levels of 350 and 400.

SOLUTION

a) We are fitting an equation of the type $y = ax^b$ to the data. Entering the data into the grapher and carrying out the regression procedure, we find that

$$a = 0.0241789574,$$
$$b = 1.527457172,$$
$$r = 0.9739361336.$$

This tells us that the equation is about $y = 0.024x^{1.527}$. The *correlation coefficient* of about 0.974 is very close to 1. This indicates that the data fit a power function fairly well, although an exponential function with $r = 0.996$ is a better fit. For illustrative purposes, though, we will continue with the power model.

b) The graph is shown at left.

c) To predict the heart attack rate for men with cholesterol levels of 350 and 400, we substitute 350 and 400 for x in the power equation:

$$y = 0.024x^{1.527} = 0.024(350)^{1.527} \approx 184,$$
$$y = 0.024x^{1.527} = 0.024(400)^{1.527} \approx 226.$$

Thus the heart attack rate is about 184 out of 10,000 for men with a cholesterol level of 350 and about 226 out of 10,000 for men with a cholesterol level of 400.

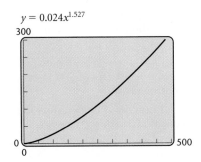
$y = 0.024x^{1.527}$

6.5 Exercise Set

1. *World Population Growth.*   In 1997, the world population was 5.8 billion. The exponential growth rate was 1.5% per year.

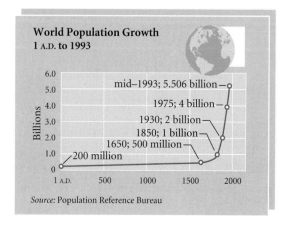

World Population Growth
1 A.D. to 1993

mid–1993; 5.506 billion
1975; 4 billion
1930; 2 billion
1850; 1 billion
1650; 500 million
200 million

Source: Population Reference Bureau

a) Find the exponential growth function.
b) Predict the population of the world in 2000 and in 2010.
c) When will the world population be 8 billion?
d) Find the doubling time.

2. *Population Growth of Rabbits.*   Under ideal conditions, a population of rabbits has an exponential growth rate of 11.7% per day. Consider an initial population of 100 rabbits.

a) Find the exponential growth function.
b) What will the population be after 7 days?
c) Find the doubling time.
d) Graph the function.

3. *Population Growth.*   Complete the following table.

POPULATION	GROWTH RATE, k	DOUBLING TIME, T
a) Mexico	3.5% per year	
b) Europe		69.31 yr
c) Oil reserves	10% per year	
d) Coal reserves	4% per year	
e) Alaska		24.8 yr
f) Central America		19.8 yr

4. *Female Olympic Athletes.*   In 1985, the number of female athletes participating in Summer Olympic-Type Games was 500. In 1996, about 3600 participated in the Summer Olympics in Atlanta. Assuming the exponential model applies:

a) Find the value of k ($P_0 = 500$), and write the function.
b) Estimate the number of female athletes in the Summer Olympics of 2000.

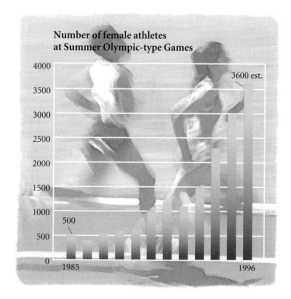

Number of female athletes at Summer Olympic-type Games

3600 est.

500

1985   1996

5. *Population Growth of the Virgin Islands.*   The population of the U.S. Virgin Islands has a growth rate of 2.6% per year. In 1990, the population was 512,000. The land area of the Virgin Islands is 3,097,600 square yards. Assuming this growth rate continues and is exponential, after how long will there be one person for every square yard of land? (*Source:* Statistical Abstract of the United States)

6. *Value of Manhattan Island.*   In 1626, Peter Minuit of the Dutch West India Company purchased Manhattan Island from the Indians for $24. Assuming an exponential rate of inflation of 8% per year, how much will Manhattan be worth in 2000?

7. *Interest Compounded Continuously.*   Suppose $10,000 is invested at an interest rate of 5.4% per year, compounded continuously.

a) Find the exponential function that describes the amount in the account after time t, in years.
b) What is the balance after 1 yr? 2 yr? 5 yr? 10 yr?
c) What is the doubling time?

8. *Interest Compounded Continuously.* Complete the following table.

INITIAL INVESTMENT AT $t = 0$, P_0	INTEREST RATE, k	DOUBLING TIME, T	AMOUNT AFTER 5 YR
a) $35,000	6.2%		
b) $5000			$ 7,130.90
c)	8.4%		$11,414.71
d)		11 yr	$17,539.32

9. *Carbon Dating.* A mummy discovered in the pyramid Khufu in Egypt has lost 46% of its carbon-14. Determine its age.

10. *Carbon Dating.* The statue of Zeus at Olympia in Greece is one of the Seven Wonders of the World. It is made of gold and ivory. The ivory was found to have lost 35% of its carbon-14. Determine the age of the statue.

11. *Radioactive Decay.* Complete the following table.

RADIOACTIVE SUBSTANCE	DECAY RATE, k	HALF-LIFE, T
a) Polonium		3 min
b) Lead		22 yr
c) Iodine-131	9.6% per day	
d) Krypton-85	6.3% per year	
e) Strontium-90		25 yr
f) Uranium-238		4560 yr
g) Plutonium		23,105 yr

12. *Decline of Long-Playing Records.* The sales S of long-playing records has declined considerably in the past 10 yr because of the emergence of the cassette tape and compact disc. In 1983, 205 million LP records were sold and in 1993, 1.2 million records were sold. (*Source:* Recording Industry Association of America) Assuming the sales are decreasing according to the exponential-decay model:

a) Find the value k, and write an exponential function that describes the number of long-playing records sold after time t, in years.

b) Estimate the sales of LP records in the year 2000.

c) In what year (theoretically) will only 1 long-playing record be sold?

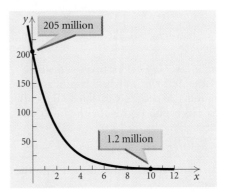

13. *Decline in Beef Consumption.* In 1985, the average annual consumption of beef B was about 80 lb per person. In 1996, it was about 67 lb per person. Assuming consumption is decreasing according to the exponential-decay model:

a) Find the value k, and write an equation that describes beef consumption after time t, in years.

b) Estimate the consumption of beef in the year 2000.

c) After how many years (theoretically) will the average annual consumption of beef be 20 lb per person?

14. *The Value of Eddie Murray's Baseball Card.* The collecting of baseball cards has become a popular hobby. The card shown here is a photograph of Eddie Murray in his rookie season of 1978.

In 1983, the value of the card was $7.75 and in 1987, its value was $27.00. (*Source: Sport Collectors Digest*) Assuming that the value of the card has grown exponentially:

a) Find the value k and determine the exponential growth function, assuming $V_0 = 7.75$.

b) Estimate the value of the card in 1995 and in 2000. Check your answer for 1995 with a baseball card dealer.

c) What is the doubling time for the value of the card?

d) After how long will the value of the card be $2000?

15. *Spread of an Epidemic.*   In a town whose population is 2000, a disease creates an epidemic. The number of people N infected t days after the disease has begun is given by the function

$$N(t) = \frac{2000}{1 + 19.9e^{-0.6t}}.$$

a) How many are initially infected with the disease $(t = 0)$?

b) Find the number infected after 2 days, 5 days, 8 days, 12 days, and 16 days.

c) Graph the function.

16. *Acceptance of Seat Belt Laws.*   In recent years, many states have passed mandatory seat belt laws. The total number of states N that have passed a seat belt law t years after 1984 is given by the function

$$N(t) = \frac{50}{1 + 22e^{-0.6t}}.$$

(*Source:* National Highway Traffic Safety Administration)

a) How many states had passed the law in 1984? ($t = 0$ corresponds to 1984.)

b) Find the number of states that had passed the law by 1988, 1992, 1996, and 2002.

c) Graph the function.

d) If the function were to continue to be appropriate, would all 50 states ever pass the law? Explain.

17. *Limited Population Growth in a Lake.*   A lake is stocked with 400 fish of a new variety. The size of the lake, the availability of food, and the number of other fish restrict growth in the lake to a *limiting value* of 2500. The population of fish in the lake after time t, in months, is given by the function

$$P(t) = \frac{2500}{1 + 5.25e^{-0.32t}}.$$

a) Find the population after 0, 1, 5, 10, 15, and 20 months.

b) Graph the function.

For each of the following scatterplots, determine which, if any, of these functions might be used as a model for the data.

a) *Quadratic,* $f(x) = ax^2 + bx + c$
b) *Polynomial, not quadratic*

c) *Exponential,* $f(x) = ab^x$, or Be^{kx}, $k > 0$
d) *Exponential,* $f(x) = ab^x$, or Be^{-kx}, $k > 0$
e) *Logarithmic,* $f(x) = a + b \ln x$
f) *Logistic,* $f(x) = \dfrac{a}{1 + be^{-kx}}$

18. 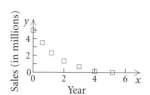   **19.**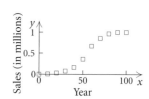

20. 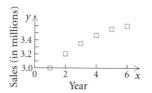   **21.**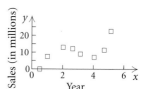

22. 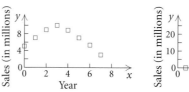   **23.**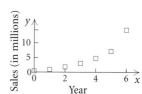

24. *Opinion on Capital Punishment.*   The following table contains factual data from a recent survey of college freshmen regarding their opinion of capital punishment.

x, YEAR	PERCENTAGE OF COLLEGE FRESHMEN AGREEING THAT CAPITAL PUNISHMENT SHOULD BE ABOLISHED, y
0. 1971	56
8. 1979	32
13. 1984	24
18. 1989	21
23. 1994	18

Source: UCLA Higher Education Research Institute.

Using a grapher:

a) Create a scatterplot of the data. Determine whether an exponential function appears to fit the data.

b) Use regression to fit an exponential function $y = ab^x$, or ae^{-kx}, to the data, where $x =$ the number of years after 1971.

c) Use the function to predict the percent of college freshmen in 2000 agreeing that capital punishment should be abolished.
d) In what year would only 1% agree that capital punishment should be abolished?

25. *Total Revenue of Microsoft, Inc.*

x, YEAR	TOTAL REVENUE, y (IN BILLIONS)
0. 1989	$0.8
1. 1990	1.1
2. 1991	1.8
3. 1992	2.7
4. 1993	3.75

Source: Microsoft Annual Report.

Using a grapher:

a) Create a scatterplot of the data. Determine whether the data seem to fit an exponential function.
b) Use regression to fit an exponential function $y = ab^x$, or ae^{kx}, to the data, where $x =$ the number of years after 1989.
c) Use the function to predict the total revenue of Microsoft, Inc., in 2000.
d) In what year would the revenue be $20 billion?

26. *Forgetting.* In an art class, students were tested at the end of the course on a final exam. Then they were retested with an equivalent test at subsequent time intervals. Their scores after time t, in months, are given in the following table.

TIME, t (IN MONTHS)	SCORE, y
1	84.9%
2	84.6%
3	84.4%
4	84.2%
5	84.1%
6	83.9%

Using a grapher:

a) Use regression to fit a logarithmic function $y = a + b \ln x$ to the data.
b) Use the function to predict test scores after 8, 10, 24, and 36 months.
c) After how long will the test scores fall below 82%?

27. *Video Rentals.* Video rental spending has been on the increase. The data in the following table shows the average amount of video rental spending per family in recent years.

x, YEAR	VIDEO RENTAL SPENDING PER FAMILY, y
1. 1991	$113
2. 1992	114
3. 1993	119
4. 1994	122
5. 1995	129

Source: Media Group Research.

a) Use the REGRESSION feature on a grapher to fit both an exponential and a power function to the data, where $x = 1$ corresponds to 1991.
b) Graph each function.
c) Use each function in part (a) to predict the video rental spending in 2005.
d) Discuss the relative merits of using each function as a predictor.

28. *Effect of Advertising.* A company introduces a new software product on a trial run in a city. They advertised the product on television and found the following data relating the percentage P of people who bought the product after x ads were run.

NUMBER OF ADS, x	PERCENTAGE WHO BOUGHT, P
0	0.2
10	0.7
20	2.7
30	9.2
40	27
50	57.6
60	83.3
70	94.8
80	98.5
90	99.6

a) Use the REGRESSION feature on a grapher to fit a logistic function

$$P(x) = \frac{a}{1 + be^{-kx}}$$

to the data.
b) What percent will buy the product when 55 ads are run? 100 ads?
c) Find an asymptote for the graph. Interpret the asymptote in terms of the advertising situation.

Skill Maintenance

Find the missing lengths in each right triangle.

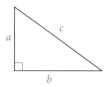

29. $b = 1$, $c = 2$ **30.** $a = 1$, $b = 1$

31. $a = 47$, $b = 34$ **32.** $b = \sqrt{13}$, $c = 200$

Synthesis

33. ◆ Browse through some newspapers or magazines until you find some data and/or a graph that seem as though they can be fit to an exponential function. Make a case for why such a fit is appropriate. Then fit an exponential function to the data and make some predictions.

34. ◆ *Atmospheric Pressure.*   Atmospheric pressure P at an altitude a is given by

$$P = P_0 e^{-0.00005a},$$

where P_0 = the pressure at sea level $\approx 14.7 \text{ lb/in}^2$ (pounds per square inch). Explain how a barometer, or some device for measuring atmospheric pressure, can be used to find the height of a skyscraper.

35. *Present Value.*   Following the birth of a child, a parent wants to make an initial investment P_0 that will grow to $50,000 for the child's education at age 18. Interest is compounded continuously at 7%. What should the initial investment be? Such an amount is called the **present value** of $50,000 due 18 yr from now.

36. *Present Value.*   Referring to Exercise 35:

a) Solve $P = P_0 e^{kt}$ for P_0.

b) Find the present value of $50,000 due 18 yr from now at interest rate 6.4%.

37. *Supply and Demand.*   The supply and demand for the sale of a certain type of VCR are given by

$$S(p) = 480e^{-0.003p} \quad \text{and} \quad D(p) = 150e^{0.004p},$$

where $S(p)$ = the number of VCRs that the company is willing to sell at price p and $D(p)$ = the quantity that the public is willing to buy at price p. Find p, called the **equilibrium price**, such that $D(p) = S(p)$.

38. *Carbon Dating.*   Recently, while digging in Chaco Canyon, New Mexico, archeologists found corn pollen that was 4000 yr old. This was evidence that

Native Americans had been cultivating crops in the Southwest centuries earlier than scientists had thought. What percent of the carbon-14 had been lost from the pollen? (*Source: American Anthropologist*)

39. *Newton's Law of Cooling.*   Suppose a body with temperature T_1 is placed in surroundings with temperature T_0 different from that of T_1. The body will either cool or warm to temperature $T(t)$ after time t, in minutes, where

$$T(t) = T_0 + |T_1 - T_0|e^{-kt}.$$

A cup of coffee with temperature 105°F is placed in a freezer with temperature 0°F. After 5 min, the temperature of the coffee is 70°F. What will its temperature be after 10 min?

40. *When Was the Murder Committed?*   The police discover the body of a math professor. Critical to solving the crime is determining when the murder was committed. The coroner arrives at the murder scene at 12:00 P.M. She immediately takes the temperature of the body and finds it to be 94.6°. She then takes the temperature 1 hr later and finds it to be 93.4°. The temperature of the room is 70°. When was the murder committed? (Use Newton's law of cooling in Exercise 39.)

41. *Electricity.*   The formula

$$i = \frac{V}{R}[1 - e^{-(R/L)t}]$$

occurs in the theory of electricity. Solve for t.

42. *The Beer–Lambert Law.*   A beam of light enters a medium such as water or smog with initial intensity I_0. Its intensity decreases depending on the thickness (or concentration) of the medium. The intensity I at a depth (or concentration) of x units is given by

$$I = I_0 e^{-\mu x}.$$

The constant μ (the Greek letter "mu") is called the **coefficient of absorption**, and it varies with the medium. For sea water, $\mu = 1.4$.

a) What percentage of light intensity I_0 remains at a depth of sea water that is 1 m? 3 m? 5 m? 50 m?

b) Plant life cannot exist below 10 m. What percentage of I_0 remains at 10 m?

43. Given that $y = ae^x$, take the natural logarithm on both sides. Let $Y = \ln y$. Consider Y as a function of x. What kind of function is Y?

44. Given that $y = ax^b$, take the natural logarithm on both sides. Let $Y = \ln y$ and $X = \ln x$. Consider Y as a function of X. What kind of function is Y?

CHAPTER

6 Summary and Review

Important Properties and Formulas

Exponential Function:	$f(x) = a^x$
The Number $e = 2.7182818284\ldots$	
Logarithmic Function:	$f(x) = \log_a x$
A Logarithm is an Exponent:	$\log_a x = y \longleftrightarrow x = a^y$
The Change-of-Base Formula:	$\log_b M = \dfrac{\log_a M}{\log_a b}$
The Product Rule:	$\log_a MN = \log_a M + \log_a N$
The Power Rule:	$\log_a M^p = p \log_a M$
The Quotient Rule:	$\log_a \dfrac{M}{N} = \log_a M - \log_a N$
Other Properties:	$\log_a a = 1, \qquad \log_a 1 = 0,$
	$\log_a a^x = x, \qquad a^{\log_a x} = x$
Base–Exponent Property:	$a^x = a^y \longleftrightarrow x = y$
Exponential Growth Model:	$P(t) = P_0 e^{kt}$
Exponential Decay Model:	$P(t) = P_0 e^{-kt}$
Interest Compounded Continuously:	$P(t) = P_0 e^{kt}$
Limited Growth:	$P(t) = \dfrac{a}{1 + be^{-kt}}$

REVIEW EXERCISES

In Exercises 1–6, match the equation with one of figures
(a)–(f), which follow. If needed, use a grapher.

a)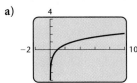

b)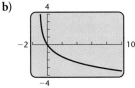

e)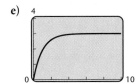

f)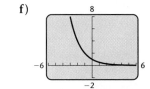

c)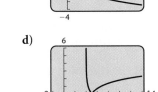

d)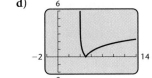

1. $f(x) = e^{x-3}$ **2.** $f(x) = \log_3 x$

3. $y = -\log_3 (x + 1)$ **4.** $y = \left(\frac{1}{2}\right)^x$

5. $f(x) = 3(1 - e^{-x}), \ x \geq 0$

6. $f(x) = |\ln (x - 4)|$

7. Convert to an exponential equation:
$$\log_4 x = 2.$$

8. Convert to a logarithmic equation:
$$e^x = 80.$$

Solve. Use any method.

9. $\log_4 x = 2$

10. $3^{1-x} = 9^{2x}$

11. $e^x = 80$

12. $4^{2x-1} - 3 = 61$

13. $\log_{16} 4 = x$

14. $\log_x 125 = 3$

15. $\log_2 x + \log_2 (x - 2) = 3$

16. $\log (x^2 - 1) - \log (x - 1) = 1$

17. $\log x^2 = \log x$

18. $e^{-x} = 0.02$

Express as a single logarithm and simplify if possible.

19. $3 \log_b x - 4 \log_b y + \frac{1}{2} \log_b z$

20. $\ln (x^3 - 8) - \ln (x^2 + 2x + 4) + \ln (x + 2)$

Express in terms of sums and differences of logarithms.

21. $\ln \sqrt[4]{wr^3}$ **22.** $\log \sqrt[3]{\dfrac{M^2}{N}}$

Given that $\log_a 2 = 0.301$, $\log_a 5 = 0.699$, *and* $\log_a 6 = 0.778$, *find each of the following.*

23. $\log_a 3$

24. $\log_a 50$

25. $\log_a \frac{1}{5}$

26. $\log_a \sqrt[3]{5}$

Simplify.

27. $\ln e^{-5k}$ **28.** $\log_5 5^{-6t}$

29. How long will it take an investment to double itself if it is invested at 5.4%, compounded continuously?

30. The population of a city doubled in 30 yr. What was the exponential growth rate?

31. How old is a skeleton that has lost 27% of its carbon-14?

32. The hydrogen ion concentration of milk is 2.3×10^{-6}. What is the pH?

33. What is the loudness, in decibels, of a sound whose intensity is $1000I_0$?

34. *The Population of Brazil.* The population of Brazil was 52 million in 1959, and the exponential growth rate was 2.8% per year. (*Source:* U.S. Bureau of the Census, World Population Profile)

a) Find the exponential growth function.
b) What will the population be in 2000? in 2020?
c) When will the population be 300 million?
d) What was the doubling time?

35. *Toll-free 800 Numbers.* The use of toll-free 800 numbers has grown exponentially. In 1967, there were 7 million such calls and in 1991, there were 10.2 billion such calls. (*Source:* Federal Communication Commission)

a) Find the exponential growth rate k.
b) Find the exponential growth function.
c) Graph the exponential growth function.
d) How many toll-free 800 number calls will be placed in 1998? in 2005?
e) In what year will 20 billion such calls be placed?

36. *Walking Speed.* The average walking speed w, in feet per second, of a person living in a city of population P, in thousands, is given by the function
$$w(P) = 0.37 \ln P + 0.05.$$

a) The population of Austin, Texas, is 466,000. Find the average walking speed.
b) A city's population has an average walking speed of 3.4 ft/sec. Find the population.

37. *Multimedia Personal Computers.* The following table contains estimated data regarding the number of multimedia personal computers installed in homes in various years.

x, YEAR	MULTIMEDIA PERSONAL COMPUTERS, y (IN MILLIONS)
0. 1992	2
1. 1993	4.0
2. 1994	8.0
3. 1995	16.0
4. 1996	22.5
5. 1997	31.4

Source: Piper Joffray Research.

Using a grapher:

a) Create a scatterplot of the data. Determine whether the data seem to fit an exponential function.
b) Use regression to fit an exponential function $y = ab^x$, or ae^{kx}, to the data, where $x = $ the number of years after 1992.

c) Use the function to predict the number of multimedia personal computers in homes in 2000.

d) In how many years will there be 100 million such computers in homes?

Synthesis

38. ◈ Suppose that you were trying to convince a fellow student that

$$\log_2 (x + 5) \neq \log_2 x + \log_2 5.$$

Give as many explanations as you can.

39. ◈ Explain how the graph of $f(x) = e^x$ could be used to graph the function $g(x) = \ln x - 1$.

Solve.

40. $|\log_4 x| = 3$

41. $\log x = \ln x$

42. $5^{\sqrt{x}} = 625$

43. a) Use a grapher to graph $f(x) = 5e^{-x} \ln x$ in the viewing window $[-1, 10, -5, 5]$.

b) Estimate the relative maximum and the minimum values. Use a MAX–MIN feature if such exists on your grapher.

44. Find the domain: $f(x) = \log_3 (\ln x)$.

45. Find the points of intersection of the graphs of the following equations:

$$y = 5x^2 e^{-x}, \qquad y = 2 - e^{-x^2}.$$

Appendixes

A.1
Solving Equations

- *Solve linear, quadratic, rational, and radical equations and equations with absolute value.*
- *Solve a formula for a given variable.*

An **equation** is a statement that two expressions are equal. To **solve** an equation in one variable is to find all the values of the variable that make the equation true. Each of these numbers is a **solution** of the equation. The set of all solutions of an equation is its **solution set**. Equations that have the same solution set are called **equivalent equations**.

Linear and Quadratic Equations

> A *linear equation in one variable* is an equation that is equivalent to one of the form $ax + b = 0$, where a and b are real numbers and $a \neq 0$.
>
> A *quadratic equation* is an equation that is equivalent to one of the form $ax^2 + bx + c = 0$, where a, b, and c are real numbers and $a \neq 0$.

The following principles allow us to solve many linear and quadratic equations.

> **Equation-Solving Principles**
>
> For any real numbers a, b, and c:
>
> **The Addition Principle:** If $a = b$ is true, then $a + c = b + c$ is true.
>
> **The Multiplication Principle:** If $a = b$ is true, then $ac = bc$ is true.
>
> **The Principle of Zero Products:** If $ab = 0$ is true, then $a = 0$ or $b = 0$, and if $a = 0$ or $b = 0$, then $ab = 0$.
>
> **The Principle of Square Roots:** If $x^2 = k$, then $x = \sqrt{k}$ or $x = -\sqrt{k}$.

Example 1 Solve: $3(7 - 2x) = 14 - 8(x - 1)$.

╷╴╴ ALGEBRAIC SOLUTION

We have

$3(7 - 2x) = 14 - 8(x - 1)$

$21 - 6x = 14 - 8x + 8$ Using the distributive property

$21 - 6x = 22 - 8x$ Combining like terms

$21 + 2x = 22$ Using the addition principle to add $8x$ on both sides

$2x = 1$ Using the addition principle to add -21 or subtract 21 on both sides

$x = \frac{1}{2}.$ Using the multiplication principle to multiply by $\frac{1}{2}$ or divide by 2 on both sides

CHECK: $\dfrac{3(7 - 2x) = 14 - 8(x - 1)}{}$

$3\left(7 - 2 \cdot \frac{1}{2}\right)$? $14 - 8\left(\frac{1}{2} - 1\right)$ Substituting $\frac{1}{2}$ for x

$3(7 - 1)$ | $14 - 8\left(-\frac{1}{2}\right)$

$3 \cdot 6$ | $14 + 4$

18 | 18 TRUE

The solution is $\frac{1}{2}$. The set of all solutions, the solution set, is $\left\{\frac{1}{2}\right\}$.

We can also use the TABLE feature of a grapher—set in ASK mode, if available— to check the answer. We let $y_1 = 3(7 - 2x)$ and $y_2 = 14 - 8(x - 1)$. When we enter .5 for x, we see that Y_1 and Y_2 are both 18.

X	Y₁	Y₂
.5	18	18

X = .5

╷╴╴ GRAPHICAL SOLUTION

We can use a grapher to solve this equation. Graph $y_1 = 3(7 - 2x)$ and $y_2 = 14 - 8(x - 1)$.

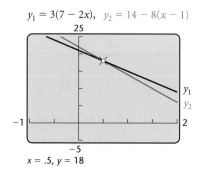

$y_1 = 3(7 - 2x), \quad y_2 = 14 - 8(x - 1)$

$x = .5, y = 18$

Using the TRACE and ZOOM features or the INTERSECT feature, we find that the first coordinate of the point of intersection is .5, or $\frac{1}{2}$. The solution set is $\left\{\frac{1}{2}\right\}$.

We can also use the SOLVE feature. It might be necessary to first write the equation as

$3(7 - 2x) - (14 - 8(x - 1)) = 0.$

The grapher returns the value .5, or $\frac{1}{2}$.

Note that the equation in Example 1 is not an identity, because it is not true for *all* values of x in its domain. In fact, we see from the graph above that $y_1 = 3(7 - 2x)$ and $y_2 = 14 - 8(x - 1)$ coincide, or intersect, only for $x = \frac{1}{2}$.

Example 2 Solve each of the following.

a) $2x^2 - x = 3$

b) $2x^2 - 10 = 0$

SOLUTION

a) We solve $2x^2 - x = 3$ both algebraically and graphically.

r··· *ALGEBRAIC SOLUTION*

We have

$$2x^2 - x = 3$$

$$2x^2 - x - 3 = 0 \qquad \text{Subtracting 3 on both sides}$$

$$(2x - 3)(x + 1) = 0 \qquad \text{Factoring}$$

$$2x - 3 = 0 \quad or \quad x + 1 = 0$$
$$\text{Using the principle of zero products}$$

$$x = \tfrac{3}{2} \quad or \qquad x = -1.$$

Both numbers check. We can use the TABLE feature of a grapher, set in ASK mode, to confirm this. Let $y = 2x^2 - x$. The y-values in the table should be 3 for both $x = 1.5$ and $x = -1$.

X	Y₁	
1.5	3	
−1	3	

X =

The solution set is $\left\{\tfrac{3}{2}, -1\right\}$.

r··· *GRAPHICAL SOLUTION*

Graph $y_1 = 2x^2 - x$ and $y_2 = 3$.

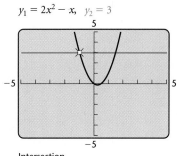

Intersection
$x = -1 \quad y = 3$

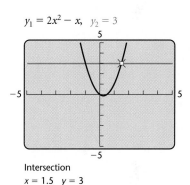

Intersection
$x = 1.5 \quad y = 3$

We use the TRACE and ZOOM features or the INTERSECT feature and find that the solutions are -1 and 1.5. We can also use the SOLVE feature to get this result. The solution set is $\{-1, 1.5\}$.

b) We solve $2x^2 - 10 = 0$ algebraically:

$$2x^2 - 10 = 0$$

$$2x^2 = 10 \qquad \text{Adding 10 on both sides}$$

$$x^2 = 5 \qquad \text{Dividing by 2 on both sides}$$

$$x = \sqrt{5} \quad or \quad x = -\sqrt{5}. \qquad \text{Using the principle of square roots}$$

Both numbers check. The solution set is $\{\sqrt{5}, -\sqrt{5}\}$, or $\{\pm\sqrt{5}\}$. Use the TABLE feature of a grapher, set in ASK mode, to confirm this. Using a grapher, we find that $x \approx 2.236$ or $x \approx -2.236$.

None of the preceding methods would yield the *exact* solutions to a quadratic equation like $x^2 - 6x - 10 = 0$. If we wish to do that, we can use a procedure called *completing the square* and then use the principle of square roots.

Example 3 Solve

$$x^2 - 6x - 10 = 0$$

by completing the square.

SOLUTION Our goal is to find an equivalent equation of the form $x^2 + bx + c = d$ in which $x^2 + bx + c$ is a perfect square. This is accomplished as follows:

$x^2 - 6x - 10 = 0$

$x^2 - 6x \qquad = 10$ Adding 10

$x^2 - 6x + 9 = 10 + 9$ Adding 9 to complete the square: $\frac{1}{2}(-6) = -3$ and $(-3)^2 = 9$

$x^2 - 6x + 9 = 19.$

Because $x^2 - 6x + 9$ is a perfect square, we are able to write it as the square of a binomial. We can then use the principle of square roots to finish the solution:

$(x - 3)^2 = 19$ Factoring

$x - 3 = \pm\sqrt{19}$ Using the principle of square roots

$x = 3 \pm \sqrt{19}.$ Adding 3

The exact solutions are $3 + \sqrt{19}$ and $3 - \sqrt{19}$, or simply $3 \pm \sqrt{19}$ (read "three plus or minus $\sqrt{19}$"). The solution set is $\{3 + \sqrt{19}, 3 - \sqrt{19}\}$, or $\{3 \pm \sqrt{19}\}$. Decimal approximations can be found on a grapher. ▬

Example 4 Solve: $2x^2 - 1 = 3x$.

SOLUTION

$2x^2 - 1 = 3x$

$2x^2 - 3x - 1 = 0$ Subtracting $3x$; we are unable to factor the result.

$2x^2 - 3x \qquad = 1$ Adding 1

$x^2 - \dfrac{3}{2}x \qquad = \dfrac{1}{2}$ Dividing by 2

$x^2 - \dfrac{3}{2}x + \dfrac{9}{16} = \dfrac{1}{2} + \dfrac{9}{16}$ Completing the square: $\frac{1}{2}\left(-\frac{3}{2}\right) = -\frac{3}{4}$ and $\left(-\frac{3}{4}\right)^2 = \frac{9}{16}$

$\left(x - \dfrac{3}{4}\right)^2 = \dfrac{17}{16}$ Factoring and simplifying

$x - \dfrac{3}{4} = \pm\dfrac{\sqrt{17}}{4}$ Using the principle of square roots and the quotient rule for radicals

$x = \dfrac{3 \pm \sqrt{17}}{4}$ Adding $\frac{3}{4}$

The solution set is $\left\{\dfrac{3 \pm \sqrt{17}}{4}\right\}$. ▬

To solve a quadratic equation by completing the square:

1. Isolate the terms with variables on one side of the equation and arrange them in descending order.
2. Divide by the coefficient of the squared term if that coefficient is not 1.
3. Complete the square by taking half the coefficient of the first-degree term and adding its square on both sides of the equation.
4. Express one side of the equation as the square of a binomial.
5. Use the principle of square roots.
6. Solve for the variable.

Using the Quadratic Formula

Because completing the square works for *any* quadratic equation, it can be used to solve the general quadratic equation $ax^2 + bx + c = 0$ for x. The result will be a formula that can be used to solve any quadratic equation quickly.

Consider any quadratic equation in standard form:

$$ax^2 + bx + c = 0, \quad a > 0.$$

If a is negative, we first multiply on both sides by -1. Then we solve by completing the square. As we carry out the steps, compare them with those of Example 4.

$$ax^2 + bx = -c \qquad \text{Adding } -c$$

$$x^2 + \frac{b}{a}x = -\frac{c}{a} \qquad \text{Dividing by } a, \text{ since } a > 0$$

Half of $\dfrac{b}{a}$ is $\dfrac{b}{2a}$ and $\left(\dfrac{b}{2a}\right)^2 = \dfrac{b^2}{4a^2}$. Thus we add $\dfrac{b^2}{4a^2}$:

$$x^2 + \frac{b}{a}x + \frac{b^2}{4a^2} = -\frac{c}{a} + \frac{b^2}{4a^2} \qquad \text{Adding } \frac{b^2}{4a^2} \text{ to complete the square}$$

$$\left(x + \frac{b}{2a}\right)^2 = -\frac{4ac}{4a^2} + \frac{b^2}{4a^2} \qquad \begin{array}{l}\text{Factoring and finding a common} \\ \text{denominator:} \\ -\dfrac{c}{a} = -\dfrac{4a}{4a} \cdot \dfrac{c}{a} = -\dfrac{4ac}{4a^2}\end{array}$$

$$\left(x + \frac{b}{2a}\right)^2 = \frac{b^2 - 4ac}{4a^2}$$

$$x + \frac{b}{2a} = \pm\frac{\sqrt{b^2 - 4ac}}{2a} \qquad \begin{array}{l}\text{Using the principle of square roots} \\ \text{and the quotient rule for radicals;} \\ \text{since } a > 0, \sqrt{4a^2} = 2a.\end{array}$$

$$x = \frac{-b \pm \sqrt{b^2 - 4ac}}{2a}. \qquad \text{Adding } -\frac{b}{2a}$$

Program

QUADFORM: This program solves a quadratic equation, giving real and complex solutions. (See the Graphing Calculator Manual that accompanies this text.)

> **The Quadratic Formula**
>
> The solutions of $ax^2 + bx + c = 0$, $a \neq 0$, are given by
>
> $$x = \frac{-b \pm \sqrt{b^2 - 4ac}}{2a}.$$

Example 5 Solve $3x^2 + 2x = 7$. Find exact solutions and approximate solutions rounded to the nearest thousandth.

We show both algebraic and graphical solutions. Note that only the algebraic approach yields the exact solutions.

ALGEBRAIC SOLUTION

After finding standard form, we are unable to factor, so we identify a, b, and c in order to use the quadratic formula:

$$3x^2 + 2x - 7 = 0;$$
$$a = 3, \quad b = 2, \quad c = -7.$$

We then use the quadratic formula:

$$x = \frac{-b \pm \sqrt{b^2 - 4ac}}{2a}$$

$$= \frac{-2 \pm \sqrt{2^2 - 4(3)(-7)}}{2(3)} \qquad \text{Substituting}$$

$$= \frac{-2 \pm \sqrt{4 + 84}}{6} = \frac{-2 \pm \sqrt{88}}{6}$$

$$= \frac{-2 \pm \sqrt{4 \cdot 22}}{6} = \frac{-2 \pm 2\sqrt{22}}{6}$$

$$= \frac{2}{2} \cdot \frac{-1 \pm \sqrt{22}}{3} = \frac{-1 \pm \sqrt{22}}{3}.$$

The exact solutions are

$$\frac{-1 - \sqrt{22}}{3} \quad \text{and} \quad \frac{-1 + \sqrt{22}}{3}.$$

Using the scientific keys on a grapher, we approximate the solutions to be -1.897 and 1.230.

GRAPHICAL SOLUTION

To solve $3x^2 + 2x = 7$, or $3x^2 + 2x - 7 = 0$, we first graph the function $f(x) = 3x^2 + 2x - 7$. Then we look for points where the graph crosses the x-axis. It appears that there are two possible zeros, one near -2 and one near 1. We can use TRACE and ZOOM to approximate these zeros, or we can use a SOLVE or POLY feature.

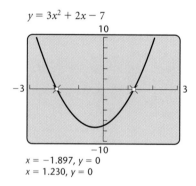

$$y = 3x^2 + 2x - 7$$

$x = -1.897, y = 0$
$x = 1.230, y = 0$

We get the approximate zeros -1.896805 and 1.2301386, or -1.897 and 1.230, rounded to three decimal places. The zeros of the function are the solutions of the equation $3x^2 + 2x = 7$.

Rational and Radical Equations

Equations containing rational expressions are called **rational equations**. Solving such equations involves multiplying on both sides by the LCD.

Example 6 Solve: $\dfrac{x^2}{x-3} = \dfrac{9}{x-3}$.

SOLUTION   The LCD is $x - 3$.

$$(x - 3) \cdot \frac{x^2}{x - 3} = (x - 3) \cdot \frac{9}{x - 3}$$

$$x^2 = 9$$

$$x = -3 \quad or \quad x = 3 \qquad \text{Using the principle of square roots}$$

The possible solutions are 3 and -3. We check.

CHECK: For 3:

$$\frac{x^2}{x-3} = \frac{9}{x-3}$$

$$\frac{3^2}{3-3} \ ? \ \frac{9}{3-3}$$

$$\frac{9}{0} \ \bigg| \ \frac{9}{0} \qquad \text{UNDEFINED}$$

For -3:

$$\frac{x^2}{x-3} = \frac{9}{x-3}$$

$$\frac{(-3)^2}{-3-3} \ ? \ \frac{9}{-3-3}$$

$$\frac{9}{-6} \ \bigg| \ \frac{9}{-6} \qquad \text{TRUE}$$

Since division by 0 is undefined, 3 is not a solution. Note that 3 is not in the domain of $x^2/(x-3)$ or $9/(x-3)$. (See the table below.) The number -3 checks, so it is a solution. The solution set is $\{-3\}$.

$$y_1 = \frac{x^2}{x-3}, \ y_2 = \frac{9}{x-3}$$

X	Y₁	Y₂
0	0	−3
1	−.5	−4.5
2	−4	−9
3	ERROR	ERROR
4	16	9
5	12.5	4.5
6	12	3

X = 3

When we use the multiplication principle to multiply (or divide) both sides of an equation by an expression with a variable, we might not obtain an equivalent equation. We must check possible solutions by substituting in the original equation.

A **radical equation** is an equation in which variables appear in one or more radicands. For example,

$$\sqrt{2x - 5} - \sqrt{x - 3} = 1$$

is a radical equation. The following principle is needed to solve such equations.

The Principle of Powers

For any positive integer n:

If $a = b$ is true, then $a^n = b^n$ is true.

Example 7 Solve: $5 + \sqrt{x + 7} = x$.

We first isolate the radical and then use the principle of powers.

$$5 + \sqrt{x + 7} = x$$
$$\sqrt{x + 7} = x - 5 \qquad \text{Subtracting 5 on both sides}$$
$$(\sqrt{x + 7})^2 = (x - 5)^2 \qquad \text{Using the principle of powers; squaring both sides}$$
$$x + 7 = x^2 - 10x + 25$$
$$0 = x^2 - 11x + 18$$
$$0 = (x - 9)(x - 2)$$
$$x - 9 = 0 \quad \text{or} \quad x - 2 = 0$$
$$x = 9 \quad \text{or} \qquad x = 2$$

The possible solutions are 9 and 2.

CHECK: For 9:

$$\frac{5 + \sqrt{x + 7} = x}{5 + \sqrt{9 + 7} \; ? \; 9}$$
$$5 + \sqrt{16}$$
$$5 + 4$$
$$9 \;\big|\; 9 \quad \text{TRUE}$$

For 2:

$$\frac{5 + \sqrt{x + 7} = x}{5 + \sqrt{2 + 7} \; ? \; 2}$$
$$5 + \sqrt{9}$$
$$5 + 3$$
$$8 \;\big|\; 2 \quad \text{FALSE}$$

Since 9 checks but 2 does not, the only solution is 9. The solution set is {9}.

Graph $y_1 = 5 + \sqrt{x + 7}$ and $y_2 = x$.

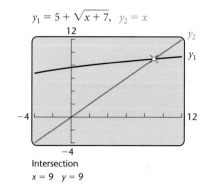

$y_1 = 5 + \sqrt{x + 7}, \quad y_2 = x$

Intersection
$x = 9 \quad y = 9$

Using the TRACE and ZOOM features or the INTERSECT feature, we see that the only solution is 9. The solution set is {9}.

We can also use the SOLVE feature to get this result. It might be necessary, however, to first write the equation as

$$5 + \sqrt{x + 7} - x = 0.$$

When we raise both sides of an equation to an even power, the resulting equation can have solutions that the original equation does not. This is because the converse of the principle of powers is not necessarily true. That is, if $a^n = b^n$ is true, we do not know that $a = b$ is true. For example, $(-2)^2 = 2^2$, but $-2 \neq 2$. Thus, as we see in Example 7, it is necessary to check the possible solutions in the original equation when using the principle of powers to raise both sides of an equation to an even power.

When a radical equation has two radical terms on one side, we isolate one of them and then use the principle of powers. If, after doing so, a radical term remains, we repeat these steps.

Example 8 Solve: $\sqrt{x - 3} + \sqrt{x + 5} = 4$.

SOLUTION

$$\sqrt{x - 3} = 4 - \sqrt{x + 5} \qquad \text{Isolating one radical}$$
$$(\sqrt{x - 3})^2 = (4 - \sqrt{x + 5})^2 \qquad \text{Using the principle of powers; squaring both sides}$$

$$x - 3 = 16 - 8\sqrt{x + 5} + (x + 5)$$
$$x - 3 = 21 - 8\sqrt{x + 5} + x \quad \text{Combining like terms}$$
$$-24 = -8\sqrt{x + 5} \quad \text{Isolating the remaining radical}$$
$$3 = \sqrt{x + 5} \quad \text{Dividing by } -8 \text{ on both sides}$$
$$3^2 = (\sqrt{x + 5})^2 \quad \text{Using the principle of powers;}$$
$$\text{squaring both sides}$$
$$9 = x + 5$$
$$4 = x$$

The number 4 checks and is the solution. Use graphs or the TABLE feature of a grapher to confirm this. The solution set is {4}.

Equations with Absolute Value

Recall that the absolute value of a number is its distance from 0 on the number line. We use this concept to solve equations with absolute value.

> For $a > 0$:
>
> $$|x| = a \text{ is equivalent to } x = -a \text{ or } x = a.$$

Example 9 Solve each of the following.

a) $|x| = 5$

b) $|x - 3| = 2$

SOLUTION

a) We solve $|x| = 5$ both algebraically and graphically.

ALGEBRAIC SOLUTION

We have

$$|x| = 5$$
$$x = -5 \quad or \quad x = 5.$$

Writing an equivalent statement

The solution set is $\{-5, 5\}$. Use the TABLE feature of a grapher to confirm this.

GRAPHICAL SOLUTION

Graph $y_1 = |x|$ and $y_2 = 5$ and find the first coordinates of the points of intersection using TRACE and ZOOM or INTERSECT. We can also use the SOLVE feature to get this result.

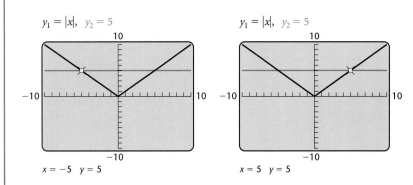

$y_1 = |x|, \ y_2 = 5$

$x = -5 \quad y = 5$

$y_1 = |x|, \ y_2 = 5$

$x = 5 \quad y = 5$

The solution set is $\{-5, 5\}$.

b) We solve $|x - 3| = 2$ algebraically:

$$|x - 3| = 2$$
$$x - 3 = -2 \quad or \quad x - 3 = 2 \qquad \text{Writing an equivalent statement}$$
$$x = 1 \qquad or \qquad x = 5. \qquad \text{Adding 3}$$

The solution set is $\{1, 5\}$.

When $a = 0$, $|x| = a$ is equivalent to $x = 0$. Note that for $a < 0$, $|x| = a$ has no solution, because the absolute value of an expression is never negative. We can use a graph to illustrate the last statement for a specific value of a. For example, let $a = -3$. Graph $y_1 = |x|$ and $y_2 = -3$.

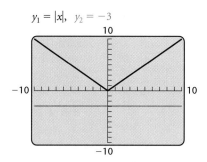

$$y_1 = |x|, \quad y_2 = -3$$

The graphs do not intersect. Thus the equation $|x| = -3$ has no solution. The solution set is the **empty set**, denoted $\varnothing$.

Formulas

A **formula** is an equation that can be used to *model* a situation. For example, the formula $P = 2l + 2w$ gives the perimeter of a rectangle with length l and width w. The equation-solving principles presented earlier can be used to solve a formula for a given variable.

Example 10   Solve $P = 2l + 2w$ for l.

SOLUTION

$$P = 2l + 2w$$
$$P - 2w = 2l \qquad \text{Subtracting } 2w \text{ on both sides}$$
$$\frac{P - 2w}{2} = l \qquad \text{Dividing by 2 on both sides}$$

The formula $l = \dfrac{P - 2w}{2}$ can be used to determine a rectangle's length if we are given its perimeter and its width.

A.1 Exercise Set

Solve using a grapher.

1. $4x + 5 = 21$

2. $2y - 1 = 3$

3. $y + 1 = 2y - 7$

4. $5 - 4x = x - 13$

Solve algebraically. Check your answers on a grapher.

5. $5x - 2 + 3x = 2x + 6 - 4x$

6. $5x - 17 - 2x = 6x - 1 - x$

7. $7(3x + 6) = 11 - (x + 2)$

8. $4(5y + 3) = 3(2y - 5)$

9. $(2x - 3)(3x - 2) = 0$

10. $(5x - 2)(2x + 3) = 0$

11. $3x^2 + x - 2 = 0$

12. $10x^2 - 16x + 6 = 0$

13. $2x^2 = 6x$

14. $18x + 9x^2 = 0$

15. $3y^3 - 5y^2 - 2y = 0$

16. $3t^3 + 2t = 5t^2$

17. $7x^3 + x^2 - 7x - 1 = 0$
(*Hint:* Factor by grouping.)

18. $3x^3 + x^2 - 12x - 4 = 0$
(*Hint:* Factor by grouping.)

Solve by completing the square to obtain exact solutions. Check your answers on a grapher.

19. $x^2 + 6x = 7$

20. $x^2 + 8x = -15$

21. $x^2 = 8x - 9$

22. $x^2 = 22 + 10x$

23. $3x^2 + 5x - 2 = 0$

24. $2x^2 - 5x - 3 = 0$

25. $6x + 1 = 4x^2$

26. $3x^2 + 5x = 3$

27. $2x^2 - 4 = 5x$

28. $4x^2 - 2 = 3x$

Solve. Use any method, but obtain exact solutions. Check your answers on a grapher.

29. $x^2 - 2x = 15$

30. $x^2 + 4x = 5$

31. $5m^2 + 3m = 2$

32. $2y^2 - 3y - 2 = 0$

33. $3x^2 + 6 = 10x$

34. $3t^2 + 8t + 3 = 0$

35. $5t^2 - 8t = 3$

36. $2t^2 - 5t = 1$

37. $\dfrac{1}{4} + \dfrac{1}{5} = \dfrac{1}{t}$

38. $\dfrac{1}{3} - \dfrac{5}{6} = \dfrac{1}{x}$

39. $\dfrac{x + 2}{4} - \dfrac{x - 1}{5} = 15$

40. $\dfrac{t + 1}{3} - \dfrac{t - 1}{2} = 1$

41. $\dfrac{1}{2} + \dfrac{2}{x} = \dfrac{1}{3} + \dfrac{3}{x}$

42. $\dfrac{1}{t} + \dfrac{1}{2t} + \dfrac{1}{3t} = 5$

43. $\dfrac{3x}{x + 2} + \dfrac{6}{x} = \dfrac{12}{x^2 + 2x}$

44. $\dfrac{5x}{x - 4} - \dfrac{20}{x} = \dfrac{80}{x^2 - 4x}$

45. $\dfrac{4}{x^2 - 1} - \dfrac{2}{x - 1} = \dfrac{3}{x + 1}$

46. $\dfrac{3y + 5}{y^2 + 5y} + \dfrac{y + 4}{y + 5} = \dfrac{y + 1}{y}$

47. $\dfrac{490}{x^2 - 49} = \dfrac{5x}{x - 7} - \dfrac{35}{x + 7}$

48. $\dfrac{3}{m + 2} + \dfrac{2}{m} = \dfrac{4m - 4}{m^2 - 4}$

49. $\dfrac{1}{x - 6} - \dfrac{1}{x} = \dfrac{6}{x^2 - 6x}$

50. $\dfrac{8}{x^2 - 4} = \dfrac{x}{x - 2} - \dfrac{2}{x + 2}$

51. $\dfrac{8}{x^2 - 2x + 4} = \dfrac{x}{x + 2} + \dfrac{24}{x^3 + 8}$

52. $\dfrac{18}{x^2 - 3x + 9} - \dfrac{x}{x + 3} = \dfrac{81}{x^3 + 27}$

53. $\sqrt{3x - 4} = 1$

54. $\sqrt[3]{2x + 1} = -5$

55. $\sqrt[4]{x^2 - 1} = 1$

56. $\sqrt{m + 1} - 5 = 8$

57. $\sqrt{y - 1} + 4 = 0$

58. $\sqrt[5]{3x + 4} = 2$

59. $\sqrt[3]{6x + 9} + 8 = 5$

60. $\sqrt{6x + 7} = x + 2$

61. $\sqrt{x - 3} + \sqrt{x + 2} = 5$

62. $\sqrt{x} - \sqrt{x - 5} = 1$

63. $\sqrt{3x - 5} + \sqrt{2x + 3} + 1 = 0$

64. $\sqrt{2m - 3} = \sqrt{m + 7} - 2$

65. $\sqrt{x} - \sqrt{3x - 3} = 1$

66. $\sqrt{2x + 1} - \sqrt{x} = 1$

67. $\sqrt{2y - 5} - \sqrt{y - 3} = 1$

68. $\sqrt{4p + 5} + \sqrt{p + 5} = 3$

69. $x^{1/3} = -2$

70. $t^{1/5} = 2$

71. $t^{1/4} = 3$

72. $m^{1/2} = -7$

Solve.

73. $|x| = 7$

74. $|x| = 4.5$

75. $|x| = -10.7$

76. $|x| = -\frac{3}{5}$

77. $|x - 1| = 4$

78. $|x - 7| = 5$

79. $|3x| = 1$

80. $|5x| = 4$

81. $|x| = 0$

82. $|6x| = 0$

83. $|3x + 2| = 1$

84. $|7x - 4| = 8$

85. $\left|\frac{1}{2}x - 5\right| = 17$

86. $\left|\frac{1}{3}x - 4\right| = 13$

87. $|x - 1| + 3 = 6$

88. $|x + 2| - 5 = 9$

Solve.

89. $A = \dfrac{1}{2}bh$, for b
(Area of a triangle)

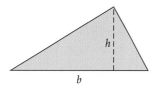

90. $A = \pi r^2$, for π
(Area of a circle)

91. $P = 2l + 2w$, for w
(Perimeter of a rectangle)

92. $A = P + Prt$, for P
(Simple interest)

93. $\dfrac{P_1 V_1}{T_1} = \dfrac{P_2 V_2}{T_2}$, for T_1
(A chemistry formula for gases)

94. $A = \dfrac{1}{2}h(b_1 + b_2)$, for h
(Area of a trapezoid)

95. $\dfrac{1}{R} = \dfrac{1}{R_1} + \dfrac{1}{R_2}$, for R_2
(Resistance)

96. $A = P(1 + i)^2$, for i
(Compound interest)

97. $\dfrac{1}{F} = \dfrac{1}{m} + \dfrac{1}{p}$, for p
(A formula from optics)

98. $\dfrac{1}{F} = \dfrac{1}{m} + \dfrac{1}{p}$, for F
(A formula from optics)

Synthesis

99. ◈ Explain why it is necessary to check the possible solutions of a rational equation.

100. ◈ Explain in your own words why it is necessary to check the possible solutions when the principle of powers is used to solve an equation.

101. ◈ Use a graphical argument to explain why the equation $x^2 - 6x + 9 = 1 - x^4$ has no solution.

Solve.

102. $(x + 1)^3 = (x - 1)^3 + 26$

103. $(x - 2)^3 = x^3 - 2$

104. $\dfrac{x + 3}{x + 2} - \dfrac{x + 4}{x + 3} = \dfrac{x + 5}{x + 4} - \dfrac{x + 6}{x + 5}$

105. $(x - 3)^{2/3} = 2$

106. $\sqrt{15 + \sqrt{2x + 80}} = 5$

107. $\sqrt{x + 5} + 1 = \dfrac{6}{\sqrt{x + 5}}$

108. $x^{2/3} = x + 1$

A.2

Distance, Midpoints, and Circles

- *Find the distance between two points in the plane and the midpoint of a segment.*
- *Find an equation of a circle with a given center and radius, and given an equation of a circle, find the center and the radius.*
- *Graph equations of circles.*

We have seen that graphs can provide a useful way of modeling real-world situations. In carpentry, surveying, engineering, and other fields, it is often necessary to determine distances and midpoints and to produce accurately drawn circles.

The Distance Formula

Suppose that a conservationist needs to determine the distance across an irregularly shaped pond. One way in which the conservationist might proceed is to measure two legs of a right triangle that is situated as shown below. The Pythagorean theorem, $a^2 + b^2 = c^2$, can then be used to find the length of the hypotenuse.

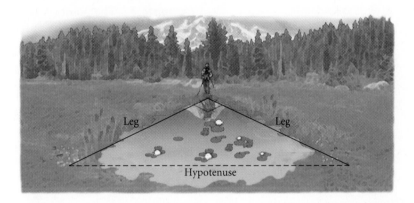

A similar strategy is used to find the distance between two points in a plane. For two points (x_1, y_1) and (x_2, y_2), we can draw a right triangle in which the legs have lengths $|x_2 - x_1|$ and $|y_2 - y_1|$.

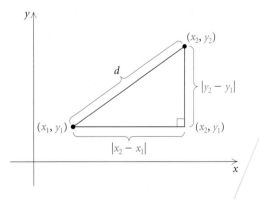

Using the Pythagorean theorem, we have

$$d^2 = |x_2 - x_1|^2 + |y_2 - y_1|^2.$$

Because we are squaring, parentheses can replace the absolute value symbols:

$$d^2 = (x_2 - x_1)^2 + (y_2 - y_1)^2.$$

Taking the principal square root, we obtain the distance formula.

The Distance Formula

The *distance d* between any two points (x_1, y_1) and (x_2, y_2) is given by

$$d = \sqrt{(x_2 - x_1)^2 + (y_2 - y_1)^2}.$$

The subtraction of the x-coordinates can be done in any order, as can the subtraction of the y-coordinates. Although we derived the distance formula by considering two points not on a horizontal or a vertical line, the distance formula holds for *any* two points.

Example 1 The point $(-2, 5)$ is on a circle that has $(3, -1)$ as its center. Find the length of the radius of the circle.

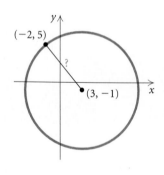

SOLUTION Since the length of the radius is the distance from the center to a point on the circle, we substitute into the distance formula:

$$r = \sqrt{[3 - (-2)]^2 + [-1 - 5]^2} \quad \text{Either point can serve as } (x_1, y_1).$$

$$= \sqrt{5^2 + (-6)^2} = \sqrt{61} \approx 7.810. \quad \text{Rounded to the nearest thousandth}$$

The circle's radius is approximately 7.8.

Midpoints of Segments

The distance formula can be used to develop a way of determining the *midpoint* of a segment when the endpoints are known. We state the formula and leave its proof to the exercises.

The Midpoint Formula

If the endpoints of a segment are (x_1, y_1) and (x_2, y_2), then the coordinates of the *midpoint* are

$$\left(\frac{x_1 + x_2}{2}, \frac{y_1 + y_2}{2} \right).$$

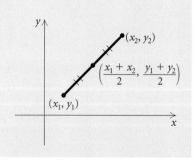

Note that we obtain the coordinates of the midpoint by averaging the coordinates of the endpoints. This is an easy way to remember the midpoint formula.

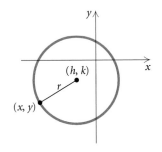

Example 2 The diameter of a circle connects two points $(2, -3)$ and $(6, 4)$ on the circle. Find the coordinates of the center of the circle.

SOLUTION Using the midpoint formula, we obtain

$$\left(\frac{2 + 6}{2}, \frac{-3 + 4}{2}\right), \quad \text{or} \quad \left(\frac{8}{2}, \frac{1}{2}\right), \quad \text{or} \quad \left(4, \frac{1}{2}\right).$$

The coordinates of the center are $\left(4, \frac{1}{2}\right)$.

Circles

A **circle** is the set of all points in a plane that are a fixed distance r, the radius, from a center (h, k). Thus if a point (x, y) is to be r units from the center, we must have

$$r = \sqrt{(x - h)^2 + (y - k)^2}. \qquad \text{Using the distance formula}$$

Squaring both sides gives an equation of a circle.

The Equation of a Circle

The equation, in standard form, of a circle with center (h, k) and radius r is

$$(x - h)^2 + (y - k)^2 = r^2.$$

Example 3 Find an equation of the circle having radius 5 and center $(3, -7)$.

SOLUTION Using the standard form, we have

$$[x - 3]^2 + [y - (-7)]^2 = 5^2 \qquad \text{Substituting}$$
$$(x - 3)^2 + (y + 7)^2 = 25.$$

Example 4 Graph the circle: $(x + 5)^2 + (y - 2)^2 = 16$.

SOLUTION We write the equation in standard form to determine the center and the radius:

$$[x - (-5)]^2 + [y - 2]^2 = 4^2.$$

The center is $(-5, 2)$ and the radius is 4. We locate the center and draw the circle using a compass.

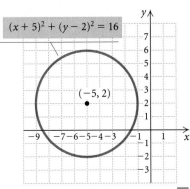

Circles can also be graphed on a grapher.

Example 5   Graph the circle: $x^2 + y^2 = 16$.

SOLUTION   We first solve $x^2 + y^2 = 16$ for y:

$$x^2 + y^2 = 16$$
$$y^2 = 16 - x^2 \qquad \text{Solving for } y^2$$
$$y = \pm\sqrt{16 - x^2}. \qquad \text{Solving for } y$$

We graph both equations $y_1 = \sqrt{16 - x^2}$ and $y_2 = -\sqrt{16 - x^2}$ using the same set of axes and a square viewing window. There are graphers that can draw graphs with a given center and radius directly from the DRAW menu. A square window will still be necessary. Some graphers have the ability to graph an equation like $x^2 + y^2 = 16$ without first solving for y.

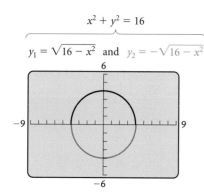

$x^2 + y^2 = 16$

$y_1 = \sqrt{16 - x^2}$ and $y_2 = -\sqrt{16 - x^2}$

A.2 | Exercise Set

Find the distance between each pair of points. Give an exact answer and, where appropriate, an approximation to three decimal places.

1. (4, 6) and (5, 9)

2. (−3, 7) and (2, 11)

3. (6, −1) and (9, 5)

4. (−4, −7) and (−1, 3)

5. $(\sqrt{3}, -\sqrt{5})$ and $(-\sqrt{6}, 0)$

6. $(-\sqrt{2}, 1)$ and $(0, \sqrt{7})$

7. The points (−3, −1) and (9, 4) are the endpoints of the diameter of a circle. Find the length of the radius of the circle.

8. The point (0, 1) is on a circle that has center (−3, 5). Find the length of the diameter of the circle.

Use the distance formula and the Pythagorean theorem to determine whether each set of points could be vertices of a right triangle.

9. (−4, 5), (6, 1), and (−8, −5)

10. (−3, 1), (2, −1), and (6, 9)

11. The points (−3, 4), (0, 5), and (3, −4) are all on the circle $x^2 + y^2 = 25$. Show that these three points are vertices of a right triangle.

12. The points (−3, 4), (2, −1), (5, 2), and (0, 7) are vertices of a quadrilateral. Show that the quadrilateral is a rectangle. (*Hint:* Show that the quadrilateral's opposite sides are the same length and that the two diagonals are the same length.)

13–18. Find the midpoint of each segment having the endpoints given in Exercises 1–6, respectively.

19. Graph the rectangle described in Exercise 12. Then determine the coordinates of the midpoint for each of the four sides. Are the midpoints vertices of a rectangle?

20. Graph the square with vertices (−5, −1), (7, −6), (12, 6), and (0, 11). Then determine the midpoint for each of the four sides. Are the midpoints vertices of a square?

21. The points $(\sqrt{7}, -4)$ and $(\sqrt{2}, 3)$ are endpoints of the diameter of a circle. Determine the center of the circle.

22. The points $(-3, \sqrt{5})$ and $(1, \sqrt{2})$ are endpoints of the diagonal of a square. Determine the center of the square.

In Exercises 23 and 24, how would you change the window so the circle is not distorted? Answers may vary.

23.

$(x + 3)^2 + (y - 2)^2 = 36$

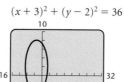

24.

$(x - 4)^2 + (y + 5)^2 = 49$

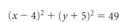
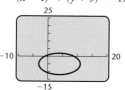

Find an equation for a circle satisfying the given conditions.

25. Center (2, 3), radius of length $\frac{5}{3}$

26. Center (4, 5), diameter of length 8.2

27. Center (−1, 4), passes through (3, 7)

28. Center (6, −5), passes through (1, 7)

29. The points $(7, 13)$ and $(-3, -11)$ are at either end of a diameter.

30. The points $(-9, 4)$, $(-2, 5)$, $(-8, -3)$, and $(-1, -2)$ are vertices of an inscribed square.

31. Center $(-2, 3)$, tangent (touching at one point) to the y-axis

32. Center $(4, -5)$, tangent to the x-axis

Find the center and the radius of each circle. Then graph each circle using a square viewing rectangle.

33. $x^2 + y^2 = 4$

34. $x^2 + y^2 = 81$

35. $x^2 + (y - 3)^2 = 16$
(*Hint:* Solving for y, we get $y = 3 \pm \sqrt{16 - x^2}$.)

36. $(x + 2)^2 + y^2 = 100$

37. $(x - 1)^2 + (y - 5)^2 = 36$

38. $(x - 7)^2 + (y + 2)^2 = 25$

39. $(x + 4)^2 + (y + 5)^2 = 9$

40. $(x + 1)^2 + (y - 2)^2 = 64$

Synthesis

Find the distances between each pair of points and find the midpoint of the segment having the given points as endpoints.

41. $(a, \sqrt{a})$ and $(a + h, \sqrt{a + h})$

42. $\left(a, \dfrac{1}{a}\right)$ and $\left(a + h, \dfrac{1}{a + h}\right)$

43. ◈ Explain how the Pythagorean theorem is used to develop the equation of a circle in standard form.

44. ◈ Explain how you could find the coordinates of a point $\frac{7}{8}$ of the way from point A to point B.

Find an equation of a circle satisfying the given conditions.

45. Center $(2, -7)$ with an area of 36π square units

46. Center $(-5, 8)$ with a circumference of 10π units

47. *Swimming Pool.* A swimming pool is being constructed in the corner of a yard, as shown. Before installation, the contractor needs to know measurements a_1 and a_2. Find them.

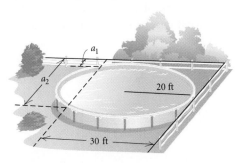

48. *An Arch of a Circle in Carpentry.* Ace Carpentry needs to cut an arch for the top of an entranceway. The arch needs to be 8 ft wide and 2 ft high. To draw the arch, the carpenters will use a stretched string with chalk attached at an end as a compass.

a) Using a coordinate system, locate the center of the circle.

b) What radius should the carpenters use to draw the arch?

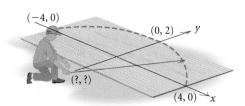

*Determine whether each of the following points lies on the **unit circle**, $x^2 + y^2 = 1$.*

49. $\left(\dfrac{\sqrt{3}}{2}, -\dfrac{1}{2}\right)$

50. $(0, -1)$

51. $\left(-\dfrac{\sqrt{2}}{2}, \dfrac{\sqrt{2}}{2}\right)$

52. $\left(\dfrac{1}{2}, -\dfrac{\sqrt{3}}{2}\right)$

53. Find the point on the y-axis that is equidistant from the points $(-2, 0)$ and $(4, 6)$.

54. Consider any right triangle with base b and height h, situated as shown. Show that the midpoint of the hypotenuse P is equidistant from the three vertices of the triangle.

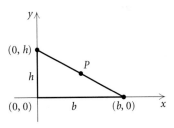

55. Prove the midpoint formula by showing that:

a) $\left(\dfrac{x_1 + x_2}{2}, \dfrac{y_1 + y_2}{2}\right)$ is equidistant from the points (x_1, y_1) and (x_2, y_2); and

b) the distance from (x_1, y_1) to the midpoint plus the distance from (x_2, y_2) to the midpoint equals the distance from (x_1, y_1) to (x_2, y_2).

A.3
Data Analysis, Curve Fitting, and Regression

- *Analyze a set of data to determine whether it can be modeled by a linear, quadratic, cubic, or quartic function.*
- *Use regression to fit a curve to a set of data; then use this model to make predictions.*

Mathematical Models

When a real-world problem can be described in mathematical language, we have a **mathematical model**. For example, the natural numbers constitute a mathematical model for situations in which counting is essential. Situations in which algebra can be brought to bear often require the use of functions.

Mathematical models are abstracted from real-world situations. Procedures within the mathematical model then give results that allow one to predict what will happen in that real-world situation. If the predictions are inaccurate or the results of experimentation do not conform to the model, the model needs to be changed or discarded.

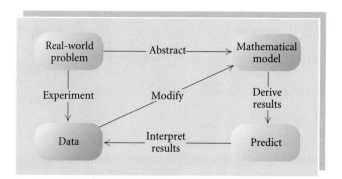

Mathematical modeling can be an ongoing process. For example, finding a mathematical model that will enable an accurate prediction of population growth is not a simple problem. Any population model that one might devise would need to be reshaped as further information is acquired.

Curve Fitting

We will develop and use many kinds of mathematical models in this text. There are many functions that can be used as models. Let's look at four of them.

Constant function:
$y_1 = b$

Linear function:
$y_2 = mx + b$

Squaring function:
$y_3 = x^2$

Quadratic function:
$y_4 = ax^2 + bx + c, a > 0$

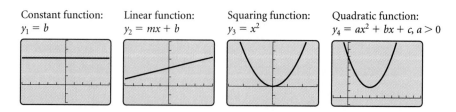

In general, we try to find a function that fits, as well as possible, observations (data), theoretical reasoning, and common sense. We call this **curve fitting**; it is one aspect of mathematical modeling.

Let's first look at some data and related graphs or scatterplots. Do any of the above functions seem to fit either set of data?

Life Expectancy of Women

YEAR, x	LIFE EXPECTANCY, y (IN YEARS)	SCATTERPLOT
0. 1900	49.1	
1. 1910	53.7	
2. 1920	56.3	
3. 1930	61.4	
4. 1940	65.3	
5. 1950	70.9	
6. 1960	73.2	
7. 1970	74.8	
8. 1980	77.5	
9. 1990	78.6	

Life expectancy of women

It appears that the data points can be represented or fitted by a straight line.

The graph is linear.

Source: Statistical Abstract of the United States.

Number of Cellular Phones

YEAR, x	NUMBER OF CELLULAR PHONES, y (IN MILLIONS)	SCATTERPLOT
0. 1985	0.2	
1. 1986	0.4	
2. 1987	1.0	
3. 1988	1.8	
4. 1989	2.8	
5. 1990	4.1	
6. 1991	6.2	
7. 1992	8.6	
8. 1993	13.0	
9. 1994	19.3	

Cellular phone boom

It appears that the data points cannot be represented by a straight line.

The graph is nonlinear.

Source: Cellular Telecommunications Industry Association.

Looking at the scatterplots, we see that the life expectancy data seem to be rising in a manner to suggest that a linear function might fit, although a "perfect" straight line cannot be drawn through the data points. A linear function does not seem to fit the cellular phone data. Part of a quadratic function might be close. In fact, it can be modeled by an exponential function that we will consider in Chapter 6.

The Regression Line

We now consider a method to fit a linear function to a set of data: **linear regression**. Although discussion leading to a complete understanding of this method belongs in a statistics course, we present the procedure here because we can carry it out easily using technology. The grapher gives us the powerful capability to find linear models and to make predictions using them.

Example 1 *Predicting the Life Expectancy of Women.*   Consider the data just presented on the life expectancy of women.

a) Fit a regression line to the data using a REGRESSION feature on a grapher.

b) Use the linear model to predict the life expectancy of women in the year 2000.

SOLUTION

a) Using linear regression, we fit a linear equation to the data. We use a grapher and enter the data into a list, like the following. We select the L_1 list to be the independent variable (x) and the L_2 list to be the dependent variable (y). The grapher can then create a scatterplot of the data, as shown on the right.

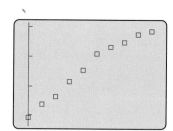

L1	L2	L3
0	49.1	------
1	53.7	
2	56.3	
3	61.4	
4	65.3	
5	70.9	
6	73.2	

L2(7) = 73.2

Consider the data points $(0, 49.1)$, $(1, 53.7)$, $(2, 56.3)$, ..., $(9, 78.6)$ on the graph. We want to fit a **regression line**,

$$y = mx + b,$$

to these data points. We use the REGRESSION feature, LINREG(AX + B), to find the regression line. The result is

$$y = 3.4x + 50.7.$$ Regression line

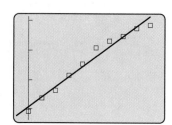

We can then graph the regression line on the same graph as the scatterplot.

b) To predict the life expectancy of women in the year 2000, we substitute the corresponding x-value, $x = 10$, into the formula for the regression function:

$$y = 3.4(10) + 50.7 = 84.7.$$ In 1980, $x = 8$; in 1990, $x = 9$. Thus in 2000, $x = 10$.

We can also use the grapher to do the calculation. Thus we estimate the life expectancy of women in the year 2000 to be 84.7 yr.   ▬

The understanding behind the development of the regression line is best left to a course in calculus and/or statistics.

The Correlation Coefficient

On some graphers, a constant r in $[-1, 1]$, called the **coefficient of linear correlation**, appears with the equation of the regression line. Though we cannot develop a formula for calculating r in this text, keep in mind that r is a number for which $-1 \leq r \leq 1$. It is used to describe the strength of the linear relationship between x and y. The closer $|r|$ is to 1, the better the correlation. A positive value of r also indicates that a positive slope and an increasing function exist. A negative value of r indicates a negative slope and a decreasing function. For the life expectancy data just discussed, $r = 0.9861$, which indicates a very good linear correlation. The following scatterplots summarize the interpretation of a correlation coefficient.

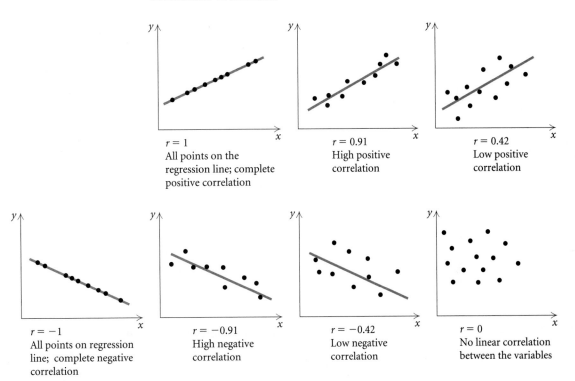

$r = 1$
All points on the
regression line; complete
positive correlation

$r = 0.91$
High positive
correlation

$r = 0.42$
Low positive
correlation

$r = -1$
All points on regression
line; complete negative
correlation

$r = -0.91$
High negative
correlation

$r = -0.42$
Low negative
correlation

$r = 0$
No linear correlation
between the variables

Keep in mind that a high linear correlation coefficient does not necessarily indicate a "cause-and-effect" connection between the variables. We might be able to calculate a high positive correlation between stock prices and rainfall in India, but before we place our life savings in the market, further analysis and common sense should be applied!

Polynomial Curve Fitting

By looking at an input–output table, we can tell whether the data fit a polynomial function. Let's consider an example:

x	y	FIRST DIFFERENCE	SECOND DIFFERENCE	THIRD DIFFERENCE
-11	-1194			
		912		
-7	-282		-640	
		272		384
-3	-10		-256	
		16		384
1	6		128	
		144		384
5	150		512	
		656		384
9	806		896	
		1552		
13	2358			

In the first column, we see that the differences of the x-values are always the same, 4. Next, we take the first differences of the y-values, but these are not constant. We continue, taking the second differences; these are still not constant. The third differences, however, are all constant—the number 384. The following theorem tells us that the data in this table fit a third-degree, or cubic, polynomial.

The Polynomial Difference Theorem

A function f is a polynomial function of degree n if and only if for any set of x-values that differ by the same number, the nth differences of the corresponding y-values are the same nonzero constant.

Thus we can look at an input–output table and know for sure whether the data fit a polynomial function. Unfortunately, real-world data do not always make such a perfect fit. We can still find quadratic, cubic, and quartic polynomials that fit the data approximately. The remainder of this section presumes the use of a grapher that does quadratic, cubic, and quartic regression.

As we have moved through this text, we have developed a "stable" of functions that can serve as models for many applications. Let's now add quadratic, cubic, and quartic functions to this stable.

Linear Function:
$y_1 = mx + b$

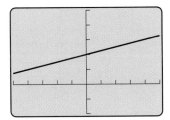

Quadratic Function:
$y_2 = ax^2 + bx + c, a > 0$

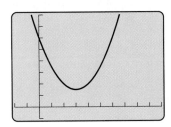

Quadratic Function:
$y_3 = ax^2 + bx + c, a < 0$

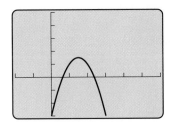

Cubic Function:
$y_4 = ax^3 + bx^2 + cx + d$

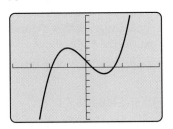

Quartic Function:
$y_5 = ax^4 + bx^3 + cx^2 + dx + e$

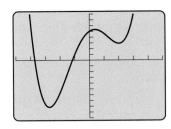

Now let's consider some real-world data. How can we decide which function might fit the data? Our simple way is to graph the data and look over the result. Simply stated, we "eyeball" the situation. Then we use a grapher to find a regression equation and make predictions.

Example 2 *Hours of Sleep versus Death Rate.* In a study by Dr. Harold J. Morowitz of Yale University, data were gathered that showed the relationship between the death rate of men and the average number of hours per day that the men slept.

Death Rate Related to Sleep

AVERAGE NUMBER OF HOURS OF SLEEP, x	DEATH RATE PER 100,000 MALES, y
5	1121
6	805
7	626
8	813
9	967

Source: "Hiding in the Hammond Report," *Hospital Practice*, by Harold J. Morowitz.

a) Make a scatterplot of the data.

b) Determine which, if any, of the functions seems to fit the data.

c) Use a grapher to fit the function to the data. Graph the equation using the same axes as the scatterplot.

d) Predict the death rate of males who sleep 4 hr, 5.5 hr, 7.5 hr, and 10 hr.

SOLUTION

a) We make a scatterplot of the data as shown on the left below.

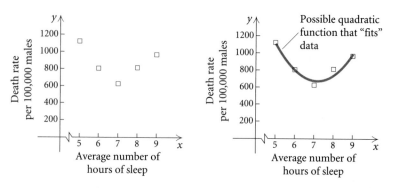

b) Note that the rate drops and then rises. This suggests that a quadratic function might fit the data. See the graph on the right above.

c) Using the quadratic REGRESSION feature, we get the following quadratic function:

$$f(x) = y = 93.28571429x^2 - 1336x + 5460.828571.$$

Depending on the accuracy required, we might round these coefficients, though it is easy to keep this equation in the grapher.

d) We then compute the outputs:

$$f(4) \approx 1609.4, \qquad f(7.5) \approx 688.1,$$
$$f(5.5) \approx 934.7, \qquad f(10) \approx 1429.4.$$

These can also be found with the TABLE feature set in ASK mode, after copying the quadratic function to the $y=$ screen. The results assert, for instance, that the death rate is about 688 per 100,000 for males who sleep 7.5 hr per night.

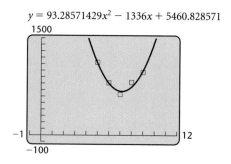

$$y = 93.28571429x^2 - 1336x + 5460.828571$$

One must always be aware that a model such as the one in Example 2 has its limitations.

Interactive Discovery

In the model of Example 2, find and interpret $f(0)$ and $f(24)$. Argue whether such computations have meaning in the real world. What about $f(-1)$ and $f(25)$?

A.3 Exercise Set

In Exercises 1–4, determine whether the graph might be modeled by a linear function.

1.

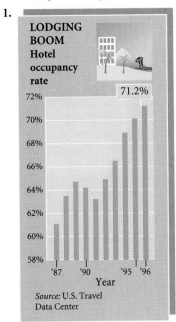

LODGING BOOM
Hotel occupancy rate

71.2%

72%
70%
68%
66%
64%
62%
60%
58%

'87   '90   '95 '96
Year

Source: U.S. Travel Data Center

2.

OIL DEMAND
In millions of barrels a day

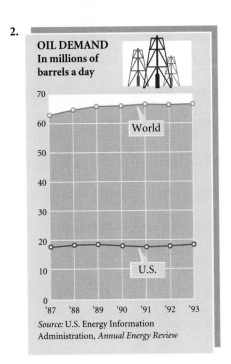

70
60   World
50
40
30
20   U.S.
10
0
'87 '88 '89 '90 '91 '92 '93

Source: U.S. Energy Information Administration, *Annual Energy Review*

3.

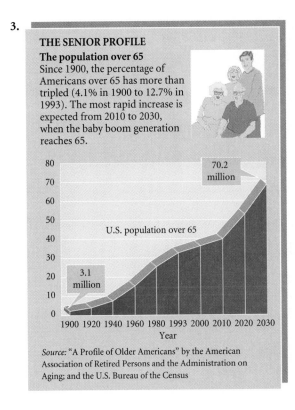

THE SENIOR PROFILE
The population over 65
Since 1900, the percentage of Americans over 65 has more than tripled (4.1% in 1900 to 12.7% in 1993). The most rapid increase is expected from 2010 to 2030, when the baby boom generation reaches 65.

80
70   70.2 million
60
50
40   U.S. population over 65
30
20   3.1
10   million
0
1900 1920 1940 1960 1980 1993 2000 2010 2020 2030
Year

Source: "A Profile of Older Americans" by the American Association of Retired Persons and the Administration on Aging; and the U.S. Bureau of the Census

4.

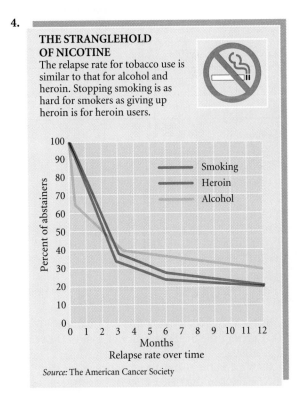

THE STRANGLEHOLD OF NICOTINE
The relapse rate for tobacco use is similar to that for alcohol and heroin. Stopping smoking is as hard for smokers as giving up heroin is for heroin users.

100
90   Smoking
80   Heroin
70   Alcohol
60
50
40
30
20
10
0
0 1 2 3 4 5 6 7 8 9 10 11 12
Months
Relapse rate over time

Percent of abstainers

Source: The American Cancer Society

In Exercises 5–8, determine whether the scatterplot might be fit by a linear model.

5.

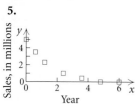

6.

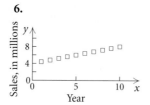

7.

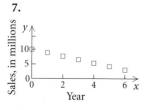

8.

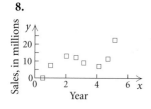

Exercises 9–14 involve creating regression lines on a grapher.

9.
Life Expectancy of Men

YEAR, x	LIFE EXPECTANCY, y (IN YEARS)	SCATTERPLOT
0. 1900	49.6	
1. 1910	50.2	
2. 1920	54.6	Life expectancy of men
3. 1930	58.0	
4. 1940	60.9	
5. 1950	65.3	
6. 1960	66.6	
7. 1970	67.1	
8. 1980	69.9	
9. 1990	71.8	

Source: Statistical Abstract of the United States.

a) Fit a regression line to the data and use it to predict the life expectancy of men in the year 2000.
b) What is the correlation coefficient for the regression line? How close a fit is the regression line?

10.
Cases of Skin Cancers in the United States

YEAR, x	NUMBER OF CASES, y (IN THOUSANDS)	SCATTERPLOT
0. 1988	5800	Skin cancer on the rise
1. 1989	6000	
2. 1990	6300	
3. 1991	6500	
4. 1992	6700	

Source: American Cancer Society.

a) Fit a regression line to the data and use it to predict the number of cases of skin cancer in the United States in 1998, 2000, and 2010.
b) What is the correlation coefficient for the regression line? How good a predictor is the regression line?

11.   *The Cost of Tuition at State Universities*

COLLEGE YEAR	ESTIMATED TUITION
0. (1997–1998)	$ 9,838
1. (1998–1999)	10,424
2. (1999–2000)	11,054
3. (2000–2001)	11,717
4. (2001–2002)	12,420

Source: College Board, Senate Labor Committee.

a) Fit a regression line to the data and use it to predict the cost of college tuition in 2004–2005, 2006–2007, and 2010–2011.
b) What is the correlation coefficient for the regression line? How close a fit is the regression line?
c) A college president asserts that in the college year 2006–2007, the cost of tuition will be $16,619. How well does your estimate compare?

12.   *Book Buying Growth in the United States*

YEAR, x	BOOK SALES, y (IN BILLIONS)
0. 1992	$21
1. 1993	23
2. 1994	24
3. 1995	25
4. 1996	26

Source: Book Industry Trends 1995.

a) Fit a regression line to the data and use it to predict book sales in 1998, 2000, and 2010.
b) What is the correlation coefficient for the regression line? How close a fit is the regression line?
c) A bookstore owner asserts that in the year 2000, book sales will be $45 billion. How well does your estimate stack up to this?

13. *Maximum Heart Rate.* A person who is exercising should not exceed his or her maximum heart rate, which is determined on the basis of that person's sex, age, and resting heart rate. The following table relates resting heart rate and maximum heart rate for a 20-yr-old man.

RESTING HEART RATE, H (IN BEATS PER MINUTE)	MAXIMUM HEART RATE, M (IN BEATS PER MINUTE)
50	166
60	168
70	170
80	172

Source: American Heart Association.

a) Fit a regression line $M = mH + b$ to the data.
b) Predict a maximum heart rate if the resting heart rate is 40, 65, 76, and 84.
c) What is the correlation coefficient? How confident are you about using the regression line as a predictor?

14. *Study Time versus Grades.* A math instructor asked her students to keep track of how much time each spent studying a chapter on functions in her algebra–trigonometry course. She collected the information together with test scores from that chapter's test. The data are listed in the table below.

STUDY TIME, x (IN HOURS)	TEST GRADE, y (IN PERCENT)
23	81
15	85
17	80
9	75
21	86
13	80
16	85
11	93

a) Fit a regression line to the data.
b) Predict a student's score if he or she studies 24 hr, 6 hr, and 18 hr.

c) What is the correlation coefficient? How confident are you about using the regression line as a predictor?

For the scatterplots and graphs in Exercises 15–23, determine which, if any, of the following functions might be used as a model for the data.
a) *Linear,* $f(x) = mx + b$
b) *Quadratic,* $f(x) = ax^2 + bx + c,\ a > 0$
c) *Quadratic,* $f(x) = ax^2 + bx + c,\ a < 0$
d) *Polynomial, not quadratic or linear*

15.
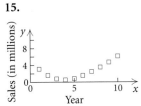

16.
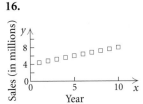

17.
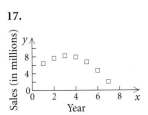

18.
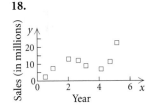

19.
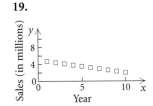

20.
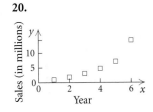

21.
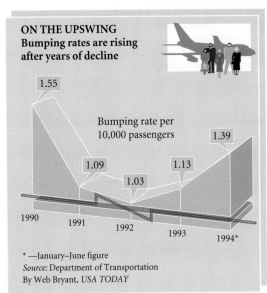

22.

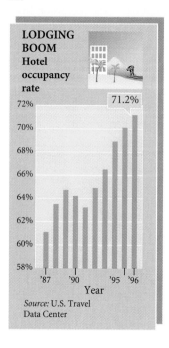

LODGING
BOOM
Hotel
occupancy
rate

71.2%

72%
70%
68%
66%
64%
62%
60%
58%

'87   '90   '95 '96
Year

Source: U.S. Travel
Data Center

23.

DRIVER FATALITIES BY AGE
Number of licensed drivers per 100,000 who died in
motor vehicle accidents in 1990. The fatality rates for
both the 70–79 group and 80+ age group were lower than
for the 15- to 24-year-olds.

Number of driver deaths per 100,000

35
30 — 28
25
20
15 — 15   15
10 —   10
5
0

15–24 25–39 40–69 70–79 80+
Ages

25

Source: National Highway Traffic Administration

*For each table of data in Exercises 24–27, answer the
following questions.*

a) *Is the change in the inputs x the same?*
b) *Find the first differences of the y-values. Is the change
in the outputs y the same?*
c) *Find the second, third, and fourth differences, as
needed. Is any set of differences nonzero and
constant?*
d) *Does a quadratic, cubic, or quartic function fit the
data?*
e) *If so, fit a function to the data using the* REGRESSION
feature of a grapher.
f) *Then find the missing data.*

24.

x	y
1	7
6	47
11	137
16	277
21	467
26	707
31	997
36	?
?	2657

25.

x	y
−11	1204
−7	292
−3	20
1	4
5	−140
9	−796
13	−2348
17	?
21	?

26.

x	y
10	37
12	42.6
14	47.4
16	51.4
18	54.6
20	57
22	58.6
24	?
?	45.1

27.

x	y
−8	3447
−5	369
−2	−27
1	−9
4	99
7	1917
10	9009
12	?
?	−19.94

28. *Airline Passenger Bumping.* Use the data in the
graph of Exercise 21.

a) Use the REGRESSION feature of a grapher to fit a
quadratic function to the data.
b) Predict the rate of passenger bumping in 1998,
2000, and 2005.

Shoe Size and Life Expectancy. *The data in the
following table are the result of a Swedish study relating
one's life expectancy to the size of one's feet. Use these
data in Exercises 29 and 30.*

SHOE SIZE	LIFE EXPECTANCY FOR WOMEN (IN YEARS)	SHOE SIZE	LIFE EXPECTANCY FOR MEN (IN YEARS)
4	69	5	66
5	76	6	69
6	82	7	69
7	84	8	72
8	75	9	75
9	72	10	77
10	70	11	82
11	69	12	79
		13	72
		14	69

Source: *Orthopedic Quarterly.*

29. a) Use the REGRESSION feature of a grapher to fit a quadratic function to the data for women.
  b) Estimate the life expectancy of women with shoe sizes of $5\frac{1}{2}$ and $8\frac{1}{2}$.

30. a) Use the REGRESSION feature of a grapher and fit a quadratic function to the data for men.
  b) Estimate the life expectancy of men with shoe sizes of $10\frac{1}{2}$ and $15\frac{1}{2}$.

31. *Prices of Personal Computers.* The price P of a personal computer has varied greatly in recent years. The data in the table relate price P, in dollars, to time t, in years, where $t = 1$ corresponds to 1981.

YEAR, t	AVERAGE PRICE P OF A PERSONAL COMPUTER
1981 ($t = 1$)	$2290
1982 ($t = 2$)	1500
1983 ($t = 3$)	1400
1984 ($t = 4$)	1850
1985 ($t = 5$)	2540
1986 ($t = 6$)	2400
1987 ($t = 7$)	1720
1988 ($t = 8$)	1860
1989 ($t = 9$)	1930
1990 ($t = 10$)	2100
1991 ($t = 11$)	1820
1992 ($t = 12$)	1640

  a) Find a cubic function $y = ax^3 + bx^2 + cx + d$ that fits the data.

  b) Graph the cubic function of part (a).
  c) Use the cubic function to predict the price of a personal computer in 1998.
  d) Find a quartic polynomial function $y = ax^4 + bx^3 + bx^2 + cx + d$ that fits the data.
  e) Graph the quartic function of part (d).
  f) Use the quartic function to predict the price of a personal computer in 1998.

32. *Damage to the Ozone Layer.* The concentration of chlorine compounds in the stratosphere serves as an indicator of damage to the ozone layer. The data in the following table show the relationship of estimated chlorine concentration in the atmosphere, in parts per billion (ppb), to the year.

YEAR	CHLORINE CONCENTRATION (IN PARTS PER BILLION)
1985	2.5
1995	3.3
2010	3.9
2035	3.7

Source: Adapted by U.S. EPA by NRDC Earth Action Guide, "Saving the Ozone."

  a) Use the REGRESSION feature on a grapher to find a linear, a quadratic, a cubic, and a quartic function to fit the data.
  b) It is estimated that in 2060 the chlorine concentration will be 3.3 ppb. Which function in part (a) would best make this prediction?
  c) Use the answer to part (b) to predict the chlorine concentration in 2085.

Synthesis

33. ◈ Why does a correlation coefficient of 1 not guarantee the existence of a cause-and-effect relationship?

34. ◈ Conduct your own experiment. Use some kind of measuring device and gather data relating body height to arm length. Fit a regression line to the data and find the correlation coefficient. How confident would you be about using these data to predict Shaquille O'Neal's arm length, knowing that his height is 7'0"?

35. *Planet Diameters versus Distance from the Sun.* The following table lists the diameters of the planets and their distances from the sun.

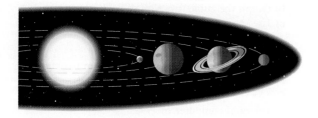

PLANET	DIAMETER, d (IN KILOMETERS)	DISTANCE FROM SUN, D (IN ASTRONOMICAL UNITS)*
Mercury	5,000	0.4
Venus	12,000	0.7
Earth	13,000	1.0
Mars	7,000	1.6
Jupiter	143,000	5.2
Saturn	120,000	10.0
Uranus	51,000	19.6
Pluto	5,000	30.8
Neptune	49,000	40.8

*1 astronomical unit (A.U.) = 150 million km.
Source: Encyclopedia Britannica.

a) Find a regression line $D = md + b$, expressing D as a function of d. What is the correlation coefficient?

b) Using a space probe, astronomers discover a new planet that has a diameter of 180,000 km. How confident would you be of using the function of part (a) to predict the distance of this planet from the sun?

36. *Ted Williams and the War Years.* Ted Williams played for the Boston Red Sox from 1939–1960. Many credit him with being the greatest hitter of all time. Unfortunately, his career totals are somewhat less impressive than others because his career was interrupted from 1943–1945 for World War II and from 1952–1953 for the Korean War. Some assert that if he had been playing during the war years, he would have broken Hank Aaron's home run record of 755 and runs batted in (RBI) record of 2297. Below are Williams' statistics.

YEAR, x	HOME RUNS, H	RBIs, R
1939	31	145
1940	23	113
1941	37	120
1942	36	137
1943	0*	0*
1944	0*	0*
1945	0*	0*
1946	38	123
1947	32	114
1948	25	127
1949	43	159
1950	28	97
1951	30	126
1952	1[†]	3[†]
1953	13[†]	34[†]
1954	29	89
1955	28	83
1956	24	82
1957	38	87
1958	26	85
1959	10	43
1960	29	72

*World War II
[†]Korean War

a) Excluding all data from the war years, fit a linear regression equation $H = mx + b$ to the data regarding the number of home runs. Then use the equation to predict how many home runs Williams would have hit in each of the war years. What is the correlation coefficient? How confident are you of your prediction?

b) Would Williams have broken Aaron's home run record?

c) Excluding all data from the war years, fit a linear regression equation $R = mx + b$ to the data regarding the number of RBIs. Then use the equation to predict how many RBIs Williams would have had in each of the war years. What is the correlation coefficient? How confident are you of your prediction?

d) Would Williams have broken Aaron's RBI record?

For each of the two input–output tables in Exercises 37 and 38:

a) Fit a cubic function to the existing data. Then use that equation to find the missing values.
b) Fit a quartic function to the existing data. Then use that equation to find the missing values.

37.

x	y
−4	?
−3	7
−2	11
−1	33
0	45
1	19
?	23
3	46
4	?

38.

x	y
20	12.4
?	24.3
30	28.5
40	19.6
?	39.2
50	78.4
60	196.8
70	793.6
80	?

39. *The Effect of Advertising on Movie Revenue.* The following table contains actual data pertaining to the box office revenue of certain movies together with the amount of money that was spent to advertise each movie. We want to explore whether box office revenue is directly related to advertising expenditures.

MOVIE	ADVERTISING BUDGET (IN MILLIONS)	BOX OFFICE REVENUE (IN MILLIONS)
The Lion King	$23.3	$300.4
Forrest Gump	25.0	298.5
True Lies	20.5	146.3
The Santa Clause	19.4	137.8
The Flintstones	10.6	130.6
Clear & Present Danger	17.9	121.8
Speed	17.2	121.2
The Mask	11.2	118.8
Maverick	13.8	101.6
Interview with the Vampire	15.4	100.7

Source: The Hollywood Reporter.

a) Using a ZOOMSTAT feature, make a scatterplot of the data.
b) Analyze the scatterplot and try to decide whether a linear, quadratic, cubic, or quartic equation might fit the data.
c) Use the REGRESSION feature on a grapher to find a quadratic, a cubic, and a quartic function to fit the data. Assume that advertising is the independent variable.
d) Plot each regression equation with the scatterplot. Examine the results and try to decide which function fits best.
e) You are about to make a blockbuster movie called *A Day in the Life of a Math Professor.* You decide to spend $30 million to advertise the film. Use each of the three functions found in part (c) to predict the box office revenue from the movie.

Answers

Introduction to Graphs and Graphers

1.

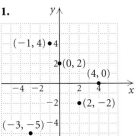

3.

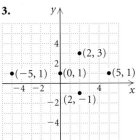

19.

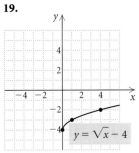

21.

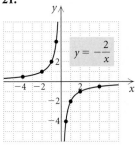

23. (f) **25.** (c) **27.** (b) **29.** (d)

5. Yes; no **7.** No; yes **9.** No; yes

11.

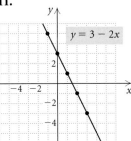

13.

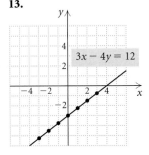

31.

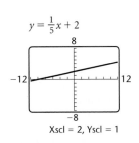

$y = \frac{1}{5}x + 2$

Xscl = 2, Yscl = 1

33.

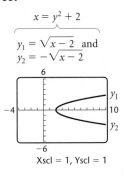

$x = y^2 + 2$

$y_1 = \sqrt{x-2}$ and
$y_2 = -\sqrt{x-2}$

Xscl = 1, Yscl = 1

15.

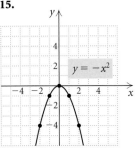

17.

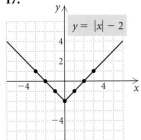

35.

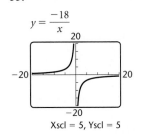

$y = \frac{-18}{x}$

Xscl = 5, Yscl = 5

37.

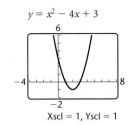

$y = x^2 - 4x + 3$

Xscl = 1, Yscl = 1

A-1

39.

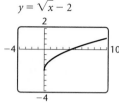

$y = \sqrt{x} - 2$

Xscl = 1, Yscl = 0.5

41.

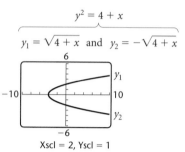

$y^2 = 4 + x$

$y_1 = \sqrt{4 + x}$ and $y_2 = -\sqrt{4 + x}$

Xscl = 2, Yscl = 1

43. (a) Does not show curvature; (b) best; (c) tick marks too close; (d) does not show curvature; ticks too close
45. (a) Does not show curvature; (b) best; (c) does not show curvature as well as (b); (d) cannot see the graph
47. Error, error, .6245, 1, 1.261, 1.4697, 1.6462; −32, −40, −53.33, −80, −160, error, 160   **49.** Yes   **51.** No
53. Yes   **55.** No   **57.** Yes   **59.** −8   **61.** −0.882
63. −5.8   **65.** 20.376   **67.** −0.104   **69.** ±1.959
71. 1.489, 5.673   **73.** −2, 3
75. (−4.378, 0), (2.545, 0), (−1.167, 0)
77. (−0.929, −3.388), (1.080, −2.240), (2.848, −1.229)
79. (−2.809, −2.528), (0, 0), (2.809, 2.528)
81. Answer depends on the grapher and its features.
83. $y = x^2 + 1$; x: ±9, ±11; y: 26

Chapter 1

EXERCISE SET 1.1

1. Yes   **3.** Yes   **5.** No   **7.** Yes   **9.** Yes   **11.** Yes
13. No   **15.** Function; domain: {2, 3, 4}; range: {10, 15, 20}
17. Not a function; domain: {−7, −2, 0}; range: {3, 1, 4, 7}
19. Function; domain: {−2, 0, 2, 4, −3}; range: {1}
21. $f(-1) = 2$; $f(0) = 0$; $f(1) = -2$
23. (a) 1; (b) 6; (c) 22; (d) $3x^2 + 2x + 1$; (e) $3t^2 - 4t + 2$; (f) $6a + 3h - 2$
25. (a) 8; (b) −8; (c) $-x^3$; (d) $27y^3$; (e) $8 + 12h + 6h^2 + h^3$; (f) $3a^2 + 3ah + h^2$
27. (a) $\dfrac{1}{8}$; (b) 0; (c) does not exist; (d) $\dfrac{81}{53}$; (e) $\dfrac{x + h - 4}{x + h + 3}$; (f) $\dfrac{7}{(x + h + 3)(x + 3)}$
29. 0; does not exist; does not exist as a real number; $\dfrac{1}{\sqrt{3}}$ or $\dfrac{\sqrt{3}}{3}$ **31.** All real numbers   **33.** $\{x \mid x \neq 0\}$

35. $\left\{ x \mid x \geq -\frac{4}{7} \right\}$ **37.** $\{x \mid x \neq 5 \text{ and } x \neq -1\}$
39. $\{x \mid -3 \leq x \leq 3\}$
41. Domain: all real numbers; range: $[0, \infty)$
43. Domain: $[-3, 3]$; range: $[0, 3]$
45. Domain: all real numbers; range: all real numbers
47. Domain: $(-\infty, 7]$; range: $[0, \infty)$
49. Domain: $(-\infty, 0) \cup (0, \infty)$; range: $(-\infty, 0) \cup (0, \infty)$
51. No   **53.** Yes   **55.** Yes   **57.** No
59. Function; domain: $[0, 5]$; range: $[0, 3]$
61. Function; domain: $[-2\pi, 2\pi]$; range: $[-1, 1]$
63. (a) The domain of y_1 does not include the intervals $(-\infty, -1.2]$ and $[3.2, \infty)$. The range of y_1 appears to be $[0, 2]$. The domain of y_2 does not include 3.2. The range of y_2 appears to be all real numbers. (b) The largest output of y_1 seems to be 2, and the smallest 0. There does not appear to be a largest or smallest output of y_2.
65. 645 m; 0 m   **67.** $\left\{\frac{15}{2}\right\}$ **69.** $\left\{\frac{53}{2}\right\}$ **71.** ◈
73. $\dfrac{1}{\sqrt{x + h} + \sqrt{x}}$ **75.** $f(x) = x$, $g(x) = x + 1$
77. 35

EXERCISE SET 1.2

1. −1.532, −0.347, 1.879   **3.** −1.414, 0, 1.414   **5.** −1, 0, 1
7. 1.5, 9.5   **9.** −2, 3
11. All numbers in the interval $[-1, 2]$
13. (a) $[-5, 1]$; (b) $[3, 5]$; (c) $[1, 3]$
15. (a) $[-3, -1]$, $[3, 5]$; (b) $[1, 3]$; (c) $[-5, -3)$
17. Increasing: $[0, \infty)$; decreasing: $(-\infty, 0]$; relative minimum: 0 at $x = 0$
19. Increasing: $(-\infty, 0]$; decreasing: $[0, \infty)$; relative maximum: 5 at $x = 0$
21. Increasing: $[3, \infty)$; decreasing: $(-\infty, 3]$; relative minimum: 1 at $x = 3$
23. Increasing: $[1, 3]$; decreasing: $(-\infty, 1]$, $[3, \infty)$; relative maximum: −4 at $x = 3$; relative minimum: −8 at $x = 1$
25. Increasing: $[-1.552, 0]$, $[1.552, \infty)$; decreasing: $(-\infty, -1.552]$, $[0, 1.552]$; relative maximum: 4.07 at $x = 0$; relative minima: −2.314 at $x = -1.552$, −2.314 at $x = 1.552$
27. (a)

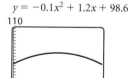

$y = -0.1x^2 + 1.2x + 98.6$

(b) 102.2; (c) 6 days after the illness began, the temperature is 102.2°F.
29. Increasing: $[-1, 1]$; decreasing: $(-\infty, -1]$, $[1, \infty)$; relative maximum: 4 at $x = 1$; relative minimum: −4 at $x = -1$
31. Increasing: $[-1.414, 1.414]$; decreasing: $[-2, -1.414]$, $[1.414, 2]$; relative maximum: 2 at $x = 1.414$; relative minimum: −2 at $x = -1.414$

33.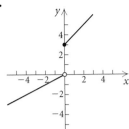

35.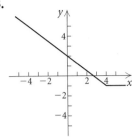

47.
$$y = \begin{cases} \sqrt[3]{x + 27}, & \text{for } x < 1, \\ \left| 2 - \dfrac{x}{5} \right|, & \text{for } x \geq 1 \end{cases}$$

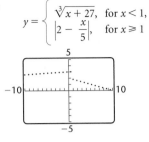

49. $d(t) = \sqrt{(120t)^2 + (400)^2}$ **51.** $A(x) = 24x - x^2$

53. $A(w) = 10w - \dfrac{w^2}{2}$ **55.** $d(s) = \dfrac{14}{s}$

57. (a) $V(x) = 4x^3 - 48x^2 + 144x$, or $4x(x - 6)^2$;
(b) $\{x \mid 0 < x < 6\}$;
(c) $y = 4x^3 - 48x^2 + 144x$ **(d)** 8 cm by 8 cm by 2 cm

37.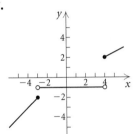

39.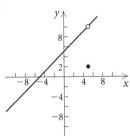

59. (a) $A(x) = x\sqrt{256 - x^2}$; **(b)** $\{x \mid 0 < x < 16\}$;
(c)

$y = x\sqrt{256 - x^2}$

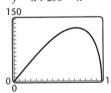

41.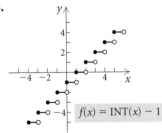

$f(x) = \text{INT}(x)$

(d) 11.314 ft by 11.314 ft   **61.** 7   **63.** ◈
65. Increasing: $[-1.933, 1.933]$; decreasing: $(-\infty, -1.933]$, $[1.933, \infty)$; relative maximum: 58.242 at $x = 1.933$; relative minimum: -58.242 at $x = -1.933$
67. Increasing: $[-5, -2]$, $[4, \infty)$; decreasing: $(-\infty, -5]$, $[-2, 4]$; relative maximum: 560 at $x = -2$; relative minima: 425 at $x = -5$, -304 at $x = 4$
69. (a) C

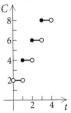

43.

$f(x) = \text{INT}(x) - 1$

(b) $C(t) = 2(\text{INT}(t) + 1)$, $t > 0$
71. $\{x \mid -5 \leq x < -4 \text{ or } 5 \leq x < 6\}$
73. (a) $h(r) = \dfrac{30 - 5r}{3}$; **(b)** $V(r) = \pi r^2 \left(\dfrac{30 - 5r}{3} \right)$;
(c) $V(h) = \pi h \left(\dfrac{30 - 3h}{5} \right)^2$

45.
$$y = \begin{cases} \sqrt[3]{x}, & \text{for } x \leq -1, \\ x^2 - 3x, & \text{for } -1 < x < 4, \\ \sqrt{x - 4}, & \text{for } x \geq 4 \end{cases}$$

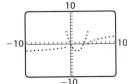

EXERCISE SET 1.3

1. (a) Yes; **(b)** yes; **(c)** yes   **3. (a)** Yes; **(b)** no; **(c)** no
5. $-\frac{3}{5}$ **7.** $\frac{1}{5}$ **9.** $-\frac{5}{3}$ **11.** Not defined   **13.** $2a + h$
15. 6.7% grade; $y = 0.067x$
17. $\frac{3}{5}$; $(0, -7)$
19. $-\frac{1}{2}$; $(0, 5)$
21. $-\frac{3}{2}$; $(0, 5)$
23. $\frac{4}{3}$; $(0, -7)$
25. $y = \frac{2}{9}x + 4$ **27.** $y = -4x - 7$
29. $y = -4.2x + \frac{3}{4}$ **31.** $y = \frac{2}{9}x + \frac{19}{3}$
33. $y = 3x - 5$ **35.** $y = -\frac{3}{5}x - \frac{17}{5}$
37. $y = -3x + 2$ **39.** $y = -\frac{1}{2}x + \frac{7}{2}$
41. Parallel   **43.** Perpendicular
45. $y = \frac{2}{7}x + \frac{29}{7}$; $y = -\frac{7}{2}x + \frac{31}{2}$
47. $y = -0.3x - 2.1$; $y = \frac{10}{3}x + \frac{70}{3}$
49. $y = -\frac{3}{4}x + \frac{1}{4}$; $y = \frac{4}{3}x - 6$ **51.** $x = 3$; $y = -3$
53. (a) $W(h) = 3.5h - 110$; **(b)** 107 lb;
(c) $y = 3.5x - 110$

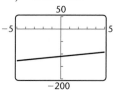

(d) $\{h \mid h > 31.43\}$, or $(31.43, \infty)$
55. (a) 115, 75, 135, 179; **(b)** $y = 2x + 115$

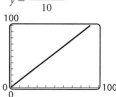

(c) Below $-57.5°$, stopping distance is negative; above $32°$, ice doesn't form.
57. (a) $\frac{11}{10}$. For each mile per hour faster that the car travels, it takes $\frac{11}{10}$ ft longer to stop; **(b)** 6, 11.5, 22.5, 55.5, 72;
(c)
$$y = \frac{11x + 5}{10}$$

(d) $\{r \mid r > 0\}$, or $(0, \infty)$. If r is allowed to be 0, the function says that a stopped car travels $\frac{1}{2}$ ft before stopping.

59. $\dfrac{\text{Height of corn in feet}}{\text{Number of weeks after planting}}$

61. $C(t) = 60 + 40t$; $y = 60 + 40x$

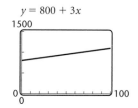

63. $C(x) = 800 + 3x$; $y = 800 + 3x$

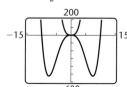

65. 16   **67.** 40   **69.** $a^2 + 3a$ **71.** ◆   **73.** -7.75
75. $C(F) = \frac{5}{9}(F - 32)$; $F(C) = \frac{9}{5}C + 32$ **77.** False
79. False   **81.** $f(x) = x + b$

EXERCISE SET 1.4

1. x-axis, no; y-axis, yes; origin, no
3. x-axis, yes; y-axis, no; origin, no
5. x-axis, no; y-axis, no; origin, yes
7. x-axis, no; y-axis, yes; origin, no
9. x-axis, no; y-axis, no; origin, no
11. x-axis, no; y-axis, yes; origin, no
13. x-axis, no; y-axis, no; origin, yes
15. x-axis, no; y-axis, no; origin, yes
17. x-axis, yes; y-axis, yes; origin, yes
19. x-axis, no; y-axis, yes; origin, no
21. x-axis, yes; y-axis, yes; origin, yes
23. x-axis, no; y-axis, no; origin, no
25. x-axis, no; y-axis, no; origin, yes   **27.** Even   **29.** Odd
31. Neither   **33.** Odd   **35.** Even   **37.** Odd   **39.** Even
41. Odd   **43.** Neither   **45.** Odd   **47.** Neither   **49.** Even
51. $3a^2 - 32a + 85$ **53.** $3a^2 + 22a + 40$ **55.** ◆
57. Odd   **59.** Neither   **61.** Odd
63. x-axis, yes; y-axis, yes; origin, yes
65. x-axis, yes; y-axis, no; origin, no
67.

$$y_1 = \frac{1}{2}x^4 - 5x^3 + 2,$$
$$y_2 = \frac{1}{2}x^4 + 5x^3 + 2$$

The graph of $f(x)$ is the reflection of the graph of $g(x)$ across the y-axis.

EXERCISE SET 1.5

1. Start with the graph of $g(x) = x^2$. Shift it up 1 unit.

$$y = x^2 + 1$$

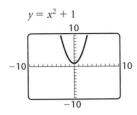

3. Start with the graph of $g(x) = 1/x$. Shift it up 3 units.

$$y = \frac{1}{x} + 3$$

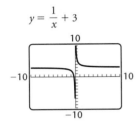

5. Start with the graph of $g(x) = \sqrt{x}$. Shift it down 5 units.

$$y = \sqrt{x} - 5$$

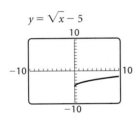

7. Start with the graph of $g(x) = x^2$. Reflect it across the x-axis.

$$y = -x^2$$

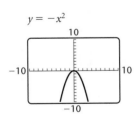

9. Start with the graph of $g(x) = |x|$. Shift it right 3 units.

$$y = |x - 3|$$

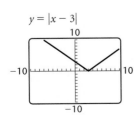

11. Start with the graph of $f(x) = x^3$. Shift it left 5 units.

$$y = (x + 5)^3$$

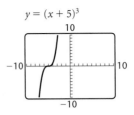

13. Start with the graph of $f(x) = x^2$. Shift it left 1 unit.

$$y = (x + 1)^2$$

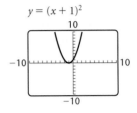

15. Start with the graph of $g(x) = \sqrt{x}$. Shrink it vertically by multiplying each y-coordinate by $\frac{1}{2}$.

$$y = \frac{1}{2}\sqrt{x}$$

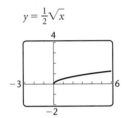

17. Start with the graph of $f(x) = 1/x$. Stretch it vertically by multiplying each y-coordinate by 2.

$$y = \frac{2}{x}$$

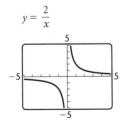

19. Start with the graph of $g(x) = \sqrt{x}$. Stretch it vertically by multiplying each y-coordinate by 3, then shift it down 5 units.

$$y = 3\sqrt{x} - 5$$

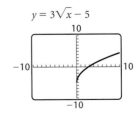

21. Start with the graph of $g(x) = x^2$. Shrink it vertically by multiplying each y-coordinate by $\frac{1}{2}$, then shift it right 3 units.

$$y = \frac{1}{2}(x - 3)^2$$

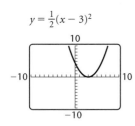

23. Start with the graph of $f(x) = |x|$. Stretch it horizontally by multiplying each x-coordinate by 3, then shift it down 4 units.

$$y = \left|\frac{1}{3}x\right| - 4$$

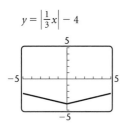

25. Start with the graph of $g(x) = x^2$. Shift it left 5 units and down 4 units.

$$y = (x + 5)^2 - 4$$

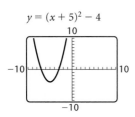

27. Start with the graph of $g(x) = x^2$. Shift it right 5 units, shrink it vertically by multiplying each y-coordinate by $\frac{1}{4}$, and reflect it across the x-axis.

$$y = -\frac{1}{4}(x - 5)^2$$

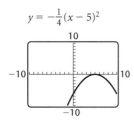

29. Start with the graph of $g(x) = 1/x$. Shift it left 3 units and up 2 units. Use DOT mode.

$$y = \frac{1}{x + 3} + 2$$

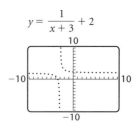

31. Start with the graph of $g(x) = x^3$. Shift it left 4 units and up 3 units.

$$y = (x + 4)^3 + 3$$

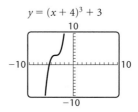

33. Start with the graph of $g(x) = \sqrt{x}$. Reflect it across the y-axis; then shift it up 5 units.

$$y = \sqrt{-x} + 5$$

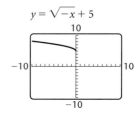

35. Start with the graph of $g(x) = x^2$. Shift it left 4 units, stretch it vertically by multiplying each y-coordinate by 3, and then shift it down 3 units.

$$y = 3(x + 4)^2 - 3$$

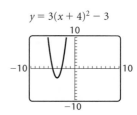

37. Start with the graph of $g(x) = x^3$. Shift it right 3 units, shrink it vertically by multiplying each y-coordinate by 0.43, and then shift it up 2.4 units.

$$y = 0.43(x - 3)^3 + 2.4$$

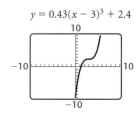

39. Start with the graph of $f(x) = x^2$. Shift it left 5.2 units, stretch it vertically by multiplying each y-coordinate by 2.8, and then shift it up 1.1 units.

$y = 2.8(x + 5.2)^2 + 1.1$

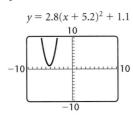

41. $f(x) = -(x - 8)^2$ **43.** $f(x) = |x + 7| + 2$

45. $f(x) = \dfrac{1}{2x} - 3$ **47.** $f(x) = -(x - 3)^2 + 4$

49. $f(x) = \sqrt{-(x + 2)} - 1$ **51.** $f(x) = 0.83(x + 4)^3$

53.

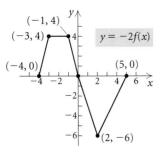

$y = -2f(x)$

55.

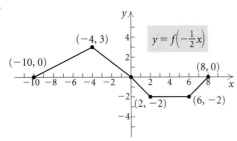

$y = f\left(-\dfrac{1}{2}x\right)$

57.

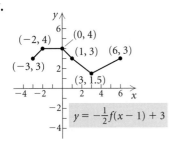

$y = -\dfrac{1}{2}f(x - 1) + 3$

59. (f)   **61.** (f)   **63.** (d)   **65.** (c)   **67.** Use a grapher.
69. Use a grapher.
71. $f(-x) = 2(-x)^4 - 35(-x)^3 + 3(-x) - 5 = 2x^4 + 35x^3 - 3x - 5 = g(x)$ **73.** $g(x) = x^3 - 3x^2 + 2$
75. $k(x) = (x + 1)^3 - 3(x + 1)^2$ **77.** 6   **79.** 16
81. ◈   **83.** $g(x) = 7 - \sqrt{x - 4}$
85. Start with the graph of $g(x) = \text{INT}(x)$. Shift it right $\frac{1}{2}$ unit. Domain: all real numbers; range: all integers.

87. Start with the graph of $g(x) = \text{INT}(x)$. Shift it left 2 units, stretch it vertically by multiplying each y-coordinate by 3, and shift it up 1 unit. Domain: all real numbers; range: all integers that are 1 more than a multiple of 3.
89. Start with the graph of $f(x) = 1/x$. Reflect across the x-axis the portion of the graph for which the y-coordinates are negative. Domain: all real numbers except 0; range: $(0, \infty)$
91. Start with the graph of $g(x) = |x|$. Shift it down 5 units; then reflect across the x-axis the portion of the graph for which the y-coordinates are negative. Domain: all real numbers; range: $[0, \infty)$ **93.** $(3, 8)$; $(3, 6)$; $\left(\frac{3}{2}, 4\right)$

EXERCISE SET 1.6

1. (a) $(4, -4)$; (b) $x = 4$; (c) minimum value: -4
3. (a) $\left(\frac{7}{2}, -\frac{1}{4}\right)$; (b) $x = \frac{7}{2}$; (c) minimum value: $-\frac{1}{4}$
5. (a) $\left(-\frac{3}{2}, \frac{7}{2}\right)$; (b) $x = -\frac{3}{2}$; (c) minimum value: $\frac{7}{2}$
7. (a) $\left(\frac{1}{2}, \frac{3}{2}\right)$; (b) $x = \frac{1}{2}$; (c) maximum value: $\frac{3}{2}$
9. (f)   **11.** (b)   **13.** (h)   **15.** (c)
17. (a) $(3, -4)$; (b) minimum value: -4; (c) $[-4, \infty)$; (d) increasing: $[3, \infty)$; decreasing: $(-\infty, 3]$
19. (a) $(-1, -18)$; (b) minimum value: -18; (c) $[-18, \infty)$; (d) increasing: $[-1, \infty)$; decreasing: $(-\infty, -1]$
21. (a) $\left(5, \frac{9}{2}\right)$; (b) maximum value: $\frac{9}{2}$; (c) $\left(-\infty, \frac{9}{2}\right]$; (d) increasing: $(-\infty, 5]$; decreasing: $[5, \infty)$
23. (a) $(-1, 2)$; (b) minimum value: 2; (c) $[2, \infty)$; (d) increasing: $[-1, \infty)$; decreasing: $(-\infty, -1]$
25. \$3240   **27.** 3.5 in.   **29.** $b = h = 10$ cm
31. 4800 yd^2 **33.** \$797 when $x = 40$
35. 3.36 ft by 3.36 ft   **37.** 6   **39.** 16   **41.** ◈
43. -236.25
45.

$y = (|x| - 5)^2 - 3$

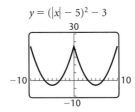

EXERCISE SET 1.7

1. 33   **3.** 78   **5.** 5   **7.** 2   **9.** 0   **11.** 0
13. (a) Domain of f, all real numbers; domain of g, $[-4, \infty)$; domain of $f + g$, $f - g$, and fg, $[-4, \infty)$; domain of ff, all real numbers; domain of f/g, $(-4, \infty)$; domain of g/f, $[-4, 3) \cup (3, \infty)$; domain of $f \circ g$, $[-4, \infty)$; domain of $g \circ f$, $[-1, \infty)$; (b) $(f + g)(x) = x - 3 + \sqrt{x + 4}$; $(f - g)(x) = x - 3 - \sqrt{x + 4}$; $(fg)(x) = (x - 3)\sqrt{x + 4}$; $(ff)(x) = x^2 - 6x + 9$; $(f/g)(x) = (x - 3)/\sqrt{x + 4}$; $(g/f)(x) = \sqrt{x + 4}/(x - 3)$; $(f \circ g)(x) = \sqrt{x + 4} - 3$; $(g \circ f)(x) = \sqrt{x + 1}$

15. (a) Domain of f, g, $f + g$, $f - g$, fg, and ff, all real numbers; domain of f/g, all real numbers except -3 and $\frac{1}{2}$; domain of g/f, all real numbers except 0; domain of $f \circ g$ and $g \circ f$, all real numbers;
(b) $(f + g)(x) = x^3 + 2x^2 + 5x - 3$;
$(f - g)(x) = x^3 - 2x^2 - 5x + 3$;
$(fg)(x) = 2x^5 + 5x^4 - 3x^3$; $(ff)(x) = x^6$;
$(f/g)(x) = x^3/(2x^2 + 5x - 3)$;
$(g/f)(x) = (2x^2 + 5x - 3)/x^3$; $(f \circ g)(x) = (2x^2 + 5x - 3)^3$;
$(g \circ f)(x) = 2x^6 + 5x^3 - 3$
17. $(f + g)(x) = 2x^2 - 2$; $(f - g)(x) = -6$;
$(fg)(x) = x^4 - 2x^2 - 8$; $(f/g)(x) = (x^2 - 4)/(x^2 + 2)$;
$(f \circ g)(x) = x^4 + 4x^2$
19. $(f + g)(x) = \sqrt{x - 7} + x^2 - 25$;
$(f - g)(x) = \sqrt{x - 7} - x^2 + 25$;
$(fg)(x) = \sqrt{x - 7}\,(x^2 - 25)$; $(f/g)(x) = \sqrt{x - 7}/(x^2 - 25)$;
$(f \circ g)(x) = \sqrt{x^2 - 32}$
21. Domain of F: $[0, 9]$; domain of G: $[3, 10]$; domain of $F + G$: $[3, 9]$
23. $\{x \mid 3 \le x \le 9 \text{ and } x \neq 6 \text{ and } x \neq 8\}$, or $[3, 6) \cup (6, 8) \cup (8, 9]$
25.

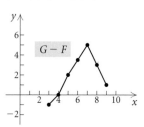

27. $(f \circ g)(x) = (g \circ f)(x) = x$
29. $(f \circ g)(x) = (g \circ f)(x) = x$
31. $(f \circ g)(x) = 20$; $(g \circ f)(x) = 0.05$
33. $(f \circ g)(x) = |x|$; $(g \circ f)(x) = x$
35. $(f \circ g)(x) = (g \circ f)(x) = x$
37. $(f \circ g)(x) = x^3 - 2x^2 - 4x + 6$;
$(g \circ f)(x) = x^3 - 5x^2 + 3x + 8$
39. $f(x) = x^5$; $g(x) = 4 + 3x$
41. $f(x) = \dfrac{1}{x}$; $g(x) = (x - 2)^4$
43. $f(x) = \dfrac{x - 1}{x + 1}$; $g(x) = x^3$
45. $f(x) = x^6$; $g(x) = \dfrac{2 + x^3}{2 - x^3}$
47. $f(x) = \sqrt{x}$; $g(x) = \dfrac{x - 5}{x + 2}$
49. $f(x) = x^3 - 5x^2 + 3x - 1$; $g(x) = x + 2$
51. (a) $P(x) = -0.4x^2 + 57x - 13$; **(b)** $R(100) = 2000$; $C(100) = 313$; $P(100) = 1687$; **(c)** Use a grapher.
53. (a) $r(t) = 3t$; **(b)** $A(r) = \pi r^2$; **(c)** $(A \circ r)(t) = 9\pi t^2$; the area of the ripple in terms of time t **55.** $[-1, 2]$
57. -21 **59.** ◈

61. **63.**

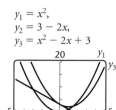

$y_1 = x^2$,
$y_2 = 3 - 2x$,
$y_3 = x^2 - 2x + 3$

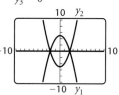

$y_1 = 5 - x^2$,
$y_2 = x^2 - 5$,
$y_3 = 0$

65. Domain of $F \circ G$, $[3, 10]$; domain of $G \circ G$, $[5, 8]$; domain of $F \circ F$, $[0, 6] \cup [8, 9]$ **67.** -2 **69.** $\dfrac{x - 1}{x}$; x
71. True **73.** True
75. $E(-x) = \dfrac{f(-x) + f(-(-x))}{2} = \dfrac{f(-x) + f(x)}{2} = E(x)$
77. (a) $E(x) + O(x) = \dfrac{f(x) + f(-x)}{2} + \dfrac{f(x) - f(-x)}{2} = \dfrac{2f(x)}{2} = f(x)$; **(b)** $f(x) = \dfrac{-22x^2 + \sqrt{x} + \sqrt{-x} - 20}{2} + \dfrac{8x^3 + \sqrt{x} - \sqrt{-x}}{2}$

EXERCISE SET 1.8

1. $\{(8, 7), (8, -2), (-4, 3), (-8, 8)\}$
3. $\{(-1, -1), (4, -3)\}$ **5.** $x = 4y - 5$ **7.** $y^3x = -5$
9.

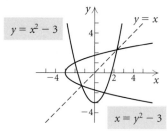

$y = x^2 - 3$
$y = x$
$x = y^2 - 3$

11.

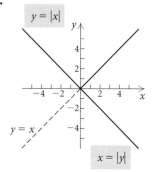

$y = |x|$
$y = x$
$x = |y|$

13. Yes **15.** No **17.** No **19.** Yes **21.** Yes **23.** No
25. No **27.** Yes

29. $y_1 = 0.8x + 1.7,$
$y_2 = \dfrac{x - 1.7}{0.8}$

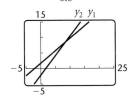

Domain and range of both f and f^{-1}: all real numbers

31. $y_1 = \dfrac{1}{2}x - 4,$
$y_2 = 2x + 8$

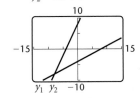

Domain and range of both f and f^{-1}: all real numbers

33. $y_1 = \sqrt{x - 3},$
$y_2 = x^2 + 3, x \geq 0$

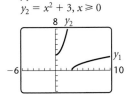

Domain of f: $[3, \infty]$; range of f: $[0, \infty)$; domain of f^{-1}: $[0, \infty)$; range of f^{-1}: $[3, \infty)$

35. $y_1 = x^2 - 4, x \geq 0;$ $y_2 = \sqrt{4 + x}$

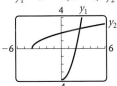

Domain of f: $[0, \infty)$; range of f: $[-4, \infty)$; domain of f^{-1}: $[-4, \infty)$; range of f^{-1}: $[0, \infty)$

37. $y_1 = (3x - 9)^3,$ $y_2 = \dfrac{\sqrt[3]{x} + 9}{3}$

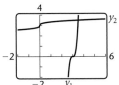

Domain and range of both f and f^{-1}: all real numbers

39. (a) One-to-one; **(b)** $f^{-1}(x) = x - 4$

41. (a) One-to-one; **(b)** $f^{-1}(x) = \dfrac{1}{2}x$

43. (a) One-to-one; **(b)** $f^{-1}(x) = \dfrac{4}{x} - 7$

45. (a) One-to-one; **(b)** $f^{-1}(x) = \dfrac{3x + 4}{x - 1}$

47. (a) One-to-one; **(b)** $f^{-1}(x) = \sqrt[3]{x} + 1$

49. (a) Not one-to-one

51. (a) One-to-one; **(b)** $f^{-1}(x) = \sqrt{\dfrac{x + 2}{5}}, x \geq 18$

53. (a) One-to-one; **(b)** $f^{-1}(x) = x^2 - 1, x \geq 0$

55. $\dfrac{1}{3}x$ **57.** $-x$ **59.** $x^3 + 5$

61. $(f^{-1} \circ f)(x) = f^{-1}(f(x))$
$= f^{-1}\!\left(\dfrac{7}{8}x\right)$
$= \dfrac{8}{7}\!\left(\dfrac{7}{8}x\right) = x;$
$(f \circ f^{-1})(x) = f(f^{-1}(x))$
$= f\!\left(\dfrac{8}{7}x\right)$
$= \dfrac{7}{8}\!\left(\dfrac{8}{7}x\right) = x$

63. $(f^{-1} \circ f)(x) = f^{-1}\!\left(\dfrac{1 - x}{x}\right)$
$= \dfrac{1}{\dfrac{1 - x}{x} + 1}$
$= \dfrac{1}{\dfrac{1}{x}} = x;$
$(f \circ f^{-1})(x) = f\!\left(\dfrac{1}{x + 1}\right)$
$= \dfrac{1 - \dfrac{1}{x + 1}}{\dfrac{1}{x + 1}}$
$= \dfrac{\dfrac{x + 1 - 1}{x + 1}}{\dfrac{1}{x + 1}} = x$

65. 5; a

67. (a) 36, 40, 44, 52, 60; **(b)** $g^{-1}(x) = \dfrac{x}{2} - 12$;

(c) 8, 10, 14, 18

69. (a) 0.5, 11.5, 22.5, 55.5, 72;

(b), (d) $y_1 = \dfrac{11x + 5}{10},$ $y_2 = \dfrac{10x - 5}{11}$

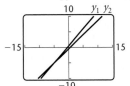

(c) $D^{-1}(r) = \dfrac{10r - 5}{11}$; the speed, in miles per hour, that the car is traveling when the reaction distance is r feet

71.

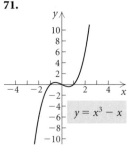

$y = x^3 - x$

73.

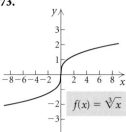

$f(x) = \sqrt[3]{x}$

75. ◆   **77.** Yes   **79.** No
81. Answers may vary. $f(x) = 3/x,\ f(x) = 1 - x,\ f(x) = x$

REVIEW EXERCISES, CHAPTER 1

1. [1.1] Not a function; domain: {3, 5, 7}; range: {1, 3, 5, 7}
2. [1.1] Function; domain: {−2, 0, 1, 2, 7}; range: {−7, −4, −2, 2, 7}   **3.** [1.1] No   **4.** [1.1] Yes
5. [1.1] All real numbers   **6.** [1.1] $\left(-\infty, \frac{7}{3}\right]$
7. [1.1] $\{x \mid x \neq 5 \text{ and } x \neq 1\}$
8. [1.1] $\{x \mid x \neq -4 \text{ and } x \neq 4\}$
9. [1.1] Domain: [−5, 3]; range: [0, 4]
10. [1.1] Domain: all real numbers; range: [−5, ∞)
11. [1.1] Domain: all real numbers; range: all real numbers
12. [1.1] Domain: all real numbers; range: [−19, ∞)
13. [1.1] **(a)** −3; **(b)** 9; **(c)** $a^2 - 3a - 1$; **(d)** $2x + h - 1$
14. [1.1] No   **15.** [1.1] Yes   **16.** [1.1] No   **17.** [1.1] Yes
18. [1.2]   **19.** [1.2]

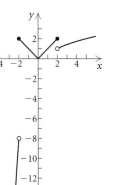

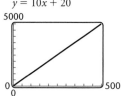

20. [1.2]

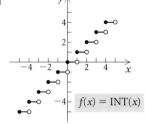

$f(x) = \text{INT}(x)$

21. [1.2]

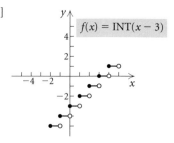

$f(x) = \text{INT}(x - 3)$

22. [1.2] **(a)** −7.029; **(b)** increasing: $(-\infty, -0.860], [6.339, \infty)$; decreasing: $[-0.860, 6.339]$; **(c)** relative maximum: $(-0.860, 710.094)$; relative minimum: $(6.339, 504.836)$
23. [1.2] **(a)** −5.732, −2.268; **(b)** increasing: $(-\infty, -4]$; decreasing: $[-4, \infty)$; **(c)** relative maximum: $(-4, 15)$
24. [1.2] **(a)** −5, 0, 3; **(b)** increasing: $[-3.589, 2.089]$; decreasing: $[-5, -3.589], [2.089, 3]$; **(c)** relative maximum: $(2.089, 4.247)$; relative minimum: $(-3.589, -8.755)$
25. [1.2] $A(x) = 2x\sqrt{4 - x^2}$
26. [1.3] **(a)** Yes; **(b)** no; **(c)** no, strictly speaking, but data might be modeled by a linear regression function.
27. [1.3] **(a)** Yes; **(b)** yes; **(c)** yes   **28.** [1.3] −2; (0, −7)
29. [1.3] $y + 4 = -\frac{2}{3}(x - 0)$ **30.** [1.3] $y + 1 = 3(x + 2)$
31. [1.3] $y - 1 = \frac{1}{3}(x - 4)$, or $y + 1 = \frac{1}{3}(x + 2)$
32. [1.3] Parallel   **33.** [1.3] Neither
34. [1.3] Perpendicular   **35.** [1.3] $y = -\frac{2}{3}x - \frac{1}{3}$
36. [1.3] $y = \frac{3}{2}x - \frac{5}{2}$
37. [1.3] $C(t) = 25 + 20t$; $y = 25 + 20x$ $145

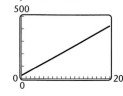

38. [1.3] **(a)** 70°C, 220°C, 10,020°C;
(b)

$y = 10x + 20$

(c) [0, 5600]
39. [1.4] x-axis, yes; y-axis, yes; origin, yes
40. [1.4] x-axis, yes; y-axis, yes; origin, yes
41. [1.4] x-axis, no; y-axis, no; origin, no
42. [1.4] x-axis, no; y-axis, yes; origin, no
43. [1.4] x-axis, no; y-axis, no; origin, yes
44. [1.4] x-axis, no; y-axis, yes; origin, no
45. [1.4] Odd   **46.** [1.4] Even   **47.** [1.4] Even
48. [1.4] Neither   **49.** [1.4] Odd

50. [1.4] Even   **51.** [1.4] Even   **52.** [1.4] Odd
53. [1.5] $f(x) = (x + 3)^2$
54. [1.5] $f(x) = -\sqrt{x - 3} + 4$
55. [1.5] $f(x) = 2|x - 3|$
56. [1.5]   **57.** [1.5]

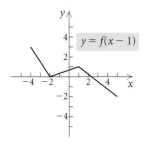

58. [1.5]   **59.** [1.5]

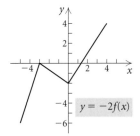   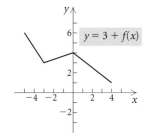

60. [1.6] **(a)** $\left(\frac{3}{8}, -\frac{7}{16}\right)$; **(b)** $x = \frac{3}{8}$; **(c)** maximum: $-\frac{7}{16}$;
(d) $\left(-\infty, -\frac{7}{16}\right]$ **61.** [1.6] **(a)** $(1, -2)$; **(b)** $x = 1$;
(c) minimum: -2;   **(d)** $[-2, \infty)$ **62.** [1.6] (d)
63. [1.6] (c)   **64.** [1.6] (b)   **65.** [1.6] (a)
66. [1.6] $35 - 5\sqrt{33}$ ft, or about 6.3 ft
67. [1.6] 6 ft by 6 ft
68. [1.6] $\dfrac{15 - \sqrt{115}}{2}$ cm, or about 2.1 cm
69. [1.7] **(a)** Domain of f, $\{x \,|\, x \neq 0\}$; domain of g, all real numbers; domain of $f + g$, $f - g$, and fg, $\{x \,|\, x \neq 0\}$;
domain of f/g, $\left\{x \,\middle|\, x \neq 0 \text{ and } x \neq \dfrac{3}{2}\right\}$; domain of $f \circ g$,
$\left\{x \,\middle|\, x \neq \dfrac{3}{2}\right\}$; domain of $g \circ f$, $\{x \,|\, x \neq 0\}$;
(b) $(f + g)(x) = \dfrac{4}{x^2} + 3 - 2x$; $(f - g)(x) = \dfrac{4}{x^2} - 3 + 2x$;
$fg(x) = \dfrac{12}{x^2} - \dfrac{8}{x}$; $(f/g)(x) = \dfrac{4}{x^2(3 - 2x)}$;
$(f \circ g)(x) = \dfrac{4}{(3 - 2x)^2}$; $(g \circ f)(x) = 3 - \dfrac{8}{x^2}$
70. [1.7] **(a)** Domain of f, g, $f + g$, $f - g$, and fg, all real
numbers; domain of f/g, $\left\{x \,\middle|\, x \neq \dfrac{1}{2}\right\}$; domain of $f \circ g$ and
$g \circ f$, all real numbers; **(b)** $(f + g)(x) = 3x^2 + 6x - 1$;
$(f - g)(x) = 3x^2 + 2x + 1$; $fg(x) = 6x^3 + 5x^2 - 4x$;

$(f/g)(x) = \dfrac{3x^2 + 4x}{2x - 1}$; $(f \circ g)(x) = 12x^2 - 4x - 1$;
$(g \circ f)(x) = 6x^2 + 8x - 1$
71. [1.7] $P(x) = -0.5x^2 + 105x - 6$
72. [1.7] $f(x) = \sqrt{x}$, $g(x) = 5x + 2$. Answers may vary.
73. [1.7] $f(x) = 4x^2 + 9$, $g(x) = 5x - 1$. Answers may vary.
74. [1.8] $\{(-2.7, 1.3), (-3, 8), (3, -5), (-3, 6), (-5, 7)\}$
75. [1.8] **(a)** $x = 3y^2 + 2y - 1$; **(b)** $0.8y^3 - 5.4x^2 = 3y$
76. [1.8] **(a)** Yes; **(b)** $f^{-1}(x) = x^2 + 6$, $x \geq 0$
77. [1.8] **(a)** Yes; **(b)** $f^{-1}(x) = \sqrt[3]{x} + 8$ **78.** [1.8] **(a)** No
79. [1.8] 657
80. [1.1], [1.7] ◈ **(a)** $4x^3 - 2x + 9$;
(b) $4x^3 + 24x^2 + 46x + 35$; **(c)** $4x^3 - 2x + 42$. **(a)** Adds 2
to each function value; **(b)** adds 2 to each input before
finding a function value; **(c)** adds the output for 2 to the
output for x
81. [1.5] ◈ In the graph of $y = f(cx)$, the constant c
stretches or shrinks the graph of $y = f(x)$ horizontally. The
constant c in $y = cf(x)$ stretches or shrinks the graph of
$y = f(x)$ vertically. For $y = f(cx)$, the x-coordinates of
$y = f(x)$ are divided by c; for $y = cf(x)$, the y-coordinates
of $y = f(x)$ are multiplied by c.
82. [1.8] ◈ The inverse of a function $f(x)$ is written $f^{-1}(x)$,
whereas $[f(x)]^{-1}$ means $\dfrac{1}{f(x)}$.
83. [1.8] No   **84.** [1.1] $\{x \,|\, x < 0\}$
85. [1.1] $\{x \,|\, x \neq -3 \text{ and } x \neq 0 \text{ and } x \neq 3\}$
86. [1.4] Let $f(x)$ and $g(x)$ be odd functions. Then by
definition, $f(-x) = -f(x)$, or $f(x) = -f(-x)$, and
$g(-x) = -g(x)$, or $g(x) = -g(-x)$. Thus,
$(f + g)(x) = f(x) + g(x) = -f(-x) + [-g(-x)] = -[f(-x) + g(-x)] = -(f + g)(-x)$ and $f + g$ is odd.
87. [1.5] Reflect the graph of $y = f(x)$ across the x-axis and
then across the y-axis.

Chapter 2

EXERCISE SET 2.1

1. $\sin \phi = \dfrac{15}{17}$, $\cos \phi = \dfrac{8}{17}$, $\tan \phi = \dfrac{15}{8}$, $\csc \phi = \dfrac{17}{15}$,
$\sec \phi = \dfrac{17}{8}$, $\cot \phi = \dfrac{8}{15}$
3. $\sin \alpha = \dfrac{\sqrt{3}}{2}$, $\cos \alpha = \dfrac{1}{2}$, $\tan \alpha = \sqrt{3}$, $\csc \alpha = \dfrac{2\sqrt{3}}{3}$,
$\sec \alpha = 2$, $\cot \alpha = \dfrac{\sqrt{3}}{3}$
5. $\sin \phi = \dfrac{7\sqrt{65}}{65}$, $\cos \phi = \dfrac{4\sqrt{65}}{65}$, $\tan \phi = \dfrac{7}{4}$,
$\csc \phi = \dfrac{\sqrt{65}}{7}$, $\sec \phi = \dfrac{\sqrt{65}}{4}$, $\cot \phi = \dfrac{4}{7}$

7. $\cos \theta = \dfrac{7}{25}$, $\tan \theta = \dfrac{24}{7}$, $\csc \theta = \dfrac{25}{24}$, $\sec \theta = \dfrac{25}{7}$,

$\cot \theta = \dfrac{7}{24}$

9. $\sin \phi = \dfrac{2\sqrt{5}}{5}$, $\cos \phi = \dfrac{\sqrt{5}}{5}$, $\csc \phi = \dfrac{\sqrt{5}}{2}$, $\sec \phi = \sqrt{5}$,

$\cot \phi = \dfrac{1}{2}$

11. $\sin \theta = \dfrac{2}{3}$, $\cos \theta = \dfrac{\sqrt{5}}{3}$, $\tan \theta = \dfrac{2\sqrt{5}}{5}$, $\sec \theta = \dfrac{3\sqrt{5}}{5}$,

$\cot \theta = \dfrac{\sqrt{5}}{2}$ **13.** $\dfrac{\sqrt{2}}{2}$ **15.** 2   **17.** $\dfrac{\sqrt{3}}{3}$ **19.** $\dfrac{1}{2}$

21. 127.3 ft   **23.** 9.72°   **25.** 35.83°   **27.** 3.03°
29. 49.65°   **31.** 17°36′   **33.** 83°1′30″   **35.** 11°45′
37. 47°49′36″   **39.** 0.6293   **41.** 0.0737   **43.** 1.2765
45. 0.7621   **47.** 0.9336   **49.** 12.4288   **51.** 1.0000
53. 1.7032   **55.** 30.8°   **57.** 12.5°   **59.** 64.4°   **61.** 46.5°
63. 25.2°   **65.** 38.6°   **67.** 45°   **69.** 60°   **71.** 45°

73. $\cos 20° = \sin 70° = \dfrac{1}{\sec 20°}$

75. $\tan 52° = \cot 38° = \dfrac{1}{\cot 52°}$

77. $\sin 25° \approx 0.4226$, $\cos 25° \approx 0.9063$, $\tan 25° \approx 0.4663$,
$\csc 25° \approx 2.366$, $\sec 25° \approx 1.103$, $\cot 25° \approx 2.145$ **79.** 1.52
81. −32.03   **83.** 27.5   **85.** ◆   **87.** 0.6534
89. Area $= \frac{1}{2}ab$. But $a = c \sin A$, so Area $= \frac{1}{2}bc \sin A$.

EXERCISE SET 2.2

1. $F = 60°$, $d = 3$, $f \approx 5.2$ **3.** $A = 22.7°$, $a \approx 52.7$,
$c \approx 136.6$ **5.** $P = 47°38′$, $n \approx 34.4$, $p \approx 25.4$
7. $B = 2°17′$, $b \approx 0.39$, $c = 9.74$
9. $A \approx 77.2°$, $B \approx 12.8°$, $a \approx 439$
11. $B = 42.42°$, $a \approx 35.7$, $b \approx 32.6$
13. $B = 55°$, $a \approx 28.0$, $c \approx 48.8$
15. $A \approx 62.4°$, $B \approx 27.6°$, $a \approx 3.56$ **17.** About 62.2 ft
19. About 2.5 ft   **21.** About 606 ft   **23.** 96.7 cm
25. 599 ft   **27.** 8 km   **29.** 275 ft   **31.** 24 km

33. $10\sqrt{2}$, or about 14.142   **35.** $\dfrac{\sqrt{3}}{2}$ **37.** $\dfrac{\sqrt{3}}{3}$ **39.** ◈

41. 3.3   **43.** Cut so that $\theta = 79.38°$ **45.** $\theta \approx 27°$

EXERCISE SET 2.3

1. III   **3.** III   **5.** I   **7.** III   **9.** II   **11.** II
13. 434°, 794°, −286°, −646°
15. 475.3°, 835.3°, −244.7°, −604.7°
17. 180°, 540°, −540°, −900°   **19.** 72.89°, 162.89°
21. 77°56′46″, 167°56′46″   **23.** 44.8°, 134.8°

25. $\sin \beta = \dfrac{5}{13}$, $\cos \beta = -\dfrac{12}{13}$, $\tan \beta = -\dfrac{5}{12}$, $\csc \beta = \dfrac{13}{5}$,

$\sec \beta = -\dfrac{13}{12}$, $\cot \beta = -\dfrac{12}{5}$

27. $\sin \phi = -\dfrac{2\sqrt{7}}{7}$, $\cos \phi = -\dfrac{\sqrt{21}}{7}$, $\tan \phi = \dfrac{2\sqrt{3}}{3}$,

$\csc \phi = -\dfrac{\sqrt{7}}{2}$, $\sec \phi = -\dfrac{\sqrt{21}}{3}$, $\cot \phi = \dfrac{\sqrt{3}}{2}$

29. $\sin \theta = -\dfrac{2\sqrt{13}}{13}$, $\cos \theta = \dfrac{3\sqrt{13}}{13}$, $\tan \theta = -\dfrac{2}{3}$

31. $\sin \theta = \dfrac{5\sqrt{41}}{41}$, $\cos \theta = \dfrac{4\sqrt{41}}{41}$, $\tan \theta = \dfrac{5}{4}$

33. $\cos \theta = -\dfrac{2\sqrt{2}}{3}$, $\tan \theta = \dfrac{\sqrt{2}}{4}$, $\csc \theta = -3$,

$\sec \theta = -\dfrac{3\sqrt{2}}{4}$, $\cot \theta = 2\sqrt{2}$

35. $\sin \theta = -\dfrac{\sqrt{5}}{5}$, $\cos \theta = \dfrac{2\sqrt{5}}{5}$, $\tan \theta = -\dfrac{1}{2}$,

$\csc \theta = -\sqrt{5}$, $\sec \theta = \dfrac{\sqrt{5}}{2}$

37. $\sin \phi = -\dfrac{4}{5}$, $\tan \phi = -\dfrac{4}{3}$, $\csc \phi = -\dfrac{5}{4}$, $\sec \phi = \dfrac{5}{3}$,

$\cot \phi = -\dfrac{3}{4}$ **39.** $-\dfrac{\sqrt{3}}{2}$ **41.** 1   **43.** 0   **45.** $-\dfrac{\sqrt{2}}{2}$

47. 2   **49.** $\sqrt{3}$ **51.** $-\dfrac{\sqrt{3}}{3}$ **53.** Undefined   **55.** −1

57. $\sqrt{3}$ **59.** $\dfrac{\sqrt{2}}{2}$ **61.** $-\sqrt{2}$ **63.** 1   **65.** 0   **67.** 0

69. 0   **71.** Positive: cos, sec; negative: sin, csc, tan, cot
73. Positive: tan, cot; negative: sin, csc, cos, sec
75. Positive: sin, csc; negative: cos, sec, tan, cot
77. Positive: all
79. $\sin 319° = -0.6561$, $\cos 319° = 0.7547$,
$\tan 319° = -0.8693$, $\csc 319° = -1.5242$, $\sec 319° = 1.3250$,
$\cot 319° = -1.1504$
81. $\sin 115° = 0.9063$, $\cos 115° = -0.4226$,
$\tan 115° = -2.1445$, $\csc 115° = 1.1034$, $\sec 115° = -2.3663$,
$\cot 115° = -0.4663$ **83.** East: 130 km; south: 75 km
85. 223 km   **87.** −1.1585   **89.** −1.4910   **91.** 0.8771
93. 0.4352   **95.** 0.9563   **97.** 2.9238   **99.** 275.4°
101. 200.1°   **103.** 288.1°   **105.** 72.6°   **107.** 44.18°
109. 66 ft/sec   **111.** ◈   **113.** 19.625 in.

EXERCISE SET 2.4

1.

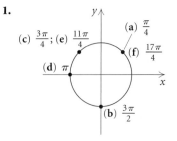

3.

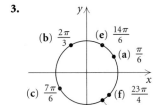

5. $M: \dfrac{2\pi}{3},\ -\dfrac{4\pi}{3};\ N: \dfrac{3\pi}{2},\ -\dfrac{\pi}{2};\ P: \dfrac{5\pi}{4},\ -\dfrac{3\pi}{4};\ Q: \dfrac{11\pi}{6},$

$-\dfrac{\pi}{6}$

7.

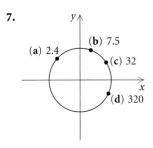

9. $\dfrac{9\pi}{4},\ -\dfrac{7\pi}{4}$ **11.** $\dfrac{19\pi}{6},\ -\dfrac{5\pi}{6}$

13. Complement: $\dfrac{\pi}{6}$; supplement: $\dfrac{2\pi}{3}$

15. Complement: $\dfrac{\pi}{8}$; supplement: $\dfrac{5\pi}{8}$ **17.** $\dfrac{5\pi}{12}$ **19.** $\dfrac{10\pi}{9}$

21. $-\dfrac{214.6\pi}{180}$ **23.** $-\pi$ **25.** 4.19   **27.** -1.05 **29.** 2.06

31. 0.02   **33.** $-135°$ **35.** $1440°$ **37.** $57.30°$

39. $134.47°$

41. $0° = 0$ radian, $30° = \dfrac{\pi}{6},\ 45° = \dfrac{\pi}{4},\ 60° = \dfrac{\pi}{3},$

$90° = \dfrac{\pi}{2},\ 135° = \dfrac{3\pi}{4},\ 180° = \pi,\ 225° = \dfrac{5\pi}{4},\ 270° = \dfrac{3\pi}{2},$

$315° = \dfrac{7\pi}{4},\ 360° = 2\pi$ **43.** 1.1, $63°$

45. $\dfrac{5\pi}{3}$, or about 5.24   **47.** $3150\ \dfrac{\text{cm}}{\text{min}}$

49. About 17,000 revolutions per hour   **51.** 1047 mph

53. 10 mph   **55.** About 133

57.

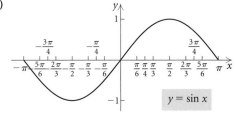

59. Domain: $\{x | x \neq -2\}$; range: $\{x | x \neq 1\}$ **61.** ◈

63. ◈   **65.** 111.7 km, 69.8 mi   **67.** (a) $5°37'30''$; (b) $19°41'15''$

69. 1.676 radians/sec   **71.** 1.46 nautical miles

EXERCISE SET 2.5

1. (a) $\left(-\dfrac{3}{4},\ -\dfrac{\sqrt{7}}{4}\right)$; (b) $\left(\dfrac{3}{4},\ \dfrac{\sqrt{7}}{4}\right)$; (c) $\left(\dfrac{3}{4},\ -\dfrac{\sqrt{7}}{4}\right)$

3. (a) $\left(\dfrac{2}{5},\ \dfrac{\sqrt{21}}{5}\right)$; (b) $\left(-\dfrac{2}{5},\ -\dfrac{\sqrt{21}}{5}\right)$; (c) $\left(-\dfrac{2}{5},\ \dfrac{\sqrt{21}}{5}\right)$

5. $\left(\dfrac{\sqrt{2}}{2},\ -\dfrac{\sqrt{2}}{2}\right)$ **7.** 0   **9.** $\sqrt{3}$ **11.** 0   **13.** $-\dfrac{\sqrt{3}}{2}$

15. 1   **17.** $\dfrac{\sqrt{3}}{2}$ **19.** $-\dfrac{\sqrt{2}}{2}$ **21.** 0   **23.** 0   **25.** 0.4816

27. 1.3065   **29.** -2.1599 **31.** 1   **33.** -1.1747 **35.** -1

37. -0.7071 **39.** 0   **41.** 0.8391

43. (a)

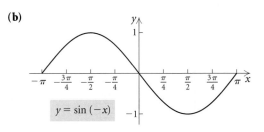

(b)

(c) same as (b); (d) the same   **45.** (a) See Exercise 43(a);

(b)

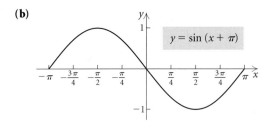

(c) same as (b); **(d)** the same

47. (a)

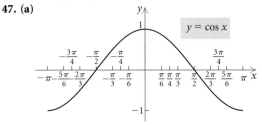

(b)

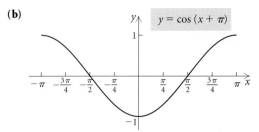

(c) same as (b); **(d)** the same   **49.** Even: cosine, secant; odd: sine, tangent, cosecant, cotangent   **51.** Positive: I, III; negative: II, IV   **53.** Positive: I, IV; negative: II, III   **55.** Stretch the graph of f vertically, then shift it down 3 units.

$$y_1 = x^2, \quad y_2 = 2x^2 - 3$$

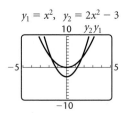

57. Shift the graph of f to the right 4 units, shrink it vertically, then shift it up 1 unit.

$$y_1 = |x|, \quad y_2 = \tfrac{1}{2}|x - 4| + 1$$

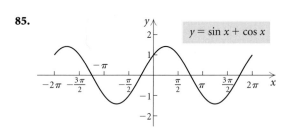

59. $y = -(x - 2)^3 - 1$ **61.** ◈   **63.** $\cos x$ **65.** $\sin x$ **67.** $\sin x$ **69.** $-\cos x$ **71.** $-\sin x$

73. (a) $\dfrac{\pi}{2} + 2k\pi,\ k \in \mathbb{Z}$; **(b)** $(2k + 1)\pi,\ k \in \mathbb{Z}$;
(c) $k\pi,\ k \in \mathbb{Z}$; **(d)** $\left(-\dfrac{3\pi}{4} + 2k\pi, \dfrac{\pi}{4} + 2k\pi\right),\ k \in \mathbb{Z}$

75. Domain: $(-\infty, \infty)$; range: $[0, 1]$; period: π;
amplitude: $\dfrac{1}{2}$ **77.** $\left[-\dfrac{\pi}{2} + 2k\pi, \dfrac{\pi}{2} + 2k\pi\right],\ k \in \mathbb{Z}$

79. $\left\{x \,\middle|\, x \neq \dfrac{\pi}{2} + k\pi,\ k \in \mathbb{Z}\right\}$ **81.** 1

83.

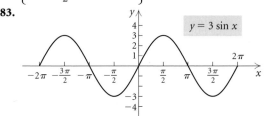

85.

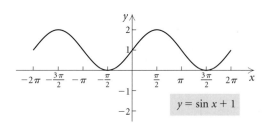

EXERCISE SET 2.6

1. Amplitude: 1; period: 2π; phase shift: 0

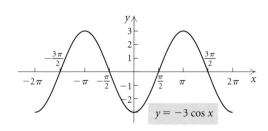

3. Amplitude: 3; period: 2π; phase shift: 0

5. Amplitude: 2; period: 4π; phase shift: 0

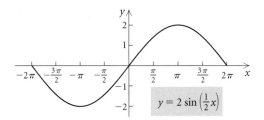

7. Amplitude: $\dfrac{1}{2}$; period: 2π; phase shift: $-\dfrac{\pi}{2}$

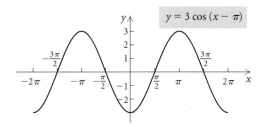

9. Amplitude: 3; period: 2π; phase shift: π

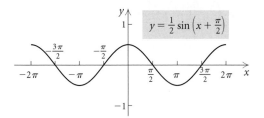

11. Amplitude: $\dfrac{1}{3}$; period: 2π; phase shift: 0

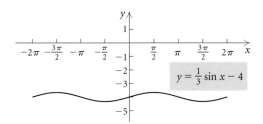

13. Amplitude: 1; period: 2π; phase shift: 0

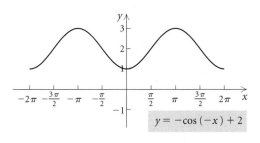

15. Amplitude: 2; period: 4π; phase shift: π

$$y = 2\cos\left(\tfrac{1}{2}x - \tfrac{\pi}{2}\right)$$

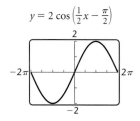

17. Amplitude: $\dfrac{1}{2}$; period: π; phase shift: $-\dfrac{\pi}{4}$

$$y = -\tfrac{1}{2}\sin\left(2x + \tfrac{\pi}{2}\right)$$

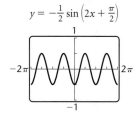

19. Amplitude: 3; period: 2; phase shift: $\dfrac{3}{\pi}$

$$y = 2 + 3\cos\left(\pi x - 3\right)$$

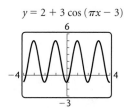

21. Amplitude: $\dfrac{1}{2}$; period: 1; phase shift: 0

$$y = -\sin\left(\tfrac{1}{2}x - \tfrac{\pi}{2}\right) + \tfrac{1}{2}$$

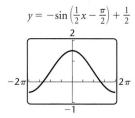

23. Amplitude: 1; period: 4π; phase shift: π

$$y = -\sin\left(\tfrac{1}{2}x - \tfrac{\pi}{2}\right) + \tfrac{1}{2}$$

25. Amplitude: 1; period: 1; phase shift: 0

$y = \cos(-2\pi x) + 2$

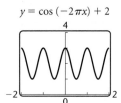

27. Amplitude: $\dfrac{1}{4}$; period: 2; phase shift: $\dfrac{4}{\pi}$

$y = -\dfrac{1}{4}\cos(\pi x - 4)$

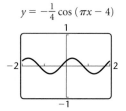

29. (b)   **31.** (h)   **33.** (a)   **35.** (f)

37.

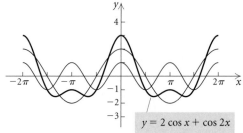

$y = 2\cos x + \cos 2x$

39.

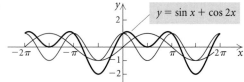

$y = \sin x + \cos 2x$

41.

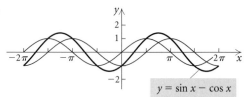

$y = \sin x - \cos x$

43.

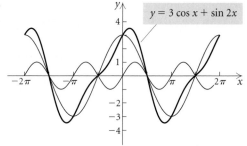

$y = 3\cos x + \sin 2x$

45.

$y = x + \sin x$

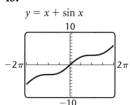

47.

$y = \cos 2x + 2x$

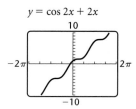

49.

$y = 4\cos 2x - 2\sin x$

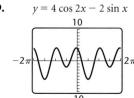

51. $\dfrac{6x + 4}{(x + 1)(x - 1)}$ **53.** $\dfrac{x - 4}{x - 1}$ **55.** ◆

57. (a) $y = 7\sin(-2.6180x + 0.5236) + 7$; (b) $10,500, $13,062

59.

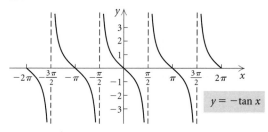

$y = -\tan x$

61.

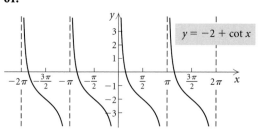

$y = -2 + \cot x$

63.

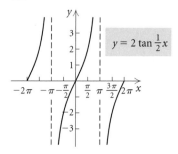

$y = 2\tan \dfrac{1}{2}x$

65.

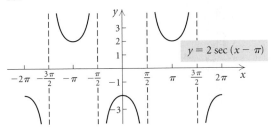

$y = 2 \sec (x - \pi)$

67.

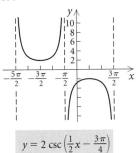

$y = 2 \csc \left(\dfrac{1}{2}x - \dfrac{3\pi}{4}\right)$

69.

$y = e^{-x/2} \cos x$

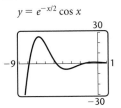

71.

$y = 0.6x^2 \cos x$

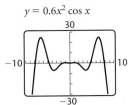

73.

$y = |\tan x|$

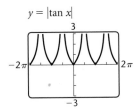

75. $-9.42, -6.28, -3.14, 3.14, 6.28, 9.42$

77. $-3.14, 0, 3.14$

79. **(a)** $y = 101.6 + 3 \sin \left(\dfrac{\pi}{8}x\right)$

(b) $104.6°, 98.6°$

81. (a) $y = 3000\left[\cos \dfrac{\pi}{45}(x - 10)\right]$

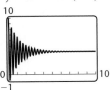

(b) amplitude: 3000; period: 90; phase shift: 10

83. (a) $y = 6e^{-0.8x} \cos (6\pi x) + 4$

(b) 4 in.

REVIEW EXERCISES, CHAPTER 2

1. [2.1] $\sin \theta = \dfrac{3\sqrt{73}}{73}$, $\cos \theta = \dfrac{8\sqrt{73}}{73}$, $\tan \theta = \dfrac{3}{8}$,

$\csc \theta = \dfrac{\sqrt{73}}{3}$, $\sec \theta = \dfrac{\sqrt{73}}{8}$, $\cot \theta = \dfrac{8}{3}$ **2.** [2.1] $\dfrac{\sqrt{2}}{2}$

3. [2.1] $\dfrac{\sqrt{3}}{3}$ **4.** [2.3] $-\dfrac{\sqrt{2}}{2}$ **5.** [2.3] $\dfrac{1}{2}$

6. [2.3] Undefined   **7.** [2.3] $-\sqrt{3}$ **8.** [2.1] $22°16'12''$

9. [2.1] $47.56°$ **10.** [2.3] 0.4452 **11.** [2.3] 1.1315

12. [2.3] 0.9498 **13.** [2.3] -0.9092 **14.** [2.3] -1.5282

15. [2.3] -0.2778 **16.** [2.3] $205.3°$ **17.** [2.3] $47.2°$

18. [2.1] $60°$ **19.** [2.1] $60°$

20. [2.1] $\sin 30.9° \approx 0.5135$, $\cos 30.9° \approx 0.8581$,

$\tan 30.9° \approx 0.5985$, $\csc 30.9° \approx 1.9474$, $\sec 30.9° \approx 1.1654$,

$\cot 30.9° \approx 1.6709$ **21.** [2.2] $b \approx 4.5$, $A \approx 58.1°$, $B \approx 31.9°$

22. [2.2] $A = 38.83°$, $b \approx 37.9$, $c \approx 48.6$ **23.** [2.2] 1748 m

24. [2.2] 14 ft   **25.** [2.3] $425°, -295°$ **26.** [2.4] $\dfrac{\pi}{3}, -\dfrac{5\pi}{3}$

27. [2.3] Complement: $76.6°$; supplement: $166.6°$

28. [2.4] Complement: $\dfrac{\pi}{3}$; supplement: $\dfrac{5\pi}{6}$

29. [2.3] $\sin \theta = \dfrac{3\sqrt{13}}{13}$, $\cos \theta = \dfrac{-2\sqrt{13}}{13}$, $\tan \theta = -\dfrac{3}{2}$,

$\csc \theta = \dfrac{\sqrt{13}}{3}$, $\sec \theta = -\dfrac{\sqrt{13}}{2}$, $\cot \theta = -\dfrac{2}{3}$

30. [2.3] $\sin \theta = -\dfrac{2}{3}$, $\cos \theta = -\dfrac{\sqrt{5}}{3}$, $\cot \theta = \dfrac{\sqrt{5}}{2}$,

$\sec \theta = -\dfrac{3\sqrt{5}}{5}$, $\csc \theta = -\dfrac{3}{2}$ **31.** [2.3] About 1743 mi

32. [2.4]

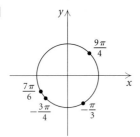

33. [2.4] II, $\dfrac{121}{150}\,\pi$, 2.534   **34.** [2.4] IV, $-\dfrac{\pi}{6}$, -0.524

35. [2.4] 270°   **36.** [2.4] 171.89°   **37.** [2.4] $\dfrac{7\pi}{4}$, or 5.5 cm

38. [2.4] 2.25, 129°   **39.** [2.4] About 37.9 ft/min

40. [2.4] 497,829 radians/hr

41. [2.5] $\left(\dfrac{3}{5},\dfrac{4}{5}\right),\ \left(-\dfrac{3}{5},-\dfrac{4}{5}\right),\ \left(-\dfrac{3}{5},\dfrac{4}{5}\right)$

42. [2.5] -1 **43.** [2.5] 1   **44.** [2.5] $-\dfrac{\sqrt{3}}{2}$ **45.** [2.5] $\dfrac{1}{2}$

46. [2.5] $\dfrac{\sqrt{3}}{3}$ **47.** [2.5] -1 **48.** [2.5] -0.9056

49. [2.5] 0.9218   **50.** [2.5] Undefined   **51.** [2.5] 4.3813

52. [2.5] -6.1685 **53.** [2.5] 0.8090

54. [2.5]

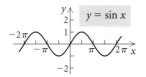

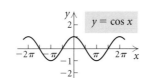

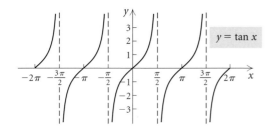

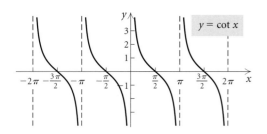

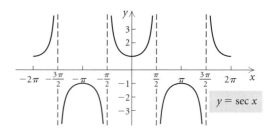

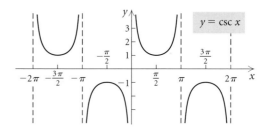

55. [2.5] Period of sin, cos, sec, csc: 2π; period of tan, cot: π

56. [2.5]

FUNCTION	DOMAIN	RANGE
Sine	$(-\infty, \infty)$	$[-1, 1]$
Cosine	$(-\infty, \infty)$	$[-1, 1]$
Tangent	$\left\{x \mid x \neq \dfrac{\pi}{2} + k\pi,\ k \in \mathbb{Z}\right\}$	$(-\infty, \infty)$

57. [2.3]

FUNCTION	I	II	III	IV
Sine	+	+	−	−
Cosine	+	−	−	+
Tangent	+	−	+	−

58. [2.6] Amplitude: 1; period: 2π; phase shift: $-\dfrac{\pi}{2}$

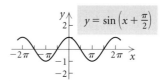

59. [2.6] Amplitude: $\frac{1}{2}$; period: π; phase shift: $\frac{\pi}{4}$

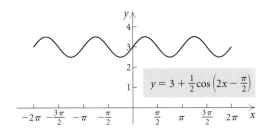

$y = 3 + \frac{1}{2}\cos\left(2x - \frac{\pi}{2}\right)$

60. [2.6] (d)   **61.** [2.6] (a)   **62.** [2.6] (c)   **63.** [2.6] (b)
64. [2.6]

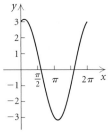

$y = 3\cos x + \sin x$

65. [2.1], [2.4] ◈ Both degrees and radians are units of angle measure. A degree is defined to be $\frac{1}{360}$ of one complete positive revolution. Degree notation has been in use since Babylonian times. Radians are defined in terms of intercepted arc length on a circle, with one radian being the measure of the angle for which the arc length equals the radius. There are 2π radians in one complete revolution.
66. [2.5] ◈ The graph of the cosine function is shaped like a continuous wave, with "high" points at $y = 1$ and "low" points at $y = -1$. The maximum value of the cosine function is 1, and it occurs at all points where $x = 2k\pi$, $k \in \mathbb{Z}$.
67. [2.6] ◈ When x is very large or very small, the amplitude of the function becomes small. The dimensions of the window must be adjusted to be able to see the shape of the graph. Also, when x is close to 0, the function is undefined, but this may not be obvious from the graph.   **68.** [2.5] ◈ No; $\sin x$ is never greater than 1.
69. [2.5] All values
70. [2.6] Domain $(-\infty, \infty)$; range $[-3, 3]$; period 4π

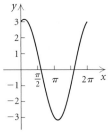

$y = 3\sin\dfrac{x}{2}$

71. [2.6] $y_2 = 2\sin\left(x + \dfrac{\pi}{2}\right) - 2$

72. [2.6] The domain consists of the intervals $\left(-\dfrac{\pi}{2} + 2k\pi, \dfrac{\pi}{2} + 2k\pi\right)$, $k \in \mathbb{Z}$.

73. [2.3] $\cos x = -0.7890$, $\tan x = -0.7787$, $\cot x = -1.2842$, $\sec x = -1.2674$, $\csc x = 1.6276$

Chapter 3

EXERCISE SET 3.1

1. $\sin^2 x - \cos^2 x$ **3.** $\sin y + \cos y$
5. $1 - 2\sin\phi\cos\phi$ **7.** $\sin^3 x + \csc^3 x$
9. $\cos x (\sin x + \cos x)$
11. $(\sin x + \cos x)(\sin x - \cos x)$
13. $(2\cos x + 3)(\cos x - 1)$
15. $(\sin x + 3)(\sin^2 x - 3\sin x + 9)$ **17.** $\tan x$
19. $\sin x + 1$ **21.** $\dfrac{2\tan t + 1}{3\tan t + 1}$ **23.** 1
25. $\dfrac{5\cot\phi}{\sin\phi + \cos\phi}$ **27.** $\dfrac{1 + 2\sin s + 2\cos s}{\sin^2 s - \cos^2 s}$
29. $\dfrac{5(\sin\theta - 3)}{3}$ **31.** $\sin x \cos x$
33. $\sqrt{\cos\alpha}\,(\sin\alpha - \cos\alpha)$ **35.** $1 - \sin y$
37. $\dfrac{\sqrt{\sin x \cos x}}{\cos x}$ **39.** $\dfrac{\sqrt{2}\cot y}{2}$ **41.** $\dfrac{\cos x}{\sqrt{\sin x \cos x}}$
43. $\dfrac{1 + \sin y}{\cos y}$ **45.** $\cos\theta = \dfrac{\sqrt{a^2 - x^2}}{a}$, $\tan\theta = \dfrac{x}{\sqrt{a^2 - x^2}}$
47. $\sin\theta = \dfrac{\sqrt{x^2 - 9}}{x}$, $\cos\theta = \dfrac{3}{x}$ **49.** $\sin\theta\tan\theta$
51. $\dfrac{\sqrt{6} - \sqrt{2}}{4}$ **53.** $\dfrac{\sqrt{3} + 1}{1 - \sqrt{3}}$, or $-2 - \sqrt{3}$
55. $\dfrac{\sqrt{6} + \sqrt{2}}{4}$ **57.** $\sin 59° \approx 0.8572$
59. $\tan 52° \approx 1.2799$ **61.** 0   **63.** $-\dfrac{7}{25}$ **65.** -1.5790
67. $2\sin\alpha\cos\beta$ **69.** $\cos u$ **71.** All real numbers
73. 1.9417 **75.** ◈   **77.** 0°; the lines are parallel
79. $\dfrac{3\pi}{4}$, or 135°   **81.** 22.83°
83. $\dfrac{\cos(x + h) - \cos x}{h}$
$$= \frac{\cos x \cos h - \sin x \sin h - \cos x}{h}$$
$$= \frac{\cos x \cos h - \cos x}{h} - \frac{\sin x \sin h}{h}$$
$$= \cos x\left(\frac{\cos h - 1}{h}\right) - \sin x\left(\frac{\sin h}{h}\right)$$
85. Let $x = \dfrac{\pi}{5}$. Then $\dfrac{\sin 5x}{x} = \dfrac{\sin\pi}{\pi/5} = 0 \neq \sin 5$. Answer may vary.

87. Let $\alpha = \dfrac{\pi}{4}$. Then $\cos(2\alpha) = \cos\dfrac{\pi}{2} = 0$, but

$2\cos\alpha = 2\cos\dfrac{\pi}{4} = \sqrt{2}$. Answer may vary.

89. Let $x = \dfrac{\pi}{6}$. Then $\dfrac{\cos 6x}{\cos x} = \dfrac{\cos\pi}{\cos\dfrac{\pi}{6}} = \dfrac{-1}{\sqrt{3}/2} \neq 6$. Answer

may vary.

91. $\dfrac{6 - 3\sqrt{3}}{9 + 2\sqrt{3}} \approx 0.0645$ **93.** 168.7°

95. $\cos 2\theta = \cos^2\theta - \sin^2\theta$, or $1 - 2\sin^2\theta$, or $2\cos^2\theta - 1$

97. $\tan\left(x + \dfrac{\pi}{4}\right) = \dfrac{\tan x + \tan\dfrac{\pi}{4}}{1 - \tan x \tan\dfrac{\pi}{4}} = \dfrac{1 + \tan x}{1 - \tan x}$

99. $\sin(\alpha + \beta) + \sin(\alpha - \beta) = \sin\alpha\cos\beta + \cos\alpha\sin\beta + \sin\alpha\cos\beta - \cos\alpha\sin\beta = 2\sin\alpha\cos\beta$

EXERCISE SET 3.2

1. (a) $\tan\dfrac{3\pi}{10} \approx 1.3763$, $\csc\dfrac{3\pi}{10} \approx 1.2361$, $\sec\dfrac{3\pi}{10} \approx 1.7013$,

$\cot\dfrac{3\pi}{10} \approx 0.7266$; **(b)** $\sin\dfrac{\pi}{5} \approx 0.5878$, $\cos\dfrac{\pi}{5} \approx 0.8090$,

$\tan\dfrac{\pi}{5} \approx 0.7266$, $\csc\dfrac{\pi}{5} \approx 1.7013$, $\sec\dfrac{\pi}{5} \approx 1.2361$,

$\cot\dfrac{\pi}{5} \approx 1.3763$

3. (a) $\cos\theta = -\dfrac{2\sqrt{2}}{3}$, $\tan\theta = -\dfrac{\sqrt{2}}{4}$, $\csc\theta = 3$,

$\sec\theta = -\dfrac{3\sqrt{2}}{4}$, $\cot\theta = -2\sqrt{2}$; **(b)** $\sin\left(\dfrac{\pi}{2} - \theta\right) = -\dfrac{2\sqrt{2}}{3}$,

$\cos\left(\dfrac{\pi}{2} - \theta\right) = \dfrac{1}{3}$, $\tan\left(\dfrac{\pi}{2} - \theta\right) = -2\sqrt{2}$,

$\csc\left(\dfrac{\pi}{2} - \theta\right) = -\dfrac{3\sqrt{2}}{4}$, $\sec\left(\dfrac{\pi}{2} - \theta\right) = 3$,

$\cot\left(\dfrac{\pi}{2} - \theta\right) = -\dfrac{\sqrt{2}}{4}$; **(c)** $\sin\left(\theta - \dfrac{\pi}{2}\right) = \dfrac{2\sqrt{2}}{3}$,

$\cos\left(\theta - \dfrac{\pi}{2}\right) = \dfrac{1}{3}$, $\tan\left(\theta - \dfrac{\pi}{2}\right) = 2\sqrt{2}$,

$\csc\left(\theta - \dfrac{\pi}{2}\right) = \dfrac{3\sqrt{2}}{4}$, $\sec\left(\theta - \dfrac{\pi}{2}\right) = 3$,

$\cot\left(\theta - \dfrac{\pi}{2}\right) = \dfrac{\sqrt{2}}{4}$

5. $\sec\left(x + \dfrac{\pi}{2}\right) = -\csc x$ **7.** $\tan\left(x - \dfrac{\pi}{2}\right) = -\cot x$

9. $\sin 2\theta = \dfrac{24}{25}$, $\cos 2\theta = -\dfrac{7}{25}$, $\tan 2\theta = -\dfrac{24}{7}$; II

11. $\sin 2\theta = \dfrac{24}{25}$, $\cos 2\theta = -\dfrac{7}{25}$, $\tan 2\theta = -\dfrac{24}{7}$; II

13. $\sin 2\theta = -\dfrac{120}{169}$, $\cos 2\theta = \dfrac{119}{169}$, $\tan 2\theta = -\dfrac{120}{119}$; IV

15. $\cos 4x = 1 - 8\sin^2 x \cos^2 x$, or $\cos^4 x - 6\sin^2 x \cos^2 x + \sin^4 x$, or $8\cos^4 x - 8\cos^2 x + 1$

17. $\dfrac{\sqrt{2 + \sqrt{3}}}{2}$ **19.** $\dfrac{\sqrt{2 + \sqrt{2}}}{2}$ **21.** $2 + \sqrt{3}$

23. 0.6421   **25.** 0.1735

27. (d); $\dfrac{\cos 2x}{\cos x - \sin x} = \dfrac{\cos^2 x - \sin^2 x}{\cos x - \sin x}$

$= \dfrac{(\cos x + \sin x)(\cos x - \sin x)}{\cos x - \sin x}$

$= \cos x + \sin x$

29. (d); $\dfrac{\sin 2x}{2\cos x} = \dfrac{2\sin x \cos x}{2\cos x} = \sin x$ **31.** $\cos x$ **33.** 1

35. $\cos 2x$ **37.** 8   **39.** $\sin^2 x$ **41.** $-\cos^2 x$ **43.** ◈

45. $\sin 141° \approx 0.6293$, $\cos 141° \approx -0.7772$, $\tan 141° \approx -0.8097$, $\csc 141° \approx 1.5891$, $\sec 141° \approx -1.2867$, $\cot 141° \approx -1.2350$

47. $-\cos x(1 + \cot x)$ **49.** $\cot^2 y$

51. $\sin\theta = -\dfrac{15}{17}$, $\cos\theta = -\dfrac{8}{17}$, $\tan\theta = \dfrac{15}{8}$

53. (a) 9.80359 m/sec²; **(b)** 9.80180 m/sec²;
(c) $g = 9.78049(1 + 0.005264\sin^2\phi + 0.000024\sin^4\phi)$;
(d) g is greatest at 90°N and 90°S (at the poles); g is least at 0° (at the equator).

EXERCISE SET 3.3

1.

$\sec x - \sin x \tan x$	$\cos x$
$\dfrac{1}{\cos x} - \sin x \cdot \dfrac{\sin x}{\cos x}$	
$\dfrac{1 - \sin^2 x}{\cos x}$	
$\dfrac{\cos^2 x}{\cos x}$	
$\cos x$	

3.

$\dfrac{1 - \cos x}{\sin x}$	$\dfrac{\sin x}{1 + \cos x}$
	$\dfrac{\sin x}{1 + \cos x} \cdot \dfrac{1 - \cos x}{1 - \cos x}$
	$\dfrac{\sin x (1 - \cos x)}{1 - \cos^2 x}$
	$\dfrac{\sin x (1 - \cos x)}{\sin^2 x}$
	$\dfrac{1 - \cos x}{\sin x}$

5.

$$\frac{1+\tan\theta}{1-\tan\theta}+\frac{1+\cot\theta}{1-\cot\theta} \qquad\Big|\qquad 0$$

$$\frac{1+\dfrac{\sin\theta}{\cos\theta}}{1-\dfrac{\sin\theta}{\cos\theta}}+\frac{1+\dfrac{\cos\theta}{\sin\theta}}{1-\dfrac{\cos\theta}{\sin\theta}}$$

$$\frac{\dfrac{\cos\theta+\sin\theta}{\cos\theta}}{\dfrac{\cos\theta-\sin\theta}{\cos\theta}}+\frac{\dfrac{\sin\theta+\cos\theta}{\sin\theta}}{\dfrac{\sin\theta-\cos\theta}{\sin\theta}}$$

$$\frac{\cos\theta+\sin\theta}{\cos\theta}\cdot\frac{\cos\theta}{\cos\theta-\sin\theta}+\frac{\sin\theta+\cos\theta}{\sin\theta}\cdot\frac{\sin\theta}{\sin\theta-\cos\theta}$$

$$\frac{\cos\theta+\sin\theta}{\cos\theta-\sin\theta}+\frac{\sin\theta+\cos\theta}{\sin\theta-\cos\theta}$$

$$\frac{\cos\theta+\sin\theta}{\cos\theta-\sin\theta}-\frac{\cos\theta+\sin\theta}{\cos\theta-\sin\theta}$$

$$0$$

7.

$$\frac{\cos^2\alpha+\cot\alpha}{\cos^2\alpha-\cot\alpha} \qquad\Big|\qquad \frac{\cos^2\alpha\tan\alpha+1}{\cos^2\alpha\tan\alpha-1}$$

$$\frac{\cos^2\alpha+\dfrac{\cos\alpha}{\sin\alpha}}{\cos^2\alpha-\dfrac{\cos\alpha}{\sin\alpha}} \qquad\Big|\qquad \frac{\cos^2\alpha\,\dfrac{\sin\alpha}{\cos\alpha}+1}{\cos^2\alpha\,\dfrac{\sin\alpha}{\cos\alpha}-1}$$

$$\frac{\cos\alpha\left(\cos\alpha+\dfrac{1}{\sin\alpha}\right)}{\cos\alpha\left(\cos\alpha-\dfrac{1}{\sin\alpha}\right)} \qquad\Big|\qquad \frac{\sin\alpha\cos\alpha+1}{\sin\alpha\cos\alpha-1}$$

$$\frac{\cos\alpha+\dfrac{1}{\sin\alpha}}{\cos\alpha-\dfrac{1}{\sin\alpha}}$$

$$\frac{\dfrac{\sin\alpha\cos\alpha+1}{\sin\alpha}}{\dfrac{\sin\alpha\cos\alpha-1}{\sin\alpha}}$$

$$\frac{\sin\alpha\cos\alpha+1}{\sin\alpha\cos\alpha-1}$$

9.

$$\frac{2\tan\theta}{1+\tan^2\theta} \qquad\Big|\qquad \sin 2\theta$$

$$\frac{2\tan\theta}{\sec^2\theta} \qquad\Big|\qquad 2\sin\theta\cos\theta$$

$$\frac{2\sin\theta}{\cos\theta}\cdot\frac{\cos^2\theta}{1}$$

$$2\sin\theta\cos\theta$$

11.

$$\begin{array}{c|c} 1-\cos5\theta\cos3\theta-\sin5\theta\sin3\theta & 2\sin^2\theta \\ 1-[\cos5\theta\cos3\theta+\sin5\theta\sin3\theta] & 1-\cos2\theta \\ 1-\cos(5\theta-3\theta) & \\ 1-\cos2\theta & \end{array}$$

13.

$$\begin{array}{c|c} 2\sin\theta\cos^3\theta+2\sin^3\theta\cos\theta & \sin 2\theta \\ 2\sin\theta\cos\theta(\cos^2\theta+\sin^2\theta) & 2\sin\theta\cos\theta \\ 2\sin\theta\cos\theta & \end{array}$$

15.

$$\frac{\tan x-\sin x}{2\tan x} \qquad\Big|\qquad \sin^2\frac{x}{2}$$

$$\frac{1}{2}\left[\frac{\dfrac{\sin x}{\cos x}-\sin x}{\dfrac{\sin x}{\cos x}}\right] \qquad\Big|\qquad \frac{1-\cos x}{2}$$

$$\frac{1}{2}\,\frac{\dfrac{\sin x-\sin x\cos x}{\cos x}\cdot\dfrac{\cos x}{\sin x}}{} \qquad\Big|\qquad \frac{1-\cos x}{2}$$

$$\frac{1-\cos x}{2}$$

17.

$$\sin(\alpha+\beta)\sin(\alpha-\beta) \qquad\Big|\qquad \sin^2\alpha-\sin^2\beta$$

$$\left(\begin{array}{c}\sin\alpha\cos\beta+\\ \cos\alpha\sin\beta\end{array}\right)\left(\begin{array}{c}\sin\alpha\cos\beta-\\ \cos\alpha\sin\beta\end{array}\right) \qquad\Big|\qquad \begin{array}{c}1-\cos^2\alpha-\\(1-\cos^2\beta)\end{array}$$

$$\sin^2\alpha\cos^2\beta-\cos^2\alpha\sin^2\beta \qquad\Big|\qquad \cos^2\beta-\cos^2\alpha$$

$$\begin{array}{c}\cos^2\beta(1-\cos^2\alpha)-\\ \cos^2\alpha(1-\cos^2\beta)\end{array}$$

$$\begin{array}{c}\cos^2\beta-\cos^2\alpha\cos^2\beta-\\ \cos^2\alpha+\cos^2\alpha\cos^2\beta\end{array}$$

$$\cos^2\beta-\cos^2\alpha$$

19.

$$\begin{array}{c|c} \tan\theta(\tan\theta+\cot\theta) & \sec^2\theta \\ \tan^2\theta+\tan\theta\cot\theta & \\ \tan^2\theta+1 & \\ \sec^2\theta & \end{array}$$

21.

$$\frac{1+\cos^2 x}{\sin^2 x} \qquad\Big|\qquad 2\csc^2 x-1$$

$$\frac{1}{\sin^2 x}+\frac{\cos^2 x}{\sin^2 x}$$

$$\csc^2 x+\cot^2 x$$

$$\csc^2 x+\csc^2 x-1$$

$$2\csc^2 x-1$$

23.

$$\frac{1+\sin x}{1-\sin x}+\frac{\sin x-1}{1+\sin x} \qquad\Big|\qquad 4\sec x\tan x$$

$$\frac{(1+\sin x)^2-(1-\sin x)^2}{1-\sin^2 x} \qquad\Big|\qquad 4\cdot\frac{1}{\cos x}\cdot\frac{\sin x}{\cos x}$$

$$\frac{(1+2\sin x+\sin^2 x)-(1-2\sin x+\sin^2 x)}{\cos^2 x} \qquad\Big|\qquad \frac{4\sin x}{\cos^2 x}$$

$$\frac{4\sin x}{\cos^2 x}$$

25.

$\cos^2 \alpha \cot^2 \alpha$	$\cot^2 \alpha - \cos^2 \alpha$
$(1 - \sin^2 \alpha) \cot^2 \alpha$	
$\cot^2 \alpha - \sin^2 \alpha \cdot \dfrac{\cos^2 \alpha}{\sin^2 \alpha}$	
$\cot^2 \alpha - \cos^2 \alpha$	

27.

$2 \sin^2 \theta \cos^2 \theta + \cos^4 \theta$	$1 - \sin^4 \theta$
$\cos^2 \theta \, (2 \sin^2 \theta + \cos^2 \theta)$	$(1 + \sin^2 \theta)(1 - \sin^2 \theta)$
$\cos^2 \theta \, (\sin^2 \theta + \sin^2 \theta + \cos^2 \theta)$	$(1 + \sin^2 \theta)(\cos^2 \theta)$
$\cos^2 \theta \, (\sin^2 \theta + 1)$	

29.

$\dfrac{1 + \sin x}{1 - \sin x}$	$(\sec x + \tan x)^2$
$\dfrac{1 + \sin x}{1 - \sin x} \cdot \dfrac{1 + \sin x}{1 + \sin x}$	$\left(\dfrac{1}{\cos x} + \dfrac{\sin x}{\cos x} \right)^2$
$\dfrac{(1 + \sin x)^2}{1 - \sin^2 x}$	$\dfrac{(1 + \sin x)^2}{\cos^2 x}$
$\dfrac{(1 + \sin x)^2}{\cos^2 x}$	

31. B;

$\dfrac{\cos x + \cot x}{1 + \csc x}$	$\cos x$
$\dfrac{\dfrac{\cos x}{1} + \dfrac{\cos x}{\sin x}}{1 + \dfrac{1}{\sin x}}$	
$\dfrac{\sin x \cos x + \cos x}{\sin x} \cdot \dfrac{\sin x}{\sin x + 1}$	
$\dfrac{\cos x \, (\sin x + 1)}{\sin x + 1}$	
$\cos x$	

33. A;

$\sin x \cos x + 1$	$\dfrac{\sin^3 x - \cos^3 x}{\sin x - \cos x}$
	$\dfrac{(\sin x - \cos x)(\sin^2 x + \sin x \cos x + \cos^2 x)}{\sin x - \cos x}$
	$\sin^2 x + \sin x \cos x + \cos^2 x$
	$\sin x \cos x + 1$

35. C;

$\dfrac{\dfrac{1}{\cot x \sin^2 x}}{\dfrac{\cos x}{\sin x} \cdot \sin^2 x}$	$\tan x + \cot x$
$\dfrac{1}{\cos x \sin x}$	$\dfrac{\sin x}{\cos x} + \dfrac{\cos x}{\sin x}$
	$\dfrac{\sin^2 x + \cos^2 x}{\cos x \sin x}$
	$\dfrac{1}{\cos x \sin x}$

37. (a), (d)

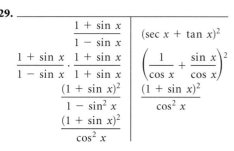

$f^{-1}(x) = \dfrac{x + 2}{3}$

(b) yes; **(c)** $f^{-1}(x) = \dfrac{x + 2}{3}$

39. (a), (d)

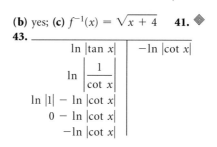

$f(x) = x^2 - 4, x \geq 0$

$f^{-1}(x) = \sqrt{x + 4}$

(b) yes; **(c)** $f^{-1}(x) = \sqrt{x + 4}$ **41.** ◈

43.

$\ln |\tan x|$	$-\ln |\cot x|$
$\ln \left| \dfrac{1}{\cot x} \right|$	
$\ln |1| - \ln |\cot x|$	
$0 - \ln |\cot x|$	
$-\ln |\cot x|$	

45. $\log (\cos x - \sin x) + \log (\cos x + \sin x)$
$= \log [(\cos x - \sin x)(\cos x + \sin x)]$
$= \log (\cos^2 x - \sin^2 x) = \log \cos 2x$

47. $\dfrac{1}{\omega C(\tan \theta + \tan \phi)} = \dfrac{1}{\omega C \left(\dfrac{\sin \theta}{\cos \theta} + \dfrac{\sin \phi}{\cos \phi} \right)}$

$= \dfrac{1}{\omega C \left(\dfrac{\sin \theta \cos \phi + \sin \phi \cos \theta}{\cos \theta \cos \phi} \right)}$

$= \dfrac{\cos \theta \cos \phi}{\omega C \sin (\theta + \phi)}$

EXERCISE SET 3.4

1. $-\dfrac{\pi}{3}$, $-60°$ **3.** $\dfrac{\pi}{4}$, $45°$ **5.** $\dfrac{\pi}{4}$, $45°$ **7.** 0, $0°$

9. $\dfrac{\pi}{6}$, $30°$ **11.** $\dfrac{\pi}{6}$, $30°$ **13.** $-\dfrac{\pi}{6}$, $-30°$

15. $-\dfrac{\pi}{6}$, $-30°$ **17.** $\dfrac{\pi}{2}$, $90°$ **19.** $\dfrac{\pi}{3}$, $60°$

21. 0.352, $20.2°$ **23.** 1.292, $74.0°$ **25.** 2.946, $168.8°$
27. -0.160, $-9.2°$ **29.** 0.829, $47.5°$ **31.** -0.960, $-55.0°$
33. $\sin^{-1}$: $[-1, 1]$; $\cos^{-1}$: $[-1, 1]$; $\tan^{-1}$: $(-\infty, \infty)$

35. $\theta = \sin^{-1}\left(\dfrac{2000}{d}\right)$ **37.** 0.3 **39.** $\dfrac{\pi}{4}$ **41.** $\dfrac{\pi}{5}$

43. $-\dfrac{\pi}{3}$ **45.** $\dfrac{1}{2}$ **47.** 1 **49.** $\dfrac{\pi}{3}$ **51.** $\dfrac{\sqrt{11}}{33}$

53. $\dfrac{a}{\sqrt{a^2 + 9}}$ **55.** $\dfrac{\sqrt{q^2 - p^2}}{p}$ **57.** $\dfrac{p}{3}$ **59.** $\dfrac{\sqrt{3}}{2}$

61. $-\dfrac{\sqrt{2}}{10}$ **63.** $xy + \sqrt{(1 - x^2)(1 - y^2)}$ **65.** 0.9861

67. $\left\{0, \dfrac{5}{2}\right\}$ **69.** $\{\pm2, \pm3i\}$ **71.** $\{27\}$ **73.** ◆   **75.** ◆

77.

	$\dfrac{\pi}{2}$
$\sin^{-1} x + \cos^{-1} x$	
$\sin\,(\sin^{-1} x + \cos^{-1} x)$	$\sin\dfrac{\pi}{2}$
$[\sin\,(\sin^{-1} x)][\cos\,(\cos^{-1} x)] +$ $[\cos\,(\sin^{-1} x)][\sin\,(\cos^{-1} x)]$	1
$x \cdot x + \sqrt{1 - x^2} \cdot \sqrt{1 - x^2}$	
$x^2 + 1 - x^2$	
	1

79.

$\sin^{-1} x$	$\tan^{-1}\dfrac{x}{\sqrt{1 - x^2}}$
$\sin\,(\sin^{-1} x)$	$\sin\left(\tan^{-1}\dfrac{x}{\sqrt{1 - x^2}}\right)$
x	x

81.

$\arcsin x$	$\arccos\sqrt{1 - x^2}$
$\sin\,(\arcsin x)$	$\sin\,(\arccos\sqrt{1 - x^2})$
x	x

83. $\theta = \arctan\dfrac{y + h}{x} - \arctan\dfrac{y}{x}$; $38.7°$

EXERCISE SET 3.5

1. $\dfrac{\pi}{6} + 2k\pi$, $\dfrac{11\pi}{6} + 2k\pi$, or $30° + k \cdot 360°$, $330° + k \cdot 360°$

3. $\dfrac{2\pi}{3} + k\pi$, or $120° + k \cdot 180°$ **5.** $98.09°$, $261.91°$

7. $\dfrac{4\pi}{3}$, $\dfrac{5\pi}{3}$ **9.** $\dfrac{\pi}{4}$, $\dfrac{3\pi}{4}$, $\dfrac{5\pi}{4}$, $\dfrac{7\pi}{4}$ **11.** $\dfrac{\pi}{6}$, $\dfrac{5\pi}{6}$, $\dfrac{3\pi}{2}$

13. $\dfrac{\pi}{6}$, $\dfrac{\pi}{2}$, $\dfrac{3\pi}{2}$, $\dfrac{11\pi}{6}$ **15.** $109.47°$, $120°$, $240°$, $250.53°$

17. 0, $\dfrac{\pi}{4}$, $\dfrac{3\pi}{4}$, π, $\dfrac{5\pi}{4}$, $\dfrac{7\pi}{4}$ **19.** $139.81°$, $220.19°$

21. $37.22°$, $169.35°$, $217.22°$, $349.35°$ **23.** 0, π, $\dfrac{7\pi}{6}$, $\dfrac{11\pi}{6}$

25. 0, $\dfrac{\pi}{2}$, π, $\dfrac{3\pi}{2}$ **27.** 0, π **29.** $\dfrac{3\pi}{4}$, $\dfrac{7\pi}{4}$

31. $\dfrac{2\pi}{3}$, $\dfrac{4\pi}{3}$, $\dfrac{3\pi}{2}$ **33.** $\dfrac{\pi}{4}$, $\dfrac{3\pi}{4}$, $\dfrac{5\pi}{4}$, $\dfrac{7\pi}{4}$ **35.** $\dfrac{\pi}{12}$, $\dfrac{5\pi}{12}$

37. 0.967, 1.853, 4.109, 4.994 **39.** $\dfrac{2\pi}{3}$, $\dfrac{4\pi}{3}$

41. 1.114, 2.773 **43.** 0.515 **45.** 0.422, 1.756
47. $B = 35°$, $b \approx 140.7$, $c \approx 245.4$ **49.** 36 **51.** ◆
53. $\dfrac{\pi}{3}$, $\dfrac{2\pi}{3}$, $\dfrac{4\pi}{3}$, $\dfrac{5\pi}{3}$ **55.** $\dfrac{\pi}{3}$, $\dfrac{4\pi}{3}$ **57.** 0
59. $e^{3\pi/2 + 2k\pi}$, where k (an integer) ≤ -1
61. 1.24 days, 6.76 days   **63.** $16.5°$N   **65.** 1 **67.** 0.1923

REVIEW EXERCISES, CHAPTER 3

1. [3.1] $\csc^2 x$ **2.** [3.1] 1 **3.** [3.1] $\tan^2 y - \cot^2 y$
4. [3.1] $\dfrac{(\cos^2 x + 1)^2}{\cos^2 x}$ **5.** [3.1] $\csc x\,(\sec x - \csc x)$
6. [3.1] $(3\sin y + 5)(\sin y - 4)$
7. [3.1] $(10 - \cos u)(100 + 10\cos u + \cos^2 u)$ **8.** [3.1] 1
9. [3.1] $\dfrac{1}{2}\sec x$ **10.** [3.1] $\dfrac{3\tan x}{\sin x - \cos x}$
11. [3.1] $\dfrac{3\cos y + 3\sin y + 2}{\cos^2 y - \sin^2 y}$ **12.** [3.1] 1
13. [3.1] $\dfrac{1}{4}\cot x$ **14.** [3.1] $\sin x + \cos x$
15. [3.1] $\dfrac{\cos x}{1 - \sin x}$ **16.** [3.1] $\dfrac{\cos x}{\sqrt{\sin x}}$ **17.** [3.1] $3\sec \theta$
18. [3.1] $\cos x \cos\dfrac{3\pi}{2} - \sin x \sin\dfrac{3\pi}{2}$
19. [3.1] $\dfrac{\tan 45° - \tan 30°}{1 + \tan 45° \tan 30°}$
20. [3.1] $\cos\,(27° - 16°)$, or $\cos 11°$ **21.** [3.1] $\dfrac{-\sqrt{6} - \sqrt{2}}{4}$
22. [3.1] $2 - \sqrt{3}$ **23.** [3.1] -0.3745 **24.** [3.2] $-\sin x$
25. [3.2] $\sin x$ **26.** [3.2] $-\cos x$

27. [3.2] **(a)** $\sin \alpha = -\dfrac{4}{5}$, $\tan \alpha = \dfrac{4}{3}$, $\cot \alpha = \dfrac{3}{4}$,

$\sec \alpha = -\dfrac{5}{3}$, $\csc \alpha = -\dfrac{5}{4}$; **(b)** $\sin\left(\dfrac{\pi}{2} - \alpha\right) = -\dfrac{3}{5}$,

$\cos\left(\dfrac{\pi}{2} - \alpha\right) = -\dfrac{4}{5}$, $\tan\left(\dfrac{\pi}{2} - \alpha\right) = \dfrac{3}{4}$,

$\cot\left(\dfrac{\pi}{2} - \alpha\right) = \dfrac{4}{3}$, $\sec\left(\dfrac{\pi}{2} - \alpha\right) = -\dfrac{5}{4}$,

$\csc\left(\dfrac{\pi}{2}-\alpha\right)=-\dfrac{5}{3}$; **(c)** $\sin\left(\alpha+\dfrac{\pi}{2}\right)=-\dfrac{3}{5}$,

$\cos\left(\alpha+\dfrac{\pi}{2}\right)=\dfrac{4}{5}$, $\tan\left(\alpha+\dfrac{\pi}{2}\right)=-\dfrac{3}{4}$,

$\cot\left(\alpha+\dfrac{\pi}{2}\right)=-\dfrac{4}{3}$, $\sec\left(\alpha+\dfrac{\pi}{2}\right)=\dfrac{5}{4}$,

$\csc\left(\alpha+\dfrac{\pi}{2}\right)=-\dfrac{5}{3}$ **28.** [3.2] $-\sec x$

29. [3.2] $\tan 2\theta=\dfrac{24}{7}$, $\cos 2\theta=\dfrac{7}{25}$, $\sin 2\theta=\dfrac{24}{25}$; I

30. [3.2] $\dfrac{\sqrt{2-\sqrt{2}}}{2}$

31. [3.2] $\sin 2\beta=0.4261$, $\cos\dfrac{\beta}{2}=0.9940$, $\cos 4\beta=0.6369$

32. [3.2] $\cos x$ **33.** [3.2] 1   **34.** [3.2] $\sin 2x$
35. [3.2] $\tan 2x$

36. [3.3]

$\dfrac{1-\sin x}{\cos x}$	$\dfrac{\cos x}{1+\sin x}$
$\dfrac{1-\sin x}{\cos x}\cdot\dfrac{\cos x}{\cos x}$	$\dfrac{\cos x}{1+\sin x}\cdot\dfrac{1-\sin x}{1-\sin x}$
$\dfrac{\cos x-\sin x\cos x}{\cos^2 x}$	$\dfrac{\cos x-\sin x\cos x}{1-\sin^2 x}$
	$\dfrac{\cos x-\sin x\cos x}{\cos^2 x}$

37. [3.3]

$\dfrac{1+\cos 2\theta}{\sin 2\theta}$	$\cot\theta$
$\dfrac{1+2\cos^2\theta-1}{2\sin\theta\cos\theta}$	$\dfrac{\cos\theta}{\sin\theta}$
$\dfrac{\cos\theta}{\sin\theta}$	

38. [3.3]

$\dfrac{\tan y+\sin y}{2\tan y}$	$\cos^2\dfrac{y}{2}$
$\dfrac{1}{2}\left[\dfrac{\dfrac{\sin y+\sin y\cos y}{\cos y}}{\dfrac{\sin y}{\cos y}}\right]$	$\dfrac{1+\cos y}{2}$
$\dfrac{1}{2}\left[\dfrac{\sin y\,(1+\cos y)}{\cos y}\cdot\dfrac{\cos y}{\sin y}\right]$	
$\dfrac{1+\cos y}{2}$	

39. [3.3]

$\dfrac{\sin x-\cos x}{\cos^2 x}$	$\dfrac{\tan^2 x-1}{\sin x+\cos x}$
	$\dfrac{\dfrac{\sin^2 x}{\cos^2 x}-1}{\sin x+\cos x}$
	$\dfrac{\dfrac{\sin^2 x-\cos^2 x}{\cos^2 x}\cdot\dfrac{1}{\sin x+\cos x}}{}$
	$\dfrac{\sin x-\cos x}{\cos^2 x}$

40. [3.3] B;

$\csc x-\cos x\cot x$	$\sin x$
$\dfrac{1}{\sin x}-\cos x\dfrac{\cos x}{\sin x}$	
$\dfrac{1-\cos^2 x}{\sin x}$	
$\dfrac{\sin^2 x}{\sin x}$	
$\sin x$	

41. [3.3] D;

$\dfrac{1}{\sin x\cos x}-\dfrac{\cos x}{\sin x}$	$\dfrac{\sin x\cos x}{1-\sin^2 x}$
$\dfrac{1}{\sin x\cos x}-\dfrac{\cos^2 x}{\sin x\cos x}$	$\dfrac{\sin x\cos x}{\cos^2 x}$
$\dfrac{1-\cos^2 x}{\sin x\cos x}$	$\dfrac{\sin x}{\cos x}$
$\dfrac{\sin^2 x}{\sin x\cos x}$	
$\dfrac{\sin x}{\cos x}$	

42. [3.3] A;

$\dfrac{\cot x-1}{1-\tan x}$	$\dfrac{\csc x}{\sec x}$
$\dfrac{\dfrac{\cos x}{\sin x}-\dfrac{\sin x}{\sin x}}{\dfrac{\cos x}{\cos x}-\dfrac{\sin x}{\cos x}}$	$\dfrac{\dfrac{1}{\sin x}}{\dfrac{1}{\cos x}}$
$\dfrac{\cos x-\sin x}{\sin x}\cdot\dfrac{\cos x}{\cos x-\sin x}$	$\dfrac{1}{\sin x}\cdot\dfrac{\cos x}{1}$
$\dfrac{\cos x}{\sin x}$	$\dfrac{\cos x}{\sin x}$

43. [3.3] C;

$$\frac{\dfrac{\cos x + 1}{\sin x} + \dfrac{\sin x}{\cos x + 1}}{\dfrac{(\cos x + 1)^2 + \sin^2 x}{\sin x\,(\cos x + 1)}} \left| \begin{array}{c} \dfrac{2}{\sin x} \end{array} \right.$$

$$\begin{array}{c} \dfrac{\cos^2 x + 2\cos x + 1 + \sin^2 x}{\sin x\,(\cos x + 1)} \\[2mm] \dfrac{2\cos x + 2}{\sin x\,(\cos x + 1)} \\[2mm] \dfrac{2(\cos x + 1)}{\sin x\,(\cos x + 1)} \\[2mm] \dfrac{2}{\sin x} \end{array}$$

44. [3.4] $-\dfrac{\pi}{6}$, $-30°$ **45.** [3.4] $\dfrac{\pi}{6}$, $30°$ **46.** [3.4] $\dfrac{\pi}{4}$, $45°$

47. [3.4] 0, $0°$ **48.** [3.4] 1.792, $102.7°$

49. [3.4] 0.398, $22.8°$ **50.** [3.4] $\dfrac{1}{2}$ **51.** [3.4] $\dfrac{\sqrt{3}}{3}$

52. [3.4] $\dfrac{\pi}{7}$ **53.** [3.4] $\dfrac{\sqrt{2}}{2}$ **54.** [3.4] $\dfrac{3}{\sqrt{b^2 + 9}}$

55. [3.4] $-\dfrac{7}{25}$

56. [3.5] $\dfrac{3\pi}{4} + 2k\pi$, $\dfrac{5\pi}{4} + 2k\pi$, or $135° + k \cdot 360°$,

$225° + k \cdot 360°$ **57.** [3.5] $\dfrac{\pi}{3} + k\pi$, or $60° + k \cdot 180°$

58. [3.5] $\dfrac{\pi}{6}, \dfrac{5\pi}{6}, \dfrac{7\pi}{6}, \dfrac{11\pi}{6}$

59. [3.5] $\dfrac{\pi}{4}, \dfrac{\pi}{2}, \dfrac{3\pi}{4}, \dfrac{5\pi}{4}, \dfrac{3\pi}{2}, \dfrac{7\pi}{4}$ **60.** [3.5] $\dfrac{2\pi}{3}, \pi, \dfrac{4\pi}{3}$

61. [3.5] 0, π **62.** [3.5] $\dfrac{\pi}{4}, \dfrac{3\pi}{4}, \dfrac{5\pi}{4}, \dfrac{7\pi}{4}$

63. [3.5] 0, $\dfrac{\pi}{2}$, π, $\dfrac{3\pi}{2}$ **64.** [3.5] $\dfrac{7\pi}{12}, \dfrac{23\pi}{12}$

65. [3.5] 0.864, 2.972, 4.006, 6.114
66. [3.5] -4.488, -2.074, 4.917 **67.** [3.5] -0.515
68. [3.3] ◈
(a) $2\cos^2 x - 1 = \cos 2x = \cos^2 x - \sin^2 x$
$= 1 \cdot (\cos^2 x - \sin^2 x)$
$= (\cos^2 x + \sin^2 x)(\cos^2 x - \sin^2 x)$
$= \cos^4 x - \sin^4 x;$
(b) $\cos^4 x - \sin^4 x = (\cos^2 x + \sin^2 x)(\cos^2 x - \sin^2 x)$
$= 1 \cdot (\cos^2 x - \sin^2 x)$
$= \cos^2 x - \sin^2 x = \cos 2x$
$= 2\cos^2 x - 1;$
(c)

$$\begin{array}{c|c} 2\cos^2 x - 1 & \cos^4 x - \sin^4 x \\ \cos 2x & (\cos^2 x + \sin^2 x)(\cos^2 x - \sin^2 x) \\ & 1 \cdot (\cos^2 x - \sin^2 x) \\ & \cos^2 x - \sin^2 x \\ & \cos 2x \end{array}$$

Answer may vary. Method 2 may be the more efficient because it involves straightforward factorization and simplification. Method 1(a) requires a "trick" such as multiplying by a particular expression equivalent to 1.
69. [3.4] ◈ The ranges of the inverse trigonometric functions are restricted in order that they might be functions.
70. [3.1] ◈ The expression $\tan(x + 450°)$ can be simplified using the sine and cosine sum formulas, but cannot be simplified using the tangent sum formula because $\tan 450°$ is undefined.
71. [3.1] $108.4°$
72. [3.1]
$\cos(u + v) = \cos u \cos v - \sin u \sin v$
$$= \cos u \cos v - \cos\left(\frac{\pi}{2} - u\right)\cos\left(\frac{\pi}{2} - v\right)$$
73. [3.2] $\cos^2 x$

74. [3.2] $\sin\theta = \sqrt{\dfrac{1}{2} + \dfrac{\sqrt{6}}{5}}$; $\cos\theta = \sqrt{\dfrac{1}{2} - \dfrac{\sqrt{6}}{5}}$;

$\tan\theta = \sqrt{\dfrac{5 + 2\sqrt{6}}{5 - 2\sqrt{6}}}$

75. [3.3] $\ln e^{\sin t} = \log_e e^{\sin t} = \sin t$

76. [3.4]

$y = \sec^{-1} x$

77. [3.4] Let $x = \dfrac{\sqrt{2}}{2}$. Then $\tan^{-1}\dfrac{\sqrt{2}}{2} \approx 0.6155$ and

$\dfrac{\sin^{-1}\dfrac{\sqrt{2}}{2}}{\cos^{-1}\dfrac{\sqrt{2}}{2}} = \dfrac{\dfrac{\pi}{4}}{\dfrac{\pi}{4}} = 1.$ **78.** [3.5] $\dfrac{\pi}{2}, \dfrac{3\pi}{2}$

Chapter 4

EXERCISE SET 4.1

1. $A = 121°$, $a \approx 33$, $c \approx 14$ **3.** $B \approx 57.4°$, $C \approx 86.1°$, $c \approx 40$, or $B \approx 122.6°$, $C \approx 20.9°$, $c \approx 14$
5. $B \approx 44°24'$, $A \approx 74°26'$, $a \approx 33.3$
7. $A = 110.36°$, $a \approx 5$ mi, $b \approx 3$ mi
9. $B \approx 83.78°$, $A \approx 12.44°$, $a \approx 12.30$ yd
11. $B \approx 14.7°$, $C \approx 135.0°$, $c \approx 28.04$ cm
13. No solution   **15.** $B = 125.27°$, $b \approx 302$ m, $c \approx 138$ m
17. 8.2 ft^2 **19.** 12 yd^2 **21.** 596.98 ft^2 **23.** 787 ft^2
25. About 12.86 ft, or 12 ft, 10 in.   **27.** About 51 ft
29. From A: about 35 mi; from B: about 66 mi
31. About 60 mi

33. 1.348, 77.2°   **35.** 18.24°   **37.**

39. Use the formula for the area of a triangle and the law of sines.

$$K = \frac{1}{2}bc \sin A \quad \text{and} \quad b = \frac{c \sin B}{\sin C},$$

$$\text{so} \quad K = \frac{c^2 \sin A \sin B}{2 \sin C}.$$

$$K = \frac{1}{2}ab \sin C \quad \text{and} \quad b = \frac{a \sin B}{\sin A},$$

$$\text{so} \quad K = \frac{a^2 \sin B \sin C}{2 \sin A}.$$

$$K = \frac{1}{2}bc \sin A \quad \text{and} \quad c = \frac{b \sin C}{\sin B},$$

$$\text{so} \quad K = \frac{b^2 \sin A \sin C}{2 \sin B}.$$

41.

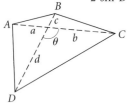

For the quadrilateral $ABCD$, we have:

$$\text{Area} = \frac{1}{2}bd \sin \theta + \frac{1}{2}ac \sin \theta$$

$$+ \frac{1}{2}ad(\sin 180° - \theta) + \frac{1}{2}bc \sin (180° - \theta)$$

$$= \frac{1}{2}(bd + ac + ad + bc) \sin \theta$$

$$= \frac{1}{2}(a + b)(c + d) \sin \theta$$

$$= \frac{1}{2}d_1 d_2 \sin \theta,$$

where $d_1 = a + b$ and $d_2 = c + d$.

EXERCISE SET 4.2

1. $a \approx 15$, $B \approx 24°$, $C \approx 126°$
3. $A \approx 36.18°$, $B \approx 43.53°$, $C \approx 100.29°$
5. $b \approx 75$ m, $A \approx 94°51'$, $C \approx 12°29'$
7. $A \approx 24.15°$, $B \approx 30.75°$, $C \approx 125.10°$
9. No solution
11. $A \approx 79.93°$, $B \approx 53.55°$, $C \approx 46.52°$
13. $c \approx 45.17$ mi, $A \approx 89.3°$, $B \approx 42.0°$
15. $a \approx 13.9$ in., $B \approx 36.127°$, $C \approx 90.417°$
17. Law of sines; $C = 98°$, $a \approx 96.7$, $c \approx 101.9$
19. Law of cosines; $A \approx 73.71°$, $B \approx 51.75°$, $C \approx 54.54°$
21. Cannot be solved
23. Law of cosines; $A \approx 33.71°$, $B \approx 107.08°$, $C \approx 39.21°$
25. About 367 ft   **27.** About 1.5 mi

29. About 37 nautical mi   **31.** About 912 km
33. (a) About 16 ft; **(b)** about 122 ft^2
35. About 4.7 cm   **37.** 5   **39.** $\frac{\sqrt{2}}{2}$ **41.** $-\frac{1}{2}$ **43.**
45. About 9386 ft
47. $A = \frac{1}{2}a^2 \sin \theta$; when $\theta = 90°$

EXERCISE SET 4.3

1. $\sqrt{17}i$ **3.** $7i$ **5.** $-9i$ **7.** $6 - 2\sqrt{21}i$
9. $-2\sqrt{19} + 5\sqrt{5}i$ **11.** $-\sqrt{55}$ **13.** $\frac{5}{4}i$ **15.** $2 + 11i$
17. $5 - 12i$ **19.** $5 + 9i$ **21.** $5 + 4i$ **23.** $5 + 7i$
25. $13 - i$ **27.** $-11 + 16i$ **29.** $35 + 14i$ **31.** $-14 + 23i$
33. 41 **35.** $12 + 16i$ **37.** $-45 - 28i$
39. $\frac{-4\sqrt{3} + 10}{41} + \frac{5\sqrt{3} + 8}{41}i$ **41.** $-\frac{14}{13} + \frac{5}{13}i$
43. $\frac{15}{146} + \frac{33}{146}i$ **45.** $-\frac{1}{2} + \frac{1}{2}i$ **47.** $-\frac{1}{2} - \frac{13}{2}i$
49. $-i$ **51.** $-i$ **53.** 1 **55.** i **57.** 625 **59.** $\{\pm\sqrt{10}i\}$
61. $\{-4 \pm 3i\}$ **63.** $\left\{-\frac{1}{2} \pm \frac{\sqrt{7}}{2}i\right\}$ **65.** $\left\{\frac{1}{10} \pm \frac{\sqrt{39}}{10}i\right\}$
67. Since $b^2 - 4ac > 0$, there are no imaginary solutions.
69. Since $b^2 - 4ac < 0$, there are imaginary solutions.
71. Since $b^2 - 4ac > 0$, there are no imaginary solutions.
73. $(x + 4)^2$ **75.** $(x - 5)^2$ **77.**   **79.** $\frac{12}{5} - \frac{1}{5}i$
81. $\frac{8}{29} + \frac{9}{29}i$ **83.** True   **85.** True   **87.** $a^2 + b^2$

EXERCISE SET 4.4

1. 5;

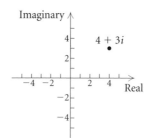

3. 1;

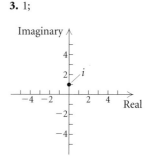

5. $\sqrt{17}$;

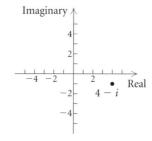

7. 3;

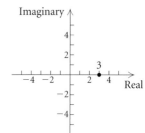

9. $3 - 3i$; $3\sqrt{2}\left(\cos \dfrac{7\pi}{4} + i \sin \dfrac{7\pi}{4}\right)$, or
$3\sqrt{2}(\cos 315° + i \sin 315°)$

11. $4i$; $4\left(\cos \dfrac{\pi}{2} + i \sin \dfrac{\pi}{2}\right)$, or $4(\cos 90° + i \sin 90°)$

13. $\sqrt{2}\left(\cos \dfrac{7\pi}{4} + i \sin \dfrac{7\pi}{4}\right)$, or $\sqrt{2}(\cos 315° + i \sin 315°)$

15. $3\left(\cos \dfrac{3\pi}{2} + i \sin \dfrac{3\pi}{2}\right)$, or $3(\cos 270° + i \sin 270°)$

17. $2\left(\cos \dfrac{\pi}{6} + i \sin \dfrac{\pi}{6}\right)$, or $2(\cos 30° + i \sin 30°)$

19. $\dfrac{2}{5}(\cos 0 \ + \ i \sin 0)$, or $\dfrac{2}{5}(\cos 0° + i \sin 0°)$

21. $\dfrac{3\sqrt{3}}{2} + \dfrac{3}{2}i$ **23.** $-10i$ **25.** $2 + 2i$ **27.** $2i$

29. $4(\cos 42° + i \sin 42°)$ **31.** $11.25(\cos 56° + i \sin 56°)$

33. 4 **35.** $-i$ **37.** $6 + 6\sqrt{3}i$ **39.** $-2i$

41. $8(\cos \pi + i \sin \pi)$ **43.** $8\left(\cos \dfrac{3\pi}{2} + i \sin \dfrac{3\pi}{2}\right)$

45. $\dfrac{27}{2} + \dfrac{27\sqrt{3}}{2}i$ **47.** $-4 + 4i$ **49.** -1

51. $-\dfrac{\sqrt{2}}{2} + \dfrac{\sqrt{2}}{2}i, \dfrac{\sqrt{2}}{2} - \dfrac{\sqrt{2}}{2}i$

53. $2(\cos 157.5° + i \sin 157.5°), 2(\cos 337.5° + i \sin 337.5°)$

55. $\dfrac{\sqrt{3}}{2} + \dfrac{1}{2}i, -\dfrac{\sqrt{3}}{2} + \dfrac{1}{2}i, -i$

57. $\sqrt[3]{4}(\cos 110° + i \sin 110°), \sqrt[3]{4}(\cos 230° + i \sin 230°),$
$\sqrt[3]{4}(\cos 350° + i \sin 350°)$

59. $2, 2i, -2, -2i$;

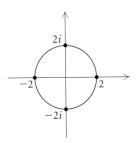

61. $\cos 36° + i \sin 36°, \cos 108° + i \sin 108°, -1,$
$\cos 252° + i \sin 252°, \cos 324° + i \sin 324°$;

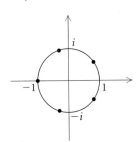

63. $\sqrt[10]{8}, \sqrt[10]{8}(\cos 36° + i \sin 36°), \sqrt[10]{8}(\cos 72° + i \sin 72°),$
$\sqrt[10]{8}(\cos 108° + i \sin 108°), \sqrt[10]{8}(\cos 144° + i \sin 144°), -\sqrt[10]{8},$
$\sqrt[10]{8}(\cos 216° + i \sin 216°), \sqrt[10]{8}(\cos 252° + i \sin 252°),$
$\sqrt[10]{8}(\cos 288° + i \sin 288°), \sqrt[10]{8}(\cos 324° + i \sin 324°)$

65. $\dfrac{\sqrt{3}}{2} + \dfrac{1}{2}i, i, -\dfrac{\sqrt{3}}{2} + \dfrac{1}{2}i, -\dfrac{\sqrt{3}}{2} - \dfrac{1}{2}i, -i, \dfrac{\sqrt{3}}{2} - \dfrac{1}{2}i$

67. $1, -\dfrac{1}{2} + \dfrac{\sqrt{3}}{2}i, -\dfrac{1}{2} - \dfrac{\sqrt{3}}{2}i$

69. $\cos 67.5° + i \sin 67.5°, \cos 157.5° + i \sin 157.5°,$
$\cos 247.5° + i \sin 247.5°, \cos 337.5° + i \sin 337.5°$

71. $\dfrac{\sqrt{3}}{2} + \dfrac{1}{2}i, i, -\dfrac{\sqrt{3}}{2} + \dfrac{1}{2}i, -\dfrac{\sqrt{3}}{2} - \dfrac{1}{2}i, -i, \dfrac{\sqrt{3}}{2} - \dfrac{1}{2}i$

73. $15°$ **75.** $\dfrac{11\pi}{6}$ **77.** $3\sqrt{5}$ **79.** ◈

81. $-\dfrac{1 + \sqrt{3}}{2} + \dfrac{1 + \sqrt{3}}{2}i, -\dfrac{1 - \sqrt{3}}{2} + \dfrac{1 - \sqrt{3}}{2}i$

83. $\cos \theta - i \sin \theta$

85. $z = a + bi, |z| = \sqrt{a^2 + b^2}; \bar{z} = a - bi,$
$|\bar{z}| = \sqrt{a^2 + (-b)^2} = \sqrt{a^2 + b^2}, \therefore |z| = |\bar{z}|$

87. $|(a + bi)^2| = |a^2 - b^2 + 2abi|$
$\qquad = \sqrt{(a^2 - b^2)^2 + 4a^2b^2}$
$\qquad = \sqrt{a^4 + 2a^2b^2 + b^4} = a^2 + b^2,$
$|a + bi|^2 = (\sqrt{a^2 + b^2})^2 = a^2 + b^2$

89. $\dfrac{z}{w} = \dfrac{r_1(\cos \theta_1 + i \sin \theta_1)}{r_2(\cos \theta_2 + i \sin \theta_2)}$
$\qquad = \dfrac{r_1}{r_2}(\cos (\theta_1 - \theta_2) + i \sin (\theta_1 - \theta_2)),$
$\left|\dfrac{z}{w}\right| = \sqrt{\left[\dfrac{r_1}{r_2}\cos (\theta_1 - \theta_2)\right]^2 + \left[\dfrac{r_1}{r_2}\sin (\theta_1 - \theta_2)\right]^2}$
$\qquad = \sqrt{\dfrac{r_1^2}{r_2^2}} = \dfrac{|r_1|}{|r_2|};$
$|z| = \sqrt{(r_1 \cos \theta_1)^2 + (r_1 \sin \theta_1)^2}$
$\qquad = \sqrt{r_1^2} = |r_1|;$
$|w| = \sqrt{(r_2 \cos \theta_2)^2 + (r_2 \sin \theta_2)^2}$
$\qquad = \sqrt{r_2^2} = |r_2|;$
Then $\left|\dfrac{z}{w}\right| = \dfrac{|r_1|}{|r_2|} = \dfrac{|z|}{|w|}.$

91.

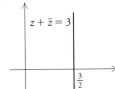

$z + \bar{z} = 3$

$\dfrac{3}{2}$

EXERCISE SET 4.5

1. Yes   **3.** No   **5.** Yes   **7.** No   **9.** No   **11.** Yes
13. 55N, 55°   **15.** 929N, 19°   **17.** 57.0, 38°   **19.** 18.4, 37°

21. 20.9, 58°   **23.** 68.3, 18°   **25.** 11 ft/sec, 63°
27. 174 nautical mi, S15°E   **29.** 60°
31. 70.7 east; 70.7 south
33. Horizontal: 215.17 mph forward; vertical: 65.78 mph up
35. Horizontal: 390 lb forward; vertical: 675.5 lb up
37. Northerly: 115 km/h; westerly: 164 km/h
39. Perpendicular: 90.6 lb; Parallel: 42.3 lb   **41.** 48.1 lb
43. $\dfrac{\sqrt{3}}{2}$ **45.** $\dfrac{\sqrt{2}}{2}$ **47.** ◈
49. (a) (4.950, 4.950); (b) (1.978, −0.950)

EXERCISE SET 4.6

1. $\langle -9, 5\rangle$; $\sqrt{106}$ **3.** $\langle -3, 6\rangle$; $3\sqrt{5}$ **5.** $\langle 4, 0\rangle$; 4   **7.** $\sqrt{37}$
9. $\langle 4, -5\rangle$ **11.** $\sqrt{257}$ **13.** $\langle -9, 9\rangle$ **15.** $\langle 41, -38\rangle$
17. $\sqrt{261} - \sqrt{65}$ **19.** $\langle -1, -1\rangle$ **21.** $\langle -8, 14\rangle$ **23.** 1
25. −34
27. 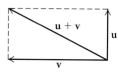   **29.**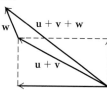

31. (a) $\mathbf{w} = \mathbf{u} + \mathbf{v}$; (b) $\mathbf{v} = \mathbf{w} - \mathbf{u}$
33. $\left\langle -\dfrac{5}{13}, \dfrac{12}{13}\right\rangle$ **35.** $\left\langle \dfrac{1}{\sqrt{101}}, -\dfrac{10}{\sqrt{101}}\right\rangle$
37. $\left\langle -\dfrac{1}{\sqrt{17}}, -\dfrac{4}{\sqrt{17}}\right\rangle$ **39.** $\mathbf{w} = -4\mathbf{i} + 6\mathbf{j}$
41. $\mathbf{s} = 2\mathbf{i} + 5\mathbf{j}$ **43.** $-7\mathbf{i} + 5\mathbf{j}$ **45.** (a) $3\mathbf{i} + 29\mathbf{j}$;
(b) $\langle 3, 29\rangle$ **47.** (a) $4\mathbf{i} + 16\mathbf{j}$; (b) $\langle 4, 16\rangle$ **49.** $\mathbf{j}$, or $\langle 0, 1\rangle$
51. $-\dfrac{1}{2}\mathbf{i} - \dfrac{\sqrt{3}}{2}\mathbf{j}$, or $\left\langle -\dfrac{1}{2}, -\dfrac{\sqrt{3}}{2}\right\rangle$ **53.** 248°   **55.** 63°
57. 50°   **59.** $|\mathbf{u}| = 3$; $\theta = 45°$ **61.** 1; 120°   **63.** 144.2°
65. 14.0°   **67.** 101.3°
69.
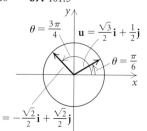
$$\theta = \frac{3\pi}{4} \qquad \mathbf{u} = \frac{\sqrt{3}}{2}\mathbf{i} + \frac{1}{2}\mathbf{j}$$
$$\theta = \frac{\pi}{6}$$
$$\mathbf{u} = -\frac{\sqrt{2}}{2}\mathbf{i} + \frac{\sqrt{2}}{2}\mathbf{j}$$

71. $\mathbf{u} = -\dfrac{\sqrt{2}}{2}\mathbf{i} - \dfrac{\sqrt{2}}{2}\mathbf{j}$ **73.** $\mathbf{u} = -\dfrac{\sqrt{10}}{10}\mathbf{i} + \dfrac{3\sqrt{10}}{10}\mathbf{j}$
75. $\sqrt{13}\left(\dfrac{2\sqrt{13}}{13}\mathbf{i} - \dfrac{3\sqrt{13}}{13}\mathbf{j}\right)$

77.
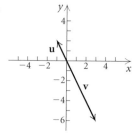

79. 174 nautical mi, S17°E   **81.** 60°
83. 500 lb on left, 866 lb on right
85. Cable: 224-lb tension, boom: 167-lb compression
87. $\mathbf{u} + \mathbf{v} = \langle u_1, u_2\rangle + \langle v_1, v_2\rangle$
$= \langle u_1 + v_1, u_2 + v_2\rangle$
$= \langle v_1 + u_1, v_2 + u_2\rangle$
$= \langle v_1, v_2\rangle + \langle u_1, u_2\rangle$
$= \mathbf{v} + \mathbf{u}$
89. (−0.286, 1.143)   **91.** (2, 5), (−3, −10).   **93.** ◈
95. (a) $\cos\theta = \dfrac{\mathbf{u} \cdot \mathbf{v}}{|\mathbf{u}||\mathbf{v}|} = \dfrac{0}{|\mathbf{u}||\mathbf{v}|}$, $\therefore \cos\theta = 0$ and $\theta = 90°$.
(b) Answers may vary. $\mathbf{u} = \langle 2, -3\rangle$ and $\mathbf{v} = \langle -3, -2\rangle$;
$\cos\theta = \dfrac{0}{\sqrt{13}\,\sqrt{13}} = 0$; $\therefore \cos\theta = 0$ and $\theta = 90°$.
97. $\dfrac{3}{5}\mathbf{i} - \dfrac{4}{5}\mathbf{j}$, $-\dfrac{3}{5}\mathbf{i} + \dfrac{4}{5}\mathbf{j}$ **99.** (5, 8)

REVIEW EXERCISES, CHAPTER 4

1. [4.2] $A \approx 153°$, $B \approx 18°$, $C \approx 9°$
2. [4.1] $A = 118°$, $a \approx 37$ in., $c \approx 24$ in.
3. [4.1] $B = 14°50'$, $a \approx 2523$ m, $c \approx 1827$ m
4. [4.1] No solution   **5.** [4.1] 33 m²   **4.** [4.1] 13.72 ft²
7. [4.1] 67 ft²   **8.** [4.2] 92.1°, 33.0°, 54.8°
9. [4.1] 419 ft   **10.** [4.2] About 650 km   **11.** [4.3] $-4i\sqrt{3}$
12. [4.3] −21   **13.** [4.3] $1 + i$ **14.** [4.3] $-5 + 3i$
15. [4.3] $27 + 18i$ **16.** [4.3] 65   **17.** [4.3] $21 - 20i$
18. [4.3] $\dfrac{4 - 7i}{13}$ **19.** [4.3] i **20.** [4.3] −1
21. [4.3] $-2 \pm 4i$
22. [4.4] $\sqrt{29}$;   **23.** [4.4] 4;

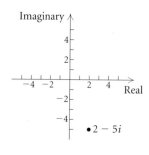

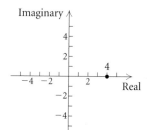

24. [4.4] 2;

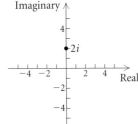

25. [4.4] $\sqrt{10}$;

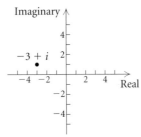

26. [4.4] $\sqrt{2}\left(\cos\dfrac{\pi}{4} + i\sin\dfrac{\pi}{4}\right)$, or $\sqrt{2}(\cos 45° + i\sin 45°)$

27. [4.4] $4\left(\cos\dfrac{3\pi}{2} + i\sin\dfrac{3\pi}{2}\right)$, or $4(\cos 270° + i\sin 270°)$

28. [4.4] $10\left(\cos\dfrac{5\pi}{6} + i\sin\dfrac{5\pi}{6}\right)$, or $10(\cos 150° + i\sin 150°)$

29. [4.4] $\dfrac{3}{4}(\cos 0 + i\sin 0)$, or $\dfrac{3}{4}(\cos 0° + i\sin 0°)$

30. [4.4] $2 + 2\sqrt{3}i$ **31.** [4.4] 7   **32.** [4.4] $-\dfrac{5}{2} + \dfrac{5\sqrt{3}}{2}i$

33. [4.4] $1 - \sqrt{3}i$ **34.** [4.4] $1 + \sqrt{3} + (-1 + \sqrt{3})i$

35. [4.4] $-i$ **36.** [4.4] $2i$ **37.** [4.4] $3\sqrt{3} + 3i$

38. [4.4] $8(\cos 180° + i\sin 180°)$

39. [4.4] $4(\cos 7\pi + i\sin 7\pi)$ **40.** [4.4] $-8i$

41. [4.4] $-\dfrac{1}{2} - \dfrac{\sqrt{3}}{2}i$

42. [4.4] $\sqrt[4]{2}\left(\cos\dfrac{3\pi}{8} + i\sin\dfrac{3\pi}{8}\right)$, $\sqrt[4]{2}\left(\cos\dfrac{11\pi}{8} + i\sin\dfrac{11\pi}{8}\right)$

43. [4.4] $\sqrt[3]{6}(\cos 110° + i\sin 110°)$, $\sqrt[3]{6}(\cos 230° + i\sin 230°)$, $\sqrt[3]{6}(\cos 350° + i\sin 350°)$

44. [4.4] $3, 3i, -3, -3i$

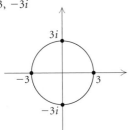

45. [4.4] $1, \cos 72° + i\sin 72°, \cos 144° + i\sin 144°$, $\cos 216° + i\sin 216°, \cos 288° + i\sin 288°$

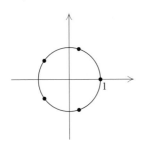

46. [4.4] $\cos 22.5° + i\sin 22.5°, \cos 112.5° + i\sin 112.5°$, $\cos 202.5° + i\sin 202.5°, \cos 292.5° + i\sin 292.5°$

47. [4.4] $\dfrac{1}{2} + \dfrac{\sqrt{3}}{2}i, -1, \dfrac{1}{2} - \dfrac{\sqrt{3}}{2}i$

48. [4.5] 13.7, 71°   **49.** [4.5] 98.7, 15°

50. [4.5]   **51.** [4.5]

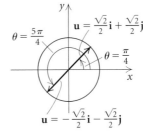

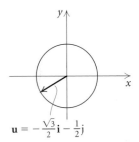

52. [4.5] 666.7 N, 36°   **53.** [4.5] 29 km/h, 329°
54. [4.5] 102.4 nautical mi, S43°E   **55.** [4.6] $\langle -4, 3\rangle$
56. [4.6] $\langle 2, -6\rangle$ **57.** [4.6] $\sqrt{61}$ **58.** [4.6] $\langle 10, -21\rangle$
59. [4.6] $\langle 14, -64\rangle$ **60.** [4.6] $5 + \sqrt{116}$ **61.** [4.6] 14
62. [4.6] $\left\langle -\dfrac{3}{\sqrt{10}}, -\dfrac{1}{\sqrt{10}}\right\rangle$ **63.** [4.6] $-9i + 4j$
64. [4.6] 194.0°   **65.** [4.6] $\sqrt{34}; \theta = 211.0°$
66. [4.6] 111.8°   **67.** [4.6] 74.5°   **68.** [4.6] $34i - 55j$
69. [4.6] $i - 12j$ **70.** [4.6] $5\sqrt{2}$
71. [4.6] $3\sqrt{65} + \sqrt{109}$ **72.** [4.6] $-5i + 5j$
73. [4.6]   **74.** [4.6]

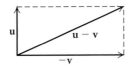

75. [4.6] $\sqrt{10}\left(\dfrac{3\sqrt{10}}{10}i - \dfrac{\sqrt{10}}{10}j\right)$

76. [4.1] ◈ When only the lengths of the sides of a triangle are known, the law of sines cannot be used to find the first angle because there will be two unknowns in the equation. To use the law of sines, you must know the measure of at least one angle and the length of the side opposite it.

77. [4.1], [4.2] ◈ A triangle has no solution when a sine or cosine value found is less than -1 or greater than 1. A triangle also has no solution if the sum of the angle measures calculated is greater than 180°. A triangle has only one solution if only one possible answer is found, or if one of the possible answers has an angle sum greater than 180°. A triangle has two solutions when two possible answers are found and neither results in an angle sum greater than 180°.

78. [4.6] $\dfrac{36}{13}i + \dfrac{15}{13}j$ **79.** [4.1] 50.52°, 129.48°

Chapter 5

EXERCISE SET 5.1

1. (f)   **3.** (b)   **5.** (d)

7. V: $(0, 0)$; F: $(0, 5)$; D: $y = -5$

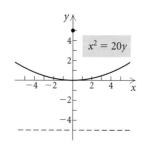

9. V: $(0, 0)$; F: $\left(-\frac{3}{2}, 0\right)$; D: $x = \frac{3}{2}$

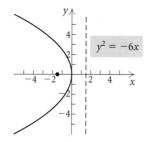

11. V: $(0, 0)$; F: $(0, 1)$; D: $y = -1$

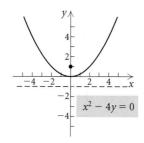

13. V: $(0, 0)$; F: $\left(\frac{1}{8}, 0\right)$; D: $x = -\frac{1}{8}$

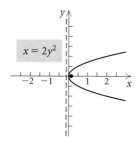

15. $y^2 = 16x$ **17.** $x^2 = -4\pi y$

19. $(y - 2)^2 = 14\left(x + \frac{1}{2}\right)$

21. V: $(-2, 1)$; F: $\left(-2, -\frac{1}{2}\right)$; D: $y = \frac{5}{2}$

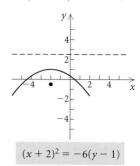

23. V: $(-1, -3)$; F: $\left(-1, -\frac{7}{2}\right)$; D: $y = -\frac{5}{2}$

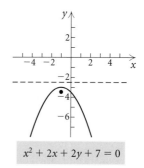

25. V: $(0, -2)$; F: $\left(0, -1\frac{3}{4}\right)$; D: $y = -2\frac{1}{4}$

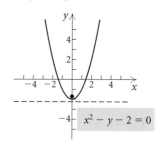

27. V: $(-2, -1)$; F: $\left(-2, -\frac{3}{4}\right)$; D: $y = -1\frac{1}{4}$

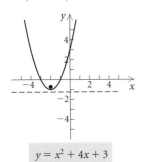

$$y = x^2 + 4x + 3$$

29. V: $\left(5\frac{3}{4}, \frac{1}{2}\right)$; F: $\left(6, \frac{1}{2}\right)$; D: $x = 5\frac{1}{2}$

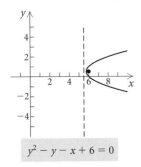

$$y^2 - y - x + 6 = 0$$

31. (a) $y^2 = 16x$; **(b)** $3\frac{33}{64}$ ft   **33.** $\frac{2}{3}$ ft, or 8 in.

35. $(x + 5)^2$ **37.** $(1, -2)$; 3   **39.** ◈

41. $(x + 1)^2 = -4(y - 2)$

43. V: $(0.867, 0.348)$; F: $(0.867, -0.191)$; D: $y = 0.887$

45. 10 ft, 11.6 ft, 16.4 ft, 24.4 ft, 35.6 ft, 50 ft

EXERCISE SET 5.2

1. (b)   **3.** (d)   **5.** (a)

7. $(7, -2)$; 8

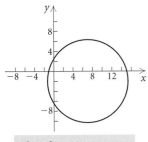

$$x^2 + y^2 - 14x + 4y = 11$$

9. $(-2, 3)$; 5

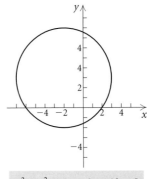

$$x^2 + y^2 + 4x - 6y - 12 = 0$$

11. $(-3, 5)$; $\sqrt{34}$

13. $\left(\frac{9}{2}, -2\right)$; $\frac{5\sqrt{5}}{2}$

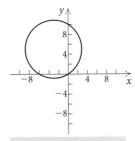

$$x^2 + y^2 + 6x - 10y = 0$$

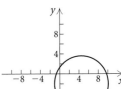

$$x^2 + y^2 - 9x = 7 - 4y$$

15. (c)   **17.** (d)

19. V: $(2, 0)$, $(-2, 0)$;

   F: $(\sqrt{3}, 0)$, $(-\sqrt{3}, 0)$

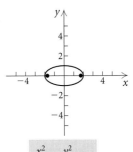

$$\frac{x^2}{4} + \frac{y^2}{1} = 1$$

21. V: $(0, 4)$, $(0, -4)$;

   F: $(0, \sqrt{7})$, $(0, -\sqrt{7})$

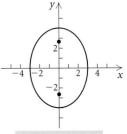

$$16x^2 + 9y^2 = 144$$

23. V: $(-\sqrt{3}, 0)$, $(\sqrt{3}, 0)$;
F: $(-1, 0)$, $(1, 0)$

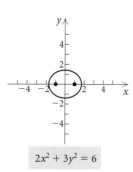

$$2x^2 + 3y^2 = 6$$

25. V: $\left(-\dfrac{1}{2}, 0\right)$, $\left(\dfrac{1}{2}, 0\right)$;

F: $\left(-\dfrac{\sqrt{5}}{6}, 0\right)$, $\left(\dfrac{\sqrt{5}}{6}, 0\right)$

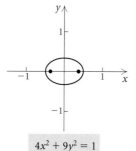

$$4x^2 + 9y^2 = 1$$

27. $\dfrac{x^2}{49} + \dfrac{y^2}{40} = 1$ **29.** $\dfrac{x^2}{25} + \dfrac{y^2}{64} = 1$ **31.** $\dfrac{x^2}{9} + \dfrac{y^2}{5} = 1$

33. C: $(1, 2)$; V: $(4, 2)$, $(-2, 2)$; F: $(1 + \sqrt{5}, 2)$, $(1 - \sqrt{5}, 2)$

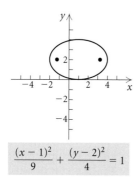

$$\dfrac{(x - 1)^2}{9} + \dfrac{(y - 2)^2}{4} = 1$$

35. C: $(-3, 5)$; V: $(-3, 11)$, $(-3, -1)$; F: $(-3, 5 + \sqrt{11})$, $(-3, 5 - \sqrt{11})$

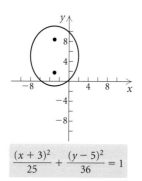

$$\dfrac{(x + 3)^2}{25} + \dfrac{(y - 5)^2}{36} = 1$$

37. C: $(-2, 1)$;
V: $(-10, 1)$, $(6, 1)$;
F: $(-6, 1)$, $(2, 1)$

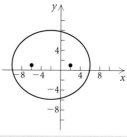

$$3(x + 2)^2 + 4(y - 1)^2 = 192$$

39. C: $(2, -1)$;
V: $(-1, -1)$, $(5, -1)$;
F: $(2 + \sqrt{5}, -1)$,
$(2 - \sqrt{5}, -1)$

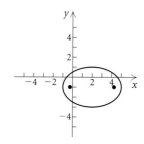

$$4x^2 + 9y^2 - 16x + 18y - 11 = 0$$

41. C: $(1, 1)$; V: $(1, 3)$, $(1, -1)$;
F: $(1, 1 + \sqrt{3})$, $(1, 1 - \sqrt{3})$

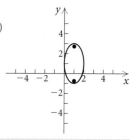

$$4x^2 + y^2 - 8x - 2y + 1 = 0$$

43. Example 2; $\dfrac{3}{5} < \dfrac{\sqrt{12}}{4}$ **45.** $\dfrac{x^2}{15} + \dfrac{y^2}{16} = 1$

47. $\dfrac{x^2}{2500} + \dfrac{y^2}{144} = 1$ **49.** 2×10^6 mi

51.

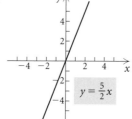

$$y = \dfrac{5}{2}x$$

53.

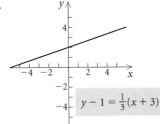

$$y - 1 = \frac{1}{3}(x + 3)$$

55. ◈   **57.** $\dfrac{(x - 3)^2}{4} + \dfrac{(y - 1)^2}{25} = 1$

59. $\dfrac{x^2}{9} + \dfrac{y^2}{484/5} = 1$

61. C: $(2.003, -1.005)$; V: $(-1.017, -1.005), (5.023, -1.005)$

63. About 9.1 ft

EXERCISE SET 5.3

1. (b)   **3.** (c)   **5.** (a)   **7.** $\dfrac{y^2}{9} - \dfrac{x^2}{16} = 1$ **9.** $\dfrac{x^2}{4} - \dfrac{y^2}{9} = 1$

11. C: $(0, 0)$; V: $(2, 0), (-2, 0)$; F: $(2\sqrt{2}, 0), (-2\sqrt{2}, 0)$;
A: $y = x, y = -x$

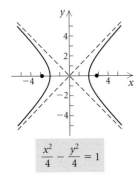

$$\frac{x^2}{4} - \frac{y^2}{4} = 1$$

13. C: $(2, -5)$; V: $(-1, -5), (5, -5)$; F: $(2 - \sqrt{10}, -5)$,
$(2 + \sqrt{10}, -5)$; A: $y = -\dfrac{x}{3} - \dfrac{13}{3}, y = \dfrac{x}{3} - \dfrac{17}{3}$

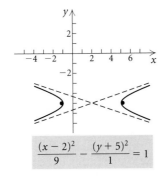

$$\frac{(x - 2)^2}{9} - \frac{(y + 5)^2}{1} = 1$$

15. C: $(-1, -3)$; V: $(-1, -1), (-1, -5)$;
F: $(-1, -3 + 2\sqrt{5}), (-1, -3 - 2\sqrt{5})$; A: $y = \frac{1}{2}x - \frac{5}{2}$,
$y = -\frac{1}{2}x - \frac{7}{2}$

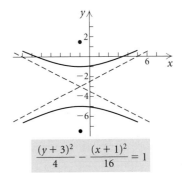

$$\frac{(y + 3)^2}{4} - \frac{(x + 1)^2}{16} = 1$$

17. C: $(0, 0)$; V: $(-2, 0), (2, 0)$; F: $(-\sqrt{5}, 0), (\sqrt{5}, 0)$;
A: $y = -\frac{1}{2}x, y = \frac{1}{2}x$

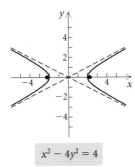

$$x^2 - 4y^2 = 4$$

19. C: $(0, 0)$; V: $(0, -3), (0, 3)$; F: $(0, -3\sqrt{10}), (0, 3\sqrt{10})$;
A: $y = \frac{1}{3}x, y = -\frac{1}{3}x$

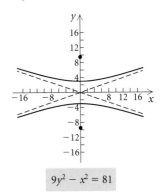

$$9y^2 - x^2 = 81$$

21. C: $(0, 0)$; V: $(-\sqrt{2}, 0)$, $(\sqrt{2}, 0)$; F: $(-2, 0)$, $(2, 0)$;
A: $y = x$, $y = -x$

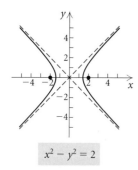

$$x^2 - y^2 = 2$$

23. C: $(0, 0)$; V: $\left(0, -\dfrac{1}{2}\right)$, $\left(0, \dfrac{1}{2}\right)$; F: $\left(0, -\dfrac{\sqrt{2}}{2}\right)$,
$\left(0, \dfrac{\sqrt{2}}{2}\right)$; A: $y = x$, $y = -x$

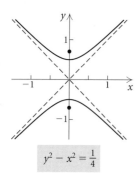

$$y^2 - x^2 = \tfrac{1}{4}$$

25. C: $(1, -2)$; V: $(0, -2)$, $(2, -2)$; F: $(1 - \sqrt{2}, -2)$,
$(1 + \sqrt{2}, -2)$; A: $y = -x - 1$, $y = x - 3$

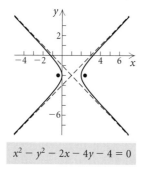

$$x^2 - y^2 - 2x - 4y - 4 = 0$$

27. C: $\left(\dfrac{1}{3}, 3\right)$; V: $\left(-\dfrac{2}{3}, 3\right)$, $\left(\dfrac{4}{3}, 3\right)$; F: $\left(\dfrac{1}{3} - \sqrt{37}, 3\right)$,
$\left(\dfrac{1}{3} + \sqrt{37}, 3\right)$; A: $y = 6x + 1$, $y = -6x + 5$

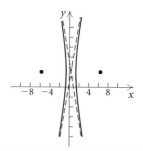

$$36x^2 - y^2 - 24x + 6y - 41 = 0$$

29. C: $(3, 1)$; V: $(3, 3)$, $(3, -1)$; F: $(3, 1 + \sqrt{13})$,
$(3, 1 - \sqrt{13})$; A: $y = \tfrac{2}{3}x - 1$, $y = -\tfrac{2}{3}x + 3$

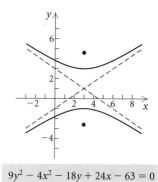

$$9y^2 - 4x^2 - 18y + 24x - 63 = 0$$

31. C: $(1, -2)$; V: $(2, -2)$, $(0, -2)$; F: $(1 + \sqrt{2}, -2)$,
$(1 - \sqrt{2}, -2)$; A: $y = x - 3$, $y = -x - 1$

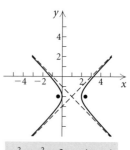

$$x^2 - y^2 - 2x - 4y = 4$$

33. C: $(-3, 4)$; V: $(-3, 10)$, $(-3, -2)$; F: $(-3, 4 + 6\sqrt{2})$, $(-3, 4 - 6\sqrt{2})$; A: $y = x + 7$, $y = -x + 1$

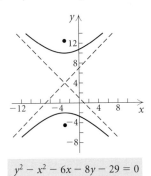

$$y^2 - x^2 - 6x - 8y - 29 = 0$$

35. Example 3; $\dfrac{\sqrt{5}}{1} > \dfrac{5}{4}$ **37.** $\dfrac{x^2}{9} - \dfrac{(y - 7)^2}{16} = 1$

39. $\dfrac{y^2}{25} - \dfrac{x^2}{11} = 1$ **41.** $(6, -1)$ **43.** $(2, -1)$ **45.** ◈

47. $\dfrac{(y + 5)^2}{9} - (x - 3)^2 = 1$

49. C: $(-1.460, -0.957)$; V: $(-2.360, -0.957)$, $(-0.560, -0.957)$; A: $y = -1.429x - 3.043$, $y = 1.429x + 1.129$

51. $\dfrac{x^2}{345.96} - \dfrac{y^2}{22{,}154.04} = 1$

EXERCISE SET 5.4

1. (e)   **3.** (c)   **5.** (b)   **7.** $(-4, -3)$, $(3, 4)$
9. $(0, 2)$, $(3, 0)$ **11.** $(-5, 0)$, $(4, 3)$, $(4, -3)$
13. $(3, 0)$, $(-3, 0)$ **15.** $(0, -3)$, $(4, 5)$ **17.** $(-2, 1)$
19. $(3, 4)$, $(-3, -4)$, $(4, 3)$, $(-4, -3)$
21. $\left(\dfrac{6\sqrt{21}}{7}, \dfrac{4i\sqrt{35}}{7}\right)$, $\left(\dfrac{6\sqrt{21}}{7}, -\dfrac{4i\sqrt{35}}{7}\right)$,
$\left(-\dfrac{6\sqrt{21}}{7}, \dfrac{4i\sqrt{35}}{7}\right)$, $\left(-\dfrac{6\sqrt{21}}{7}, -\dfrac{4i\sqrt{35}}{7}\right)$
23. $(3, 2)$, $\left(4, \dfrac{3}{2}\right)$
25. $\left(\dfrac{5 + \sqrt{70}}{3}, \dfrac{-1 + \sqrt{70}}{3}\right)$, $\left(\dfrac{5 - \sqrt{70}}{3}, \dfrac{-1 - \sqrt{70}}{3}\right)$
27. $(\sqrt{2}, \sqrt{14})$, $(-\sqrt{2}, \sqrt{14})$, $(\sqrt{2}, -\sqrt{14})$, $(-\sqrt{2}, -\sqrt{14})$
29. $(1, 2)$, $(-1, -2)$, $(2, 1)$, $(-2, -1)$
31. $\left(\dfrac{15 + \sqrt{561}}{8}, \dfrac{11 - 3\sqrt{561}}{8}\right)$,
$\left(\dfrac{15 - \sqrt{561}}{8}, \dfrac{11 + 3\sqrt{561}}{8}\right)$
33. $\left(\dfrac{7 - \sqrt{33}}{2}, \dfrac{7 + \sqrt{33}}{2}\right)$, $\left(\dfrac{7 + \sqrt{33}}{2}, \dfrac{7 - \sqrt{33}}{2}\right)$

35. $(3, 2)$, $(-3, -2)$, $(2, 3)$, $(-2, -3)$
37. $\left(\dfrac{5 - 9\sqrt{15}}{20}, \dfrac{-45 + 3\sqrt{15}}{20}\right)$,
$\left(\dfrac{5 + 9\sqrt{15}}{20}, \dfrac{-45 - 3\sqrt{15}}{20}\right)$ **39.** $(3, -5)$, $(-1, 3)$
41. $(8, 5)$, $(-5, -8)$ **43.** $(3, 2)$, $(-3, -2)$
45. $(2, 1)$, $(-2, -1)$, $(1, 2)$, $(-1, -2)$
47. $\left(4 + \dfrac{3i\sqrt{6}}{2}, -4 + \dfrac{3i\sqrt{6}}{2}\right)$, $\left(4 - \dfrac{3i\sqrt{6}}{2}, -4 - \dfrac{3i\sqrt{6}}{2}\right)$
49. $(3, \sqrt{5})$, $(-3, -\sqrt{5})$, $(\sqrt{5}, 3)$, $(-\sqrt{5}, -3)$
51. $\left(\dfrac{8\sqrt{5}}{5}i, \dfrac{3\sqrt{105}}{5}\right)$, $\left(\dfrac{8\sqrt{5}}{5}i, -\dfrac{3\sqrt{105}}{5}\right)$,
$\left(-\dfrac{8\sqrt{5}}{5}i, \dfrac{3\sqrt{105}}{5}\right)$, $\left(-\dfrac{8\sqrt{5}}{5}i, -\dfrac{3\sqrt{105}}{5}\right)$

53. $(2, 1)$, $(-2, -1)$, $\left(-i\sqrt{5}, \dfrac{2i\sqrt{5}}{5}\right)$, $\left(i\sqrt{5}, -\dfrac{2i\sqrt{5}}{5}\right)$

55. 6 cm by 8 cm   **57.** 4 in. by 5 in.   **59.** 1 m by $\sqrt{3}$ m
61. 30 yd by 75 yd   **63.** 16 ft, 24 ft   **65.** 120°, 300°
67. 15°, 105°, 195°, 285°   **69.** ◈

71. $(x - 2)^2 + (y - 3)^2 = 1$ **73.** $\dfrac{x^2}{4} + y^2 = 1$

75. $\left(x + \dfrac{5}{13}\right)^2 + \left(y - \dfrac{32}{13}\right)^2 = \dfrac{5365}{169}$

77. There is no number x such that $\dfrac{x^2}{a^2} - \dfrac{\left(\dfrac{b}{a}x\right)^2}{b^2} = 1$,

because the left side simplifies to $\dfrac{x^2}{a^2} - \dfrac{x^2}{a^2}$, which is 0.
79. $\left(\dfrac{1}{2}, \dfrac{1}{4}\right)$, $\left(\dfrac{1}{2}, -\dfrac{1}{4}\right)$, $\left(-\dfrac{1}{2}, \dfrac{1}{4}\right)$, $\left(-\dfrac{1}{2}, -\dfrac{1}{4}\right)$
81. Factor: $x^3 + y^3 = (x + y)(x^2 - xy + y^2)$. We know
that $x + y = 1$, so $(x + y)^2 = x^2 + 2xy + y^2 = 1$, or
$x^2 + y^2 = 1 - 2xy$. We also know that $xy = 1$, so
$x^2 + y^2 = 1 - 2 \cdot 1 = -1$. Then
$x^3 + y^3 = 1 \cdot (-1 - 1) = -2$.
83. $(2, 4)$, $(4, 2)$ **85.** $(3, -2)$, $(-3, 2)$, $(2, -3)$, $(-2, 3)$
87. $(0.965, 4402.33)$, $(-0.965, -4402.33)$
89. $(2.112, -0.109)$, $(-13.041, -13.337)$
91. $(400, 1.431)$, $(-400, 1.431)$, $(400, -1.431)$, $(-400, -1.431)$

EXERCISE SET 5.5

1. $(0, -2)$ **3.** $(1, \sqrt{3})$ **5.** $(\sqrt{2}, 0)$ **7.** $(\sqrt{3}, 1)$
9. Ellipse or circle   **11.** Hyperbola   **13.** Parabola
15. Hyperbola   **17.** Ellipse or circle

19.

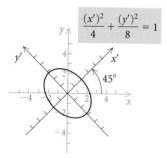

$$\frac{(x')^2}{4} + \frac{(y')^2}{8} = 1$$

21.

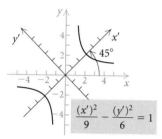

$$\frac{(x')^2}{9} - \frac{(y')^2}{6} = 1$$

23.

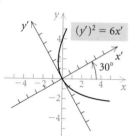

$$(y')^2 = 6x'$$

25.

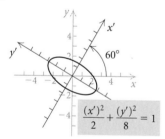

$$\frac{(x')^2}{2} + \frac{(y')^2}{8} = 1$$

27.

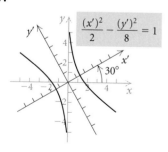

$$\frac{(x')^2}{2} - \frac{(y')^2}{8} = 1$$

29.

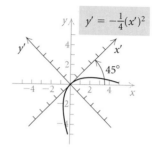

$$y' = -\frac{1}{4}(x')^2$$

31.

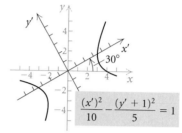

$$\frac{(x')^2}{10} - \frac{(y'+1)^2}{5} = 1$$

33.

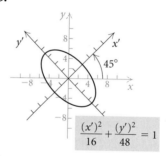

$$\frac{(x')^2}{16} + \frac{(y')^2}{48} = 1$$

35.

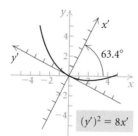

$$(y')^2 = 8x'$$

37.

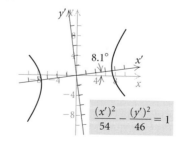

$$\frac{(x')^2}{54} - \frac{(y')^2}{46} = 1$$

39. $\dfrac{2\pi}{3}$ **41.** $60°$ **43.** $\sqrt{29}$ **45.** ◈

47. $x = x' \cos\theta - y' \sin\theta,\ y = x' \sin\theta + y' \cos\theta$

49. $A' + C' = A\cos^2\theta + B\sin\theta\cos\theta + C\sin^2\theta$
$\qquad\qquad + A\sin^2\theta - B\sin\theta\cos\theta + C\cos^2\theta$
$\qquad\quad = A(\sin^2\theta + \cos^2\theta) + C(\sin^2\theta + \cos^2\theta)$
$\qquad\quad = A + C$

EXERCISE SET 5.6

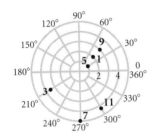

13. A: $(4, 30°)$, $(4, 390°)$, $(-4, 210°)$; B: $(5, 300°)$, $(5, -60°)$, $(-5, 120°)$; C: $(2, 150°)$, $(2, 510°)$, $(-2, 330°)$; D: $(3, 225°)$, $(3, -135°)$, $(-3, 45°)$; answers may vary

15. $(3, 270°)$, $\left(3, \dfrac{3\pi}{2}\right)$ **17.** $(6, 300°)$, $\left(6, \dfrac{5\pi}{3}\right)$

19. $(8, 330°)$, $\left(8, \dfrac{11\pi}{6}\right)$ **21.** $(2, 225°)$, $\left(2, \dfrac{5\pi}{4}\right)$

23. $(7.616, 66.8°)$, $(7.616, 1.166)$

25. $(4.643, 132.9°)$, $(4.643, 2.320)$

27. $\left(\dfrac{5}{2}, \dfrac{5\sqrt{3}}{2}\right)$ **29.** $\left(-\dfrac{3\sqrt{2}}{2}, -\dfrac{3\sqrt{2}}{2}\right)$

31. $\left(-\dfrac{3}{2}, -\dfrac{3\sqrt{3}}{2}\right)$ **33.** $(-1, \sqrt{3})$ **35.** $(2.19, -2.05)$

37. $(1.30, -3.99)$ **39.** $r(3\cos\theta + 4\sin\theta) = 5$

41. $r\cos\theta = 5$ **43.** $r = 6$ **45.** $r^2\cos^2\theta = 25r\sin\theta$

47. $r^2\sin^2\theta - 5r\cos\theta - 25 = 0$ **49.** $x^2 + y^2 = 25$

51. $y = 2$ **53.** $y^2 = -6x + 9$

55. $x^2 - 9x + y^2 - 7y = 0$ **57.** $x = 5$

59.

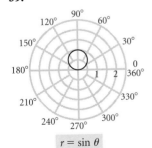
$r = \sin\theta$

61.

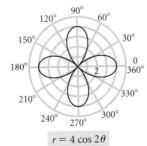
$r = 4\cos 2\theta$

63. (d)   **65.** (g)   **67.** (j)   **69.** (b)   **71.** (e)   **73.** (k)

75.

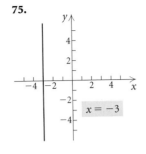
$x = -3$

77.

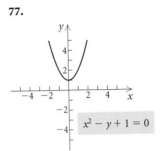
$x^2 - y + 1 = 0$

79.

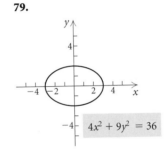
$4x^2 + 9y^2 = 36$

81. ◈   **83.** $y^2 = -4x + 4$

85.

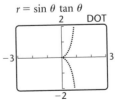
$r = \sin\theta\tan\theta$

87.

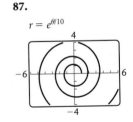
$r = e^{\theta/10}$

89.

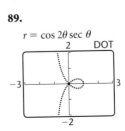
$r = \cos 2\theta\sec\theta$

91.

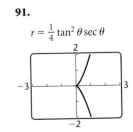
$r = \dfrac{1}{4}\tan^2\theta\sec\theta$

EXERCISE SET 5.7

1. b   **3.** a   **5.** d

7. (a) Parabola; (b) vertical, 1 unit to the right of the pole;
(c) $\left(\frac{1}{2}, 0\right)$; (d)

$$r = \frac{1}{1 + \cos \theta}$$

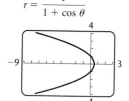

9. (a) Hyperbola; (b) horizontal, $\frac{3}{2}$ units below the pole;
(c) $\left(-3, \frac{\pi}{2}\right), \left(1, \frac{3\pi}{2}\right)$; (d)

$$r = \frac{15}{5 - 10 \sin \theta}$$

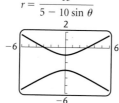

11. (a) Ellipse; (b) vertical, $\frac{8}{3}$ units to the left of the pole;
(c) $\left(\frac{8}{3}, 0\right), \left(\frac{8}{9}, \pi\right)$; (d)

$$r = \frac{8}{6 - 3 \cos \theta}$$

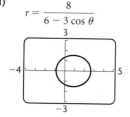

13. (a) Hyperbola; (b) horizontal, $\frac{4}{3}$ units above the pole;
(c) $\left(\frac{4}{5}, \frac{\pi}{2}\right), \left(-4, \frac{3\pi}{2}\right)$; (d)

$$r = \frac{20}{10 + 15 \sin \theta}$$

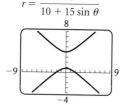

15. (a) Ellipse; (b) vertical, 3 units to the right of the pole;
(c) $(1, 0), (3, \pi)$; (d)

$$r = \frac{9}{6 + 3 \cos \theta}$$

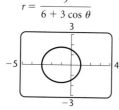

17. (a) Parabola; (b) horizontal, $\frac{3}{2}$ units below the pole;
(c) $\left(\frac{3}{4}, \frac{3\pi}{2}\right)$; (d)

$$r = \frac{3}{2 - 2 \sin \theta}$$

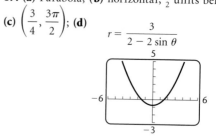

19. (a) Ellipse; (b) vertical, 4 units to the left of the pole;
(c) $(4, 0), \left(\frac{4}{3}, \pi\right)$; (d)

$$r = \frac{4}{2 - \cos \theta}$$

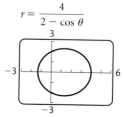

21. (a) Hyperbola; (b) horizontal, $\frac{7}{10}$ units above the pole;
(c) $\left(\frac{7}{12}, \frac{\pi}{2}\right), \left(-\frac{7}{8}, \frac{3\pi}{2}\right)$; (d)

$$r = \frac{7}{2 + 10 \sin \theta}$$

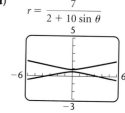

23. $y^2 + 2x - 1 = 0$ **25.** $x^2 - 3y^2 - 12y - 9 = 0$
27. $27x^2 + 36y^2 - 48x - 64 = 0$
29. $4x^2 - 5y^2 + 24y - 16 = 0$
31. $3x^2 + 4y^2 + 6x - 9 = 0$ **33.** $4x^2 - 12y - 9 = 0$
35. $3x^2 + 4y^2 - 8x - 16 = 0$

37. $4x^2 - 96y^2 + 140y - 49 = 0$ **39.** $r = \dfrac{6}{1 + 2 \sin \theta}$

41. $r = \dfrac{4}{1 + \cos \theta}$ **43.** $r = \dfrac{1}{1 - \frac{1}{2} \cos \theta}$, or $r = \dfrac{2}{2 - \cos \theta}$

45. $r = \dfrac{\frac{15}{4}}{1 + \frac{3}{4} \sin \theta}$, or $r = \dfrac{15}{4 + 3 \sin \theta}$

47. $r = \dfrac{8}{1 - 4 \sin \theta}$

49. $f(t) = (t - 3)^2 + 4$, or $t^2 - 6t + 13$
51. $f(t - 1) = (t - 4)^2 + 4$, or $t^2 - 8t + 20$ **53.** ◈

55. $r = \dfrac{1.5 \times 10^8}{1 + \sin \theta}$

EXERCISE SET 5.8

1. $y = 12x - 7,\ -\dfrac{1}{2} \le x \le 3$

3. $y = \sqrt[3]{x} - 4,\ -1 \le x \le 1000$ **5.** $x = y^4,\ 0 \le x \le 16$

7. $y = \dfrac{1}{x},\ 1 \le x \le 5$ **9.** $y = \dfrac{1}{4}(x + 1)^2,\ -7 \le x \le 5$

11. $y = \dfrac{1}{x},\ x > 0$ **13.** $x^2 + y^2 = 9,\ -3 \le x \le 3$

15. $x^2 + \dfrac{y^2}{4} = 1,\ -1 \le x \le 1$ **17.** $y = \dfrac{1}{x},\ x \ge 1$

19. $(x - 1)^2 + (y - 2)^2 = 4,\ -1 \le x \le 3$ **21.** 0.7071

23. -0.2588 **25.** 0.7265   **27.** 0.5460

29. Answers may vary. $x = t,\ y = 4t - 3;\ x = \dfrac{t}{4} + 3,$

$y = t + 9$

31. Answers may vary. $x = t,\ y = (t - 2)^2 - 6t;\ x = t + 2,$
$y = t^2 - 6t - 12$

33. (a) About 36.1 ft; **(b)** about 196.6 ft; **(c)** about 3 sec

35. About 73.1 ft

37. $x = 2(t - \sin t),\ y = 2(1 - \cos t);\ 0 \le t \le 4\pi$

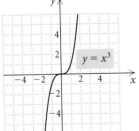

39. $x = t - \sin t,\ y = 1 - \cos t;\ -2\pi \le t \le 2\pi$

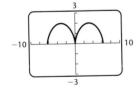

41.

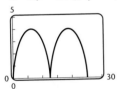

$y = x^3$

43.

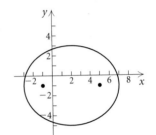

$f(x) = \sqrt{x - 2}$

45.   **47.** $x = 3 \cos t,\ y = -3 \sin t$

REVIEW EXERCISES, CHAPTER 5

1. [5.1] (d)   **2.** [5.2] (a)   **3.** [5.2] (e)   **4.** [5.3] (g)
5. [5.2] (b)   **6.** [5.2] (f)   **7.** [5.1] (h)   **8.** [5.3] (c)
9. [5.1] $x^2 = -6y$ **10.** [5.1] F: $(-3, 0)$; V: $(0, 0)$; D: $x = 3$
11. [5.1] V: $(-5, 8)$; F: $\left(-5, \dfrac{15}{2}\right)$; D: $y = \dfrac{17}{2}$
12. [5.2] C: $(2, -1)$; V: $(-3, -1),\ (7, -1)$; F: $(-1, -1)$,
$(5, -1)$;

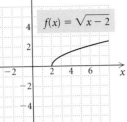

$16x^2 + 25y^2 - 64x + 50y - 311 = 0$

13. [5.2] $\dfrac{x^2}{9} + \dfrac{y^2}{16} = 1$

14. [5.3] C: $\left(-2, \dfrac{1}{4}\right)$; V: $\left(0, \dfrac{1}{4}\right),\ \left(-4, \dfrac{1}{4}\right)$;

F: $\left(-2 + \sqrt{6}, \dfrac{1}{4}\right),\ \left(-2 - \sqrt{6}, \dfrac{1}{4}\right)$;

A: $y - \dfrac{1}{4} = \dfrac{\sqrt{2}}{2}(x + 2),\ y - \dfrac{1}{4} = -\dfrac{\sqrt{2}}{2}(x + 2)$

15. [5.1] 0.167 ft   **16.** [5.4] $(-8\sqrt{2}, 8),\ (8\sqrt{2}, 8)$

17. [5.4] $\left(3, \dfrac{\sqrt{29}}{2}\right),\ \left(-3, \dfrac{\sqrt{29}}{2}\right),\ \left(3, -\dfrac{\sqrt{29}}{2}\right),$

$\left(-3, -\dfrac{\sqrt{29}}{2}\right)$ **18.** [5.4] $(7, 4)$ **19.** [5.4] $(2, 2),\ \left(\dfrac{32}{9}, -\dfrac{10}{9}\right)$

20. [5.4] $(0, -3),\ (2, 1)$
21. [5.4] $(4, 3),\ (4, -3),\ (-4, 3),\ (-4, -3)$
22. [5.4] $(-\sqrt{3}, 0),\ (\sqrt{3}, 0),\ (-2, 1),\ (2, 1)$
23. [5.4] $\left(-\dfrac{3}{5}, \dfrac{21}{5}\right),\ (3, -3)$
24. [5.4] $(6, 8),\ (6, -8),\ (-6, 8),\ (-6, -8)$
25. [5.4] $(2, 2),\ (-2, -2),\ (2\sqrt{2}, \sqrt{2}),\ (-2\sqrt{2}, -\sqrt{2})$

26. [5.4] 7, 4   **27.** [5.4] 7 m by 12 m   **28.** [5.4] 4, 8
29. [5.4] 32 cm, 20 cm   **30.** [5.4] 11 ft, 3 ft
31. [5.5]

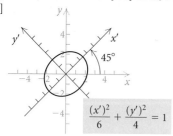

$$\frac{(x')^2}{6} + \frac{(y')^2}{4} = 1$$

32. [5.5]

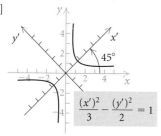

$$\frac{(x')^2}{3} - \frac{(y')^2}{2} = 1$$

33. [5.5]

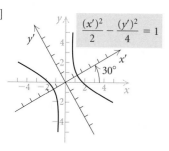

$$\frac{(x')^2}{2} - \frac{(y')^2}{4} = 1$$

34. [5.5]

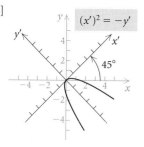

$$(x')^2 = -y'$$

35. [5.6] A: (5, 120°), (5, 480°), (−5, 300°); B: (3, 210°),
(−3, 30°), (−3, 390°); C: (4, 60°), (4, 420°), (−4, 240°); D:
(1, 300°), (1, −60°), (−1, 120°); answers may vary

36. [5.6] (8, 135°), $\left(8, \dfrac{3\pi}{4}\right)$ **37.** [5.6] (5, 270°), $\left(5, \dfrac{3\pi}{2}\right)$

38. [5.6] (5.385, 111.8°), (5.385, 1.951)
39. [5.6] (4.964, 147.8°), (4.964, 2.579)

40. [5.6] $\left(\dfrac{3\sqrt{2}}{2}, \dfrac{3\sqrt{2}}{2}\right)$ **41.** [5.6] $(3, 3\sqrt{3})$

42. [5.6] (1.93, −0.52)   **43.** [5.6] (−1.86, −1.35)
44. [5.6] $r(5 \cos\theta - 2\sin\theta) = 6$ **45.** [5.6] $r\sin\theta = 3$
46. [5.6] $r = 3$ **47.** [5.6] $r^2 \sin^2\theta - 4r\cos\theta - 16 = 0$
48. [5.6] $x^2 + y^2 = 36$ **49.** [5.6] $x^2 + 2y = 1$
50. [5.6] $y^2 - 6x = 9$ **51.** [5.6] $x^2 - 2x + y^2 - 3y = 0$
52. [5.6] (b)   **53.** [5.6] (d)   **54.** [5.6] (a)   **55.** [5.6] (c)

56. [5.7]

$$r = \frac{6}{3 - 3\sin\theta}$$

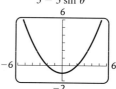

Horizontal directrix 6 units below the pole;
vertex: $\left(1, \dfrac{3\pi}{2}\right)$

57. [5.7]

$$r = \frac{8}{2 + 4\cos\theta}$$

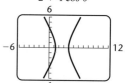

Vertical directrix 2 units to the right of the pole;
vertices: $\left(0, \dfrac{4}{3}\right)$, $(-4, \pi)$

58. [5.7]

$$r = \frac{4}{2 - \cos\theta}$$

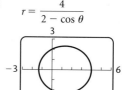

Vertical directrix 4 units to the left of the pole;
vertices: $(0, 4)$, $\left(\dfrac{4}{3}, \pi\right)$

59. [5.7]

$$r = \frac{18}{9 + 6\sin\theta}$$

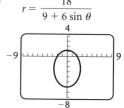

Horizontal directrix 3 units above the pole;
vertices: $\left(\dfrac{6}{5}, \dfrac{\pi}{2}\right)$, $\left(6, \dfrac{3\pi}{2}\right)$

60. [5.7] $x^2 - 12y - 36 = 0$

61. [5.7] $4x^2 - y^2 - 16x + 16 = 0$

62. [5.7] $3x^2 + 4y^2 - 8x - 16 = 0$

63. [5.7] $9x^2 + 5y^2 + 24y - 36 = 0$

64. [5.7] $r = \dfrac{1}{1 + \frac{1}{2}\cos\theta}$, or $r = \dfrac{2}{2 + \cos\theta}$

65. [5.7] $r = \dfrac{18}{1 - 3\sin\theta}$ **66.** [5.7] $r = \dfrac{4}{1 - \cos\theta}$

67. [5.7] $r = \dfrac{6}{1 + 2\sin\theta}$

68. [5.8] $x = t, \; y = 2 + t, \; -3 \le t \le 3$

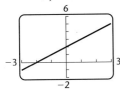

$y = 2 + x, \; -3 \le x \le 3, \; -1 \le y \le 5$

69. [5.8] $x = \sqrt{t}, \; y = t - 1, \; 0 \le t \le 9$

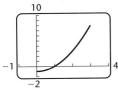

$y = x^2 - 1, \; 0 \le x \le 3, \; -1 \le y \le 8$

70. [5.8] $x = 2\cos t, \; y = 2\sin t, \; 0 \le t \le 2\pi$

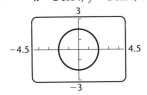

$x^2 + y^2 = 1$

71. [5.8] $x = 3\sin t, \; y = \cos t, \; 0 \le t \le 2\pi$

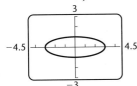

$\dfrac{x^2}{9} + y^2 = 1$

72. [5.8] Answers may vary. $x = t, \; y = 2t - 3; \; x = t + 1,$
$y = 2t - 1$

73. [5.8] Answers may vary. $x = t, \; y = t^2 + 4; \; x = t - 2,$
$y = t^2 - 4t + 8$

74. [5.4] ◆ An algebraic solution of a nonlinear system of
equations may be preferable to a graphical solution if there
are complex-number solutions or if any of the equations is
difficult to enter on a grapher. However, there are nonlinear
systems that are difficult or impossible to solve using
algebraic methods. Approximations of the real-number
solutions of many of those systems may be found using
graphical methods.

75. [5.2] ◆ The equation of a circle can be written as

$$\dfrac{(x - h)^2}{a^2} + \dfrac{(y - k)^2}{b^2} = 1,$$

where $a = b = r$, the radius of the circle. In an ellipse,
$a > b$, so a circle is not a special type of ellipse.

76. [5.2] $x^2 + \dfrac{y^2}{9} = 1$ **77.** [5.4] $\dfrac{8}{7}, \dfrac{7}{2}$

78. [5.2], [8.4] $(x - 2)^2 + (y - 1)^2 = 100$

79. [5.3] $\dfrac{x^2}{778.41} - \dfrac{y^2}{39{,}221.59} = 1$

80. [5.6] $y^2 = 4x + 4$

Chapter 6

EXERCISE SET 6.1

1. 54.5982   **3.** 0.0856

5.

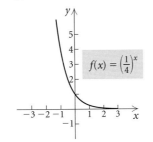

$f(x) = 3^x$

7.

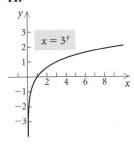

$f(x) = 6^x$

9.

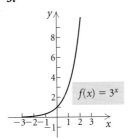

$f(x) = \left(\dfrac{1}{4}\right)^x$

11.

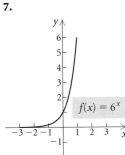

$x = 3^y$

13.

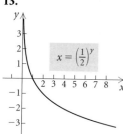

$x = \left(\frac{1}{2}\right)^y$

15.

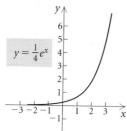

$y = \frac{1}{4}e^x$

17.

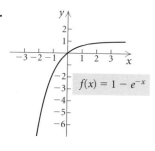

$f(x) = 1 - e^{-x}$

19. Shift the graph of $y = 2^x$ left 1 unit.

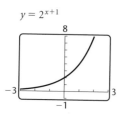

$y = 2^{x+1}$

21. Shift the graph of $y = 2^x$ down 3 units.

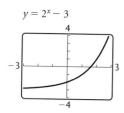

$y = 2^x - 3$

23. Reflect the graph of $y = 3^x$ across the y-axis, then across the x-axis, and then shift it up 4 units.

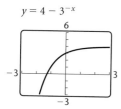

$y = 4 - 3^{-x}$

25. Shift the graph of $y = \left(\frac{3}{2}\right)^x$ right 1 unit.

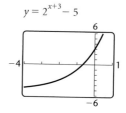

$y = \left(\frac{3}{2}\right)^{x-1}$

27. Shift the graph of $y = 2^x$ left 3 units, and then down 5 units.

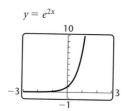

$y = 2^{x+3} - 5$

29. Shrink the graph of $y = e^x$ horizontally.

$y = e^{2x}$

31. Shift the graph of $y = e^x$ left 1 unit, then reflect it across the y-axis.

$y = e^{-x+1}$

33. Reflect the graph of $y = e^x$ across the y-axis, then across the x-axis, then shift it up 1 unit, and then stretch it vertically.

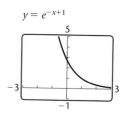

$y = 2(1 - e^{-x}), x > 0$

35. (a) 1,475,789; **(b)** 5,669,391;
(c) $y = 100,000(1.4)^x$ **(d)** 6.8 yr

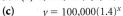
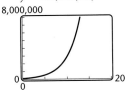

37. (a) 350,000; 233,333; 69,136; 6070;
(b) $y = 350,000\left(\frac{2}{3}\right)^x$ **(c)** after about 13 yr

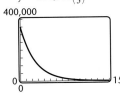

39. (a) $5800, $4640, $3712, $1900.54, $622.77;
(b) $y = 5800(0.8)^x$ **(c)** after 11 yr

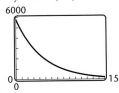

41. (a) About 50.0 billion ft³, 62.0 billion ft³, 78.2 billion ft³; **(b)** $y = 46.6(1.018)^x$ **(c)** After about 39 yr

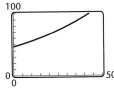

43. (a) About 63%; **(b)** $y = 100(1 - e^{-0.04x})$

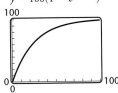

(c) after 58 days

**45. (c)   47. (a)   49. (l)   51. (g)   53. (i)   55. (k)
57. (m)   59.** (1.481, 4.090)
61. (−0.402, −1.662), (1.051, 2.722)   **63.** {4.448}
65. (0, ∞)   **67.** {2.294, 3.228}   **69.** 1   **71.** 1000   **73.** 2
75. ◈   **77.** ◈   **79.** $y = 1$

81. (a) $y = e^{-x^2}$

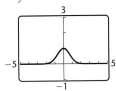

(b) none; **(c)** relative maximum: 1 at $x = 0$

EXERCISE SET 6.2

1.   **3.**

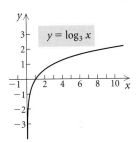

5. 4   **7.** 3   **9.** −3   **11.** −2   **13.** 0   **15.** 1
17. $\log_{10} 1000 = 3$ **19.** $\log_8 2 = \frac{1}{3}$ **21.** $\log_e t = 3$
23. $\log_e 7.3891 = 2$ **25.** $\log_p 3 = k$ **27.** $5^1 = 5$
29. $10^{-2} = 0.01$ **31.** $e^{3.4012} = 30$ **33.** $a^{-x} = M$
35. $a^x = T^3$ **37.** 0.4771   **39.** 2.7259   **41.** −0.2441
43. Does not exist   **45.** 0.6931   **47.** 6.6962
49. Does not exist
51. 3.3219   **53.** −0.2614   **55.** 0.7384
57. Translate the graph of $y = \log_2 x$ left 3 units. Domain: $(-3, \infty)$; vertical asymptote: $x = -3$;

$$y = \frac{\log (x + 3)}{\log 2}$$

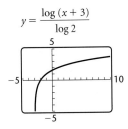

59. Translate the graph of $y = \log_3 x$ down 1 unit. Domain: $(0, \infty)$; vertical asymptote: $x = 0$;

$$y = \frac{\log x}{\log 3} - 1$$

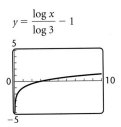

61. Stretch the graph of $y = \ln x$ vertically. Domain: $(0, \infty)$; vertical asymptote: $x = 0$;

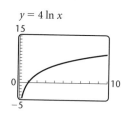

$y = 4 \ln x$

63. Reflect the graph of $y = \ln x$ across the x-axis and translate it up 2 units. Domain: $(0, \infty)$; vertical asymptote: $x = 0$;

$y = 2 - \ln x$

65.

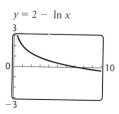

$y_1 = 3^x$,

$y_2 = \dfrac{\log x}{\log 3}$

67.

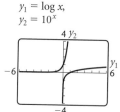

$y_1 = \log x$,
$y_2 = 10^x$

69. (a) 3.1 ft/sec; **(b)** 3.0 ft/sec; **(c)** 2.3 ft/sec; **(d)** 1.5 ft/sec
71. (a) 78%; **(b)** 67.5%, 57%;
(c) $y = 78 - 15 \log (x + 1)$, $x \geqslant 0$ **(d)** after 73 months

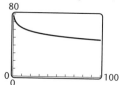

73. (a) 10^{-7}; **(b)** 4.0×10^{-6}; **(c)** 6.3×10^{-4}; **(d)** 1.6×10^{-5}
75. (a) 34 decibels; **(b)** 64 decibels; **(c)** 60 decibels;
(d) 90 decibels
77. y^{24} **79.** 10^{4t} **81.** ◈   **83.** 3   **85.** $(0, \infty)$
87. $(-\infty, 0) \cup (0, \infty)$ **89.** $\left(-\frac{5}{2}, -2\right)$ **91.** (d)   **93.** (b)

95. (a)

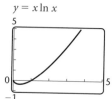

$y = x \ln x$

(b) 1; **(c)** relative minimum: -0.368 at $x = 0.368$
97. (a)

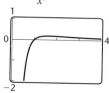

$y = \dfrac{\ln x}{x^2}$

(b) 1; **(c)** relative maximum: 0.184 at $x = 1.649$

EXERCISE SET 6.3

1. $\log_3 81 + \log_3 27$ **3.** $\log_5 5 + \log_5 125$
5. $\log_t 8 + \log_t Y$ **7.** $3 \log_b t$ **9.** $8 \log y$ **11.** $-6 \log_c K$
13. $\log_t M - \log_t 8$ **15.** $\log_a x - \log_a y$
17. $\log_a 6 + \log_a x + 5 \log_a y + 4 \log_a z$
19. $2 \log_b p + 5 \log_b q - 4 \log_b m - 9$
21. $3 \log_a x - \frac{5}{2} \log_a p - 4 \log_a q$
23. $2 \log_a m + 3 \log_a n - \frac{3}{4} - \frac{5}{4} \log_a b$ **25.** $\log_a 150$
27. $\log 100 = 2$ **29.** $\log_a x^{-5/2} y^4$, or $\log_a \dfrac{y^4}{x^{5/2}}$ **31.** $\ln x$
33. $\ln (x - 2)$ **35.** $\ln \dfrac{x}{(x^2 - 25)^3}$ **37.** $\ln \dfrac{2^{11/5} x^9}{y^8}$
39. -0.5108 **41.** -1.6094 **43.** $\frac{1}{2}$ **45.** 2.6094
47. 4.3174 **49.** 3   **51.** $|x - 4|$ **53.** $4x$ **55.** w **57.** $8t$
59. $\{4\}$ **61.** $\{5\}$ **63.** ◈   **65.** $\{4\}$ **67.** $\log_a (x^3 - y^3)$
69. $\frac{1}{2} \log_a (x - y) - \frac{1}{2} \log_a (x + y)$ **71.** 7   **73.** True
75. True   **77.** True   **79.** -2 **81.** 3
83. $\log_a \left(\dfrac{x + \sqrt{x^2 - 5}}{5} \cdot \dfrac{x - \sqrt{x^2 - 5}}{x - \sqrt{x^2 - 5}} \right)$

$$= \log_a \dfrac{5}{5(x - \sqrt{x^2 - 5})}$$

$$= -\log_a (x - \sqrt{x^2 - 5})$$

EXERCISE SET 6.4

1. $\{4\}$ **3.** $\left\{\frac{3}{2}\right\}$ **5.** $\{5.044\}$ **7.** $\left\{\frac{5}{2}\right\}$ **9.** $\left\{-3, \frac{1}{2}\right\}$
11. $\{0.959\}$ **13.** $\{6.908\}$ **15.** $\{84.191\}$ **17.** $\{-1.710\}$
19. $\{2.844\}$ **21.** $\{-1.567, 1.567\}$ **23.** $\{0.347\}$ **25.** $\{625\}$
27. $\{0.0001\}$ **29.** $\{e\}$ **31.** $\left\{\frac{22}{3}\right\}$ **33.** $\{10\}$ **35.** $\left\{\frac{1}{63}\right\}$
37. $\{5\}$ **39.** $\left\{\frac{21}{8}\right\}$ **41.** $\{0.367\}$ **43.** $\{0.621\}$

45. $\{-1.532\}$ **47.** $\{7.062\}$ **49.** $\{2.444\}$
51. $(4.093, 0.786)$ **53.** $(7.586, 6.684)$
55. $\{-3i, 3i, -2, 2\}$
57. (a) $y = 400.27x + 10,655.15$, \$14,257.6 million;
(b) $y = -58.325x^2 + 691.895x + 10,363.525$,
\$11,866.3 million
59. ◈   **61.** $\{10\}$ **63.** $\{1, e^4\}$, or $\{1, 54.598\}$ **65.** $\left\{\frac{1}{3}, 27\right\}$
67. $\{1, e^2\}$, or $\{1, 7.389\}$ **69.** $\{0, 0.431\}$ **71.** $\{-9, 9\}$
73. $\{e^{-2}, e^2\}$, or $\{0.135, 7.389\}$ **75.** $\left\{\frac{7}{4}\right\}$ **77.** $(56.598, \infty)$
79. $\{5\}$ **81.** $a = \frac{2}{3}b$ **83.** 88

EXERCISE SET 6.5

1. (a) $P(t) = 6.6e^{0.028t}$; **(b)** 7.2 billion, 9.5 billion;
(c) in 6.9 yr; **(d)** 24.8 yr
3. (a) 19.8 yr; **(b)** 1% per year; **(c)** 6.93 yr; **(d)** 17.3 yr;
(e) 2.8% per year; **(f)** 3.5% per year
5. After about 69 yr
7. (a) $P(t) = 10,000e^{0.054t}$; **(b)** \$10,555; \$11,140; \$13,100;
\$17,160; **(c)** 12.8 yr
9. About 5135 yr
11. (a) 23.1% per minute; **(b)** 3.15% per year; **(c)** 7.2 days;
(d) 11 yr; **(e)** 2.8% per year; **(f)** 0.015% per year; **(g)** 0.003%
per year
13. (a) $k \approx 0.016$; $P(t) = 80e^{-0.016t}$; **(b)** about 63 lb per
person; **(c)** 86.6 yr
15. (a) 96; **(b)** 286, 1005, 1719, 1971, 1997;
(c)
$$y = \frac{2000}{1 + 19.9e^{-0.6x}}$$

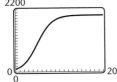

17. (a) 400, 520, 1214, 2059, 2396, 2478;
(b)
$$y = \frac{2500}{1 + 5.25e^{-0.32x}}$$

19. (f)   **21.** (b)   **23.** (c)
25. (a) Yes; **(b)** $y = 0.785(1.490)^x$, or $y = 0.785e^{0.399x}$;
(c) \$63.1 billion; **(d)** 1997

27. (a) Exponential function: $y = (107.9)(1.034)^x$, or
$y = (107.9)e^{0.033x}$; power function: $y = 110.8x^{0.077}$;
(b) $y_1 = (107.9)(1.034)^x$, $y_2 = 110.8x^{0.077}$

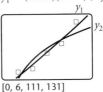

$[0, 6, 111, 131]$

(c) exponential function: \$178; power function: \$136;
(d) The exponential function has a higher correlation
coefficient and seems to be following data better.
29. $\sqrt{3}$ **31.** $\sqrt{3365}$ **33.** ◈   **35.** \$14,182.70

37. \$166.16   **39.** 46.7°F   **41.** $t = -\dfrac{L}{R}\left[\ln\left(1 - \dfrac{iR}{V}\right)\right]$

43. Linear

REVIEW EXERCISES, CHAPTER 6

1. [6.1] (c)   **2.** [6.2] (a)   **3.** [6.2] (b)   **4.** [6.1] (f)
5. [6.1] (e)   **6.** [6.2] (d)   **7.** [6.2] $4^2 = x$
8. [6.2] $\log_e 80 = x$ **9.** [6.4] $\{16\}$ **10.** [6.4] $\left\{\frac{1}{5}\right\}$
11. [6.4] $\{4.382\}$ **12.** [6.4] $\{2\}$ **13.** [6.4] $\left\{\frac{1}{2}\right\}$
14. [6.4] $\{5\}$ **15.** [6.4] $\{4\}$ **16.** [6.4] $\{9\}$ **17.** [6.4] $\{1\}$
18. [6.4] $\{3.912\}$ **19.** [6.3] $\log_b \dfrac{x^3\sqrt{z}}{y^4}$
20. [6.3] $\ln(x^2 - 4)$ **21.** [6.3] $\frac{1}{4}\ln w + \frac{1}{2}\ln r$
22. [6.3] $\frac{2}{3}\log M - \frac{1}{3}\log N$ **23.** [6.3] 0.477
24. [6.3] 1.699   **25.** [6.3] -0.699 **26.** [6.3] 0.233
27. [6.3] $-5k$ **28.** [6.3] $-6t$ **29.** [6.5] 12.8 yr
30. [6.5] 2.3%   **31.** [6.5] About 2623 yr   **32.** [6.2] 5.6
33. [6.2] 30 decibels
34. [6.5] **(a)** $P(t) = 52e^{0.028t}$; **(b)** 164 million, 287 million;
(c) in 62.6 yr; **(d)** 24.8 yr
35. [6.5] **(a)** $k = 0.304$; **(b)** $P(t) = 7e^{0.304t}$;
(c) $y = 7e^{0.304x}$

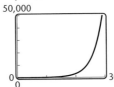

(d) 86.7 billion, 727.9 billion; **(e)** 1993

36. [6.2] **(a)** 2.3 ft/sec; **(b)** 8,553,143

37. [6.5] **(a)** Yes; **(b)** $y = (2.329)1.753^x$, or $y = 2.329e^{0.561x}$; **(c)** 207.7 million; **(d)** in 6.7 yr

38. [6.3] ◈ By the product rule, $\log_2 x + \log_2 5 = \log_2 5x$, not $\log_2 (x + 5)$. Also, substituting various numbers for x shows that both sides of the inequality are indeed unequal. You could also graph each side and show that the graphs do not coincide.

39. [6.1], [6.2] ◈ Reflect the graph of $f(x) = e^x$ across the line $y = x$ and then shift it down 1 unit.

40. [6.4] $\left\{\frac{1}{64}, 64\right\}$ **41.** [6.4] $\{1\}$ **42.** [6.4] $\{16\}$

43. [6.1], [6.2] **(a)** $y = 5e^{-x}\ln x$

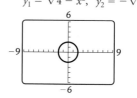

(b) Relative maximum: 0.486 at $x = 1.763$, (1.763, 0.486)

44. [6.2] $(1, \infty)$ **45.** [6.1] $(-0.393, 1.143)$, $(0.827, 1.495)$

Appendixes

EXERCISE SET A.1

1. $\{4\}$ **3.** $\{8\}$ **5.** $\left\{\frac{4}{5}\right\}$ **7.** $\left\{-\frac{3}{2}\right\}$ **9.** $\left\{\frac{3}{2}, \frac{2}{3}\right\}$ **11.** $\left\{\frac{2}{3}, -1\right\}$

13. $\{0, 3\}$ **15.** $\left\{-\frac{1}{3}, 0, 2\right\}$ **17.** $\left\{-1, 1, -\frac{1}{7}\right\}$ **19.** $\{-7, 1\}$

21. $\{4 \pm \sqrt{7}\}$ **23.** $\left\{-2, \frac{1}{3}\right\}$ **25.** $\left\{\frac{3 \pm \sqrt{13}}{4}\right\}$

27. $\left\{\frac{5 \pm \sqrt{57}}{4}\right\}$ **29.** $\{-3, 5\}$ **31.** $\left\{-1, \frac{2}{5}\right\}$

33. $\left\{\frac{5 \pm \sqrt{7}}{3}\right\}$ **35.** $\left\{\frac{4 \pm \sqrt{31}}{5}\right\}$ **37.** $\left\{\frac{20}{9}\right\}$

39. $\{286\}$ **41.** $\{6\}$ **43.** $\varnothing$ **45.** $\varnothing$ **47.** $\varnothing$

49. $\{x \mid x$ is a real number and $x \neq 0$ and $x \neq 6\}$ **51.** $\{2\}$

53. $\left\{\frac{5}{3}\right\}$ **55.** $\{\pm\sqrt{2}\}$ **57.** $\varnothing$ **59.** $\{-6\}$ **61.** $\{7\}$

63. $\varnothing$ **65.** $\{1\}$ **67.** $\{3, 7\}$ **69.** $\{-8\}$ **71.** $\{81\}$

73. $\{-7, 7\}$ **75.** $\varnothing$ **77.** $\{-3, 5\}$ **79.** $\left\{-\frac{1}{3}, \frac{1}{3}\right\}$ **81.** $\{0\}$

83. $\left\{-1, -\frac{1}{3}\right\}$ **85.** $\{-24, 44\}$ **87.** $\{-2, 4\}$ **89.** $b = \dfrac{2A}{h}$

91. $w = \dfrac{P - 2l}{2}$ **93.** $T_1 = \dfrac{P_1 V_1 T_2}{P_2 V_2}$ **95.** $R_2 = \dfrac{RR_1}{R_1 - R}$

97. $p = \dfrac{Fm}{m - F}$ **99.** ◈   **101.** ◈   **103.** $\{1\}$

105. $\{1, 5\}$ **107.** $\{-1\}$

EXERCISE SET A.2

1. $\sqrt{10}$, 3.162   **3.** $\sqrt{45}$, 6.708   **5.** $\sqrt{14 + 6\sqrt{2}}$, 4.742

7. 6.5   **9.** Yes

11. The distances between the points are $\sqrt{10}$, 10, and $\sqrt{90}$. Since $(\sqrt{10})^2 + (\sqrt{90})^2 = 10^2$, the three points are vertices

of a right triangle.   **13.** $\left(\dfrac{9}{2}, \dfrac{15}{2}\right)$ **15.** $\left(\dfrac{15}{2}, 2\right)$

17. $\left(\dfrac{\sqrt{3} - \sqrt{6}}{2}, -\dfrac{\sqrt{5}}{2}\right)$

19.

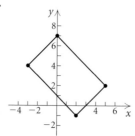

$\left(-\dfrac{1}{2}, \dfrac{3}{2}\right), \left(\dfrac{7}{2}, \dfrac{1}{2}\right), \left(\dfrac{5}{2}, \dfrac{9}{2}\right), \left(-\dfrac{3}{2}, \dfrac{11}{2}\right)$; no

21. $\left(\dfrac{\sqrt{7} + \sqrt{2}}{2}, -\dfrac{1}{2}\right)$

23. Square the window; for example, $[-12, 9, -4, 10]$

25. $(x - 2)^2 + (y - 3)^2 = \dfrac{25}{9}$

27. $(x + 1)^2 + (y - 4)^2 = 25$

29. $(x - 2)^2 + (y - 1)^2 = 169$

31. $(x + 2)^2 + (y - 3)^2 = 4$

33. $(0, 0)$; 2;

$$x^2 + y^2 = 4$$
$$y_1 = \sqrt{4 - x^2}, \ y_2 = -\sqrt{4 - x^2}$$

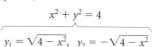

35. $(0, 3)$; 4;

$$x^2 + (y - 3)^2 = 16$$
$$y_1 = 3 + \sqrt{16 - x^2}, \ y_2 = 3 - \sqrt{16 - x^2}$$

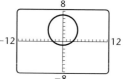

37. $(1, 5)$; 6;

$$(x - 1)^2 + (y - 5)^2 = 36$$
$$y_1 = 5 + \sqrt{36 - (x - 1)^2}, \ y_2 = 5 - \sqrt{36 - (x - 1)^2}$$

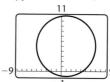

39. $(-4, -5)$; 3;

$$(x + 4)^2 + (y + 5)^2 = 9$$

$$y_1 = -5 + \sqrt{9 - (x + 4)^2}, \quad y_2 = -5 - \sqrt{9 - (x + 4)^2}$$

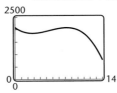

41. $\sqrt{h^2 + h + 2a - 2\sqrt{a^2 + ah}}$,

$\left(\dfrac{2a + h}{2}, \dfrac{\sqrt{a} + \sqrt{a + h}}{2}\right)$

43. ◈   **45.** $(x - 2)^2 + (y + 7)^2 = 36$

47. $a_1 = 2.7$ ft, $a_2 = 37.3$ ft   **49.** Yes   **51.** Yes

53. $(0, 4)$ **55.** Let $P_1 = (x_1, y_1)$, $P_2 = (x_2, y_2)$, and

$M = \left(\dfrac{x_1 + x_2}{2}, \dfrac{y_1 + y_2}{2}\right)$. Let $d(AB)$ denote the distance

from point A to point B.

(a) $d(P_1 M) = \sqrt{\left(\dfrac{x_1 + x_2}{2} - x_1\right)^2 + \left(\dfrac{y_1 + y_2}{2} - y_1\right)^2}$

$\qquad = \dfrac{1}{2}\sqrt{(x_2 - x_1)^2 + (y_2 - y_1)^2};$

$d(P_2 M) = \sqrt{\left(\dfrac{x_1 + x_2}{2} - x_2\right)^2 + \left(\dfrac{y_1 + y_2}{2} - y_2\right)^2}$

$\qquad = \dfrac{1}{2}\sqrt{(x_1 - x_2)^2 + (y_1 - y_2)^2}$

$\qquad = \dfrac{1}{2}\sqrt{(x_2 - x_1)^2 + (y_2 - y_1)^2} = d(P_1 M).$

(b) $d(P_1 M) + d(P_2 M) = \dfrac{1}{2}\sqrt{(x_2 - x_1)^2 + (y_2 - y_1)^2}$

$\qquad\qquad + \dfrac{1}{2}\sqrt{(x_2 - x_1)^2 + (y_2 - y_1)^2}$

$\qquad\qquad = \sqrt{(x_2 - x_1)^2 + (y_2 - y_1)^2}$

$\qquad\qquad = d(P_1 P_2).$

EXERCISE SET A.3

1. Yes   **3.** No   **5.** No   **7.** Yes

9. (a) $y = 2.6x + 49.7$, 75.7; **(b)** $r = 0.9856$, a good fit

11. (a) $y = 645.7x + 9799.2$; \$14,319.10; \$15,610.50;

\$18,193.30; **(b)** $r = 0.9994$, a good fit; **(c)** The president's

assertion is about \$1000, or 6%, higher than the estimate

from the model.

13. (a) $M = 0.2H + 156$; **(b)** 164, 169, 171, 173; **(c)** $r = 1$;

the regression line fits the data perfectly and should be a

good predictor.

15. (b)   **17.** (c)   **19.** (a)   **21.** (b)   **23.** (b)

25. (a) Yes; **(b)** no; **(c)** third differences are all -384;

(d) cubic; **(e)** $f(x) = -x^3 - x^2 + x + 5$; **(f)** $f(17) = -5180$;

$f(21) = -9676$

27. (a) Yes; **(b)** no; **(c)** fourth differences are all 1944;

(d) quartic; **(e)** $f(x) = x^4 - 10x^2 + x - 1$; $f(13) = 26{,}883$;

$x = -2.825, -1.500, 1.707,$ or 2.618 when $y = -19.94$

29. (a) $f(x) = -0.8630952381x^2 + 12.125x + 36.76785714$;

(b) 77.3 yr, 77.5 yr

31. (a) $f(x) = -2.775187775x^3 + 45.77200577x^2 -$

$190.2173752x + 2084.949495$;

(b)
$$y = -2.775187775x^3 + 45.77200577x^2 -$$
$$190.2173752x + 2084.949495$$

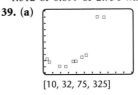

(c) $-\$2694$; **(d)** $f(x) = 1.586902681x^4 -$

$44.03465747x^3 + 400.5581051x^2 - 1316.011477x +$

3075.176768;

(e)
$$y = 1.586902681x^4 - 44.03465747x^3 +$$
$$400.5581051x^2 - 1316.011477x + 3075.176768$$

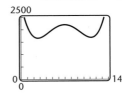

(f) \$18,944

33. ◈

35. (a) $D = 0.000008d + 11.9$, $r = 0.0282$;

(b) The model does not fit the data and would not be a

good predictor.

37. (a) $f(x) = 0.8586956522x^3 - 0.8452380952x^2 -$

$0.7546583851x + 31.10766046$; $f(-4) \approx -34.354$;

$f(4) \approx 69.522$; x is about -1.950 when y is 23;

(b) $g(x) = 1.558569182x^4 + 1.807389937x^3 -$

$15.7629717x^2 - 9.835017969x + 42.25965858$;

$g(-4) \approx 112.713$; $g(4) \approx 265.379$; x is about -3.301 or

-1.512 or 0.899 or 2.754 when $y = 23$

39. (a)

[image: scatter plot with window [10, 32, 75, 325]]

[10, 32, 75, 325]

(b) Linear seems inappropriate; quadratic, cubic, quartic

seem possible. **(c)** *Quadratic:* $y = 2.031904548x^2 -$

$59.04179598x + 527.2818092$; *cubic:* $y = -0.0273804413x^3 +$

$3.488800304x^2 - 83.76947384x + 660.2911579$; *quartic:*

$y = -0.0403607597x^4 + 2.842908639x^3 - 70.96865626x^2 +$

$749.2437238x - 2721.562456$

(d) Quadratic

[10, 32, 75, 325]

Cubic

[10, 32, 75, 325]

The quadratic and cubic equations seem to fit best. They indicate that, for an advertising budget of about $16 million or greater, box office revenue increases as the budget increases. The quartic equation indicates that revenue declines sharply as the advertising budget increases beyond about $25.5 million.

(e) $584.74 million, $547.86 million, −$49.72 million

Quartic

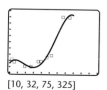

[10, 32, 75, 325]

Index of Applications

Index

Geometry

Plane Geometry

Rectangle
Area: $A = lw$
Perimeter: $P = 2l + 2w$

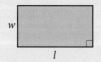

Square
Area: $A = s^2$
Perimeter: $P = 4s$

Triangle
Area: $A = \frac{1}{2}bh$

Sum of Angle Measures
$A + B + C = 180°$

Right Triangle
Pythagorean theorem
(equation):
$a^2 + b^2 = c^2$

Parallelogram
Area: $A = bh$

Trapezoid
Area: $A = \frac{1}{2}h(a + b)$

Circle
Area: $A = \pi r^2$
Circumference:
  $C = \pi d = 2\pi r$
  $\left(\frac{22}{7}\right.$ and 3.14 are different
  approximations for $\pi\left.\right)$

Solid Geometry

Rectangular Solid
Volume: $V = lwh$

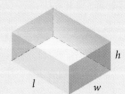

Cube
Volume: $V = s^3$

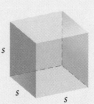

Right Circular Cylinder
Volume: $V = \pi r^2 h$
Lateral surface area:
  $L = 2\pi rh$
Total surface area:
  $S = 2\pi rh + 2\pi r^2$

Right Circular Cone
Volume: $V = \frac{1}{3}\pi r^2 h$
Lateral surface area:
  $L = \pi rs$
Total surface area:
  $S = \pi r^2 + \pi rs$
Slant height:
  $s = \sqrt{r^2 + h^2}$

Sphere
Volume: $V = \frac{4}{3}\pi r^3$
Surface area: $S = 4\pi r^2$

Algebra

Properties of Real Numbers

Commutative: $a + b = b + a; \quad ab = ba$

Associative: $a + (b + c) = (a + b) + c;$
$a(bc) = (ab)c$

Additive Identity: $a + 0 = 0 + a = a$

Additive Inverse: $-a + a = a + (-a) = 0$

Multiplicative Identity: $a \cdot 1 = 1 \cdot a = a$

Multiplicative Inverse: $a \cdot \dfrac{1}{a} = 1, \ a \neq 0$

Distributive: $a(b + c) = ab + ac$

Exponents and Radicals

$a^m \cdot a^n = a^{m+n}$ $\qquad \dfrac{a^m}{a^n} = a^{m-n}$

$(a^m)^n = a^{mn}$ $\qquad (ab)^m = a^m b^m$

$\left(\dfrac{a}{b}\right)^m = \dfrac{a^m}{b^m}$ $\qquad a^{-n} = \dfrac{1}{a^n}$

$\sqrt[n]{a} \cdot \sqrt[n]{b} = \sqrt[n]{ab}, \ a, b \geq 0$

$\sqrt[n]{\dfrac{a}{b}} = \dfrac{\sqrt[n]{a}}{\sqrt[n]{b}}$

$\sqrt[n]{a^m} = (\sqrt[n]{a})^m = a^{m/n}$

The Distance Formula

The distance from (x_1, y_1) to (x_2, y_2) is given by

$$d = \sqrt{(x_2 - x_1)^2 + (y_2 - y_1)^2}.$$

Midpoint Formula

The midpoint of the line segment from (x_1, y_1) to (x_2, y_2) is given by

$$\left(\dfrac{x_1 + x_2}{2}, \dfrac{y_1 + y_2}{2}\right).$$

Special-Product Formulas

$(a + b)(a - b) = a^2 - b^2$

$(a + b)^2 = a^2 + 2ab + b^2$

$(a - b)^2 = a^2 - 2ab + b^2$

$(a + b)^3 = a^3 + 3a^2b + 3ab^2 + b^3$

$(a - b)^3 = a^3 - 3a^2b + 3ab^2 - b^3$

Factoring Formulas

$a^2 - b^2 = (a + b)(a - b)$

$a^2 + 2ab + b^2 = (a + b)^2$

$a^2 - 2ab + b^2 = (a - b)^2$

$a^3 + b^3 = (a + b)(a^2 - ab + b^2)$

$a^3 - b^3 = (a - b)(a^2 + ab + b^2)$

Interval Notation

$(a, b) = \{x | a < x < b\}$ $\qquad [a, b] = \{x | a \leq x \leq b\}$

$(a, b] = \{x | a < x \leq b\}$ $\qquad [a, b) = \{x | a \leq x < b\}$

$(-\infty, a) = \{x | x < a\}$ $\qquad (a, \infty) = \{x | x > a\}$

$(-\infty, a] = \{x | x \leq a\}$ $\qquad [a, \infty) = \{x | x \geq a\}$

Absolute Value

$|a| \geq 0$ $\qquad\qquad |-a| = |a|$

$|ab| = |a| \cdot |b|$ $\qquad \left|\dfrac{a}{b}\right| = \dfrac{|a|}{|b|}, \ b \neq 0$

For $a > 0$,

$$|x| = a \rightarrow x = -a \quad \text{or} \quad x = a,$$
$$|x| < a \rightarrow -a < x < a,$$
$$|x| > a \rightarrow x < -a \quad \text{or} \quad x > a.$$

(Algebra continued)

Algebra (continued)

Equation-Solving Principles

$a = b \rightarrow a + c = b + c$

$a = b \rightarrow ac = bc$

$a = b \rightarrow a^n = b^n$

$ab = 0 \leftrightarrow a = 0 \quad \text{or} \quad b = 0$

$x^2 = k \rightarrow x = \sqrt{k} \quad \text{or} \quad x = -\sqrt{k}$

Inequality-Solving Principles

$a < b \rightarrow a + c < b + c$

$a < b \text{ and } c > 0 \rightarrow ac < bc$

$a < b \text{ and } c < 0 \rightarrow ac > bc$

Formulas Involving Lines

The slope of the line containing points (x_1, y_1) to (x_2, y_2) is given by

$$m = \frac{y_2 - y_1}{x_2 - x_1}.$$

Slope–intercept equation: $\quad y = f(x) = mx + b$

Horizontal line: $\quad y = b \quad \text{or} \quad f(x) = b$

Vertical line: $\quad x = a$

Point–slope equation: $\quad y - y_1 = m(x - x_1)$

Two-point equation: $\quad y - y_1 = \dfrac{y_2 - y_1}{x_2 - x_1}(x - x_1)$

The Quadratic Formula

The solutions of $ax^2 + bx + c = 0$, $a \neq 0$, are given by

$$x = \frac{-b \pm \sqrt{b^2 - 4ac}}{2a}.$$

Compound Interest Formulas

Compounded n times per year: $\quad A = P\left(1 + \dfrac{i}{n}\right)^{nt}$

Compounded continuously: $\quad P(t) = P_0 e^{kt}$

Properties of Exponential and Logarithmic Functions

$\log_a x = y \leftrightarrow x = a^y \qquad a^x = a^y \leftrightarrow x = y$

$\log_a MN = \log_a M + \log_a N \qquad \log_a M^p = p \log_a M$

$\log_a \dfrac{M}{N} = \log_a M - \log_a N$

$\log_b M = \dfrac{\log_a M}{\log_a b}$

$\log_a a = 1 \qquad\qquad \log_a 1 = 0$

$\log_a a^x = x \qquad\qquad a^{\log_a x} = x$

Conic Sections

Circle: $\qquad (x - h)^2 + (y - k)^2 = r^2$

Ellipse: $\qquad \dfrac{(x - h)^2}{a^2} + \dfrac{(y - k)^2}{b^2} = 1,$

$\qquad\qquad \dfrac{(x - h)^2}{b^2} + \dfrac{(y - k)^2}{a^2} = 1$

Parabola: $\qquad (x - h)^2 = 4p(y - k),$

$\qquad\qquad (y - k)^2 = 4p(x - h)$

Hyperbola: $\qquad \dfrac{(x - h)^2}{a^2} - \dfrac{(y - k)^2}{b^2} = 1,$

$\qquad\qquad \dfrac{(y - k)^2}{a^2} - \dfrac{(x - h)^2}{b^2} = 1$